CHICAGO PUBLIC LIBRARY
HAROLD WASHINGTON LIBRARY CENTER

W9-ABC-935

QK
157
.M63
Cop. 4

Mohlenbrock, Robert
H.

Guide to the
vascular flora of
Illinois

DATE		

Business/Science/Technology
Division

© THE BAKER & TAYLOR CO.

Guide to the Vascular Flora of Illinois

Robert H. Mohlenbrock

Southern Illinois University Press
Carbondale and Edwardsville

Feffer & Simons, Inc.
London and Amsterdam

QK
157
.M63
cop.4

Library of Congress Cataloging in Publication Data

Mohlenbrock, Robert H
  Guide to the vascular flora of Illinois.

  Includes index.
  1. Botany—Illinois. I. Title.
QK157.M63   581.9'773   75-22414
ISBN 0-8093-0704-9
ISBN 0-8093-0756-1 pbk.

Copyright © 1975 by Southern Illinois University Press
All rights reserved
Printed in the United States of America
Designed by Gary Gore

BUSINESS/SCIENCE/TECHNOLOGY DIVISION
THE CHICAGO PUBLIC LIBRARY

JAN 2 0 1978

Contents

Contents

Preface

THIS book is intended to be a field guide to the vascular flora of Illinois. It provides keys to 3,047 taxa of ferns, gymnosperms, and flowering plants in the state. For each taxon there is given a common name if one is in general usage in Illinois, followed by an indication of flowering time for angiospermous plants. A statement of habitat is provided for each taxon, as well as a general comment on its distribution in Illinois. For plants considered rare and therefore known from only a very few counties, those counties are usually enumerated.

The sequence of families follows that found in the eighth edition of Asa Gray's *Manual of Botany* (Fernald, 1950), except that the Phrymaceae is placed next to the Verbenaceae in this work. I am aware that there is criticism to the Englerian sequence of families found in Gray's *Manual,* and that consideration of data derived from cytological, biochemical, and anatomical studies indicates different affinities for many of our families. Nonetheless, because systems devised by Cronquist (1968), Thorne (1968), and others are unfamiliar to the majority of field botanists, both professional and amateur, the more familiar Englerian system is employed in this work.

The nomenclature used has been derived after consultation with many floras, monographs, and revisions. Where the nomenclature adopted in this work is at variance with that found in the eighth edition of Gray's *Manual of Botany* (Fernald, 1950), the third edition of the Nathaniel Lord Britton and Addison Brown *Illustrated Flora of the Northern United States, Canada, and the British Possessions* . . . (Gleason, 1952), and the third edition of *Flora of Illinois* (Jones, 1963), the nomenclature used in these three works is indicated, followed by the abbreviations F, G, and J, respectively.

Included in this manual are all taxa known to be native in Illinois either at present or in the past, and all alien vascular plants for which there is evidence that they maintain themselves year after year without cultivation. No taxon is included in this work unless it has been verified by the writer or by Mr. Floyd Swink who is responsible for many recent discoveries in the Chicago area.

Illinois extends for nearly 380 miles north to south, more than one-third the distance from Canada to the Gulf of Mexico. Its maximum width is about 200 miles. The total number of square miles is just less than 58,000. The highest point of elevation is at Charles Mound in the extreme northwestern corner of the state, with an elevation above mean sea level of 1,241 feet. Average elevation for the state is 600 feet above sea level.

The French botanist André Michaux was apparently the first plant collector to visit Illinois when he crossed into the territory from Vincennes in 1795, traversed westward to Kaskaskia, and then southward and out of the state. In this short visit, Michaux collected fifty-two plants which were new to science. In 1818, Constantine Rafinesque botanized in the southeastern corner of the state.

The first flora for the state was prepared by Dr. S. B. Mead, a physician from Augusta, Illinois, who published, in 1846, a list of 900 species of plants known to occur in Illinois. In 1857, Increase Lapham wrote *A catalogue of the plants of the state of Illinois* in which he recorded 1,127 plants for the state. By 1876, H. N. Patterson was able to list 1,542 species of plants from Illinois.

For nearly three quarters of a century, no updated accounting of the plants of the entire state was available. In 1945, 1950, and 1963, G. N. Jones published the three editions of *Flora of Illinois* in which he listed 2,125, 2,200, and 2,400 species, respectively. Since I began work on the forty-volume "Illustrated Flora of Illinois" in 1960, more than 500 taxa of vascular plants have been added to the Illinois flora. This *Guide to the Vascular Flora of Illinois* accounts for 2,699 species, 265 lesser taxa, and 83 hybrids in Illinois. Continued study and exploration are adding several dozen taxa yearly to the Illinois flora.

I am indebted to Dr. Dan K. Evans who contributed *Carex* to this work, and to Dr. Richard P. Wunderlin who wrote the Compositae. Dr. Robert Haynes of Ohio State University has offered suggestions in *Potamogeton,* and Dr. James Richardson, Dr. Robert A. DeFilipps, and Dr. Kenneth L. Weik have helped with *Sagittaria, Juncus,* and the Lemnaceae, respectively.

I am grateful to the Illinois Nature Preserves Commission who has kindly let me use *The Natural Divisions of Illinois* in its entirety, and to Mr. John Schwegman, who was the principal author.

Mr. Floyd Swink of the Morton Arboretum has been a constant

source of inspiration for this book. Without his expertise, much would be lacking from this work.

Lastly I give much thanks to my wife, Beverly, who typed the several difficult drafts of the manuscript.

Robert H. Mohlenbrock

Southern Illinois University
August 1974

County Map of Illinois

Guide to the Vascular Flora of Illinois

1 Wisconsin Driftless Division

2 Rock River Hill Country Division
   a Freeport Section
   b Oregon Section

3 Northeastern Morainal Division
   a Morainal Section
   b Lake Michigan Dunes Section
   c Chicago Lake Plain Section
   d Winnebago Drift Section

4 Grand Prairie Division
   a Grand Prairie Section
   b Springfield Section
   c Western Section
   d Green River Lowland Section
   e Kankakee Sand Area Section

5 Upper Mississippi River and Illinois
   River Bottomlands Division
   a Illinois River Section
   b Mississippi River Section

6 Illinois River and Mississippi
   River Sand Areas Division
   a Illinois River Section
   b Mississippi River Section

7 Western Forest-Prairie Division
   a Galesburg Section
   b Carlinville Section

8 Middle Mississippi Border Division
   a Glaciated Section
   b Driftless Section

9 Southern Till Plain Division
   a Effingham Plain Section
   b Mt Vernon Hill Country Section

10 Wabash Border Division
    a Bottomlands Section
    b Southern Uplands Section
    c Vermilion River Section

11 Ozark Division
    a Northern Section
    b Central Section
    c Southern Section

12 Lower Mississippi River Bottomlands Division
    a Northern Section
    b Southern Section

13 Shawnee Hills Division
    a Greater Shawnee Hills Section
    b Lesser Shawnee Hills Section

14 Coastal Plain Division
    a Cretaceous Hills Section
    b Bottomlands Section

The Natural Divisions of Illinois

Courtesy of the Illinois Nature Preserves Commission

The Natural Divisions of Illinois

1. Introduction

The Natural Divisions of Illinois [1] is a classification of the natural environments and biotic communities of Illinois based on physiography, flora, and fauna. The general approach was first applied to Illinois by Vestal,[2] who developed a preliminary vegetation map of the state based mainly on topography, soil type, and natural vegetation.

The natural divisions of Illinois have been derived through an expansion of Vestal's approach. Factors considered in delimiting the 14 natural divisions are topography, soils, bedrock, glacial history, and the distribution of plants and animals. The divisions are divided into 33 sections based on differences that are of the same nature as those used to separate the divisions but deemed less significant. The role of each factor in developing this system is discussed in the text.

TOPOGRAPHY

Glacial history has played an important role in shaping Illinois's topography. Except for relatively small driftless areas in northwestern, west-central, and extreme southern Illinois, the topography can be characterized as a more or less dissected plain of glacial till. Rugged topography is characteristic of the driftless areas and of much of the till plain along the major river valleys. Other areas of significant relief within the till plain are the youthful moraines of northeastern Illinois and areas of thin drift and rolling topography in the Rock River valley and southeastern Illinois. The remainder of

---

[1] Used by permission of the Illinois Nature Preserves Commission. Written primarily by John Schwegman.

[2] Vestal, A. G. 1931. A Preliminary Vegetation Map of Illinois. Transactions of the Illinois State Academy of Science 23:204–17.

this till plain is an arc of older, eroded Illinoian till south and wes of the younger, less dissected Wisconsinan till.

Topography has influenced the biota of Illinois by controlling th diversity of habitats. In the glaciated regions, forest was restricted mainly to the rugged topography of stream valleys and moraine while prairie occupied most of the level uplands and some of the broad floodplains. In general, the more rugged the topography th greater the diversity of habitats. Youthful topography, such as the Wisconsinan till plain, is poorly drained, resulting in an abundance of aquatic habitats. Drainage has been improved by ditching ir most prairie regions of the till plain.

SOILS

The principal soil features influencing the recognition of divisions and sections are sand areas, soils derived from glacial till deep loess soils, and claypan soils. The description of soil conditions within the divisions is only generalized. Alluvial soils occur ir every division, but they are not usually mentioned in the text.

Some of the most important features of soils are parent material, texture, and degree of development. The general terms "heavy soils," "light soils," and "sandy soils" refer to texture. Heavy soils are predominantly clay; moisture generally moves slowly through such soils and aeration is poor. They often provide a more rigorous environment for plant growth. Light soils contain a mixture of clay, silt, and sand, are well aerated, and readily yield moisture to plants. Sandy soils are well aerated and have poor moisture-holding ability. They provide a severe environment for plants because of their droughtiness and susceptibility to wind erosion.

The parent material for most Illinois soils is loess, a wind-deposited silt. Loess deposits are deepest near the major river valleys and become thinner and of finer texture with increased distance. This mantle of loess has reduced the diversity of soil types in much of the state. Deep loess soils are generally less weathered, and more fertile than those derived from thin loess, especially in the southern third of Illinois.

Soils derived primarily from glacial till are restricted to northeastern Illinois, where geographic position has prevented deep accumulations of loess. The texture of these soils ranges from gravel to clay. The diversity of soil parent material is partly responsible for the great variety of environments and varied biota of this region.

Sand derived from glacial outwash is another important parent material that occurs as terraces along major streams throughout Illinois. In some instances the soils developed from sand support

biota distinctive enough to warrant recognition in the natural divisions. Local sand areas are not listed because they are generally
not considered to be of statewide significance.

Alluvial soils tend to be lighter in central and northern Illinois
than in southern Illinois. They frequently support distinctive biota.

Peat and bedrock are parent materials that give rise to soils
which often support unusual plants and animals. Although limited
in area, they have contributed to the recognition of some divisions
and sections.

The degree of soil development is a function of time and climate.
Older soils tend to be more strongly developed and have many of
their soluble minerals and clay particles leached from the surface
layers. The same is true of soils developed in areas of higher precipitation, as in southern Illinois. A strongly developed soil may
have a claypan formed by the accumulation of clay particles in the
subsoil. Claypan restricts penetration by roots and movement of
ground water, thus sometimes determining the dominant vegetation over wide areas. For example, the strongly developed claypan
soils of south-central Illinois determine to a great extent the distribution of post oak flatwoods. These soils are a primary factor in
the recognition of the Southern Till Plain Division.

Younger soils are derived from relatively recent deposits of till,
loess, or alluvium, or exposure of bedrock. They are not strongly
developed, are relatively rich in minerals, and lack claypans.

BEDROCK

Bedrock differences are often reflected in the topography, especially in driftless areas. Bedrock also controls the plant life to some
extent through the influence it has on the thin soils developed from
it. Bedrock type can determine the presence of caves and other
habitats.

The Ozark Division shows the effect of bedrock differences. The
Northern Section (Monroe County) and the Southern Section
(Union and Alexander counties) are both parts of the Salem Plateau
of the Ozarks and were presumably uplifted at about the same time.
The Southern Section is underlain by cherty limestone which is
resistant to erosion and has produced very rugged topography with
surface drainage. Steep cherty slopes in this section support acid-
soil plants such as shortleaf pine, azalea, and farkleberry. On the
other hand, the Northern Section is underlain by relatively pure
limestone that is less resistant to erosion and groundwater solution, producing gentler topography with some internal drainage
through sinkholes and solution cavities. Natural sinkhole ponds are

common. Most of the acid-soil plants are lacking, and the flora includes a distinctive assemblage of Ozark "limestone glade" species.

NATURAL VEGETATION

The dominant species and some of the most frequent associates in the principal plant communities of each natural division are listed in the text, but no attempt is made to characterize the total vegetation of each division. Generally the descriptions pertain to natural vegetation in Illinois about the time of settlement. At present this vegetation has been completely destroyed by man over vast sections of the state.

The striking contrast between prairie and forest vegetation that existed at the time of settlement is of little consequence to the recognition of natural divisions, as both forest and prairie or prairie-like communities occur in every division. However, the relative abundance of the prairie community was important in delimiting some divisions.

At the time of the settlement of Illinois there were considerable areas termed "barrens" by land surveyors. Apparently the barrens developed into forest or were cleared for agriculture soon after settlement. In any event, they were destroyed before being thoroughly described. They probably occurred in all the natural divisions, but they are not included in the vegetation accounts because their species composition is unknown.

FLORA

An analysis of the distribution of vascular plants in relation to physiographic provinces, bedrock types, soil types, topography, and drainage systems reveals a considerable number of distinctive species apparently restricted to certain provinces or natural features. These restricted distributions, especially for such dominant plants as trees, were important in delimiting the natural divisions. Many of these distinctive plants are listed in the descriptions of the divisions.

FAUNA

The distribution of birds in Illinois is less adequately documented than that of other classes of vertebrates. Nor does their distribution correlate as well with the natural divisions recognized here, being more heavily influenced by climate and by artificial and transitory conditions. For these reasons, distinctive species of birds are mentioned in the descriptions of only a few natural divisions.

Examination of the distribution of non-avian vertebrates in Illinois reveals many correlations with physiographic features, vegetation, watersheds, and soil types. The present work indicates considerable correlation between the natural divisions and patterns of fish distribution, and to a lesser extent with those of mammals.

The vertebrates that are considered distinctive of certain divisions and species that have influenced the location of divisional boundaries are mentioned in the descriptions of the divisions. There is no attempt to characterize the fauna of each division.

No attempt has been made to examine the distribution of invertebrates, or their relation to the natural divisions. The plains scorpion is mentioned as a distinctive feature of the Ozark Division, but this is the only exception.

CLIMATE

Climate has played a role in the delineation of some divisions and sections by its effects on the distribution of plants and animals, especially in extreme northern Illinois and southern Illinois. However, there is little direct correspondence between climatic regions and the natural divisions.

Illinois has a continental climate with hot humid summers and cold winters. This climate varies considerably from north to south in temperature and precipitation. The northern counties have lower mean annual temperatures, shorter growing seasons, and less precipitation than the southern counties.

The mean annual temperature ranges from about 47° F. in the north to about 59° F. in the south. The mean January temperature ranges from 22° F. in the north to 36° F. in the south, and the mean July temperature ranges from 73° F. in the north to 80° F. in the south. The average length of the growing season ranges from less than 160 days in northwestern Illinois to more than 200 days in the southern tip of the state.

Average annual precipitation increases from about 32 inches in the north to 47 inches in the south. Although the southern counties receive the most rainfall, much of this occurs in winter. Precipitation during the growing season is more nearly uniform throughout the state.

2. Wisconsin Driftless Division

The Wisconsin Driftless Division is part of an area extending from northwestern Illinois into Wisconsin, Iowa, and Minnesota that apparently escaped Pleistocene glaciation. This division is one of the

most maturely developed land surfaces in Illinois and is character
ized by rugged terrain that originally was mostly forested. It has the
coldest climate in the state. It contains several distinctive plants of
northern affinity and some species that may represent relicts of the
pre-ice-age flora. The division contains lead deposits.

BEDROCK

The Wisconsin Driftless Division is a maturely dissected upland of
Ordovician and Silurian limestone, dolomite, and shale. Bedrock
crops out along the major watercourses. Prominent "mounds"
capped with the more resistant dolomite are common. A mineralized zone containing deposits of lead and zinc is an important feature. Caves are known in the dolomite.

TOPOGRAPHY

The topography of the Wisconsin Driftless Division is one of rolling hills and great relief, particularly along interior stream canyons.
High erosional remnants (including Charles Mound, the highest
point in Illinois, with an elevation of 1,257 feet), are prominent features. There are loess-capped bluffs and palisades along the Mississippi River valley and ravines and bluffs throughout the division.

SOILS

The soils of this division have developed from loess or, on
steeper slopes, from loess on bedrock. The loess soils are derived
from thick deposits and are weakly to moderately developed. The
soils on bedrock are thin to moderately thick and well drained.

PLANT COMMUNITIES

FOREST   The original vegetation of the Wisconsin Driftless Division was predominantly upland hardwood forest dominated by
black oak (*Quercus velutina*) and white oak (*Quercus alba*) on dry
sites and by sugar maple (*Acer saccharum*), basswood (*Tilia americana*), and red oak (*Quercus rubra*) on mesic sites. Floodplain
forests dominated by silver maple (*Acer saccharinum*), American
elm (*Ulmus americana*), and green ash (*Fraxinus pennsylvanica* var.
subintegerrima) occupy alluvial soils of the stream valleys. Cliffs
and cool, shaded slopes of this division often support white pine
(*Pinus strobus*), Canada yew (*Taxus canadensis*), and white birch
(*Betula papyrifera*).

PRAIRIE   Dry prairie on the rolling uplands contained such species of the northern Great Plains as the plains buttercup (*Ranunculus rhomboideus*), pasque flower (*Anemone patens*), June grass
(*Koeleria macrantha*), and Wilcox's panic grass (*Panicum wilcox-
anum*), along with the dominant little bluestem (*Schizachyrium
scoparium*) and side-oats grama (*Bouteloua curtipendula*). Areas of
mesic and wet prairie were infrequent. Frink's Prairie and Jules
Prairie were two extensive prairies on the uplands. Loess hill
prairies dominated by little bluestem and side-oats grama occur on
the steep southwest-facing bluffs above the Mississippi River floodplain.

AQUATIC HABITATS

The main aquatic habitats of the Wisconsin Driftless Division are
creeks and rivers. Springs are local, and intermittent streams are
characteristic of the ravines.

SPECIAL FEATURES

The Wisconsin Driftless Division contains several distinctive
plants considered relicts of preglacial or interglacial floras, such as
jeweled shooting star (*Dodecatheon amethystinum*), sullivantia
(*Sullivantia renifolia*), and cliff goldenrod (*Solidago sciaphila*).
Cool, shaded ravines, cliffs, and river bluffs provided habitat for
relict populations of some distinctive northern plants including
woodland white violet (*Viola incognita*), bird's-eye primrose (*Primula mistassinica*), American stickseed (*Hackelia americana*), and
moschatel (*Adoxa moschatellina*). Some of these distinctive plants
are restricted to river bluffs of the Mississippi River valley, and
others grow only in interior stream canyons.

Principal Natural Features

FOREST: Dry upland, mesic upland, floodplain.
PRAIRIE: Dry, mesic, wet.
LOESS HILL PRAIRIE
BEDROCK: Outcrops of dolomite and shale, zinc and lead de-
posits, caves.
TOPOGRAPHY: Mississippi River bluffs, interior stream can-
yons, ravines, ridges and "mounds," level to rolling upland.
AQUATIC HABITATS: Creeks, rivers.
SPECIAL FEATURES: Northern and preglacial or interglacial
plants.

3. Rock River Hill Country Division

The Rock River Hill Country Division is a region of rolling topog
raphy that is drained by the Rock River. It has a thin mantle o
glacial till. Prairie formerly occupied the larger expanses of leve
uplands, but forest was equally abundant along the water courses
and in the more dissected uplands. Several distinctive plant species
occur in this division. The Freeport and Oregon sections are recog
nized on the basis of bedrock types and resultant floral differences.

GLACIAL HISTORY

The Rock River Hill Country Division is thinly mantled with glacial
drift from the Illinoian and early Wisconsinan stages of Pleistocene
glaciation. The Pecatonica lobe of the Altonian substage of the Wisconsinan glacial stage extended westward in this division into Stephenson and Ogle counties.

BEDROCK

The bedrock is primarily Ordovician and Silurian dolomite and
limestone. Outcrops of the dolomite occur throughout the division,
particularly along the streams. St. Peter sandstone underlies the
Oregon Sections and crops out frequently. There are caves in the
dolomite near the Mississippi River.

TOPOGRAPHY

The topography varies from level to rolling, with river valleys, and
"dells" or bluffs along streams throughout the division. The Oregon
Section is very rough, with bluffs, ridges, and ravines in the sandstone. The Pecatonica River meanders through a broad plain
formed by sediments of a glacial lake.

SOILS

The soils have developed primarily from moderately thick loess
and thin loess on bedrock. Soils developed from glacial outwash
occur in the major river valleys. Small areas of soils developed from
glacial till occur throughout the division.

PLANT COMMUNITIES

FOREST   The forests are similar to those of the Wisconsin Driftless Division. They occurred on slopes and areas protected from

prairie fires. The dry upland forests are dominated by black oak, white oak, bur oak (*Quercus macrocarpa*), and wild black cherry (*Prunus serotina*). Mesic sites support forests dominated by sugar maple, basswood, slippery elm (*Ulmus rubra*), and red oak. White pine, Canada yew, and yellow birch (*Betula lutea*) are occasionally found on cool, shaded bluffs. The floodplain forests are particularly well developed along the Pecatonica River and are dominated by silver maple, black willow (*Salix nigra*), cottowood (*Populus deltoides*), American elm, and ashes (*Fraxinus* spp.). The Oregon Section is heavily forested with black oak, white oak, and bur oak.

PRAIRIE   Prairies occupied much of the level to rolling uplands. The largest was known as Shannon Prairie. The dry upland prairies contained the floral elements of the northern Great Plains listed for the Wisconsin Driftless Division. Mesic prairies were the most common type and contained the species typical of the Grand Prairie, the dominant grasses being big bluestem (*Andropogon gerardii*), Indian grass (*Sorghastrum nutans*), and prairie dropseed (*Sporobolus heterolepis*). Wet prairies contained cord grass (*Spartina pectinata*), bluejoint grass (*Calamagrostis canadensis*), and big bluestem.

MARSH   Marshes in poorly drained parts of the prairies and along the streams and river floodplains are dominated by cattail (*Typha* sp.), bulrushes (*Scirpus* spp.), sedges (Cyperaceae), and occasionally common reed (*Phragmites australis*).

AQUATIC HABITATS

The major aquatic habitats of the Rock River Hill Country Division are rivers and creeks. Meander scar sloughs characterized the Pecatonica River floodplain. Springs and seepage areas occur locally, especially in the sandstone areas of the Oregon Section.

FREEPORT SECTION   The Freeport Section includes most of the Rock River Hill Country Division and is characterized by rolling hills and dolomite and limestone bedrock.

Principal Natural Features

FOREST: Dry upland, mesic upland, floodplain.
PRAIRIE: Dry, mesic, wet.
MARSH
BEDROCK: Outcrops of dolomite and limestone caves.
TOPOGRAPHY: Stream canyons and bluffs, floodplain, rolling uplands, meander scars.
AQUATIC HABITATS: Rivers, creeks, sloughs.
SPECIAL FEATURE: Northern relict plants.

OREGON SECTION   The Oregon Section is distinguished fror
the rest of the division by its sandstone bedrock and the uniqu
northern plants associated with it. These distinctive northern relict
include ground pine (*Lycopodium dendroideum*), rusty woodsi
(*Woodsia ilvensis*), oak fern (*Gymnocarpium dryopteris*), and Amer
can wintergreen (*Pyrola americana*).

Principal Natural Features

FOREST: Dry upland, mesic upland, floodplain.
PRAIRIE: Dry, mesic, wet.
BEDROCK: Sandstone outcrops.
TOPOGRAPHY: Rolling uplands, floodplain, ravines, bluffs.
AQUATIC HABITATS: Rivers, creeks.
SPECIAL FEATURE: Northern relict plants.

4. Northeastern Morainal Division

The Northeastern Morainal Division is the region of most recent
glaciation in Illinois. Glacial landforms are common features and
are responsible for the rough topography over most of the area.
Lake-bed deposits and beach sands are also frequent features. Unlike most of Illinois, the soils of this division are derived from glacial
drift rather than loess. Drainage is poorly developed, and many natural lakes are found. This division contains distinctive northern and
eastern floral elements including the bog community. Several species of animals are known in Illinois only from this area.

The sections are recognized because of differences in topography, soil, glacial history, flora, and fauna.

BEDROCK

The bedrock is primarily Ordovician and Silurian limestone and
dolomite with some shale. The bedrock is deeply buried by glacial
drift but limestone crops out along some of the streams.

GLACIAL HISTORY

The Northeastern Morainal Division is covered with deep glacial
drift from the Altonian and Woodfordian substages of the Wisconsinan glacial stage. Moraines, kames, eskers, and other glacial
landforms occur throughout the division.

TOPOGRAPHY

Moraines and morainic systems are dominant topographic fea-
ures and account for the rough, hilly, and rolling terrain of most of
ne division. There are outwash plains at the fronts of major termi-
al moraines, such as the Marengo Ridge. The Chicago lake plain
nd ancient beach ridges are prominent features of the Chicago
rea and were formed by sediments of glacial Lake Chicago. Bluffs
vere formed along Lake Michigan north of Chicago during high-
vater stages of glacial Lake Chicago.

Sand dunes are present along Lake Michigan north of Waukegan
nd east of the Sugar River in Winnebago County. The Lake Mich-
gan dunes are well developed and are associated with the beach
rea. This is the only division to have a natural beach-and-dunes association. Ridges and swales occur in the sand area north of Waukegan and in the Chicago lake plain.

SOILS

The soils are derived primarily from glacial drift, lake bed sediments, beach deposits, and peat. They range from very poorly
drained to well drained on the uplands. They are diverse in texture,
ranging from gravel and sand to silty clay loams. The many different soils are responsible for the diversity of plant communities
found in this division.

PLANT COMMUNITIES

FOREST   Bur oak and white oak dominate the dry upland forests
on moraines and other glacial landforms of this division. Many of
the forests were typical oak openings but have been heavily grazed
or have rapidly grown up to closed, dense forests due to the effec-
tive exclusion of fire. The mesic upland forests were dominated by
sugar maple, basswood, red oak, and white ash (*Fraxinus americana*). Notable northern shrubs of these mesic forests are highbush
cranberry (*Viburnum trilobum*) and red elder (*Sambucus pubens*).
Beech (*Fagus grandifolia*) occurred in a few ravines along Lake Michigan. The dominant species of the floodplain forests are silver
maple, green ash, and American elm. Some of the poorly drained
upland forests are dominated by swamp white oak (*Quercus bicolor*). Black oak and Hill's oak (*Quercus ellipsoidalis*) occur on the
sandy soils of the dunes and ridges, forming savanna-like or scrub-
oak forests. Jack pine (*Pinus banksiana*) and white pine once grew
on sandy ridges in the Chicago lake plain. Tamarack (*Larix laricina*)

occurs in some poorly drained depressions of the Valparais
morainic system. Plants essentially restricted in Illinois to th
forests of the Northeastern Morainal Division include maple-lea
viburnum (*Viburnum acerifolium*), round-lobed hepatica (*Hepatic
nobilis* var. *obtusa*), wood reed (*Cinna latifolia*), large-leaved aste
(*Aster macrophyllus*), moccasin flower (*Cypripedium acaule*), an
purple trillium (*Trillium erectum*).

PRAIRIE The presettlement vegetation of the Northeaster
Morainal Divison was about 60 percent prairie. Dry prairie on th
gravel moraines and eroded bluffs was dominated by little bluesten
and side-oats grama, and contained several species typical of th
western plains. The mesic prairie and wet prairie were dominatec
by prairie dropseed, big bluestem, Indian grass, switch grass (*Pani
cum virgatum*), cord grass, and bluejoint grass, and containec
many characteristic forbs. Extensive areas of sand prairie occurrec
in the sand areas of the Chicago lake plain, east of the Sugar Rive
in Winnebago County, and in the Lake Michigan Dunes Section
Some distinctive plants of sandy prairie and sandy open woods are
sweet-fern (*Comptonia peregrina*), speckled alder (*Alnus rugosa*)
yellow fringed orchid (*Habenaria ciliaris*), fringed gentian (*Gentiana
crinita*), and small fringed gentian (*Gentiana procera*). Colic-roo
(*Aletris farinosa*), hardhack (*Spiraea tomentosa*), and lupine (*Lu
pinus perennis*) occur in sand prairies near Lake Michigan.

FEN A fen is a type of wet prairie with an alkaline water source.
Fens are associated with calcareous springs and seeps and also
occur in swales and on low ground near lakes in areas of cal
careous ground water. They warrant recognition because of their
distinctive species composition. Some of the notable plants are
small, white lady's slipper (*Cypripedium candidum*), grass-of-Par
nassus (*Parnassia glauca*), meadow spikemoss (*Selaginella apoda*),
Ohio goldenrod (*Solidago ohioensis*), Kalm's lobelia (*Lobelia kal
mii*), shrubby cinquefoil (*Potentialla fruticosa*), low calamint (*Sa
tureja arkansana*), and white camass (*Zigadenus glaucus*).

MARSH Marshes are conspicuous plant communities of the
Northeastern Morainal Division, common because of the poorly
drained soils. Marshes are generally dominated by cattails and
bulrushes, with common reed locally abundant.

SEDGE MEADOW Many of the marshes of this division grade
into an open community where sedges dominate instead of cattails
and bulrushes, forming sedge meadows. These are often as
sociated with wet-shrub communities dominated by dogwoods
(*Cornus* spp.), willows (*Salix* spp.) and, infrequently, speckled alder
(*Alnus rugosa*).

BOG True bogs are found in Illinois only in the Northeastern

orainal Division. These have formed in poorly drained depres-
ions in the Valparaiso morainic system and contain many distinc-
ve plants such as pitcher plant (*Sarracenia purpurea*), sundew
Drosera sp.), cranberry (*Vaccinium macrocarpon*), leatherleaf
Chamaedaphne calyculata), poison sumac (*Rhus vernix*), winter-
erry (*Ilex verticillata*), and dwarf birch (*Betula pumila*). All stages
f bog succession—young, mature, and old—are represented.

AQUATIC HABITATS

The Northeastern Morainal Division is poorly drained and has
many aquatic habitats. It contains all of Illinois's glacial lakes and is
he only division on Lake Michigan. The glacial lakes are generally
of two types: those with a peat base, and those with a sand or marl
base. The biota of each type is different. Some species of water
marigold (*Bidens beckii*) and two species of bladderwort (*Utricu-
laria* spp.) are restricted to the glacial lakes of this division.

DISTINCTIVE FAUNA

The pugnose shiner, blackchin shiner, and banded killifish are
known in Illinois only from the glacial lakes of this division. The
alewife, American smelt, lake chub, and ninespine stickleback are
fishes that occur in Illinois only along the shore of Lake Michigan.
Lake trout and lake whitefish are species restricted to the deeper
waters of Lake Michigan. The common tern breeds in Illinois in only
a few locations near Lake Michigan, and the piping plover did so
until fairly recently. Breeding of the golden-winged warbler in Illinois is restricted to this division, where it has been known to hybri-
dize with the blue-winged warbler. Breeding of the Nashville warbler in Illinois is reported only from this division. Except for casual
occurrences, the large population of oldsquaws that winters on Illinois waters is restricted to Lake Michigan; and much the same is
true for wintering white-winged, common, and surf scoters. The occurrence in Illinois of several species of migrant waterbirds such as
the parasitic, pomarine, and long-tailed jaegers, the knot, and Bon-
aparte's and little gull is limited almost entirely to Lake Michigan or
its shores. Other animals restricted to this division are the spotted
turtle, blue-spotted salamander, and pigmy shrew.

MORAINAL SECTION   The Morainal Section of the Northeastern
Morainal Division encompasses the moraines and morainic systems
of the late advances of the Woodfordian substage of Wisconsinan
glaciation. This section contains most of Illinois's glacial lakes as
well as its true bogs. Glacial landforms are well represented.

Principal Natural Features

> FOREST: Dry upland, mesic upland, wet upland, floodplain tamarack swamp.
> PRAIRIE: Dry, mesic, wet.
> FEN
> MARSH
> SEDGE MEADOW
> BOG: Youthful, mature, old.
> GLACIAL LANDFORMS: Moraines, kames, eskers, drumlins kettleholes.
> TOPOGRAPHY: Rolling upland, ravines, lake bluffs.
> AQUATIC HABITATS: Rivers, creeks, glacial lakes, sloughs.

LAKE MICHIGAN DUNES SECTION This section is recognized because of the unique flora of its dunes and beaches. This flora includes some sand-binding plants that differ from those of the inland sand deposits, such as beach grass (*Ammophila breviligulata*), creeping juniper (*Juniperus horizontalis*), and bearberry (*Arctostaphylos uva-ursi*). Plant communities range from sand prairie to marsh, to scrub-oak forest. The continuing vegetational succession from shifting sand to stabilized sand results in a wide variety of plant associations.

Principal Natural Features

> FOREST: Scrub-oak.
> SAND PRAIRIE: Dry, mesic, and wet.
> FEN
> MARSH
> TOPOGRAPHY: Beach, ridges and swales, dunes.
> AQUATIC HABITATS: Creeks, Lake Michigan.

CHICAGO LAKE PLAIN SECTION The Chicago lake plain is a flat, poorly drained area of lake-bed sediments deposited by glacial Lake Chicago. It is recognized because of its special topography and physiographic history. Long ridges of shore-deposited sands are conspicuous topographic features. A few natural lakes occur near Calumet City. The original vegetation was mostly prairie and marsh, with scrub-oak forests on sandy ridges. Black gum (*Nyssa sylvatica*) and sassafras (*Sassafras albidum*) are found in some of the wet forests. Thismia (*Thismia americana*), one of the most unsual plants of the American flora, occurred only in the Chicago ake Plain Section, near Lake Calumet, but is now considered ex-nct because this area has been converted to indusry.

Principal Natural Features

FOREST: Scrub-oak, mesic upland, floodplain.
PRAIRIE: Dry, mesic, wet.
FEN
MARSH
TOPOGRAPHY: Lake plain, ridges and swales.
AQUATIC HABITATS: Lakes, creeks, Lake Michigan.

WINNEBAGO DRIFT SECTION This section encompasses the Winnebago Formation of Altonian drift. This early Wisconsinan drift s better drained than that of the Morainal Section, having fewer marshes and no glacial lakes. The original vegetation was predomi-nantly prairie, with oak openings, dry upland forest, and well-developed floodplain forests. Glacial outwash is extensive along many of the creeks and rivers. Wet prairie and marsh with large sedge meadows or "prairie bogs" occur in the sand area along Coon Creek. Dunes have formed along the east bank of the Sugar River and support sand prairie vegetation and dry upland forests of black oak and Hill's oak.

Distinctive features of this section are the extensive gravel hill prairies which once extended along the eroded east bluffs of the Rock River valley into Wisconsin. These prairies contained many western elements, including pasque flower, plains buttercup, and prairie smoke (*Geum triflorum*).

Principal Natural Features

FOREST: Dry upland, mesic upland, floodplain.
PRAIRIE: Dry, mesic, wet.
SAND PRAIRIE: Dry, mesic, wet.
MARSH
SEDGE MEADOW
BEDROCK: Outcrops of limestone and sandstone.
TOPOGRAPHY: Dunes, outwash plains, river terraces, mean-der scars.
AQUATIC HABITATS: Rivers, creeks, sloughs.

5. Grand Prairie Division

The Grand Prairie Division is a vast plain formerly occupied pr
marily by tall-grass prairie. The soils were developed from recent
deposited loess, lake-bed sediments, and outwash and are gene
ally very fertile. Natural drainage was poor, resulting in man
marshes and prairie potholes. Forest bordered the rivers and ther
were occasional groves on moraines and other prominent glacia
landforms. The sections of this division are differentiated on th
basis of soils, topography, and glacial history.

At one time bison grazed the prairies and waterfowl in grea
numbers occupied the marshes and potholes. The steel plov
brought about the rapid destruction of the vast Illinois prairies
Ditches and tile lines drained almost all of the marshes and pot
holes. The bison were gone by 1814. The abundant waterfowl were
displaced. The giant Canada goose was extirpated as a breeding
bird, and other characteristic species disappeared or became
scarce. The prairie, once seemingly limitless, is now one of the
rarest plant communities in Illinois, with only pitifully small and
often degraded patches remaining.

GLACIAL HISTORY

The Grand Prairie is a rather level, poorly drained plain of glacia
drift from the Illinoian and Wisconsinan stages of Pleistocene gla
ciation. Repeated advances and retreats of the Wisconsinan
glaciers created a series of moraines and morainic systems o
which the Shelbyville and Bloomington morainic systems are conspicuous.

BEDROCK

The bedrock, deeply buried by glacial drift, crops out only along
the larger rivers. Major outcrops of sandstone are found near Ot-
tawa along the Illinois and Fox rivers. Dolomite crops out along the
Kankakee River west of Kankakee.

TOPOGRAPHY

The topography of the Grand Prairie Division is generally level to
rolling, with the major stream valleys and the extensive systems of
moraines providing the greatest relief. Large flat expanses of lake-
bed deposits are found in Lasalle, Kendall, Will, Grundy, Livingston,
Ford, Iroquois, Kankakee, and Douglas counties. Extensive out-

ash plains and sand dunes are found in Kankakee and Iroquois ounties and in the valleys of the Green River and lower Rock iver. The major rivers have well-developed floodplains, and in 1any areas there are ravines in the bluffs.

SOILS

The soils are relatively young and high in organic content, having eveloped from a thin to moderately thick layer of loess, glacial rift, or lake-bed sediments. Soils developed from sand, muck, and 1eat exist in the Kankakee Sand Area and Green River Lowland ections. Deep loess occurs along the Illinois and the lower Sanga-10n rivers.

PLANT COMMUNITIES

FOREST   The forests of the Grand Prairie Division are generally 1ssociated with the stream valleys and moraines. On dry sites, the orests are dominated by white oak, black oak, and shagbark hick-ory (*Carya ovata*), with shingle oak (*Quercus imbricaria*) and bur oak frequent associates. On mesic sites these species are replaced by basswood, sugar maple, slippery elm, American elm, hackberry Celtis occidentalis), red oak, and white ash. Black walnut (*Juglans nigra*), bitternut hickory (*Carya cordiformis*) and, in the northern part, bigtooth aspen (*Populus grandidentata*) are common. The floodplain forests are of the silver maple-American elm-ash type. The prairie groves were influenced by recurrent fires and are gener-ally of two types: one dominated by bur oak and the other domi-1ated by American elm and hackberry. Sandy soils in the Green River Lowland and Kankakee Sand Area sections support scrub forests of black oak.

PRAIRIE   The vast prairies of Illinois were once one of its most remarkable features. They contained several hundred species of grasses and forbs and were interspersed by numerous marshes and prairie potholes. At the time of settlement, wet and mesic prairie were the most widespread plant communities of the Grand Prairie Division. Mesic prairie was dominated by big bluestem, Indian grass, prairie dropseed, switch grass, and little bluestem and con-tained many characteristic prairie plants such as leadplant (*Amor-pha canescens*), compass plant (*Silphium laciniatum*), prairie dock (*Silphium terebinthinaceum*), and rattlesnake master (*Eryngium yuccifolium*). The wet sites were dominated by cord grass, sedges, and bluejoint grass and supported such species as ironweed (*Vernonia fasciculata*), boneset (*Eupatorium perfoliatum*), swamp milkweed (*Asclepias incarnata*), and water hemlock (*Cicuta macu lata*).

Dry upland prairie occurs mainly on steep slopes along the Illinois River, on loess bluffs along the lower Sangamon River, and on gravel moraines and kames. The dominant species of the dr prairies are little bluestem and side-oats grama. They common contain such forbs as scurf pea (*Psoralea tenuiflora*), pale bearc tongue (*Penstemon pallidus*), false boneset (*Kuhnia eupatorioides* cylindrical blazing star (*Liatris cylindracea*), and fringed puccoo (*Lithospermum incisum*).

Sand prairie occurs in the Kankakee Sand Area and Green Rive Lowland sections. Little bluestem, fall witch-grass (*Leptoloma cog natum*), and sand dropseed (*Sporobolus cryptandrus*) are the im portant grasses, and goat's rue (*Tephrosia virginiana*) and spotte monarda (*Monarda punctata*) are common forbs. June grass an porcupine grass (*Stipa spartea*) are commom. The sand prairie contain several species of Great Plains affinity such as wester ragweed (*Ambrosia psilostachya*), prickly-pear cactus (*Opuntia ra finesquii*), poppy mallow (*Callirhoe triangulata*), hairy grama (*Bou teloua hirsuta*), western sunflower (*Helianthus occidentalis*), silk aster (*Aster sericeus*), and flax-leaved aster (*Aster linariifolius*).

MARSH   The formerly common marshes and prairie potholes o the Grand Prairie Division were dominated by bulrushes, sedges bur-reeds (*Sparganium* spp.), cattails, and common reed and con tained many species of aquatic and semiaquatic plants such as ar rowhead (*Sagittaria* spp.), water plantain (*Alisma subcordatum*) pondweed (*Potamogeton* spp.), pickerelweed (*Pontederia cordata*) beggar-ticks (*Bidens* spp.), and water crowfoot (*Ranunculus* spp.).

AQUATIC HABITATS

The aquatic habitats of the Grand Prairie Division are rivers creeks, and prairie potholes.

DISTINCTIVE FAUNA

Distinctive animals of the Grand Prairie Division are Blanding's turtle, western smooth green snake, western fox snake, eastern plains garter snake, Kirtland's water snake, northern lined snake, Franklin's ground squirrel, and thirteen-lined ground squirrel. None of these animals are completely restricted to this division, but all are most abundant here. Several amphibians and reptiles common outside the limits of Wisconsinan glaciation are conspicuously absent from the Grand Prairie.

GRAND PRAIRIE SECTION   The Grand Prairie Section encom-
asses the area outside the Northeastern Morainal Division that
as covered by the Woodfordian substage of the Wisconsinan
age of Pleistocene glaciation, excluding the outwash and sand
reas. The Shelbyville and Bloomington morainic systems form the
oundaries of this section. Mesic black-soil prairie, marshes, and
rairie potholes in the young, poorly-drained drift are character-
tic. Glacial landforms are common. The Kankakee mallow
(Iamna remota) and Chase aster (Aster schreberi) are endemic to
linois and restricted to small areas of the Grand Prairie Section.
he Kankakee mallow is found only on an island in the Kankakee
,iver, and Chase aster is known only from three counties near
'eoria.

Principal Natural Features

FOREST: Floodplain, mesic upland, dry upland.
PRAIRIE GROVE
PRAIRIE: Wet, mesic, dry.
MARSH
BEDROCK: Sandstone and dolomite outcrops.
TOPOGRAPHY: Level to rolling upland, floodplain, ravines,
river bluffs, lake plains, glacial landforms.
AQUATIC HABITATS: Prairie potholes, rivers, creeks.
SPECIAL FEATURES: Endemic plants.

SPRINGFIELD SECTION   The Springfield Section is part of the
llinoian drift, and the drainage system is better developed than in
he younger Wisconsinan drift of the Grand Prairie Section. This
;ection was mostly prairie in presettlement times. Deep loess de-
)osits that support dry hill prairie occur along the lower Sangamon
River. Large tracts of floodplain forest exist in the valley of the
ower Sangamon River and its tributaries.

Principle Natural Features

FOREST: Floodplain, mesic upland, dry upland.
PRAIRIE: Wet, mesic, dry.
LOESS HILL PRAIRIE
MARSH
BEDROCK: Outcrops.
TOPOGRAPHY: Level to rolling upland, floodplain, ravines,
river bluffs.
AQUATIC HABITATS: River, creeks.

WESTERN SECTION The Western Section was predominant prairie in presettlement times and therefore is in the Grand Prairi Division. It is part of the older, dissected Illinoian drift.

Principal Natural Features

FOREST: Floodplain, mesic upland, dry upland.
PRAIRIE: Wet, mesic, dry.
MARSH
BEDROCK: Outcrops.
TOPOGRAPHY: Level to rolling upland, floodplain, ravines.
AQUATIC HABITATS: Creeks.

GREEN RIVER LOWLAND SECTION The broad valley of th Green River and lower Rock River was formed by glacial melt wa ters. Much glacial outwash was deposited, and sand flats an dunes developed. This section had extensive marshes and we prairies. It has scrub-oak forests on the dry sandy ridges and flood plain forests along the rivers. Sand prairie occupied the sand flat and dunes. Most of this section has been disturbed by grazing drainage, and cultivation. There are some active dunes where th stabilizing cover has been removed.

Principal Natural Features.

FOREST: Scrub-oak, floodplain.
PRAIRIE: Wet, mesic.
SAND PRAIRIE: Wet, mesic, dry.
MARSH
TOPOGRAPHY: Outwash plain, dunes.
AQUATIC HABITATS: Rivers.

KANKAKEE SAND AREA SECTION The sand of the Kankakee Sand Area Section was deposited by the Kankakee Flood during late Wisconsinan glaciation. Sand prairie and marsh were the pre dominant vegetation of this section before the land was drained fo cultivation. Primrose violet (*Viola primulifolia*) and Carey's smart weed (*Polygonum careyi*) are restricted in Illinois to this section Scrub-oak forests occur on drier sites. The clear, well-vegetated sand-bottomed streams contain such unusual fishes as the weec shiner, iron color shiner, and least darter.

Principal Natural Features

FOREST: Scrub-oak.
SAND PRAIRIE: Wet, mesic, dry.
MARSH
TOPOGRAPHY: Outwash plain, dunes.
AQUATIC HABITATS: Creeks.

6. Upper Mississippi River and Illinois River Bottomlands Division

The Upper Mississippi River and Illinois River Bottomlands Divi-
ɔn encompasses the rivers and floodplains of the Mississippi
ver above its confluence with the Missouri River and the bot-
ᴛmlands and associated backwater lakes of the Illinois River and
 major tributaries south of LaSalle. It does not include the major
ɴd deposits, which are in a separate division. Much of the divi-
ɔn was originally forested but prairie and marsh also occurred.
ᴇ more sluggish nature of the Illinois River and its distinctive
ckwater lakes distinguishes the Illinois River Section from the
ɔper Mississippi River Section.

BEDROCK

The bedrock of the two river valleys is deeply covered by alluvial
ᴇposits.

TOPOGRAPHY

The bottomlands of the Mississippi River and the Illinois River are
ᴛaracterized by broad floodplains and gravel terraces formed by
ᴀcial flood waters.

SOILS

The soils are from recent alluvium and glacial outwash. They are
ɔorly drained, alkaline to slightly acidic, and vary from sandy to
ayey. In general they are lighter than the alluvial soils of the
ɔwer Mississippi River Bottomlands Division.

PLANT COMMUNITIES

FOREST The bottomland forests are generally dominated by
ᴠver maple, American elm, and green ash. Pin oak (*Quercus palus-*

tris) is the most important oak; pecan (*Carya illinoensis*), bur oa
sycamore (*Platanus occidentalis*), honey locust (*Gleditsia tr
canthos*), hickories (*Carya* spp.), and black walnut are freque
Black willow and river birch (*Betula nigra*) are common in the M
sissippi River Section. A few southern lowland species, includi
water locust (*Gleditsia aquatica*), overcup oak (*Quercus lyrat*
sugarberry (*Celtis laevigata*), deciduous holly (*Ilex decidua*), a
swamp privet (*Forestiera acuminata*) range into the southern part
this division.

PRAIRIE   In presettlement times mesic prairie and wet prairie c
curred in the broad bottomlands. The species composition of the
prairies was similar to that of the prairies of the Grand Prairie Di
sion.

MARSH   Marshes containing the typical marsh species are i
portant features throughout both sections.

SPRING BOGS   Spring-fed bogs with peat deposits are found
terraces along the Illinois River. These are unique to the Illino
River and the species composition differs from that of the bogs
the Northeastern Morainal Division. Distinctive plants include bla
ash (*Fraxinus nigra*), willows, poison sumac, and skunk cabba
(*Symplocarpus foetidus*).

AQUATIC HABITATS

Oxbow lakes occur in both the Illinois River valley and the Missi
sippi River valley. Backwater lakes are distinctive of the Illino
River valley, and the Illinois River is more sluggish than the Missi
sippi River. Springs are common in gravel terraces along the Illino
River. The bottomland forests are subject to prolonged periods
flooding.

DISTINCTIVE FAUNA

The fish faunas of the Illinois River and the upper Mississip
River are similar although they differ somewhat from that of the si
laden Mississippi River south of its confluence with the Missou
River.

ILLINOIS RIVER SECTION   The Illinois River Section is disti
guished from the Mississippi River Section by its distinctive bac
water lakes and differences in forest vegetation. The spring bog
along the river bluffs are a special feature.

Principal Natural Features

FOREST: Bottomland.
PRAIRIE: Wet, mesic.
MARSH
SPRING BOGS
TOPOGRAPHY: River floodplain, river terraces.
AQUATIC HABITATS: Backwater lakes, oxbow lakes, rivers.

MISSISSIPPI RIVER SECTION   This section is composed of sev-
ral distinct bottomlands along the Mississippi River, from Wiscon-
in to Calhoun County. Most of the prairies of this section have
een drained for agriculture. Forests are still found along the river
side the levees and on the river islands.

Principal Natural Features

FOREST: Bottomland.
PRAIRIE: Wet, mesic.
MARSH
TOPOGRAPHY: River floodplain, river terraces.
AQUATIC HABITATS: Oxbow lakes, rivers.

7. Illinois River and Mississippi River Sand Areas Division

The Illinois River and Mississippi River Sand Areas Division
encompasses the sand areas and dunes in the bottomlands of
the Illinois and Mississippi rivers and includes the "perched
dunes" atop the bluffs near Hanover in JoDaviess County.
Scrub-oak forest and dry sand prairie are the natural vegetation of this division. Several plant species found here are
more typical of the short-grass prairies to the west of Illinois.
Several "relict" western amphibians and reptiles are known
only from these sand areas. The two sections are distinguished because of differences in flora and fauna.

TOPOGRAPHY

The topography is generally one of level to rolling plains of sand
deposited by glacial melt waters and blown into widespread areas
east of the rivers. In many areas the sand has migrated onto the
bluffs and uplands east of the river terraces. In places, dunes 20 to

40 feet high have formed and blowouts are common in unstabiliz
sand.

SOILS

The soils are derived from sand and sandy material. Other soils
depressions surrounded by sand are also in this division. The so
are generally droughty and subject to wind erosion. Low areas a
generally wet.

PLANT COMMUNITIES

FOREST The forests of this division are limited to scrub
stands dominated by black oak in the Mississippi River Section ar
black and blackjack oaks (*Quercus marilandica*) in the Illinois Riv
Section.

PRAIRIE Sand prairie, composed of such species as little blu
stem, June grass, Indian grass, and porcupine grass, is the maj
community of this division. The sand prairie habitats range from d
to wet and include such plants as goat's rue, spotted monard
prickly-pear cactus, tubercled three-awned grass (*Aristida tuberc*
losa), poppy mallow, and fall witch-grass. The dry sites have wes
ern floral elements, including sand love-grass (*Eragrostis tr*
chodes), hairy grama and, rarely, bladderpod (*Lesquerel*
ludoviciana), and Patterson's bindweed (*Stylisma pickeringii*). Th
mesic and wet sites have prairie vegetation similar to that of th
Grand Prairie Division. There are also various plant association
related to unstabilized sand. Long-leaved calamovilfa (*Calamovilf*
longifolia) is one of the principal sand binders.

MARSH Marsh occurs in low, poorly drained areas.

DISTINCTIVE FAUNA

Distinctive animals are the bullsnake, plains hognose snake, Ill
nois mud turtle, and Illinois chorus frog. The Illinois chorus frog i
restricted to the Illinois River Section. The white-tailed jackrabbit i
found in Illinois only in the northern part of the Mississippi Rive
Section. The lark sparrow breeds most commonly in Illinois in the
sandy habitats of this division, and a northern outlier population o
the summer tanager breeds in forests of the Illinois River Section.

ILLINOIS RIVER SECTION The Illinois River Section is distin
guished from the Mississippi River Section on the basis of flora
and faunal differences.

Principal Natural Features

FOREST: Scrub-oak.
SAND PRAIRIE: Dry, mesic, wet.
MARSH
TOPOGRAPHY: Dunes, blowouts, level to rolling plain.

MISSISSIPPI RIVER SECTION   The Mississippi River Section has
oral elements absent from the Illinois River Section, including
ock spikemoss (*Selaginella rupestris*), rough-seeded rock-pink (*Ta-
num rugospermum*), and beach-heath (*Hudsonia tomentosa*).
ock spikemoss and beach-heath form large mats that stabilize
lowouts.

Principal Natural Features

FOREST: Scrub-oak.
SAND PRAIRIE: Dry, mesic, wet.
MARSH
TOPOGRAPHY: Dunes, blowouts, level to rolling plain.

8. Western Forest-Prairie Division

The Western Forest-Prairie Division is a strongly dissected glacial
ill plain of Illinoian and Kansan age. At the time of settlement,
orest was the predominant vegetation, but there was also considerable prairie on the level uplands. The prairie soils were developed
rom loess and are fertile. The two sections are geographically separated by the Illinois River valley and also have some faunal dif-
erences.

GLACIAL HISTORY

Most of the bedrock is covered by glacial drift from the Illinoian
stage of Pleistocene glaciation. There is an area of older Kansan
drift in the western parts of Pike and Adams counties.

BEDROCK

Pennsylvanian and Mississippian bedrocks of limestone, sandstone, shale, and coal crop out frequently along the major streams.

TOPOGRAPHY

The till plain is strongly dissected, with many ravines in the leve
to rolling uplands. Floodplains are developed along the majo
streams.

SOILS

Most of the soils are fairly young, having developed from four to
five feet of loess. The prairie soils are high in organic matter and
are similar to those of the Grand Prairie Division. The forest soil
are acidic and low in organic matter. Relatively small areas c
droughty, fine-textured soils have developed in till on some stee
slopes.

PLANT COMMUNITIES

FOREST   The upland forests consist of an oak-hickory associa
tion with black oak, white oak, and several species of hickory as the
dominants. Scattered sites of fine-textured soils support a post oak
blackjack oak community. This forest community also occurred or
the margins of the prairies, perhaps because of fires. The mesi
forests contain white oak, red oak, basswood, sugar maple, and
slippery elm. The floodplain forests are dominated by silver maple
American elm, ashes, and box elder (*Acer negundo*).

PRAIRIE   In presettlement times large prairies existed on the
uplands of this division. Carthage Prairie, Hancock Prairie, and
Bushnell Prairie were in the Galesburg Section; and String Prairie
and Brown's Prairie were in the Carlinville Section. The prairie veg
etation was similar to that of the Grand Prairie Division, but we
prairie was less frequent.

MARSH   Poorly drained areas with marsh vegetation are less
frequent than in the Grand Prairie Division but are similar in com
position.

AQUATIC HABITATS

The aquatic habitats of this well-drained division consist mainly
of rivers and creeks.

DISTINCTIVE FAUNA

There is some continuity of animal life between the Western
Forest-Prairie Division and the Southern Till Plain Division. The

ive-lined skink, the broad-headed skink, and the ornate box turtle are species that occur throughout the Southern Till Plain Division and range into this division but are absent from the adjoining Grand Prairie Division.

GALESBURG SECTION   The Galesburg Section is the area north of the broad Illinois River valley and is distinguished from the Carlinville Section because of its separate location. There were about equal amounts of forest and prairie in this section at the time of settlement, with the forests primarily in the well-dissected areas along tributaries of the Illinois River.

Principal Natural Features

FOREST: Dry upland, mesic upland, floodplain.
PRAIRIE: Dry, mesic, wet.
MARSH
BEDROCK: Outcrops.
TOPOGRAPHY: Level to rolling uplands, ravines, floodplain.
AQUATIC HABITATS: Rivers, creeks.

CARLINVILLE SECTION   The Carlinville Section is the area of well-dissected land southeast of the Illinois River valley. The original vegetation of this section was mostly forest, with only about 12 percent of the area prairie.

Principal Natural Features

FOREST: Dry upland, mesic upland, floodplain.
PRAIRIE: Dry, mesic, wet.
MARSH
BEDROCK: Outcrops.
TOPOGRAPHY: Level to rolling uplands, ravines, floodplain.
AQUATIC HABITATS: Creeks.

9. Middle Mississippi Border Division

The Middle Mississippi Border Division consists of a relatively narrow band of river bluffs and rugged terrain bordering the Mississippi River floodplain from Rock Island County to St. Clair County and the lower Illinois River floodplain. Forest is the predominant

natural vegetation but hill prairies are common on the west-facing
bluffs. The soils were generally developed from very deep loess.
Limestone cliffs are common features. This division is best distinguished from the river bluffs to the north and south of it by the absence of certain floral and faunal elements. The Driftless Section
was never glaciated and is distinguished from the remainder of the
division.

BEDROCK

The bedrock of the Middle Mississippi Border Division consists of
limestone and sandstone, with dolomite associated with the sandstone in Calhoun County. Outcrops and cliffs of limestone are common along the bluffs, and sandstone outcrops occur in Rock Island
and Calhoun counties. Cretaceous gravels and clays occur in Pike
and Adams counties. There are caves throughout the division in the
limestone and dolomite but they are most abundant in unglaciated
Calhoun County.

GLACIAL HISTORY

Most of the division was glaciated during the Illinoian stage of
Pleistocene glaciation. An area of older Kansan drift is in western
Hancock, Adams, and Pike counties. Calhoun County and parts of
Pike and Adams counties apparently escaped Pleistocene glacia-
tion.

TOPOGRAPHY

The Middle Mississippi Border Division is greatly dissected, particularly along the major streams where there are bluffs and ravines. Sinkhole plains are most common in the southern part of the
division. The Driftless Section is higher and has more rugged topography than the Glaciated Section.

SOILS

Most of the soils on the uplands have developed from deep, well-
drained loess. Isolated areas of heavy soils are also present, especially in the southern part of the division.

PLANT COMMUNITIES

FOREST The vegetation of the Middle Mississippi Border Division is mostly mesic and dry forests associated with the dissected

uplands. The forests of the dry sites are dominated by black oak and white oak. Sugar maple, basswood, red oak, hackberry, slippery elm, and black walnut are major components of the forests on mesic sites. Floodplain forests along the creeks contain silver maple, hickories, cottonwood, and sycamore. Post oak (*Quercus stellata*) is common on the heavy soils and near ridge tops.

PRAIRIE Prairies of the Middle Mississippi Border Division are limited to the steep slopes and ridges of deep loess atop the river bluffs. The prairies are dominated by little bluestem and side-oats grama, with purple prairie clover (*Petalostemum purpureum*) and flowering spurge (*Euphorbia corollata*) among the most frequent forbs. Scurf pea, a distinctive western plant, is also common. Stickleaf (*Mentzelia oligosperma*), another western plant, reaches its northeastern limits on the exposed limestone ledges of this division and of the Northern Section of the Ozark Division.

AQUATIC HABITATS

There are creeks throughout the division and sinkhole ponds in the Driftless Section.

DISTINCTIVE FAUNA

The dark-sided salamander and western worm snake are restricted in Illinois to the Middle Mississippi Border Division. Forested glens of this division serve as major nighttime roosting places for wintering bald eagles.

GLACIATED SECTION The topography of this section has been modified by the Illinoian and Kansan stages of Pleistocene glaciation. Limestone underlies most of the Glaciated Section and frequently forms cliffs along the river bluffs.

Principal Natural Features

FOREST: Dry upland, mesic upland, floodplain.
LOESS HILL PRAIRIE
BEDROCK: Limestone outcrops, caves.
TOPOGRAPHY: River bluffs, ravines, floodplain, sinkhole plain.
AQUATIC HABITATS: Creeks.
SPECIAL FEATURE: Roosting areas for wintering bald eagles.

DRIFTLESS SECTION The Driftless Section apparently escaped Pleistocene glaciation. Its topography is rougher than that of the

Glaciated Section, and it has many sinkholes and sinkhole pond
Except for the jeweled shooting star, the Driftless Section is n
known to harbor preglacial relict plants.

Principal Natural Features

FOREST: Dry upland, mesic upland, floodplain.
LOESS HILL PRAIRIE
BEDROCK: Limestone and sandstone outcrops, caves.
TOPOGRAPHY: River bluffs, ravines, floodplain, sinkhol
plain.
AQUATIC HABITATS: Creeks, sinkhole ponds.
SPECIAL FEATURE: Roosting areas for wintering bald eagles

10. Southern Till Plain Division

The Southern Till Plain Division encompasses most of the area c
dissected Illinoian glacial till plain south of the Shelbyville Morain
and the Sangamon River and Macoupin Creek watersheds. Bot
forest and prairie were present at the time of settlement. The soil
are relatively poor because of their high clay content and th
frequent occurrence of a "claypan" subsoil. Post oak flatwoo
forest is characteristic of the division. The two sections are distir
guished because of topographic differences.

BEDROCK

The bedrock of the Southern Till Plain Division consists of sand
stone, limestone, coal, and shale, which commonly crop out in th
eastern and southeastern parts of the division. Bedrock lies nea
the surface in the Mt. Vernon Hill Country Section.

GLACIAL HISTORY

The Illinoian stage of Pleistocene glaciation reached the south
ernmost limit of North American continental glaciation just beyon
the limits of this division. The Southern Till Plain Division is entirel
covered by Illinoian till. Glacial landforms are common only in th
northwestern part of the division.

TOPOGRAPHY

The glacial till of the Southern Till Plain Division becomes thinne
from north to south. The bedrock of the Mt. Vernon Hill Countr

ection is near the surface, accounting for its hilly and rolling to-
ography. The Effingham Plain Section is a nearly level to dis-
ected till plain. There are broad floodplains along the major
treams and there are ravines in the bluffs along the stream valleys.

SOILS

The soils on the uplands are light colored and strongly devel-
ped, with poor internal drainage. They have developed from thin
oess and till under both forest and prairie vegetation. Fragipan and
laypan layers are characteristic of the upland soils. Some of the
rairie soils have a high sodium content and are known locally as
alkaline slicks."

PLANT COMMUNITIES

FOREST   The level, poorly drained, heavy soils on the uplands of
he Southern Till Plain Division support a characteristic flatwoods
orest of post oak, swamp white oak, blackjack oak, and pin oak.
Forests on the uplands, where drainage is slightly improved, include black oak, shingle oak, mockernut hickory (*Carya tomen-
osa*), and shagbark hickory in the post oak community. The forests
on the slopes along stream valleys are dominated by white oak,
shingle oak, and black oak on the drier southern and western exposures, with hickories, white ash, basswood, sugar maple, wild
black cherry, slippery elm, and black walnut with the oaks on the
more mesic sites. Forests in the broad floodplains of the Kaskaskia
and Big Muddy rivers are dominated by silver maple, willows, sycamore, and American elm near the rivers, with pin oak, white oak,
hickories, ashes, hackberry, and honey locust on the heavier soils
farther from the rivers. Pin oak occasionally grows in nearly pure
stands over large areas of the floodplain. The floodplain forests of
the smaller streams have a higher percentage of oaks than the
floodplain forests of central and northern Illinois. Pin oak and shingle oak are dominant, with white oak, red oak, hickories, black walnut, river birch, and cottonwood occasional associates. Shumard
oak (*Quercus shumardii*) and sweet gum (*Liquidambar styraciflua*)
grow in the floodplain of the Big Muddy River but are not generally
abundant.

PRAIRIE   At the time of settlement about 40 percent of the
uplands of the Southern Till Plain Division supported prairie vegetation. Most of the prairie was of the mesic tall-grass type characteristic of the Grand Prairie Division. Twelve Mile Prairie and Look-
ing Glass Prairie were two large expanses of prairie. Mesic prairies
extended along the west side of the division almost to the limit of

glaciation but were rare in the southeastern part. Wet prairie wa
not common but did occur in parts of the Kaskaskia River flooc
plain. It is not known whether the alkaline slicks of this divisio
supported a unique prairie flora.

MARSH   Marshes were associated with the stream floodplains c
this division.

AQUATIC HABITATS

The aquatic habitats of the Southern Till Plain Division consist c
rivers, creeks, and oxbow lakes.

DISTINCTIVE FAUNA

The northern crayfish frog, northern fence lizard, ground skink
five-lined skink, and broad-headed skink are common in the South
ern Till Plain Division but are rare or absent from the Grand Prairi
Division.

EFFINGHAM PLAIN SECTION   The Effingham Plain Section is
relatively flat plain drained by the Kaskaskia River. It originally wa
mostly prairie. Post oak flatwoods are characteristic of the upland
of this section.

A few flocks of the greater prairie chicken remain within this sec
tion. Sanctuaries are being established to maintain the populatio
by providing nesting habitat.

Principal Natural Features

FOREST: Upland flatwoods, dry upland, mesic upland, anc
floodplain.

PRAIRIE: Wet, mesic, dry.

MARSH

BEDROCK: Outcrops.

TOPOGRAPHY: Level to rolling upland, dissected till plain
ravines, floodplain.

AQUATIC HABITATS: Creeks, rivers, oxbow lakes.

SPECIAL FEATURES: Greater prairie chicken population, alka
line slicks.

MT. VERNON HILL COUNTRY SECTION   The Mt. Vernon Hill
Country Section is distinguished from the Effingham Plain Section
by its rolling, hilly topography. In presettlement times uplanc
forests covered most of this section. The striped shiner and the

stoneroller, two fishes of nearly statewide distribution, are absent from most of the Mt. Vernon Hill Country Section. The broad bottomlands of the major rivers that drain the eastern part of this section are considered to be part of the Wabash Border Division.

Principal Natural Features

FOREST: Upland flatwoods, dry upland, mesic upland, floodplain.
PRAIRIE: Wet, mesic, dry.
MARSH
BEDROCK: Outcrops
TOPOGRAPHY: Rolling till plain, dissected till plain, ravines, floodplain.
AQUATIC HABITATS: Creeks, rivers, oxbow lakes.

11. Wabash Border Division

The Wabash Border Division includes the bottomlands of the Wabash River and its major tributaries, the loess-covered uplands bordering the Wabash River, and the forests of the Vermilion River, Little Vermilion River, and Crab Apple Creek. This is a region of lowland oak forests containing beech, tuliptree (*Liriodendron tulipifera*), and other trees typical of the forests to the east of Illinois. The Wabash River drainage contains several distinctive fishes. The sections are distinguished by differences in topography, glacial history, and flora and fauna.

BEDROCK

The bedrock of the Wabash Border Division consists of limestone, sandstone, coal, and shale. Small outcrops of bedrock occur along some of the streams throughout the division. Bedrock outcrops along the river bluffs are few and never form the towering cliffs characteristic of the western border of Illinois.

GLACIAL HISTORY

All but a small area in the southern part of the Wabash Border Division was subject to Pleistocene glaciation. The Vermilion River dissects younger Wisconsinan drift, whereas the remainder of the division is the older Illinoian till plain. There are areas of outwash sand along the lower Wabash River.

TOPOGRAPHY

The topography of the Wabash Border Division is relatively gentle. The Southern Uplands Section is a low, eroded till plain with bluffs above the bottomlands. The Vermilion River Section has more rugged topography resulting from erosion of its streams into Wisconsinan drift. The rivers have broad floodplains formed by glacial lakes, which include terrace deposits and many meander scars.

SOILS

The soils in the Bottomlands Section range from floodplain soils to terrace soils. The soils of the Southern Uplands Section have developed from moderately deep loess deposits. Soils of the Vermilion River Section are derived from thin loess over loamy till. Despite the diversity of soils, this division is united by the continuity of its forest vegetation and its fish and amphibian faunas.

PLANT COMMUNITIES

FOREST   The forests of the bottomlands are developed either on Recent alluvium or on Pleistocene terraces, which are generally better drained. Bottomland forests in the floodplains of the major rivers are dominated by pin oak, overcup oak, swamp white oak, swamp chestnut oak (*Quercus michauxii*), bur oak, cherrybark oak (*Quercus pagodaefolia*), and Shumard oak. Commonly associated with the oaks are sweetgum, hackberry, American elm, kingnut hickory (*Carya laciniosa*), silver maple, and pecan. The forests on the terraces contain a mixture of the tree species found on the floodplains and those found on the uplands. Shumard oak, bur oak, and sweetgum are the commonest, with pin oak and swamp white oak in poorly drained areas. The best-drained terraces contain shagbark hickory and tuliptree. Other characteristic trees are kingnut hickory, Kentucky coffeetree (*Gymnocladus dioica*), and hackberry. Along the rivers black willows, cottonwood, sycamore, and silver maple predominate. Sloughs near the Wabash River and Saline River contain bald cypress (*Taxodium distichum*) and some other southern swamp species.

The mesic forests of the Southern Uplands Section and the Vermilion River Section are dominated by white oak, red oak, and sugar maple and frequently contain beech and tuliptree. Forests on drier sites are dominated by black oak and hickories. Forests in the floodplains of the Southern Uplands Section and Vermilion River Section contain silver maple, cottonwood, willows, sycamore, and American elm.

PRAIRIE At the time of settlement some of the Southern Uplands Section supported mesic prairie, while mesic and wet prairie occurred in the Bottomlands Section. The prairies of this division were similar to those of the Grand Prairie Division.

MARSH The poorly drained bottomlands contain large areas of marsh associated with the sloughs and meander scars. These are dominated by cord grass and river bulrush (*Scirpus fluviatilis*).

AQUATIC HABITATS

The poorly drained Bottomlands Section of the Wabash Border Division contains many aquatic habitats, including rivers, oxbow lakes, and sloughs. The Southern Uplands and Vermilion River sections are better drained, and the aquatic habitats consist mainly of creeks and rivers.

DISTINCTIVE FAUNA

Fishes limited in Illinois to the Wabash Border Division are the river chub, river redhorse, mountain madtom, and greenside darter. The northern bigeye chub and bluebreast darter are known in Illinois only from the Vermilion River system, and the harlequin darter is kown only from the Embarras River. The red-backed salamander is essentially restricted in Illinois to this division.

BOTTOMLANDS SECTION The Bottomlands Section of the Wabash Border Division encompasses the bottomland forests, sloughs, marshes, and oxbow lakes in the floodplains of the Wabash River, the Ohio River, and their major tributaries. Bottomland forests are the predominant vegetation with wet prairie and marsh associated with the sloughs.

Principal Natural Features

FOREST: Floodplain, terrace, swamp.

PRAIRIE: Mesic, wet.

MARSH

TOPOGRAPHY: River floodplain, terrace deposits, meander scars.

AQUATIC HABITATS: Oxbow lakes, sloughs, rivers.

SOUTHERN UPLANDS SECTION The Southern Uplands Section contains the dry and mesic upland forests on the deep loess bluffs along the Wabash River. The upland forests of white oak, sugar maple, beech, and sweetgum are the predominant plant community. Some sandstone ravines support an unusual combination of plant species that includes some relict northern species.

Principal Natural Features

FOREST: Dry upland, mesic upland, floodplain.
PRAIRIE: Mesic, dry.
BEDROCK: Outcrops.
TOPOGRAPHY: Dissected till plain, river bluffs, ravines.
AQUATIC HABITATS: Creeks.
SPECIAL FEATURE: Relict northern plants.

VERMILION RIVER SECTION The Vermilion River Section is characterized by rugged topography and the beech-maple forests in the ravines along the Vermilion River and its tributaries. The beech-maple forests represent an important climax deciduous forest type of the northeastern United States, which is found in Illinois only in the extreme eastern and southern portions.

Principal Natural Features

FOREST: Dry upland, mesic upland, floodplain.
PRAIRIE: Dry, mesic, wet.
TOPOGRAPHY: Dissected till plain, ravines, floodplain.
AQUATIC HABITATS: Rivers, creeks.

12. Ozark Division

The Ozark Division consists of the Illinois part of the Salem Plateau of the Ozark uplift from northern Monroe County southward and includes the glaciated sandstone ravines in Randolph County. The area is mostly forested, but many hill prairies occur in the Northern Section. The division contains many Ozarkian, southern, and southwestern plants and animals that are rare or absent elsewhere in Illinois. The sections are based on differences in bedrock, topography, flora, and fauna.

BEDROCK

Most of the Ozark Division is part of the Salem Plateau of the Ozark uplift. The northern part of the division is underlain by relatively pure limestone, while the southern part is underlain by cherty limestone that is more resistant to erosion. Sandstone underlies the Central Section of the division. Bedrock crops out in all sections. Caves and sinkholes are numerous in the limestone of the Northern Section and less so in the Southern Section.

GLACIAL HISTORY

Part of the Northern Section and all of the Central Section of the Ozark Division were glaciated during the Illinoian stage of Pleistocene glaciation. The Southern Section is driftless.

TOPOGRAPHY

The topography of the Ozark Division is that of a maturely dissected plateau with steep bluffs along the Mississippi River. There are ravines and stream canyons throughout the division, especially in the sandstone of the Central Section. The Northern Section has a well-developed sinkhole plain topography.

SOILS

Most of the soils of the Ozark Division are derived from deep loess, but thin soils occur over the bedrock outcrops along the river bluffs and in interior ravines. Some of the soils of the Southern Section are derived from bedrock and are acidic.

PLANT COMMUNITIES

FOREST At the time of settlement the Ozark Division was almost entirely forested. The forests of the Northern section consist in part of red oak, sugar maple, basswood, and Ohio buckeye (*Aesculus glabra*) on the mesic sites, with white oak, black oak, and hickories on the ridge tops. Beech and tuliptree occur in the mesic forests of the Central Section. The forests of the Southern Section contain a rich assemblage of tree species including cucumbertree (*Magnolia acuminata*), blackgum, butternut (*Juglans cinerea*), black walnut, and bitternut hickory in addition to red oak, sugar maple, basswood, white oak, black oak, Ohio buckeye, beech, and tuliptree. Yellow-wood (*Cladrastis lutea*) occurs in the Southern Section in Alexander County. A mixed association of white oak, red oak, sycamore, American elm, river birch, wild black cherry, and cottonwood is found along the stream floodplains throughout the division. Stands of shortleaf pine (*Pinus echinata*) are found in the

Southern Section and as far north as Piney Creek in Randolph County.

PRAIRIE   Loess hill prairies are common on the river bluffs in the Northern Section of the Ozark Division but are rare in the Southern Section and absent from the Central Section. These prairies have a similar species composition to that of loess hill prairies of the Middle Mississippi Border Division, with little bluestem and side-oats grama dominating; but they have several species of plants that are restricted to this division in Illinois.

AQUATIC HABITATS

The Ozark Division has few aquatic habitats. Ponds occur in some of the sinkholes. Springs occur at cave entrances and at the bases of some of the bluffs. Creeks are the commonest aquatic feature of the division.

DISTINCTIVE FAUNA

The plains scorpion, eastern narrow-mouthed toad, and eastern coachwhip are restricted in Illinois to the Northern Section of the Ozark Division, while the spring cavefish, northern blacktail shiner, and scarlet snake are known only from the Southern Section. The northern flat-headed snake occurs in both the Northern and Southern sections but is found nowhere else in the State.

NORTHERN SECTION   The Northern Section is distinguished from the other sections of the Ozark Division by its limestone bedrock, numerous caves and sinkholes, unique plant and animal species, and forest composition. Plant species unique to this section are the reticulate-seeded spurge (*Euphorbia spathulata*), stiff bedstraw (*Galium virgatum*), Missouri black-eyed susan (*Rudbeckia fulgida* var. *missouriensis*), and small heliotrope (*Heliotropium tenellum*). These plants grow in hill prairies or on exposed limestone ledges.

Principal Natural Features

FOREST: Dry upland, mesic upland, floodplain.

LOESS HILL PRAIRIE

BEDROCK: Sinkholes, caves, limestone outcrops.

TOPOGRAPHY: Sinkhole plain, ravines, river bluffs, floodplain.

AQUATIC HABITATS: Sinkhole ponds, creeks, springs.

SPECIAL FEATURES: Ozark "limestone glade" plants, distinctive reptiles and amphibians.

CENTRAL SECTION   The Central Section of the Ozark Division is distinguished because of its sandstone bedrock, forest composition, and distinctive flora. Distinctive plants of this section include Harvey's buttercup (*Ranunculus harveyi*), large-flowered rock-pink (*Talinum calycinum*), and Bradley's spleenwort (*Asplenium bradleyi*). These species apparently entered Illinois from the Missouri Ozarks after Illinoian glaciation.

Principal Natural Features

FOREST: Dry upland, mesic upland, floodplain.
BEDROCK: Sandstone outcrops.
TOPOGRAPHY: Hills, ravines, stream canyons, floodplain.
AQUATIC HABITATS: Creeks.
SPECIAL FEATURE: Ozarkian floral element.

SOUTHERN SECTION   The Southern Section of the Ozark Division is distinguished by its bedrock, topography, glacial history, unique fauna, forest composition, and distinctive southern and Ozarkian flora. Black spleenwort (*Asplenium resiliens*), shortleaf pine, azalea (*Rhododendron prinophyllum*), and big-leaf snowbell-bush (*Styrax grandifolia*) are part of the distinctive floral element.

Principal Natural Features

FOREST: Dry upland, mesic upland, floodplain.
LOESS HILL PRAIRIE
BEDROCK: Outcrops.
TOPOGRAPHY: Steep ravines, river bluffs, floodplain.
AQUATIC HABITATS: Creeks, springs, sinkhole ponds.
SPECIAL FEATURES: Ozarkian floral element, spring cavefish.

13. Lower Mississippi River Bottomlands Division

The Lower Mississippi River Bottomlands Division includes the Mississippi River and its floodplain from Alton to the Thebes Gorge. The Mississippi River is muddy here due to the silt load brought in by the Missouri River. Its fish fauna contains a distinctive assemblage of silt-tolerant plains species.

The Northern Section (the American Bottom) originally contained prairies, marshes, and forest. The Southern Section was densely forested. The forests of this division contain a greater number of tree species than the forests of the Upper Mississippi River, including some southern lowland species.

TOPOGRAPHY

The broad bottomlands of the Lower Mississippi River Bottomlands Division were formed by glacial flood waters. Since the retreat of the glaciers the river has meandered through this broad floodplain, and many meander scars and oxbow lakes remain.

SOILS

The soils are generally fine textured, with areas of both sandy, well drained soils and clay soils with poor internal drainage. The soils of this division have developed from alluvium.

PLANT COMMUNITIES

FOREST   Except for areas of wet prairie and marsh in the Northern Section, the division was entirely forested in presettlement times. The bottomland forests on the light soils include silver maple, ashes, American elm, honey locust, sugarberry, and pecan. Beech, basswood, and red buckeye (*Aesculus discolor*) grew as associated species on the loamy soils in the Southern Section; but most of these soils, being better drained and fertile, have been cleared for agriculture. Bottomland forests on heavy soils of the Southern Section are dominated by pin oak, overcup oak, Shumard oak, and cherrybark oak in association with kingnut hickory, sugarberry, and sweetgum. Bottomland swamps in the Southern Section are dominated by pumpkin ash (*Fraxinus tomentosa*), swamp cottonwood (*Populus heterophylla*), Drummond's red maple (*Acer rubrum* var. *drummondii*), and water locust. Bald cypress grows in the Southern Section as far north as the southern edge of Union County. Tupelo (*Nyssa aquatica*) and some other coastal plain species are absent.

PRAIRIE   There were relatively large areas of wet and mesic prairie in the Northern Section of the Lower Mississippi River Bottomlands Division, but most have been drained for agriculture.

MARSH   The Northern Section contained large marshes dominated by river bulrush, cattail, lotus (*Nelumbo lutea*), and pickerelweed. Some of the marshes remain, even though much of the area is organized into drainage districts.

AQUATIC HABITATS

The silt-laden Mississippi River below the Missouri River provides aquatic habitats somewhat different from those of the upper Mississippi River. Oxbow lakes and sloughs are common features of the bottomlands. The springfed swamps of northwestern Union County provide a unique habitat for several species of fish.

DISTINCTIVE FAUNA

The herpetofauna of the Lower Mississippi River Bottomlands Division is similar to that of the bottomlands of the Coastal Plain Division. It includes the western cottonmouth, green water snake, green treefrog, western bird-voiced treefrog, and mole salamander. The fish fauna of this division includes several species not found elsewhere in Illinois. The bantam sunfish is found only in the springfed swamps in northwestern Union Country; and the Alabama shad, plains minnow, sturgeon chub, flathead chub, and sicklefin chub are found in the Mississippi River. The banded pigmy sunfish is known only from the springfed swamps of this division and from the bottomland swamps of the Coastal Plain Division.

NORTHERN SECTION   The Northern Section is distinguished by its forest composition, the presence of wet prairies and marshes, and the absence of the coastal plain trees of the Southern Section. The bottomlands of this section near St. Louis are called the "American Bottoms."

Principal Natural Features

FOREST: Bottomland.
PRAIRIE: Wet, mesic.
MARSH
TOPOGRAPHY: River floodplain, meander scars.
AQUATIC HABITATS: Oxbow lakes, Mississippi River.

SOUTHERN SECTION   The bottomland forests of the Southern Section contain a greater number of tree species, including some bottomland swamp species typical of the coastal plain. The composition of the forests in this section varies with the soils.

Principal Natural Features

FOREST: Bottomland on heavy soils, bottomland on light soils, bottomland swamp.

TOPOGRAPHY: River floodplain, meander scars.
AQUATIC HABITATS: Oxbow lakes, Mississippi River.
SPECIAL FEATURE: Springfed swamps.

14. Shawnee Hills Division

The Shawnee Hills extend across the southern tip of the stat
from Fountain Bluff on the Mississippi River to the Shawneetow
Hills near the mouth of the Wabash River. This unglaciated hi
country is characterized by a high east-west escarpment of sand
stone cliffs forming the Greater Shawnee Hills and a series of lowe
hills underlain by limestone and sandstone known as the Lesse
Shawnee Hills. Originally this division was mostly forested and con
siderable forest remains to the present time. There are a number o
distinctive plant species restricted to this division of Illinois.

BEDROCK

The Greater Shawnee Hills form a band along the northern edge
of the division and consist of massive Pennsylvanian sandstone
strata that dip northward toward the Illinois Basin. The range o
hills averages 10 miles wide and borders the Lesser Shawnee Hills
to the south. The Lesser Shawnee Hills are underlain by Mississip
pian limestone and sandstone, and sinkholes and caves are locall
common features. Mineralized faults containing fluorspar and zinc
silver, and other metals exist in the eastern part of the Shawnee
Hills Division. Iron deposits are found in Hardin County. There is a
dome containing an igneous rock core in western Hardin County
and outcrops of igneous rock occur in the Lesser Shawnee Hills
Section.

TOPOGRAPHY

The topography of the Shawnee Hills Division is very rugged, with
many bluffs and ravines. The north slopes of the Greater Shawnee
Hills Section are relatively gentle, but the south slopes consist of
many escarpments, cliffs, and overhanging bluffs. Streams have
eroded canyons in the sandstone. The Lesser Shawnee Hills
average about 200 feet lower than the Greater Shawnee Hills. The
Lesser Shawnee Hills have local areas of sinkhole topography.

SOILS

The soils are derived mainly from loess. Narrow bands of moderately developed deep loess soils occur along the Mississippi River in Jackson County and along the Ohio River in eastern Hardin County; however, most of the soils are derived from thinner loess and are strongly developed. Claypan and fragipan layers are frequent.

PLANT COMMUNITIES

FOREST   At the time of settlement most of the Shawnee Hills Division supported forest, and considerable land remains timbered. Most of the upland forests are dominated by white oak, black oak, and shagbark hickory with post oak, blackjack oak, scarlet oak (*Quercus coccinea*), and pignut hickory (*Carya glabra*) on dry sites. Deep mesic ravines have a forest community of red oak, beech, tuliptree, bitternut hickory, sugar maple, and white ash with black walnut, butternut, Ohio buckeye, and basswood occasional. The floodplain forests also contain sycamore, Kentucky coffeetree, sugarberry, and honey locust.

PRAIRIE   The Lesser Shawnee Hills Section contains limestone glades which support a dry prairie vegetation like that of hill prairies in western Illinois, but with the addition of such southern plants as wild blue sage (*Salvia pitcheri*) and heart-leaved tragia (*Tragia cordata*).

AQUATIC HABITATS

The Shawnee Hills Division has numerous clear rocky streams and creeks. Sinkhole ponds are found in the Lesser Shawnee Hills Section.

DISTINCTIVE FAUNA

The streams support several distinctive fishes including blackspotted topminnow, spottail darter, and stripetail darter. The latter species is restricted to extreme southeastern Illinois.

GREATER SHAWNEE HILLS SECTION   The Greater Shawnee Hills Section is distinguished by its sandstone bedrock, topography, and distinctive plants. Filmy fern (*Trichomanes boschianum*), Virginia saxifrage (*Saxifraga virginiensis*), small-flowered rock-pink (*Talinum parviflorum*), thread-leaved evening primrose (*Oenothera linifolia*), synandra (*Synandra hispidula*), French's shooting star

(*Dodecatheon frenchii*), and small-flowered alumroot (*Heuchera parviflora*) are some of the distinctive plants of this section. French's shooting star and small-flowered alumroot are abundant in their restricted habitats and may have persisted since preglacial times. Deep ravines and sandstone ledges along the larger streams support relict northern plants such as clubmosses (*Lycopodium* spp.), sphagnum (*Sphagnum* sp.), and barren strawberry (*Waldsteinia fragarioides*). Except for synandra, these distinctive plants are absent from a seemingly suitable habitat in the glaciated area of southwestern Williamson County.

Principal Natural Features

FOREST: Dry upland, mesic ravine, floodplain.
BEDROCK: Outcrops, faults.
TOPOGRAPHY: Bluffs, ravines, stream canyons, floodplain, overhanging cliffs.
AQUATIC HABITATS: Creeks.
SPECIAL FEATURE: Northern relict plants, preglacial relict plants.

LESSER SHAWNEE HILLS SECTION   The Lesser Shawnee Hills Section is distinguished by its limestone bedrock and sinkhole topography. The fluorspar deposits near Cave in Rock and Rosiclare in Hardin County are world famous. Caves are common features of the limestone bluffs. Distinctive plants of this section are wild mock-orange (*Philadelphus pubescens*) and great chickweed (*Stellaria pubera*).

Principal Natural Features

FOREST: Dry upland, mesic ravine, floodplain.
LIMESTONE GLADES
BEDROCK: Sinkholes, caves, sandstone outcrops, limestone outcrops, igneous outcrops, mineralized faults.
TOPOGRAPHY: Sinkhole plain, interior bluffs, river bluffs, ravines, floodplain.
AQUATIC HABITATS: Sinkhole ponds, creeks.
SPECIAL FEATURE: Fluorspar deposits.

15. Coastal Plain Division

The Coastal Plain Division is a region of swampy forested bottomlands and low clay and gravel hills that is the northernmost extension of the Gulf Coastal Plain Province of North America. Bald cypress-tupelo gum swamps are a unique feature of this division in Illinois as are many southern animals and plants found within them. The division encompasses the bottomlands of the Cache, Ohio, and Mississippi rivers and hills capped by Cretaceous and Tertiary sand, gravel, and clay. It has a relatively mild climate, the warmest in the state. The two sections distinguish between upland and lowland environments.

GLACIAL HISTORY

The Coastal Plain Divison was never subjected to Pleistocene glaciation, but it has been influenced by glacial floodwaters. Glacial Lake Cache was formed in the valleys of the Cache and Ohio rivers during a late stage of glaciation. Pleistocene deposits form the terraces in the bottomlands of the rivers.

BEDROCK

The Cretaceous Hills Section is composed of unconsolidated sediments of Cretaceous and Tertiary sands, gravels, and clays. The bedrock of the Bottomlands Section is deeply buried by alluvium.

TOPOGRAPHY

The Bottomlands Section of the Coastal Plain Division consists of the broad floodplain at the confluence of the Ohio and Mississippi rivers, the broad floodplain of the Cache River, and the terraces and meander scars in the floodplains. The topography of the Cretaceous Hills Section is steep to rolling.

SOILS

The soils of the uplands are derived from relatively thin loess and in a few places from gravel. Some areas in the eastern part of the Cretaceous Hills Section have gravel exposures. The soils of the Bottomlands Section range from Recent alluvium to older terrace soils. The terrace soils tend to be hardpan clays; but areas of sand occur, especially in the western part of the section. The alluvial soils are generally heavy except near the rivers.

PLANT COMMUNITIES

FOREST   The presettlement vegetation of the Coastal Plain Division was mostly forest. The upland forests of the Cretaceous Hills Section are similar to those of the Shawnee Hills with black oak, white oak, red oak, cherrybark oak, blackgum, tuliptree, shagbark hickory, and pignut hickory the common trees. In the extreme western end of these hills beech and cucumbertree are also found. A large stand of native chestnuts (*Castanea dentata*) formerly grew near Olmsted in Pulaski County. Mesic ravines along the Ohio River contain southeastern floral elements that include the silverbell (*Halesia carolina*).

The bottomland forests consist in part of Shumard oak, cherrybark oak, swamp white oak, swamp chestnut oak, pin oak, overcup oak, kingnut hickory, shagbark hickory, bitternut hickory, ashes, sweetgum, blackgum, honey locust, sugarberry, pecan, wild black cherry, and catalpa (*Catalpa speciosa*). Beech, tuliptree, and cucumbertree grow on the better-drained bottomland soils. Pin oak is the commonest tree on the heavier terrace soils, along with post oak and willow oak (*Quercus phellos*). Silver maple and American elm grow along the streams.

The bottomland swamps of this division contain an association of bald cypress, tupelo gum, swamp cottonwood, Drummond's red maple, water locust, pumpkin ash, and overcup oak. Water hickory (*Carya aquatica*) and planer-tree (*Planera aquatica*) are occasional on better-drained soils.

PRAIRIE   The Cretaceous Hills Section contained small dry prairies on the uplands of the eastern part and mesic prairies in some of the broad creek bottoms. The few persisting remnants indicate that little bluestem was the dominant species of the dry upland prairies and that big bluestem, Indian grass, and gama grass (*Tripsacum dactyloides*) were the dominant species of the mesic prairies. Prairie dropseed and cord grass are absent from the prairies of the Cretaceous Hills Section.

SOUTHERN SEEP SPRING BOG   Seep springs in the eastern end of the Cretaceous Hills Section are very acidic. They are generally dominated by sedges, royal fern (*Osmunda regalis*), lady fern (*Athyrium filix-femina*), and cinnamon fern (*Osmunda cinnamomea*) and may contain large areas of sphagnum moss. These seep springs contain a distinctive southeastern flora which includes netted chain fern (*Woodwardia areolata*), screw-stem (*Bartonia paniculata*), and incomperta sedge (*Carex incomperta*).

AQUATIC HABITATS

The aquatic habitats of the bottomlands of the Coastal Plain Division include rivers, creeks, oxbow lakes, and sloughs. The Cretaceous Hills have creeks and seep springs.

DISTINCTIVE FAUNA

The herpetofauna of the swampy lowlands of the Coastal Plain Division is similar to that of the Lower Mississippi River Bottomlands Division and includes the western cottonmouth, green water snake, green treefrog, western bird-voiced treefrog, and mole salamander. The range of the dusky salamander in Illinois is essentially restricted to the Creataceous Hills Section.

CRETACEOUS HILLS SECTION   The Cretaceous Hills Section encompasses the rolling hills of unconsolidated Cretaceous and Tertiary sands, gravels, and clays. Found in these hills are fossil beds from the Cretaceous period.

Principal Natural Features

FOREST: Dry upland, mesic upland.
PRAIRIE: Dry, mesic.
SOUTHERN SEEP SPRING BOG
BEDROCK: Cretaceous and Tertiary sands, gravels and clays, Cretaceous fossil beds.
TOPOGRAPHY: Steep to rolling hills.
AQUATIC HABITATS: Creeks, seep springs.

BOTTOMLANDS SECTION   The Bottomlands Section encompasses the bottomland forests, oxbow lakes, sloughs, and rivers of the Coastal Plain Division. This section includes the remannts of the once vast bald cypress and tupelo gum swamps along the rivers.

Principal Natural Features

FOREST: Floodplain, terrace, swamp.
TOPOGRAPHY: Terrace ridges, floodplain, meander scars.
AQUATIC HABITATS: Backwater swamps, sloughs, oxbow lakes, creeks, rivers.

General Key to Groups of Illinois Vascular Plants

1. Plants ferns, quillworts, clubmosses, or horsetails ...... Group I, p. 49
1. Plants producing seeds 2
2. Plants cacti or pitcher-like or completely non-green .... Group II, p. 50
2. Plants neither cacti, pitcher-like, nor non-green 3
3. Plants climbing or twining .... Group III, p. 50
3. Plants neither climbing nor twining, although sometimes prostrate or
   trailing 4
4. Plants woody (excluding woody vines) 5
4. Plants herbaceous 8
5. Leaves less than 3 mm wide, usually needle-like .... Group IV, p. 52
5. Leaves more than 3 mm wide, not needle-like 6
6. Leaves compound .... Group V, p. 53
6. Leaves simple 7
7. Leaves alternate .... Group VI, p. 54
7. Leaves opposite or whorled .... Group VII, p. 57
8. Plants monocotyledonous, usually with elongated, parallel-veined
   leaves, flower parts usually in 3s, and a single cotyledonn ....
   .... Group VII, p. 57
8. Plants dicotyledonous, usually with net-veined leaves, flower parts
   usually in 2s, 4s, or 5s, and two cotyledons 9
9. Plants aquatic, living in water during the entire year ... Group IX, p. 59
9. Plants terrestrial, at least during much of each year 10
10. Leaves opposite or whorled .... Group X, p. 60
10. Leaves alternate or basal 11
11. At least some of the leaves compound .... Group XI, p. 63
11. None of the leaves compound, although sometimes deeply lobed .....
   .... Group XII, p. 65

I. Plants Ferns, Quillworts, Clubmosses, or Horsetails, not Producing Seeds

1. Leaves grass-like; corm at base of leaves mostly bilobed ....
   .... 4. Isoetaceae, p. 72
1. Leaves not grass-like; corm absent 2
2. Stem conspicuously jointed and with longitudinal ridges ....
   .... 1. Equisetaceae, p. 69
2. Stem not conspicuously jointed and without longitudinal ridges 3

3. Plants true aquatics, either rooted or floating ....
3. Plants terrestrial, although occasionally in boggy situations ....
4. Plants rooted; blades shaped like a 4-leaf clover; blades and sporocarps with long stalks .... 9. Marsileaceae, p. 8.
4. Plants floating; blades very small, bilobed; blades and sporocarps sessile ....10. Salviniaceae, p. 8
5. Leaves less than 2 cm long, often scale-like ....
5. Leaves more than 2 cm long ....
6. Creeping plants with erect branches; leaves without a ligule ....
.... 2. Lycopodiaceae, p. 7
6. Creeping or tufted plants without erect branches; leaves with a ligule near the axil ....3. Selaginellaceae, p. 7
7. Sporangia borne on a branch arising from the base of the sterile portion of the blade .... 5. Ophioglossaceae, p. 7
7. Sporangia borne on the back of or on the margin of green leaves, or on wholly fertile leaves or pinnae arising directly from the rhizome, or with fertile fronds terminated by a fruiting panicle ....
8. Blades delicate, translucent, only a single layer of cells thick; sporangia borne at the base of a bristle-like projection from the margin of the blade .... 7. Hymenophyllaceae p. 7
8. Blades more firm, opaque, several layers of cells thick; sporangia not associated with a marginal bristle ....
9. Fertile and sterile portions of leaves essentially alike in form, except for a possible size differentiation ....8. Polypodiaceae, p. 7
9. Fertile and sterile portions of leaves greatly dissimilar in form 1
10. Sporangia naked and solitary (although borne in loose panicles or clusters) on modified pinnae, these borne at the apex of the leaf, or centrally, or the whole leaf fertile .... 6. Osmundaceae, p. 7
10. Sporangia covered by an indusium or, if not, then borne in well-defined sori ....8. Polypodiaceae, p. 7

II. Plants Cacti or Pitcher-like or Completely Non-green

1. Plants green ....
1. Plants non-green ....
2. Plants cacti; leaves not pitcher-like .... 110. Cactaceae, p. 32
2. Plants not cacti; leaves modified into elongated pitchers ....
.... 73. Sarraceniaceae, p. 25
3. Plants twining or creeping .....*Cuscuta* in 133. Convolvulaceae, p. 35
3. Plants erect ....
4. Stamens 1 or 3; ovary inferior ....
4. Stamens 4, 8, or 10; ovary superior ....
5. Plants up to 1.5 cm tall; stamens 3 *Thismia* in 37. Burmanniaceae, p. 18
5. Plants well over 1.5 cm tall; stamen 1 .... 38. Orchidaceae, p. 18
6. Flowers actinomorphic; stamens 8 or 10 ....122. Ericaceae, p. 34
6. Flowers zygomorphic; stamens 4 .... 144. Orobanchaceae, p. 39

III. Plants Climbing or Twining

1. Leaves simple ....
1. Leaves compound 2

. Leaves alternate · 3
. Leaves opposite or whorled · 18
. Plants with tendrils · 4
. Tendrils absent · 8
. Leaves entire, although sometimes lobed · 5
. Leaves toothed · 6
. Leaves unlobed, or with a pair of basal lobes; stems often with bristles or prickles; perianth 6-parted, undifferentiated 34. Smilacaceae, p. 185
. Leaves 3-lobed; stems neither bristly nor prickly; sepals and petals differentiated · · · · · · · · · · · · · · 108. Passifloraceae, p. 322
. Petals united to about the middle; stamens 3 153. Cucurbitaceae, p. 406
. Petals free to the base; stamens 4–5 · 7
. Flowers borne in racemes; petals falling away as flower opens · · · · · · · ·
· 100. Vitaceae, p. 309
. Flowers solitary; petals persistent during flowering · · · · · · · · · · · · · · · · · ·
· 108. Passifloraceae, p. 322
. Leaves entire, although sometimes lobed · 9
. Leaves toothed and sometimes lobed as well · · · · · · · · · · · · · · · · · · · 16
. Stems bearing ocreae at the nodes · · · · · · · · · 51. Polygonaceae, p. 212
. Stems not bearing ocreae at the nodes · 10
. Calyx united, elongated, S-shaped, with 3 lobes at the apex · · · · · · · · · · ·
· 50. Aristolochiaceae, p. 212
0. Calyx free or united, not long and S-shaped, usually with more than 3 parts · 11
1. Flowers unisexual · 12
1. Flowers bisexual · 13
2. Ovary superior; fruit a drupe · · · · · · · · · · · · · 65. Menispermaceae, p. 240
2. Ovary inferior; fruit a 3-angled capsule · · · · · · · 35. Diocoreaceae, p. 186
3. Ovary inferior; aerial roots abundant . Hedera in 119. Araliaceae, p. 331
3. Ovary superior; aerial roots sparse or absent · · · · · · · · · · · · · · · · · · · 14
4. Corolla absent; stamens 7–10; fruit a 3-angled achene · · · · · · · · · · · · · · · ·
· Brunnichia in 51. Polygonaceae, p. 212
4. Corolla 5-parted; stamens 5; fruit a berry or capsule · · · · · · · · · · · · · 15
5. Corolla deeply 5-lobed; fruit a berry · · · · · · · · · · 140. Solanaceae, p. 379
5. Corolla shallowly 5-lobed; fruit a capsule . 133. Convolvulaceae, p. 355
6. Fruit a capsule; seeds arillate · · · · · · · · · · · · · · · 93. Celastraceae, p. 304
6. Fruit a berry or drupe; seeds not arillate · 17
7. Fruit a berry; leaves coarsely serrate, often lobed; tendrils present · · · ·
· 100. Vitaceae, p. 309
7. Fruit a drupe; leaves serrulate, unlobed; tendrils absent · · · · · · · · · · · · ·
· Berchemia in 99. Rhamnaceae, p. 308
8. Leaves serrate or lobed · 19
8. Leaves entire · 20
9. None of the leaves lobed; leaves evergreen; aerial roots present; seeds arillate · 93. Celastraceae, p. 304
9. Some of the leaves lobed; leaves not evergreen; aerial roots absent; seeds without arils · · · · · · · · · · · · · · · · · · Humulus in 46. Moraceae, p. 210
0. Latex present · 21
0. Latex absent · 22
1. Fruits a pair of follicles; filaments free from each other · · · · · · · · · · · · · · ·
· 131. Apocynaceae, p. 351
1. Fruit a single follicle; filaments united into a tube · · · · · · · · · · · · · · · · · · ·
· 132. Asclepiadaceae, p. 352
2. Flowers in a loose head, sharing a common receptacle; fruit an achene
· Mikania in 155. Compositae, p. 436

22. Flowers not in a head and not sharing the same receptacle; fruit a ca
sule or berry or follicle ....
23. Perianth 6-parted; stamens 3 or 6; fruit a 3-angled capsule ....
.... 35. Dioscoreaceae, p. 1
23. Perianth composed of a 5-parted calyx and a 5-parted corolla; stamens
fruit a berry or follicle ....
24. Flowers actinomorphic; fruit a follicle *Vinca* in 131. Apocynaceae, p. 3!
24. Flowers zygomorphic; fruit a berry *Lonicera* in 149. Caprifoliaceae, p. 4
25. Leaves opposite ....
25. Leaves alternate ....
26. Leaves bifoliolate ....*Bignonia* in 142. Bignoniaceae, p. 39
26. Leaves with 3 or more leaflets ....
27. Leaves trifoliolate ....*Clematis* in 63. Ranunculaceae, p. 2
27. Leaves with 5 or more leaflets ....
28. Corolla trumpet-shaped, orange or red; sepals 5; stamens 5 ....
.... *Campsis* in 142. Bignoniaceae, p. 39
28. Corolla absent; sepals 4; stamens numerous ....
.... *Clematis* in 63. Ranunculaceae, p. 23
29. Leaves trifoliolate ....
29. Leaves with 2, 4, or 5 or more leaflets ....
30. Flowers zygomorphic; stamens 10 ....80. Leguminosae, p. 27
30. Flowers actinomorphic; stamens 5 ....
31. Petals free; aerial roots present; plants poison to the touch ....
.... 91. Anacardiaceae, p. 30
31. Petals united at the base; aerial roots absent; plants not poison to th
touch .... 140. Solanaceae, p. 37
32. Leaves with an even number of leaflets ....80. Leguminosae, p. 27
32. Leaves with an odd number of leaflets ....
33. Leaves palmately compound .... 100. Vitaceae, p. 30
33. Leaves pinnately compound ....
34. Leaves once pinnate ....
34. Leaves bi- or tri-pinnate or -ternate ....
35. Flowers zygomorphic; stamens 10 ....80. Leguminosae, p. 27
35. Flowers actinomorphic; stamens 5 ....153. Cucurbitaceae, p. 40
36. Flowers actinomorphic; petals 5; stamens 5; fruit a berry ....
.... *Ampelopsis* in 100. Vitaceae, p. 30
36. Flowers zygomorphic; petals 4; stamens 6 or 8; fruit a capsule 3
37. Stamens 8; corolla not subcordate; fruit bladdery ....
.... *Cardiospermum* in 97. Sapindaceae, p. 30
37. Stamens 6; corolla subcordate; fruit not bladdery ....
.... *Adlumia* in 69. Papaveraceae, p. 24

IV. Plants Woody; Leaves Less Than 3 mm Wide, Usually Needle-like (Scale-like in Some Genera of Cupressaceae)

1. Some or all the leaves scale-like, overlapping, up to 0.4 mm long ....
.... 14. Cupressaceae, p. 8
1. Some or all the leaves needle-like or awl-shaped ....
2. Needle-like leaves borne in fascicles of 2 or more, at least 3.5 cm long o
longer ....12. Pinaceae, p. 8
2. Needle-like leaves borne singly, up to 2 cm long ....
3. All or most of the needle-like leaves more than 1.2 cm long, usually flat

tened or with a keel; seeds either borne in a fleshy scarlet cup or in a woody cone 4
None of the needle-like leaves more than 1.2 cm long, usually awl-shaped; seeds borne in a fleshy purplish "berry" ....
.... 14. Cupressaceae, p. 84
Shrub with evergreen leaves; seeds borne in a fleshy scarlet cup ......
.... 11. Taxaceae, p. 82
Tree with deciduous leaves; seeds borne in a woody cone with shield-shaped scales .... 13. Taxodiaceae, p. 83

V. Plants Woody; Leaves Compound

Leaves opposite 2
Leaves alternate 7
Leaves trifoliolate 3
Leaves with 5 or more leaflets 4
Leaflets serrulate; flowers perfect; fruits bladdery ....
.... 94. Staphyleaceae, p. 305
Leaflets coarsely and sparsely toothed or entire; flowers unisexual; fruits a cluster of samaras ....95. Aceraceae, p. 305
Leaves palmately compound .... 96. Hippocastanaceae, p. 306
Leaves pinnately compound 5
Flowers perfect; petals 5; stamens 5; fruit fleshy ....
.... Sambucus in 149. Caprifoliaceae, p. 400
Flowers unisexual; petals 4 or absent; stamens 2–12, rarely 5; fruit a samara or pair of samaras 6
Fruit a samara; petals 4; stamens 2 .. Fraxinus in 127. Oleaceae, p. 347
Fruit a pair of samaras; petals absent; stamens 3 or more ....
.... 95. Aceraceae, p. 305
Leaves, or most of them, bi- or tri-pinnately compound 8
Leaves once compound 10
Trunk and leafstalks coarsely prickly; fruit a cluster of drupes ....
.... 119. Araliaceae, p. 331
Trunk and leafstalks without prickles, although the trunk sometimes with stout spines; fruit a legume or bladdery 9
Leaflets entire or semelate; trunk often bearing stout spines; fruit an elongated legume ....80. Leguminosae, p. 277
Leaflets coarsely dentate; trunk spineless; fruit bladdery, 3-parted .....
.... Koelreuteria in 97. Sapindaceae, p. 307
Leaves trifoliolate 11
Leaves with 5 or more leaflets 14
Plants prickly .... 79. Rosaceae, p. 262
Plants not prickly 12
Flowers zygomorphic; stamens 10 ....80. Leguminosae, p. 277
Flowers actinomorphic; stamens 3–5 13
Fruit an orbicular samara; ovary 2-locular; stigmas 2 85. Rutaceae, p. 296
Fruit a drupe; ovary 1-locular; stigmas 3 .... 91. Anacardiaceae, p. 303
Plants with thorns, spines, bristles, or prickles 15
Plants without thorns, spines, bristles, or prickles 17
Flowers zygomorphic; stamens 10; fruit a legume ....
.... 80. Leguminosae, p. 277
Flowers actinomorphic; stamens usually 4–5 or more than 10 16

16. Flowers perfect; stamens numerous; flowers often 2 cm broad
broader .... 79. Rosaceae, p. 2

16. Flowers unisexual; stamens 4–5; flowers less than 1 cm broad .....
.... 85. Rutaceae, p. 2

17. Corolla absent; all flowers unisexual; staminate flowers in catkins ..
.... 42. Juglandaceae, p. 2

17. Corolla present; some or all flowers perfect; flowers not in catkins

18. Flowers zygomorphic; fruit a legume ....80. Leguminosae, p. 2

18. Flowers actinomorphic; fruit not a legume ....

19. Leaflets mostly 5 (rarely 7) .... *Potentilla* in 79. Rosaceae, p. 2

19. Leaflets 7–31 ....

20. Ovary superior; stamens 2, 3, 5, or 10 ....

20. Ovary inferior; stamens numerous .....*Sorbus* in 79. Rosaceae, p. 2

21. Ovary simple; stamens 5; fruit a berry .... 91. Anacardiaceae, p. 3

21. Ovary 2- to 5-lobed; stamens 2, 3, or 10; fruit a samara ....
.... 86. Simaroubaceae, p. 2

VI. Plants Woody; Leaves Simple, Alternate

1. Some or all the leaves lobed ....

1. None of the leaves lobed ....

2. Latex present .... 46. Moraceae, p. 2

2. Latex absent ....

3. Flowers without petals ....

3. Flowers with petals ....

4. Leaves densely coated with white felt on the lower surface; pistilla
flowers without a calyx ....*Populus* in 40. Salicaceae, p. 1

4. Leaves without a white felt on the lower surface; pistillate flowers with
calyx ....

5. Leaves star-shaped, 5- to 7-lobed; twigs often winged ....
.... *Liquidambar* in 77. Hamamelidaceae, p. 2

5. Leaves not star-shaped; twigs unwinged ....

6. Twigs green; leaves with 2 or 3 lobes; fruit a drupe 68. Lauraceae, p. 2

6. Twigs brown or tan; leaves usually with more than 3 lobes; fruit a nut

7. Pistillate flowers in a bur-like head; leaves strongly aromatic; fruit
shiny nut subtended by 8 linear bracts ....41. Myricaceae, p. 2

7. Pistillate flowers not in a bur-like head; leaves not aromatic; fruit
acorn .... *Quercus* in 44. Fagaceae, p. 2

8. Flowers unisexual, borne in globose heads .....78. Platanaceae, p. 2

8. Flowers perfect, arranged in various inflorescences ....

9. Ovary or ovaries superior ....

9. Ovary inferior ....

10. Stamens united into a long tube .... 103. Malvaceae, p. 3

10. Stamens free from each other ....

11. Stamens inserted on the receptacle; sepals 3, usually colored like t
petals .... 66. Magnoliaceae, p. 2

11. Stamens inserted on the calyx; sepals usually 5, often green ....
.... 79. Rosaceae, p. 2

12. Styles 2 per carpel .... 76. Saxifragaceae, p. 2

12. Style 1 per carpel .... 79. Rosaceae, p. 2

13. Leaves toothed ....

13. Leaves entie ....

Corolla absent 15
Corolla present 24
Flowers unisexual 16
Flowers perfect 23
Pistillate and staminate flowers in aments, heads, or spikes 17
Pistillate flowers not in aments, heads, or spikes 21
Pistillate flowers without a calyx 18
Pistillate flowers with a calyx 19
Bracts unlobed or none .... 40. Salicaceae, p. 195
Bracts 3- to 5-lobed .... 43. Betulaceae, p. 202
Staminate and pistillate flowers in same ament; fruit a prickly bur .....
.... Castanea in 44. Fagaceae, p. 203
Staminate and pistillate aments separate; fruit not a prickly bur 20
Latex present .... 46. Moraceae, p. 209
Latex absent .... 43. Betulaceae, p. 202
Calyx absent .... 43. Betulaceae, p. 202
Calyx present 22
Styles 3–5; margins of leaves uniformly toothed; fruit a nut ....
.... 44. Fagaceae, p. 203
Style 1; margins of leaves irregularly and sparingly toothed; fruit a drupe
.... 114. Nyssaceae, p. 325
Ovary 1-locular, with 2 styles; leaves usually asymmetrical at base,
usually conspicuously serrate .... 45. Ulmaceae, p. 207
Ovary 2- to 4-locular, with 2–5 styles; leaves symmetrical at base, often
finely and obscurely serrate .... 99. Rhamnaceae, p. 308
Petals free 25
Petals united, at least at the base 35
Flowers unisexual 26
Flowers perfect 28
Leaves irregularly and sparsely toothed along the margins ....
.... 114. Nyssaceae, p. 325
Leaves regularly finely toothed along the margins 27
Flowers with a disk; sepals united at the base 92. Aquifoliaceae, p. 304
Flowers without a disk; sepals free to the base 99. Rhamnaceae, p. 308
Ovary superior 29
Ovary inferior 33
Stamens opening by terminal pores .... 22. Ericaceae, p. 340
Stamens opening lengthwise 30
Carpels 1 to several, free; sepals united with receptacle ....
.... 79. Rosaceae, p. 262
Carpels united into a compound ovary; sepals free or united only at the
base 31
Stamens 4–5; flowers and fruits not pendulous from a paddle-shaped
bract; leaves not cordate 32
Stamens numerous; flowers and fruits pendulous from a paddle-shaped
bract; leaves usually cordate .... 101. Tiliaceae, p. 310
Flowers with a disk; sepals united at the base 92. Aquifoliaceae, p. 304
Flowers with a disk; sepals free to the base ... 99. Rhamnaceae, p. 308
Carpels 1- or 2-locular, with 2 styles per carpel 34
Carpel 1-locular, with 1 style .... 79. Rosaceae, p. 262
Leaves wavy-toothed; calyx 4-parted; petals 4; stamens 8 ....
.... 77. Hamamelidaceae, p. 261
Leaves finely serrulate; calyx 5-parted; petals 5; stamens 5 ....
.... Itea in 76. Saxifragaceae, p. 258

35. Stamens united into a tube; anthers dehiscing lengthwise ....
.... 126. Styracaceae, p.
35. Stamens free; anthers dehiscing by pores at the tip 122. Ericaceae, p.
36. Corolla absent ....
36. Corolla present ....
37. Flowers unisexual ....
37. Flowers perfect ....
38. Latex present; plants often spiny; fruit globose, greenish, 4 cm or m
in diameter ....*Maclura* in 46. Moraceae, p.
38. Latex absent; plants not spiny; fruit various ....
39. Pistillate flowers in aments .... 40. Salicaceae, p.
39. Pistillate flowers not in aments ....
40. Leaves asymmetrical at the base; trunk warty ....
.... *Celtis* in 45. Ulmaceae, p.
40. Leaves symmetrical at the base; trunk not warty ....
41. Ovary inferior; leaves not aromatic when crushed ....
41. Ovary superior; leaves aromatic when crushed ...68. Lauraceae, p.
42. Fruit an acorn; styles 3 .... 44. Fagaceae, p.
42. Fruit a drupe; style 1 .... 114. Nyssaceae, p.
43. Stamens 4; leaves with peltate scales ....
.................................*Elaeagnus* in 112. Elaeagnaceae, p.
43. Stamens 8; leaves without peltate scales ...111. Thymelaeaceae, p.
44. Petals free ....
44. Petals united ....
45. Flowers unisexual .... 114. Nyssaceae, p.
45. Flowers perfect ....
46. Flowers pea-shaped; leaves heart-shaped; fruit a legume ....
....*Cercis* in 80. Leguminosae, p.
46. Flowers not pea-shaped; leaves not heart-shaped; fruit no a legume
47. Ovary superior ....
47. Ovary inferior ....
48. Plants prickly; sepals 6 ....64. Berberidaceae, p.
48. Plants not prickly; sepals mostly 3 or 5 ....
49. Stamens dehiscing by terminal pores; sepals 5; stamens 5 ....
.... 122. Ericaceae, p.
49. Stamens dehiscing lengthwise; sepals 3; stamens numerous ....
50. Petals greenish-yellow; fruit cone-like .... 66. Magnoliaceae, p.
50. Petals maroon; fruit a thickened, elongated, banana-like berry .....
.... 67. Annonaceae, p.
51. Stamens 4; fruit a drupe .... 121. Cornaceae, p.
51. Stamens 10; fruit dry ....126. Styracaceae, p.
52. Stamens usually 8 or 16; fruit a large, orange berry; calyx 4-parted ..
.... 125. Ebenaceae, p.
52. Stamens 5 or 10; fruit various; calyx 5-parted ....
53. Fertile stamens 8 or 10; ovary inferior ....
53. Fertile stamens 5; ovary superior ....
54. Anthers opening by terminal pores; fruit fleshy, several-seeded .....
.... 122. Ericaceae, p.
54. Anthers opening by longitudinal slits; fruit dry, 1-seeded ....
.... 126. Styracaceae, p.
55. Plants usually thorny; 5 sterile stamens present; anthers dehiscin
lengthwise .... 124. Sapotaceae, p.
55. Plants thornless; sterile stamens absent; anthers dehiscing by termina
pores .... 122. Ericaceae, p. 34

VII. Plants Woody; Leaves Simple, Opposite or Whorled

1. Plants parasitic, growing on the branches of trees ....
.... 49. Loranthaceae, p. 211
1. Plants not parasitic 2
2. Some or all the leaves whorled 3
2. Leaves opposite 5
3. Leaves heart-shaped, most or all of them at least 6 cm broad ....
.... *Catalpa* in 142. Bignoniaceae, p. 392
3. Leaves not heart-shaped, all of them less than 6 cm broad 4
4. Flowers white, borne in dense, globose heads; ovary inferior; corolla 4-toothed; stipules present *Cephealanthus* in 148. Rubiaceae, p. 397
4. Flowers magenta, borne in axillary clusters; ovary superior; petals 5, free; stipules absent ....*Decodon* in 113. Lythraceae, p. 324
5. Flowers borne in catkins; leaves more than four times longer than broad
.... 40. Salicaceae, p. 195
5. Flowers not borne in catkins; leaves less than four times longer than broad 6
6. Leaves palmately lobed 7
6. Leaves unlobed, or rarely pinnately lobed 8
7. Fruit a pair of samaras; ovary superior ....95. Aceraceae, p. 305
7. Fruit a drupe; ovary inferior ....149. Caprifoliaceae, p. 400
8. Leaves toothed 9
8. Leaves entire 12
9. Petals free 10
9. Petals united, or absent 11
10. Stamens 4 or 5; seeds enclosed in a red aril .. 93. Celastraceae, p. 304
10. Stamens 8–40; seeds not arillate ....76. Saxifragaceae, p. 258
11. Stamens 2; ovary superior ....127. Oleaceae, p. 347
11. Stamens 4–5; ovary inferior ....149. Caprifoliaceae, p. 400
12. Leaves heart-shaped, at least 6 cm long and 6 cm broad ....
.... *Paulownia* in 141. Scrophulariaceae, p. 383
12. Leaves not heart-shaped, never 6 cm broad and 6 cm long 13
13. Ovary superior; stamens 2 or numerous 14
13. Ovary inferior; stamens 4, 8, or 10–40 15
14. Petals united; stamens 2 ....127. Oleaceae, p. 347
14. Petals free; stamens numerous ....104. Hypericaceae, p. 314
15. Petals free or absent; stamens 4, 8, or 10–40 16
15. Petals united; stamens usually 5 ....149. Caprifoliaceae, p. 400
16. Flowers dioecious; stamens 8; petals absent ....
.... *Shepherdia* in 112. Elaeagnaceae, p. 323
16. Flowers perfect; stamens 4 or 10–40; petals present 17
17. Petals usually 5; stamens 10–40; fruit a capsule 76. Saxifragaceae, p. 258
17. Petals 4; stamens 4; fruit a drupe .... 121. Cornaceae, p. 339

VIII. Plants Monocotyledonous, Herbaceous, Usually with Elongated, Parallel-veined Leaves, Flower Parts Usually in 3s, and a Single Cotyledon

1. Plants climbing or twining (if erect, then usually with a few weak tendrils from the upper axils); leaves net-veined; flowers unisexual 2

1. Plants erect or floating in water (tendrils never present); leaves most
   parallel-veined; flowers bisexual or unisexual . . . .
2. Inflorescence umbellate; ovary superior; fruit a berry . . . .
   . . . . 34. Smilacaceae, p. 18
2. Inflorescence glomerulate or paniculate; ovary inferior; fruit a capsul
   . . . . 35. Dioscoreaceae, p. 18
3. Plants with one or two whorls of leaves . . . .
3. Plants with leaves alternate, opposite, basal, or none . . . .
4. Flowers radially symmetrical; ovary superior; stamens 6 . . . .
4. Flowers bilaterally symmetrical; ovary inferior; stamen 1
   . . . . Isotria in 38. Orchidaceae, p. 18
5. Plants never over 50 cm tall; flowers usually borne singly . . . .
   . . . . Trillium and Medeola in 33. Liliaceae, p. 17
5. Plants more than 50 cm tall; flowers usually more than one . . . .
   . . . . Lilium in 33. Liliaceae, p. 17
6. Flowers crowded together on a spadix, often subtended by a spathe
   . . . . 27. Araceae, p. 16
6. Flowers not crowded on a spadix (in Ruppia, two flowers are borne on
   spadix-like structure) . . . .
7. Plants thalloid, floating in water . . . . .28. Lemnaceae, p. 16
7. Plants with roots, stems, and leaves, aquatic or terrestrial . . . .
8. Perianth absent, or reduced to very minute scales (lodicules) or bristle
   . . . .
8. Perianth present, composed of either calyx or corolla or both (plant
   with the perianth reduced to minute scales or bristles should be sough
   under the first 8) 1
9. Each flower subtended by one or more sterile scales; plants general'
   not true aquatics 1
9. Flowers not subtended by individual scales; plants mostly aquatics . 1
10. Leaves 2-ranked; sheaths usually open; stems usually hollow, with soli
    nodes, often terete; anthers attached above the base 25. Poaceae, p. 9
10. Leaves 3-ranked (when present); sheaths closed; stems solid, with so
    nodes, often 3-angled; anthers attached at the base . . . .
    . . . . 26. Cyperaceae, p. 13
11. Plants erect; inflorescence terminal, spicate, thick; leaves very long, lir
    ear, strap-shaped . . . . 15. Typhaceae, p. 8
11. Plants not erect; free-floating or sometimes rooted in bottom mud; ir
    florescence axillary or terminal and slenderly spicate; leaves not a
    above 1
12. Leaves alternate; stamens 2 or 4; inflorescence spicate and usually te
    minal, or with flowers borne 2 per spadix 1
12. Leaves opposite; stamen 1; inflorescence not spicate, axillary 1
13. Stamens 4; flowers in a spike or head; fruit sessile, appearing as a
    achene upon drying . . . . 20. Potamogetonaceae, p. 8
13. Stamens 2; flowers on a short spadix, concealed within the leaf sheath
    fruit stipitate, drupe-like . . . . 17. Ruppiaceae, p. 8
14. Carpel one; fruit beakless .19. Najadaceae, p. 8
14. Carpels 2–4; fruit beaked . . . . .18. Zannichelliaceae, p. 8
15. Pistils simple, more than one, separate or slightly coherent at base . 1
15. Pistil one, compound 1
16. Calyx and corolla differentiated (in color and texture) 1
16. Calyx and corolla undifferentiated (i.e., similar in color and texture) . .
    . . . . 21. Juncaginaceae, p. 8
17. Inflorescence umbellate; pistils 6, coherent at base; fruit a follicle . . .
    . . . . 23. Butomaceae, p. 9

7. Inflorescence not umbellate; pistils 10 or more, free to base; fruit an achene ....22. Alismaceae, p. 90
8. Ovary superior 19
8. Ovary inferior 28
9. Calyx and corolla differentiated (in color and texture) 20
9. Calyx and corolla undifferentiated (i.e., similar in color and texture) 21
0. Flowers crowded together in a dense head; leaves basal
....29. Xyridaceae, p. 170
0. Flowers borne in cymes or umbels; leaves caulinz ....
....30. Commelinaceae, p. 170
1. Flowers unisexual 22
1. Flowers bisexual 24
2. Leaves net-veined; flowers in umbels ....34. Smilacaceae, p.185
2. Leaves parallel-veined; flowers in globose clusters, racemes, or panicles
....23
3. Perianth small, greenish; flowers aggregated in dense globose clusters; stamens 5 ....16. Sparganiaceae, p. 85
3. Perianth usually conspicuous, greenish, yellowish, white, or bronze-purple; flowers in racemes or panicles; stamens 6 .33. Liliaceae, p. 176
4. Perianth scarious ....32. Juncaceae, p. 172
4. Perianth petaloid 25
5. Stamens 3 ....31. Pontederiaceae, p. 172
5. Stamens 6 (or 4) 26
6. Stamens of different sizes ....31. Pontederiaceae, p. 172
6. Stamens all alike 27
7. Leaves evergreen, rigid; stems woody .... Yucca in 33. Liliaceae, p. 176
7. Leaves deciduous, mostly not rigid; stems herbaceous ....
....33. Liliaceae, p. 176
8. Plants growing in water 29
8. Plants growing on land 31
9. Leaves whorled ....24. Hydrocharitaceae, p. 92
9. Leaves basal, or cauline and alternate 30
0. Stamens 2, or 6–12, never 3; flowers unisexual; styles not petaloid ....
....24. Hydrocharitaceae, p. 92
0. Stamens 3; flowers bisexual; styles petaloid ....36. Iridaceae, p. 186
1. Flowers bilaterally symmetrical; stamens 1 or 2 38. Orchidaceae, p. 189
1. Flowers radially symmetrical or nearly so; stamens 3 or 6 32
2. Stamens 3; styles sometimes petaloid ....36. Iridaceae, p. 186
2. Stamens 6; styles not petaloid 33
3. Leaves reduced to scales; plants lacking chlorophyll, at most 4 cm tall
....37. Burmanniaceae, p. 189
3. Leaves blade-bearing; plants with chlorophyll, well over 4 cm tall
....33. Liliaceae, p. 176

IX. Plants True Aquatics, Spending Their Entire Life In the Water; Dicotyledons

1. Some or all of the leaves deeply divided 2
1. None of the leaves deeply divided 10
2. Flowers crowded together in a solitary head, each flower sharing a common receptacle ....Bidens in 155. Compositae, p. 409
2. Flowers not crowded, each with its own receptacle 3
3. Leaves bearing small bladders; flowers zygomorphic ....
....145. Lentibulariaceae, p. 394

3. Leaves not bearing small bladders; flowers actinomorphic ....
4. Some or all the flowers unisexual ....
4. Some or all the flowers perfect ....
5. Calyx (or involucre) 8- to 12-cleft; stamens 10–20; ovary unlobed ....
.... 59. Ceratophyllaceae, p. 23
5. Calyx 4-cleft; stamens 4–8; ovary 2- to 4- lobed 117. Haloragidaceae, p. 33
6. Pistils 2 to several, free, or pistil 1 but deeply 2- to 4-lobed ....
6. Pistil one, unlobed ....
7. Sepals 3; petals 3; stamens 3–6 ....62. Cabombaceae, p. 23
7. Sepals 4–5; petals 0, 4, or 5; stamens 3, 4, 8, or numerous ....
8. Sepals 5; petals 5; stamens numerous ....
.... Ranunculus in 63. Ranunculaceae, p. 23
8. Sepals 4; petals 0 or 4; stamens 3, 4, or 8 ..117. Haloragidaceae, p. 33
9. Sepals 5; petals 5, united at base; stamens 5 ....
.... Hottonia in 123. Primulaceae, p. 34
9. Sepals 4; petals 4, free; stamens 6 ..Neobeckia in 71. Cruciferae, p. 24
10. Leaves with conspicuous sheathing stipules (ocreae) at the base ....
.... 51. Polygonaceae, p. 21
10. Leaves without ocreae 1
11. Flowers unisexual; perianth absent; stamen 1 89. Callitrichaceae, p. 30
11. Flowers perfect; perianth present; stamens 4 or more 1
12. Leaves peltate 1
12. Leaves not peltate 1
13. Sepals 3–4; petals 3–4; stamens 12–18 ....
.... Brasenia in 62. Cabombaceae, p. 23
13. Sepals and petals indistinguishable, together totalling more than 8 segments; stamens more than 20 .... 61. Nelumbonaceae, p. 23
14. Leaves opposite; stamens 4 1
14. Leaves alternate; stamens 5 or more 1
15. Corolla absent .... Peplis in 113. Lythraceae, p. 32
15. Corolla present ....Bacopa in 141. Scrophulariaceae, p. 38
16. Petals united; stamens 5 ...Nymphoides in 130. Menyanthaceae, p. 35
16. Petals free; stamens 8 or more 1
17. Petals 4–6; stamens 8–12 ....Jussiaea in 116. Onagraceae, p. 32
17. Petals more than 6; stamens more than 12 ..60. Nymphaeaceae, p. 231

X. Dicotyledonous Herbs; Leaves Opposite or Whorled

1. Stems or leaves or both fleshy 2
1. Plants not fleshy 4
2. Leaves reduced to minute scales ....
.... Salicornia in 52. Chenopodiaceae, p. 217
2. Leaves fleshy, not scale-like 3
3. Flowers yellow or white; sepals 4–5 . Sedum in 75. Crassulaceae, p. 257
3. Flowers rose; sepals 2 .... Talinum in 57. Portulacaceae, p. 224
4. Flowers crowded in a dense head, each sharing the same receptacle .
.... 155. Compositae, p. 409
4. Each flower with its own receptacle 5
5. Leaves, or some of them, compound 6
5. Leaves simple 13
6. Leaves ternately divided 7
6. Leaves pinnately or palmately compound 8

Flowers greenish or yellowish; ovary 1, inferior .150. Adoxaceae, p. 404
Flowers white or pinkish; ovaries 4–15, superior ....
.... *Anemonella* in 63. Ranunculaceae, p. 232
Leaves palmately compound 9
Leaves pinnately compound 10
Petals 5; stamens 5; leaflets serrate ....*Panax* in 119. Araliaceae, p. 331
Petals 4; stamens 6; leaflets usually coarsely toothed ....
.... *Dentaria* in 71. Cruciferae, p. 245
Petals united; stamens 3 or 4 11
Petals free; stamens 5 12
Calyx campanulate; stamens 4 *Seymeria* in 141. Scrophulariaceae, p. 383
Calyx reduced to bristles; stamens 3 .... 151. Valerianaceae, p. 405
Erect perennials; leaves twice-pinnate; carpels 2 ....
.... *Polytaenia* in 120. Umbelliferae, p. 331
Spreading or prostrate annuals; leaves once-pinnate; carpels 4–5 or
8–12 .... 84. Zygophyllaceae, p. 296
Ovaries more than 1 .... *Anemone* in 63. Ranunculaceae, p. 232
Ovary one 14
Corolla absent 15
Corolla present 25
Flowers unisexual 16
Flowers perfect 19
Calyx absent 17
Calyx present 18
Ovary 4-lobed; flowers axillary .... 89. Callitrichaceae, p. 302
Ovary 3-lobed; flowers borne from involucres 88. Euphorbiaceae, p. 298
Stamens 3–4; calyx lobes 3–4 .... 47. Urticaceae, p. 210
Stamens 5; calyx lobes 5 .... *Iresine* in 53. Amaranthaceae, p. 221
Leaves linear; stamen 1 .... 18. Hippuridaceae, p. 330
Leaves broader; stamens more than 1 20
Calyx adnate to ovary 21
Calyx free from ovary 22
Sepals 3, maroon; stamens 12 ....50. Aristolochiaceae, p. 212
Sepals 4, greenish; stamens 4 .... *Ludwigia* in 116. Onagraceae, p. 326
Flowers subtended by a colored involucre ...54. Nyctaginaceae, p. 223
Flowers not subtended by an involucre 23
Sepals 5; stamens 5 or 10 24
Sepals 4; stamens 4 or 8 ....113. Lythraceae, p. 324
Stamens attached to the calyx; leaves whorled ...56. Aizoaceae, p. 224
Stamens attached to the receptacle; leaves opposite ....
.... 58. Caryophyllaceae, p. 225
Petals free 26
Petals united 43
Flowers unisexual; ovary 3-lobed .... 88. Euphorbiaceae, p. 298
Flowers perfect; ovary rarely 3-lobed 27
Anthers opening by terminal pores ....115. Melastomaceae, p. 325
Anthers splitting lengthwise 28
Ovary superior 29
Ovary inferior 41
Leaves large, umbrella-like, peltate, borne in pairs, or solitary ....
.... *Podophyllum* in 64. Berberidaceae, p. 240
Leaves not umbrella-like and peltate 30
Leaves palmately lobed or dissected 31
Leaves entire or toothed 32

31. Petals fimbriate at the tip; ovary not deeply lobed ....
.... *Mitella* in 76. Saxifragaceae, p. 2
31. Petals not fimbriate at the tip; ovary deeply lobed 83. Geraniaceae, p. 2
32. Leaves toothed .... *Bergia* in 105. Elatinaceae, p. 3
32. Leaves entire ....
33. Sepals 2–3 ....
33. Sepals 4–5 ....
34. Leaves in whorls; stamens 6 .... 33. Liliaceae, p. 1
34. Leaves opposite; stamens 2, 3, or 5 ....
35. Petals 5; stamens 5; leaves a single pair .... 57. Portulacaceae, p. 2
35. Petals 2–3; stamens 2–3; leaves several pairs .. 105. Elatinaceae, p. 3
36. Petals 4–5; flowers actinomorphic ....
36. Petals 3; flowers usually zygomorphic ....
37. Plants without white latex ....
37. Plants with white latex .... 88. Euphorbiaceae, p. 2
38. Ovules and seeds attached to a central column; stamens all free ...
.... 58. Caryophyllaceae, p. 2
38. Ovules and seeds not attached to a central column; stamens usua
united, at least at the base ....
39. Stamens numerous ....104. Hypericaceae, p. 3
39. Stamens 5 .... 81. Linaceae, p. 2
40. Stamens usually 8; plants glabrous or nearly so 87. Polygalaceae, p. 2
40. Stamens 5, 10, or 15; plants pubescent .... 106. Cistaceae, p. 3
41. Flowers subtended by 4 large, white bracts .... 121. Cornaceae, p. 3
41. Flowers not subtended by 4 large, white bracts ....
42. Calyx tube attached to ovary ....116. Onagraceae, p. 3
42. Calyx tube free from ovary ....113. Lythraceae, p. 3
43. Ovary superior ....
43. Ovary inferior ....
44. Stamens and corolla lobes equal in number ....
44. Stamens and corolla lobes different in number ....
45. Ovaries 2, or ovary 1 but deeply 4-lobed ....
45. Ovary 1, unlobed ....
46. Ovaries 2; latex often present ....
46. Ovary 1, deeply 4-lobed; latex absent .... 139. Labiatae, p. 36
47. Stamens free; fruit a pair of follicles ....131. Apocynaceae, p. 3
47. Stamens united into a tube; follicle solitary 132. Asclepiadaceae, p. 3
48. Leaves lobed .... *Ellisia* in 135. Hydrophyllaceae, p. 3
48. Leaves unlobed ....
49. Ovary 1-locular ....
49. Ovary with more than 1 locule ....
50. Stamens opposite the lobes of the corolla .....123. Primulaceae, p. 34
50. Stamens alternate with the lobes of the corolla 129. Gentianaceae, p. 34
51. Flowers in terminal cymes or axillary; corolla not scarious; leaves co
nected by a stipular line ....128. Loganiaceae, p. 34
51. Flowers in long-peduncled heads; corolla scarious; leaves without s
pules ....147. Plantaginaceae, p. 39
52. Corolla lobes 3; stamens usually 8 ....87. Polygalaceae, p. 29
52. Corolla lobes 5; stamens 2 or 4 ....
53. Leaves entire ....
53. Leaves toothed ....
54. Corolla more or less actinomorphic ....
54. Corolla zygomorphic ....
55. Calyx with lobes of unequal size ....

NERAL KEY TO GROUPS
63

- Calyx with lobes of equal size 57
- Ovary deeply 4-lobed .... 139. Labiatae, p. 367
- Ovary unlobed ....141. Scrophulariaceae, p. 383
- Ovary deeply 4-lobed .... 139. Labiatae, p. 367
- Ovary more or less unlobed, not deeply 4-lobed 58
- Ovary with more than 12 ovules; leaves without cystoliths ....
  ....141. Scrophulariaceae, p. 383
- Ovary with up to 12 ovules; leaves with cystoliths ....
  ....146. Acanthaceae, p. 395
- Fertile stamens 2 60
- Fertile stamens 4 62
- Ovary deeply 4-lobed .... 139. Labiatae, p. 367
- Ovary more or less unlobed, at least not deeply 4-lobed 61
- Corolla up to 1 cm long ....141. Scrophulariaceae, p. 383
- Corolla longer than 1 cm ....146. Acanthaceae, p. 395
- Ovary deeply 4-lobed .... 139. Labiatae, p. 367
- Ovary unlobed, or at least not deeply 4-lobed 63
- Fruit with 2 long, incurved beaks; upper leaves long-petiolate ....
  ....143. Martyniaceae, p. 393
- Fruit without 2 long, incurved beaks; upper leaves short-petiolate or ses-
  sile ....141. Scrophulariaceae, p. 383
- Fertile stamens 2 65
- Fertile stamens 4 66
- Ovary deeply 4-lobed .... 139. Labiatae, p. 367
- Ovary unlobed ....141. Scrophulariaceae, p. 383
- Ovary deeply 4-lobed, the style arising from between the lobes ....
  ....139. Labiatae, p. 367
- Ovary not deeply 4-lobed or, if lobed, the style arising from the tip of the
  ovary 67
- Ovules 2 or more per locule 68
- Ovule 1 per locule 69
- Fruit with 2 long, incurved beaks ....143. Martyniaceae, p. 393
- Fruit without 2 long, incurved beaks .....141. Scrophulariaceae, p. 383
- Teeth of calyx with a hooked tip; flowers reflexed after anthesis; ovary 1-
  locular ....138. Phrymaceae, p. 367
- Teeth of calyx not hooked at tip; flowers not reflexed; ovary 2- or 4-
  locular .... 137. Verbenaceae, p. 365
- Stamens 3; calyx reduced to bristles .... 151. Valerianaceae, p. 405
- Stamens 4–5; calyx not bristle-like 71
- Ovary 1-locular; flowers in dense, spiny heads 152. Dipsacaceae, p. 406
- Ovary 2- to 5-locular; flowers not in heads or, if in dense heads, not
  spiny 72
- Leaves whorled ....148. Rubiaceae, p. 397
- Leaves opposite 73
- Stipules present ....148. Rubiaceae, p. 397
- Stiples absent ....149. Caprifoliaceae, p. 400

XI. Herbs; Leaves Alternate or Basal,
At Least Some of Them Compound

1. Flowers crowded together in heads, each sharing the same receptacle
   (various clovers may be sought here, but the flowers do not share the
   same receptacle) ....155. Compositae, p. 409

1. Flowers not sharing the same receptacle ....
2. Some or all the leaves trifoliolate ....
2. Leaves not trifoliolate ....
3. Flowers papilionaceous; fruit a legume or loment ....
.... 80. Leguminosae, p. 2
3. Flowers not papilionaceous; fruit neither a legume nor loment .....
4. Flowers borne on a spadix, surrounded by a spathe 27. Araceae, p. 1
4. Spadix and spathe absent ....
5. Petals free ....
5. Petals united .... *Menyanthes* in 130. Menyanthaceae, p. 3
6. Ovary superior ....
6. Ovary inferior .... 120. Umbelliferae, p. 3
7. Calyx and corolla each 4-parted .... 70. Capparidaceae, p. 2
7. Calyx and corolla each 5-parted ....
8. Styles 5-parted at summit, united below into a beak; carpels becomin
    separate at maturity, beaked .... 82. Oxalidaceae, p. 2
8. Not as above .... 79. Rosaceae, p. 2
9. Flowers borne on a spadix, surrounded by a spathe 27. Araceae, p. 1
9. Spadix and spathe absent ....
10. Flowers zygomorphic ....
10. Flowers actinomorphic ....
11. Petals 4; stamens 6 .... 69. Papaveraceae, p. 2
11. Petals 5; stamens 5 or 10 or more ....
12. Flowers in an umbel .... 120. Umbelliferae, p. 3
12. Flowers not in an umbel ....
13. Stamens 10 or more; fruit a legume; flowers without a spur ....
.... 80. Leguminosae, p. 2
13. Stamens numerous; fruit a follicle; flower spurred ....
.... *Delphinium* in 63. Ranunculaceae, p. 23
14. Flowers unisexual ....
14. Flowers perfect ....
15. Ovaries 4–15; petals absent; calyx free from the ovaries ....
.... *Thalictrum* in 63. Ranunculaceae, p. 23
15. Ovary 1; petals present; calyx adnate to the ovary ....
.... 153. Cucurbitaceae, p. 40
16. Petals absent; stamens 3 .. *Proserpinaca* in 117. Haloragidaceae, p. 33
16. Petals present; stamens 4 or more ....
17. Petals free ....
17. Petals united ....
18. Ovary or ovaries superior ....
18. Ovary inferior ....
19. Stamens attached to the receptacle ....
19. Stamens attached to the perianth ....
20. Ovaries several; stamens usually 5 or 10 or more ....
.... 63. Ranunculaceae, p. 23
20. Ovary solitary; stamens usually 6 ....
21. Leaves palmately compound, stipulate ...... 70. Capparidaceae, p. 24
21. Leaves pinnately compound, without stipules ....71. Cruciferae, p. 24
22. Sepals 5; petals 5; stamens 5 or more ....
22. Sepals 3; petals 3; stamens 6 .... 90. Limnanthaceae, p. 30
23. Carpels sharply pointed at base; sepals awned ....
.... *Erodium* in 83. Geraniaceae, p. 29
23. Carpels not sharply pointed at base; sepals not awned ....
.... 79. Rosaceae, p. 26

XII. Dicotyledonous Herbs; Leaves Alternate or Basal, Simple

44. Stamens attached to the perianth 48
45. Pistils 2 or more, free . . . . 63. Ranunculaceae, p. 232
45. Pistil 1 46
46. Sepals 2 . . . . 69. Papaveraceae, p. 242
46. Sepals 4 or more 47
47. Flowers actinomorphic; stamens 6 . . . . 71. Cruciferae, p. 245
47. Flowers zygomorphic; stamens 12 or more . . 72. Resedaceae, p. 257
48. Pistils 2 or more, more or less free . . . . : . . . 79. Rosaceae, p. 262
48. Pistil 1 or, if more than 1, united 49
49. Flowers actinomorphic 50
49. Flowers zygomorphic 51
50. Stamens numerous, united into a column . . . . . . 103. Malvaceae, p. 311
50. Stamens 5, free or united only at the base . . . . 81. Linaceae, p. 294
51. Flowers with one spur 52
51. Flowers spurless 53
52. Petals 5, one of them spurred . . . . 107. Violaceae, p. 319
52. Petals 2, each 2-lobed . . . . 98. Balsaminaceae, p. 307
53. Petals 3; stamens 8 . . . . 87. Polygalaceae, p. 297
53. Petals 5; stamens 5–10 54
54. Corolla actinomorphic; outer sepals different in size from inner sepals
. . . . 106. Cistaceae, p. 317
54. Corolla zygomorphic; sepals all about the same size . . . .
. . . . 80. Leguminosae, p. 277
55. Stamens 20 or more; plants harshly pubescent . 109. Loasaceae, p. 322
55. Stamens 4–12; plants not harshly pubescent, sometimes glabrous . . 56
56. Stamens 4, 6, 8, or 12; petals usually 4 or 6 . . . 116. Onagraceae, p. 326
56. Stamens 5; petals 5 . . . . 120. Umbelliferae, p. 331
57. Ovary superior 58
57. Ovary inferior . . . . 154. Campanulaceae, p. 407
58. Flowers zygomorphic 59
58. Flowers actinomorphic 62
59. Stamens more numerous than the lobes of the corolla 60
59. Stamens the same number as the corolla lobes or fewer than the corolla
lobes 61
60. Petals 5; stamens 10; fruit a legume . . . . 80. Leguminosae, p. 277
60. Petals 3; stamens 8; fruit a capsule . . . . 87. Polygalaceae, p. 297
61. Fruit with 2 long, incurved horns; ovary 1-locular . . . .
. . . . 143. Martyniaceae, p. 393
61. Fruit without 2 long, incurved horns; ovary 2-locular . . . .
. . . . 141. Scrophulariaceae, p. 383
62. Stamens more numerous than the lobes of the corolla 63
62. Stamens the same number as the lobes of the corolla 64
63. Anthers 2-horned, dehiscing by pores at the tip . . . .
. . . . *Chimaphila* in 122. Ericaceae, p. 340
63. Anthers without horns, dehiscing longitudinally 103. Malvaceae, p. 311
64. Ovaries 2; latex frequently present . . . . 132. Asclepiadaceae, p. 352
64. Ovary 1; latex absent 65
65. Ovary deeply 4-lobed . . . . 136. Boraginaceae, p. 360
65. Ovary not deeply 4-lobed 66
66. Leaves reduced to minute scales *Bartonia* in 129. Gentianaceae, p. 348
66. Leaves not reduced to scales 67
67. Stigmas 3 . . . . 134. Polemoniaceae, p. 358
67. Stigmas 1 or 2 68
68. Stigma 1 69

68. Stigmas 2 70
69. Stamens opposite the petals ....123. Primulaceae, p. 344
69. Stamens alternate with the petals .... 133. Convolvulaceae, p. 355
70. Styles 2, free ....135. Hydrophyllaceae, p. 359
70. Style 1, becoming lobed at tip, but always united below 71
71. Fruit a berry or a many-seeded capsule; plants rarely prostrate ....
.... 140. Solanaceae, p. 379
71. Fruit a 4-seeded capsule; plants often prostrate ....
.... 133. Convolvulaceae, p. 355

Descriptive Flora

1. EQUISETACEAE

1. *Equisetum* [*Tourn.*] L. Horsetail, Scouring Rush

1. Aerial stems soft and flexible, dimorphic, the sterile ones profusely branched 2
1. Aerial stems hard and rigid (flexible in *E.* X *nelsonii*), monomorphic, usually unbranched (except *E. palustre, E.* X *litorale,* and *E. fluviatile*) 3
2. Ridges of stem spinulose; teeth of sheath with white margins .... 1. *E. pratense*
2. Ridges of stem not spinulose; teeth of sheath without white margins .. .. 2. *E. arvense*
3. Ridges of stem 3–6; teeth of sheaths 3; central cavity of stem absent .. .. 3. *E. scirpoides*
3. Ridges of stem usually more than 6; teeth of at least some sheaths more than 3; central cavity of stem present 4
4. Central cavity up to two-thirds the diameter of the stem 5
4. Central cavity more than two-thirds the diameter of the stem 9
5. Stems regularly unbranched above the base 6
5. Stems regularly branched above the base 8
6. Ridges of stem angled, with tubercles in two rows; teeth of sheath with a central groove throughout; stems firm, perennial 7
6. Ridges of stem rounded, with tubercles in one row; teeth of sheath with a central groove only at tip; stems flexible, annual .....6. *E.* X *nelsonii*
7. Ridges of stem strongly grooved, 4- to 10-angled; teeth of sheaths falling off early; uppermost sheath (including deciduous teeth) rarely more than 5 mm long .... 4. *E. variegatum*
7. Ridges of stem slightly grooved, 10- to 17-angled; teeth of sheaths persistent; uppermost sheath (including teeth) nearly always more than 5 mm long .... 5. *E.* X *trachyodon*
8. Sheath of first joint of branches with 5–10 teeth; stem with outer row of cavities about same size as central cavity .... 7. *E. palustre*
8. Sheath of first joint of branches with 3–4 teeth; stem with outer row of cavities considerably smaller than central cavity .... 8. *E.* X *litorale*
9. Teeth of sheaths early deciduous, basally connate 10
9. Teeth of sheaths persistent, free or basally connate 12
10. All sheaths gray .... 9. *E. hyemale*
10. Upper sheaths green, the lower sometimes gray 11
11. Stems smooth; central cavity three-fourths the diameter of the stem; cone rounded or apiculate at summit .... 10. *E. laevigatum*

69

11. Stems slightly rough; central cavity four-fifths the diameter of the stem; cone apiculate at summit ....11. *E.* X *ferrissii*
12. Teeth of sheaths free; sheaths brownish or green; stems often branched; cones not apiculate .... 12. *E. fluviatile*
12. Teeth of sheaths basally connate; sheaths gray; stems usually unbranched; cones apiculate .... 9. *E. hyemale*

1. E. pratense Ehrh. Meadow Horsetail. May–July. North-facing wooded slope, rare; JoDaviess Co.

2. E. arvense L. Common Horsetail. April–June. Railroad embankments, roadsides, fields, shores; in every co.

3. E. scirpoides Michx. Dwarf Scouring Rush. Aug.–Sept. Moist shaded woodlands, rare; Lake and McHenry cos.

4. E. variegatum Schleich. Variegated Scouring Rush. July–Oct. Lake shores and stream banks, margins of swamps, rare; Lake, Cook, Carroll, Warren, Coles, and DeWitt cos.

5. E. X trachyodon A. Br. Horsetail. June–Sept. Lake shores, very rare; Cook and Mason cos. Considered to be a hybrid between *E. variegatum* and *E. hyemale*. *E. variegatum* var. *jesupi* A. A. Eaton—F, G.

6. E. X nelsonii (A. A. Eaton) Schaffner. Horsetail. June–Sept. Sandy shores, rare; Cook Co. Considered to be a hybrid between *E. variegatum* and *E. laevigatum. E. variegatum* var. *nelsonii* A. A. Eaton—F, G.

7. E. palustre L. Marsh Horsetail. June–Sept. Banks of rivers and streams, rare; Tazewell and Kankakee cos.

8. E. X litorale Kuhl. Horsetail. June–Sept. Shores and banks, rare; Lee Co. Considered to be a hybrid between *E. fluviatile* and *E. arvense*.

9. E. hyemale L. var. affine (Engelm.) A. A. Eaton. Scouring Rush. May–Aug. Shores, banks, roadsides; in every co. *E. hyemale* L.—J. (Including *E. hiemale* var. *pseudohiemale* [Farw.] Morton and *E. hiemale* var. *elatum* [Engelm.] Morton.)

10. E. laevigatum A. Br. Smooth Scouring Rush. May–July. Open, moist, sandy areas; common in the n. half, occasional in the s. half. *E. hyemale* var. *intermedium* A. A. Eaton—F. (Including *E. kansanum* Schaffn.)

11. E. X ferrissii Clute. Intermediate Scouring Rush. May–Aug. Banks and shores; scattered throughout the n. half of the state. Considered to be a hybrid between *E. hyemale* and *E. laevigatum.*

12. E. fluviatile L. Water Horsetail. May–Aug. In water of lakes, streams, and swamps; occasional in the upper half of the state.

2. LYCOPODIACEAE

1. Lycopodium [Dill.] L. Clubmoss

1. Sporangia axillary along the stem, the leaves subtending the sporangia appearing no different from ordinary leaves; gemmae frequently present in the axils of upper leaves 2
1. Sporangia aggregated in terminal cones, the leaves subtending the

sporangia much modified from ordinary leaves (except in *L. inundatum*); gemmae absent 3

2. Leaves linear or lanceolate, broadest near base, entire; stomates present on both surfaces (with 10 X magnification); spores usually over 35μ in diameter ....
   .... 1. *L. porophilum*

2. Leaves oblanceolate, broadest near middle, serrulate or rarely entire; stomates present only on lower surfaces (with 10 X magnification); spores usually less than 35μ in diameter .... 2. *L. lucidulum*

3. Leaves subtending sporangia similar to ordinary leaves, green; sterile stems prostrate 4

3. Leaves subtending sporangia much shorter than ordinary leaves, yellow; sterile stems erect or ascending 5

4. Leaves denticulate, about 1 mm broad .... 3. *L. adpressum*

4. Leaves not denticulate, up to 0.7 mm broad .... 4. *L. inundatum*

5. Cones solitary, 5–7 mm thick; leaves 6- to 8-ranked . 5. *L. dendroideum*

5. Cones 2–4, 3–6 mm thick; leaves 4-ranked 6

6. Peduncles 3–10 cm long; leaves with appressed tips, 1–2 mm apart; sporophylls without scarious margins, ovate .... 6. *L. flabelliforme*

6. Peduncles 12–18 cm long; leaves with spreading tips, about 4 mm apart; sporophylls with scarious margins, orbicular .... 7. *L. X habereri*

1. L. porophilum Lloyd & Underw. Cliff Clubmoss. June–Sept. Moist, shaded sandstone cliffs; three n. and three se. cos. *L. selago* var. *patens* (Beauv.) Desv.—G.

2. L. lucidulum Michx. Shining Clubmoss. June–Sept. Leaves serrulate in the upper half. Shaded sandstone cliffs and ravines, rare; scattered in Ill., but apparently absent in the w. and cent. cos.

var. tryonii Mohlenbrock. June–Sept. Leaves entire. Shaded sandstone cliff, rare; Jackson Co. *L. lucidulum* var. *occidentale* (Clute) L. R. Wilson—F, G.

3. L. adpressum Lloyd & Underw. Bog Clubmoss. July–Sept. Wet ground, very rare; Winnebago Co. *L. inundatum* var. *bigelovii* Tuckerm.—F, G.

4. L. inundatum L. Bog Clubmoss. July–Sept. Bogs, rare; Cook Co.

5. L. dendroideum Michx. July–Sept. Ground Pine. Woodlands, in wet, sandy detritus of St. Peter sandstone, very rare; Ogle Co. *L. obscurum* L. var. *dendroideum* (Michx.) D. C. Eaton—F, G.

6. L. flabelliforme (Fern.) Blanch. Ground Pine. July–Sept. Sandstone ledges in Pope and Ogle cos.; adventive under conifers in Cook, DuPage, and several s. Ill. cos. *L. complanatum* var. *flabelliforme* Fern.—F, G.

7. L. X habereri House. Hybrid between *L. flabelliforme* and *L. tristachyum*. Cook Co. (in 1878).

3. SELAGINELLACEAE

1. *Selaginella* Beauv.   Spikemoss

1. Stems weak, herbaceous; leaves 4-ranked, flaccid, obtuse to acute .....
   .... 1. *S. apoda*

1. Stems wiry, evergreen; leaves spirally arranged, stiff, subulate-tipped ..
.... 2. *S. rupestris*

    1. S. apoda (L.) Fern. Small Spikemoss. June–Oct. Moist shaded
areas, grassy margins of streams; local throughout.
    2. S. rupestris (L.) Spring. Rock Spikemoss. June–Sept. Dry rocky
areas, mostly of sandstone, local; nw. Ill. and Pope and Union cos.

4. ISOETACEAE

1. Isoetes L. Quillwort

1. Sporangium punctate or striate; megaspores tuberculate; microspores
spinulose or papillose .... 1. *I. melanopoda*
1. Sporangium neither punctate nor striate; megaspores reticulate, with
narrow ridges; microspores smooth or minutely roughened ....
.... 2. *I. engelmannii*

    1. I. melanopoda Gay & Dur. Black Quillwort. May–Sept. Shallow
water of ponds and ditches; local throughout. (Including f. *pallida*
[Engelm.] Fern.)
    2. I. engelmannii A. Br. Engelmann's Quillwort. May–Oct. Shallow
water of ponds and ditches; St. Clair Co.

5. OPHIOGLOSSACEAE

1. Leaves pinnately compound or simple and deeply lobed, the venation
free; sporangia stipitate .... 1. *Botrychium*
1. Leaves simple, entire, the venation net-like; sporangia sessile ....
.... 2. *Ophioglossum*

1. Botrychium Sw. Grape Fern

1. Sterile blades long- or short-petiolate below long or short stalk of fruiting panicle, evergreen (often turning bronze in winter), coriaceous, subcoriaceous, or membranaceous 2
1. Sterile blades sessile below long stalk of fruiting panicle ....
.... 6. *B. virginianum*
2. Sterile blades ternately compound, on long petioles arising near base of
plant 3
2. Sterile blades pinnate or pinnatifid, on long or short petioles arising
midway or higher above base of plant 5
3. Ultimate segments of the leaf obtuse or rounded; blade fleshy or coriaceous, not turning bronze in winter .... 1. *B. multifidum*
3. Ultimate segments of the leaf acute or subacute; blade subcoriaceous or
membranaceous, turning bronze in winter (except *B. biternatum*) ... 4
4. Pinnules entire, crenulate, or lobed; blade subcoriaceous, turning
bronze in winter .... 2. *B. dissectum*

4. Pinnules sharply serrate; blade membranaceous, remaining green in winter .... 3. *B. biternatum*
5. Sterile blade petiolate .... 4. *B. simplex*
5. Sterile blade short-petiolate or nearly sessile ... 5. *B. matricariaefolium*

1. B. multifidum (Gmel.) Rupr. ssp. silaifolium (Presl) Clausen. Northern Grape Fern. Aug.–Sept. Rich woodlands, rare; Boone, Lake, McHenry, Ogle, Will, and Winnebago cos. *B. multifidum* (Gmel.) Rupr.—J; *B. multifidum* var. *intermedium* (D. C. Eaton) Farw.—F, G.

2. B. dissectum Spreng. Cut-leaved Grape Fern. Sept.–Nov. Blade of leaf very finely dissected. Open oak-hickory woodlands; scattered throughout the state.

var. obliquum (Muhl.) Clute. Grape Fern. Sept.–Nov. Blade of leaf shallowly divided. Open oak-hickory woodlands and pastures; throughout the state. *B. dissectum* f. *obliquum* (Muhl.) Fern.—F; *B. obliquum* Muhl.—J.

3. B. biternatum (Sav.) Underw. Grape Fern. Sept.–Nov. Open oak-hickory woodlands, rare; Jackson Co. *B. dissectum* var. *tenuifolium* (Underw.) Farw.—F, G.

4. B. simplex E. Hitchc. Grape Fern. May–June. Grassy meadow, very rare; Winnebago Co.

5. B. matricariaefolium A. Br. June–July. Grassy meadow, very rare; Winnebago Co.

6. B. virginianum (L.) Sw. Rattlesnake Fern. June–July. Dry or moist woodlands; in every co.

2. *Ophioglossum* [*Tourn.*] *L.* Adder's-tongue

1. Leaf rounded or subacute at apex, the veins forming principle areoles without secondary areoles .... 1. *O. vulgatum*
1. Leaf acute, apiculate, the veins forming principle areoles surrounding secondary areoles .... 2. *O. engelmannii*

1. O. vulgatum L. var. pycnostichum Fern. Adder's-tongue Fern. April–June. Leaf deep green, shiny; spores 37–42μ in diameter. Moist woods and shaded sandstone ledges; scattered in the n. and cent. cos., occasional in the extreme s. *O. vulgatum* L.—G, J.

var. pseudopodum (Blake) Farw. Adder's-tongue Fern. April–June. Leaf pale green, dull; spores 50–54μ in diameter. Low, rich woods, not common. *O. vulgatum* L.—G, J.

2.O. engelmannii Prantl. Adder's-tongue Fern. May–July. Limestone ledges, rare; Hardin, Jersey, Johnson, Pope, and Randolph cos.

6. OSMUNDACEAE

1. *Osmunda* [*Tourn.*] *L.* Royal Fern

1. Leaves bipinnate, the pinnules serrulate; sporangia borne on upper half of leaf .... 1. *O. regalis*
1. Leaves always pinnate-pinnatifid, the pinnules entire; sporangia borne on a separate fertile leaf or centrally on the leaf 2

2. Sporangia borne on a separate fertile leaf .... 2. *O. cinnamomea*
2. Sporangia produced centrally between sterile pinnae   3. *O. claytoniana*

    1. O. regalis L. var. spectabilis (Willd.) Gray. Royal Fern. April–June.
Swamps, woods, or, in some of the s. cos., moist, sandstone ledges
throughout the state. *O. regalis* L.—J.
    2. O. cinnamomea L. Cinnamon Fern. March–June. Swamps and
swampy woods or, rarely, on moist, sandstone ledges; in the n. one-
third of Ill.; also Pope, Johnson, and Massac cos.
    3. O. claytoniana L. Interrupted Fern. April–July. Moist, low wood-
lands or, rarely, in moist depressions along sandstone ledges; occasional in the n. half of Ill., rare in the s. half.

7. HYMENOPHYLLACEAE

1. *Trichomanes L.   Filmy Fern*

    1. T. boschianum Sturm. Filmy Fern. June–Sept. Beneath moist,
overhanging sandstone cliffs, rare; Johnson and Pope cos.

8. POLYPODIACEAE

1. Sori on the leaf margin, covered by a recurved outgrowth of the blade
1. Sori not marginal or, if so, not covered by leaf margin ....
2. Sori distinct, borne on the veins or vein tips ....
2. Sori continuous ....
3. Leaves distinctly hairy or glandular; petiole unbranched ....
3. Leaves glabrous; petiole branched at the apex .... 2. *Adiantum*
4. Blades glandular-pubescent; rhizome long-creeping, pubescent ......
.... 1. *Dennstaedtia*
4. Blades eglandular; rhizome compact, scaly .... 6. *Cheilanthes*
5. Rhizome without scales; leaves ternately pinnate; coarse ferns at least 5
cm tall, mostly in woods or in fields .... 3. *Pteridium*
5. Rhizome scaly; leaves regularly pinnate; ferns rarely over 35 cm tall,
mostly on rocks ....
6. Stipe green, weak .... 4. *Cryptogramma*
6. Stipe dark brown or purple-brown, wiry .... 5. *Pellaea*
7. Indusium absent ....
7. Indusium present, although sometimes so deeply cleft as to be difficult
to observe, and sometimes soon deciduous 1
8. Blades once-pinnatifid; leaves less than 40 cm long, usually less than 20
cm long .... 7. *Polypodium*
8. Blades bipinnatifid or pinnately divided; leaves over 40 cm long .....
9. Rachis unwinged; leaves ternate .... 11. *Gymnocarpium*
9. Rachis winged (except sometimes between the lowest pinnae); leaves
triangular, but not ternate .... 12. *Thelypteris*

10. Leaves strongly dimorphic (i.e., with the fertile quite different from the
sterile) 11
10. Leaves all alike (In *Polystichum*, the upper fertile pinnae are strongly
modified; in one *Asplenium*, the leaves are slightly dimorphic.) 13
11. Fertile leaves bipinnate, the sterile ones pinnatifid .... 9. *Onoclea*
11. Fertile leaves once-pinnate, the sterile ones once-pinnate, pinnate-pinnatifid, or pinnatifid 12
12. Veins free .... 10. *Matteuccia*
12. Veins forming a network .... 14. *Woodwardia*
13. Indusium attached at the center or the side of the sorus 14
13. Indusium attached beneath the sporangia (in *Cystopteris*, the indusium
may sometimes appear to be lateral) 19
14. Indusium attached at the center of the sorus 15
14. Indusium attached laterally 17
15. Indusium peltate; fertile pinnae apical and modified; leaf once-pinnate
.... 8. *Polystichum*
15. Indusium reniform; fertile pinnae identical to sterile pinnae; leaf pinnate-
pinnatifid to tripinnate 16
16. Rhizome slender, cord-like; leaves deciduous; petiole without scales, or
with scales at most only 5 mm long .... 12. *Thelypteris*
16. Rhizome stout; leaves more or less evergreen; petiole scaly (at least
near base), the scales over 5 mm long .... 13. *Dryopteris*
17. Sori parallel to the main vein, appearing to be in a chain ....
.... 14. *Woodwardia*
17. Sori parallel to the lateral veins 18
18. Veins reaching the margin; blades deciduous .... 15. *Athyrium*
18. Veins not reaching the margin; blades usually evergreen 16. *Asplenium*
19. Indusium separating into shreds; petioles rather densely chaffy ....
.... 17. *Woodsia*
19. Indusium hood-like, not separating into shreds; petioles sparsely chaffy,
or glabrous .... 18. *Cystopteris*

1. *Dennstaedtia* Bernh. Hay-scented Fern

1. D. punctilobula (Michx.) Moore. Hay-scented Fern. July–Oct.
Moist, shaded sandstone ravines, rare; Johnson and Pope cos; reported from Wabash Co.

2. *Adiantum* L. Maidenhair Fern

1. A. pedatum [Tourn.] L. Maidenhair Fern. June–Sept. Moist,
shaded woodlands; in every co.

3. *Pteridium* Gled. Bracken Fern

1. P. aquilinum (L.) Kuhn var. latiusculum (Desv.) Underw. Bracken
Fern. June–Sept. Margins of ultimate leaf segments hairy, the terminal segments 5–8 mm broad. Open woodlands and fields; common
throughout the state, except in the s. cos. where it is occasional. *P.
latiusculum* (Desv.) Hieron.—J.

var. pseudocaudatum (Clute) Heller. Bracken Fern. June–Sept.
Margins of ultimate leaf segments glabrous or nearly so, the terminal
segments to 4.5 mm broad. Open woodlands and fields, rare; Hardin
Co.

4. Cryptogramma R. Br. Cliffbrake Fern

1. C. stelleri (S. G. Gmel.) Prantl. Slender Cliffbrake. June–Sep
Deep, shaded, limestone ravines, uncommon; restricted to the n. one
fourth of Ill.

5. Pellaea Link Cliffbrake Fern

1. Petiole and rachis pubescent nearly throughout; fertile pinnules much
narrower than the sterile ones .... 1. *P. atropurpurea*
1. Petiole and rachis glabrous or sparsely pubescent; fertile and sterile pin
nae similar in shape .... 2. *P. glabella*

    1. *P. atropurpurea* (L.) Link. Purple Cliffbrake. June–Sept. On lime
stone cliffs, frequently under very dry conditions; in the cos. border
ing the Mississippi River in the s. half of Ill.; in the cos. bordering the
Ohio River; also Johnson and Ogle cos.
    2. *P. glabella* Mett. Smooth Cliffbrake. June–Sept. Limestone out
croppings; local in the n. cos., extending southward along the Missis
sippi and Ohio rivers. *P. atropurpurea* var. *bushii* Mack.—G.

6. Cheilanthes Sw. Lip Fern

1. Leaves densely brown-woolly beneath, to 15 cm long, tripinnate, with
7–12 (–15) pairs of pinnae; on limestone .... 1. *C. fee*
1. Leaves white-villous beneath, some usually well over 15 cm long, bipin
nate-pinnatifid, with 12–20 pairs of pinnae; chiefly on sandstone
.... 2. *C. lanosa*

    1. *C. feei* Moore. Baby Lip Fern. June–Sept. Dry, exposed limestone
cliffs; in cos. bordering the Mississippi River and the Ohio River, in
the s. half of Ill.; also JoDaviess and Carroll cos.
    2. *C. lanosa* (Michx.) D. C. Eaton. Hairy Lip Fern. June–Sept. Dry
exposed cliffs, chiefly of sandstone, local; confined to the s. one-third
of Ill. and Ogle Co. *C. vestita* (Spreng.) Sw.—F.

7. Polypodium [Tourn.] L. Polypody

1. Blades and petioles scaleless, appearing smooth, the segments minutely
toothed .... 1 *P. vulgare*
1. Lower surface of blades and petioles scaly, appearing pustular, the seg
ments entire .... 2. *P. polypodioides*

    1. *P. vulgare* L. var. *virginianum* (L.) Eaton. Common Polypody.
June–Oct. Sandstone rock, usually under some shade; occasional in
the s. one-third of Ill., local and rare in the n. one-third, absent from
the cent. one-third. *P. virginianum* L.—F, J; *P. vulgare* L.—G.
    2. *P. polypodioides* (L.) Watt var. *michauxianum* Weatherby. Gray
Polypody. June–Oct. Sandstone rocks, tree trunks, and branches, not
common; restricted to the s. one-fourth of the state. *P. polypodioides*
(L.) Watt—G, J.

8. Polystichum Roth   Christmas Fern

1. P. acrostichoides (Michx.) Schott. Christmas Fern. June–Oct. Woods, particularly in rocky soil; in every co. in the s. one-half of Ill., scattered in the n. cos.

9. Onoclea L.   Sensitive Fern

1. O. sensibilis L. Sensitive Fern. June–Oct. Moist woodlands or low, open ground, common; in nearly every co. (Including f. *obtusilobata* [Schkuhr] Gilbert.)

10. Matteuccia Todaro   Ostrich Fern

1. M. struthiopteris (L.) Todaro. Ostrich Fern. July–Oct. Rich, moist woodlands; restricted to the n. one-third of the state, usually in the w. cos. *Pteretis pensylvanica* (Willd.) Fern.—F.

11. Gymnocarpium Newm.   Oak Fern

1. G. dryopteris (L.) Newm. Oak Fern. June–Sept. Deep, moist woodlands, very rare; Ogle and St. Clair cos. *Dryopteris disjuncta* (Ledeb.) C. V. Morton—F, J.

12. Thelypteris Schmidel   Beech Fern

1. Rachis winged throughout, or only the basal pair of pinnae stalked; indusium none 2
1. Rachis not winged, the pinnae separate from it nearly to apex of blade; indusium cordate 3
2. Rachis winged only above the two basal pinna pairs, the wings not extending to the lowest pair of pinnae; blades pinnate-pinnatifid; rachis rather densely brown-scaly ....1. *T. phegopteris*
2. Rachis winged throughout; blades bipinnatifid; rachis sparsely white-scaly ....
.... 2. *T. hexagonoptera*
3. Lowest pinnae strongly reduced ....3. *T. noveboracensis*
3. Lowest pinnae only slightly reduced or not at all .... 4. *T. palustris*

1. T. phegopteris (L.) Slosson. Long Beech Fern. June–Aug. Moist, shaded woodlands in sandstone areas, local; Cass, Henderson, Jackson, LaSalle, Menard, Ogle, and Pulaski cos. *Dryopteris phegopteris* (L.) Christens.—F, J.

2. T. hexagonoptera (Michx.) Weatherby. Broad Beech Fern. June–Sept. Rich woodlands; scattered throughout the state. *Dryopteris hexagonoptera* (Michx.) Christens.—F, J.

3. T. noveboracensis (L.) Nieuwl. New York Fern. June–Sept. Moist, rarely dry, woodlands, rare; Kane, Kankakee, Monroe, Pope, and Wabash cos. *Dryopteris noveboracensis* (L.) Gray—F, J.

4. T. palustris Schott var. pubescens (Laws.) Fern. Marsh Fern. June–Oct. Marshy ground and in swamps; rather common in the n., rare in the s. *Dryopteris thelypteris* (L.) Gray—F, J.

78 POLYPODIACEAE

13. *Dryopteris* Adans. Shield Fern

1. D. carthusiana (Villars) H. P. Fuchs. Spinulose Woodfern. June–Aug. Moist, rocky woodlands, rarely in dry situations; occasional throughout the state, except for the s. cent. cos. *D. spinulosa* (O. F. Muell.) Watt.—F, J; *D. austriaca* (Jacq.) Woynar var. *spinulosa* (Muell.) Fiori—G.
2. D. X boottii (Tuckerm.) Underw. Boott's Woodfern. June–Aug. Moist, rocky woodland, very rare; LaSalle Co. Considered a hybrid between *D. intermedia* and *D. cristata*.
3. *D. intermedia* (*Muhl.*) *Gray.* Common Woodfern. June–Aug. Moist, rocky ravines and woodlands; occasional or rare in the n. half and the s. tip of Ill. *D. spinulosa* (O. F. Muell.) Watt var. *intermedia* (Muhl.) Underw.—F; *D. austriaca* (Jacq.) Woynar var. *intermedia* (Muhl.) Morton—G.
4. D. X triploidea Wherry. Woodfern. June–Sept. Moist, rocky woodland, very rare; Lake Co. Considered to be a hybrid between *D. intermedia* and *D. carthusiana*. *D. spinulosa* (O. F. Muell.) Watt—F; *D. austriaca* (Jacq.) Woynar—G.
5. D. cristata (L.) Gray. Crested Fern. June–Sept. Low, moist woodlands; occasional in the n. one-third of the state; also Pope Co.
6. D. X clintoniana (D. C. Eaton) Dowell. Clinton's Woodfern. June–Sept. Moist, rich woodland, very rare; St. Clair Co. Considered to be a hybrid between *D. cristata* and *D. goldiana*. *D. cristata* (L.) Gray var. *clintoniana* (D. C. Eaton) Underw.—F.

7. D. goldiana (Hook.) Gray. Goldie's Fern. June–Sept. Moist, shaded, often rocky woodlands, not common; scattered through the state.

8. D. celsa (Wm. Palmer) Small. Log Fern. June–Aug. Along railroad near a swamp, very rare; Johnson Co. *D. goldiana* (Hook.) Gray ssp. *celsa* Wm. Palmer—G.

9. D. X neo-wherryi W. H. Wagner. June–Aug. Rich, rocky woods, very rare; Pope Co. Considered to be a hybrid between *D. goldiana* and *D. marginalis.*

10. D. marginalis (L.) Gray. Marginal Fern. June–Oct. Rocky woodlands; common to occasional throughout the state.

14. Woodwardia Sm. Chain Fern

1. Leaves monomorphic; sterile blades once-pinnate ...... 1. *W. virginica*
1. Leaves dimorphic; sterile blades pinnatifid .... 2. *W. areolata*

1. W. virginica (L.) Sm. Chain Fern. July–Sept. Bogs, rare; Lake Co.
2. W. areolata (L.) Moore. Netted Chain Fern. July–Oct. Springy marsh in lowland woods, edge of sandstone cliffs, rare; Johnson and Pope cos.

15. Athyrium Roth Lady Fern

1. Blades once-pinnate; pinnae entire .... 1. *A. pycnocarpon*
1. Blades pinnate-pinnatifid to bi- (tri-) pinnate; pinnae pinnatifid 2
2. Blades pinnate-pinnatifid; indusium light brown (at maturity), firm, parallel to the veins; sori straight or nearly so .... 2. *A. thelypterioides*
2. Blades at least bipinnate; indusium dark brown (at maturity), membranaceous, many crossing the veins; sori more or less curved ....
.... 3. *A. filix-femina*

1. A. pycnocarpon (Spreng.) Tidestrom. Narrow-leaved Spleenwort. Aug.–Sept. Moist, shaded woodlands; occasional throughout the state, although rare in the extreme n. cos.

2. A. thelypterioides (Michx.) Desv. Silvery Spleenwort. July–Sept. Rich woodlands, occasional to common.

3. A. filix-femina (L.) Roth var. rubellum Gilb. Lady Fern. June–Sept. Fourth or fifth pair of pinnae (from the base) the largest; rachis glandular; indusium eglandular. Moist, open woodlands and borders of swamps; rather common throughout the state. *A. angustum* (Willd.) Presl—J; *A. filix-femina* var. *michauxii* (Spreng.) Farw.—F, G.

var. asplenioides (Michx.) Farw. Southern Lady Fern. June–Sept. Second or third pair of pinnae (from the base) the largest; rachis eglandular; indusium glandular. Moist, rocky woodlands, occasional in the s. cos.

16. Asplenium L. Spleenwort

1. Leaves simple, unlobed (rarely a single pair of lobe-like auricles at base); veins forming a network .... 1. *A. rhizophyllum*
1. Leaves simple and pinnatifid, or pinnate to bipinnate-pinnatifid; veins free 2
2. Rachis green throughout 3
2. Rachis partly or entirely brown, or black 6

3. Petiole green throughout .... 2. *A. ruta-muraria*
3. Petiole brown at base, sometimes throughout 4
4. Blades entirely pinnatifid, or with merely the lowest pair of pinnae distinct; spores normal .... 3. *A. pinnatifidum*
4. Blades pinnatifid above, pinnate below for at least two pairs of pinnae; spores abortive 5
5. Petiole brown throughout ....4. *A.* X *gravesii*
5. Petiole brown at base, green above ....5. *A.* X *trudellii*
6. Rachis brown below (usually about half its length), green above. 7
6. Rachis brown or black throughout, or at least for three-fourths the length 9
7. Blades once-pinnate below, pinnatifid above; spores abortive 8
7. Blades bipinnate to bipinnate-pinnatifid; spores normal .. 8. *A. bradleyi*
8. Pinnae approximate to overlapping; base of most of the pinnae truncate; outline of frond oblong-lanceolate .... 6. *A.* X *kentuckiense*
8. Pinnae separated to remote; base of most of the pinnae cuneate; outline of frond linear .... 7. *A. trichomanes* X *A. pinnatifidum*
9. Rachis brown for three-fourths its length, green above; blade pinnatifid for half its length, the tip narrowly caudate .... 9. *A.* X *ebenoides*
9. Rachis brown or black throughout; blade pinnatifid only at apex, the tip not caudate 10
10. Pinnae not auriculate .... 10. *A. trichomanes*
10. Pinnae auriculate at base of upper margin 11
11. Pinnae pairs opposite (sometimes alternate); fertile and sterile leaves similar; auricle of pinna not overlapping rachis .... 11. *A. resiliens*
11. Pinnae pairs alternate; fertile leaves more erect and taller than sterile leaves; auricle of pinna overlapping rachis .... 12. *A. platyneuron*

    1. A. rhizophyllum L. Walking Fern. May–Sept. Rocky woodlands, either on sandstone or limestone; scattered throughout the state. *Camptosorus rhizophyllus* (L.) Link—F, G, J.

    2. A. ruta-muraria L. Wall-rue Spleenwort. May–Sept. Crevices of limestone cliffs, very rare; known only from "s. Ill.," collected in the middle 1800s. *A. cryptolepis* Fern.—F.

    3. A. pinnatifidum Nutt. Pinnatifid Spleenwort. May–Sept. Crevices of sandstone cliffs; occasional in the s. tip of Ill.; also Cumberland, Fulton, and Wabash cos.

    4. A. X gravesii Maxon. Graves' Spleenwort. May–Sept. Crevices of sandstone cliffs, very rare; Union and Williamson cos. Considered to be a hybrid between *A. pinnatifidum* and *A. bradleyi*.

    5. A. X trudellii Wherry. Trudell's Spleenwort. May–Sept. Crevices of sandstone cliffs; Jackson Co. Considered to be a hybrid between *A. pinnatifidum* and *A. montanum*.

    6. A. X kentuckiense McCoy. Kentucky Spleenwort. May–Sept. Crevices of cherty cliffs; Union Co. Considered to be a hybrid between *A. pinnatifidum* and *A. platyneuron*.

    7. A. trichomanes X A. pinnatifidum. Wagner's Spleenwort. May–Sept. Crevices of cherty cliffs; Union Co. Considered to be a hybrid between *A. trichomanes* and *A. pinnatifidum*.

    8. A. bradleyi D. C. Eaton. Bradley's Spleenwort. June–Sept. Crevices of sandstone or cherty cliffs, rare; Jackson, Randolph, and Union cos.

    9. A. X ebenoides R. R. Scott. Scott's Spleenwort. May–Sept. Rocky, sandstone woodlands, rare; Alexander, Jackson, Pope, and

Union cos. Considered to be a hybrid between *A. rhizophyllum* and *A. platyneuron. Asplenosorus* X *ebenoides* (R. R. Scott) Wherry—F, G.

10. A. trichomanes L. Maidenhair Spleenwort. May–Sept. Crevices of shaded cliffs, both limestone and sandstone; occasional in the s. one-fourth of the state; also Cumberland Co.

11. A. resiliens Kunze. Black Spleenwort. May–Sept. Crevices of limestone cliffs, rare; Jackson and Union cos.

12. A. platyneuron (L.) Oakes. Ebony Spleenwort. May–Sept. Dry or moist woodlands; common in the s. half of Ill., becoming less abundant northward. (Including var. *incisum* [Howe] Robinson.)

17. Woodsia R. Br. Woodsia

1. Blades bipinnate-pinnatifid; leaves generally over 25 cm long; petiole stramineous, unjointed; scales of rhizome few . . . . 1. *W. obtusa*
1. Blades pinnate-pinnatifid; leaves generally less than 20 cm long; petiole brown, jointed below the middle; scales of rhizome numerous . . . .
. . . . 2. *W. ilvensis*

1. W. obtusa (Spreng.) Torr. Common Woodsia. May–Oct. Dry or moist rocky woods, occasionally on very exposed sandstone ledges; occasional to common, except in the e. cent. cos.

2. W. ilvensis (L.) R. Br. Rusty Woodsia. June–Oct. Sandstone cliff, very rare; Ogle Co.

18. Cystopteris Bernh. Fragile Fern

1. All veins of ultimate leaf segments running to the sinuses 1. *C. bulbifera*
1. Some or all the veins of ultimate leaf segments running to the teeth . 2
2. All veins of ultimate leaf segments running to the teeth . . . 2. *C. fragilis*
2. Some veins of ultimate leaf segments running to the teeth, some running to the sinuses 3
3. Indusium acute and toothed at the apex, eglandular . . . . . . 2. *C. fragilis*
3. Indusium truncate and entire at the apex, more or less glandular . . . . . .
. . . . 3. *C.* X *tennesseensis*

1. C. bulbifera (L.) Bernh. Bladder Fern. June–Sept. Mostly limestone cliffs and woodlands; occasional in the cos. along the Illinois and Mississippi rivers, otherwise rare.

2. C. fragilis (L.) Bernh. Fragile Fern. June–Sept. All veins of ultimate leaf segments running to the teeth; rhizome short, scaly; petiole brown for half its length. Moist woodlands; Union Co.

var. protrusa Weatherby. Fragile Fern. June–Sept. All veins of ultimate leaf segments running to the teeth; rhizome long-creeping, densely hairy; petiole stramineous, except at the base. Moist woodland; common in every co.

var. mackayi Laws. Fragile Fern. June–Sept. Moist woodlands; Jackson and LaSalle cos.

3. C. X tennesseensis Shaver. Tennessee Fragile Fern. June–Sept. Moist woodlands, rare; Champaign and Will cos.

9. MARSILEACEAE

1. *Marsilea L. Waterclover*

1. M. quadrifolia L. Waterclover. June–Dec. Introduced into pond
or lakes; scattered and uncommon.

10. SALVINIACEAE

1. *Azolla Lam. Mosquito Fern*

1. Plants less than 1 cm in diameter; leaves about 0.5 mm long; glochidi
without cross-walls .... 1. *A. carolinian*
1. Plants 1 cm in diameter or larger; leaves about 0.7 mm long or longe
glochidia with cross-walls .... 2. *A. mexican*

    1. A. caroliniana Willd. Mosquito Fern. July–Oct. Standing wate
very rare; St. Clair Co.
    2. A. mexicana Presl. Mosquito Fern. July–Oct. Ponds and standin
water, local; scattered throughout.

11. TAXACEAE

1. *Taxus L. Yew*

    1. T. canadensis Marsh. Canada Yew. Wooded hillsides, rare; cor
fined to the n. one-fourth of Ill.

12. PINACEAE

1. Leaves in fascicles of 2–5; plants evergreen .... 1. *Pinu*
1. Leaves in fascicles of 20 or more; plants deciduous .... 2. *Lari*

1. *Pinus L. Pine*

1. Leaves 5 in a fascicle .... 1. *P. strobu*
1. Leaves 2–3 in a fascicle ....
2. Leaves, or some of them, 3 in a fascicle ....
2. Leaves, or some of them, 2 in a fascicle ....
3. Leaves slender, at most 1.5 mm broad .... 2. *P. echinat*
3. Leaves stout, 1.5–3.0 mm broad ....
4. Most of the leaves more than 7 cm long; scales of cone with a spine u
to 3 mm long ....

4. Most of the leaves less than 7 cm long; scales of cone with a spine 5–6 mm long .... 5. *P. pungens*

5. Leaves up to 15 cm long; cone almost as broad as long ... 3. *P. rigida*

5. Most or all the leaves over 15 cm long; cone longer than broad .... 4. *P. taeda*

6. Leaves up to 7 cm long 7

6. Most of the leaves 7 cm long or longer 9

7. Leaves 2–4 cm long; cones shiny .... 6. *P. banksiana*

7. Some of the leaves over 4 cm long; cones not shiny 8

8. Scales of cone without a spine .... 7. *P. sylvestris*

8. Scales of cone with a spine 5–6 mm long .... 5. *P. pungens*

9. None of the leaves over 17 cm long, dark green 10

9. Some of the leaves 17–25 cm long, light green .... 4. *P. taeda*

10. Scales of cone without a spiny tip .... 8. *P. resinosa*

10. Scales of cone with a minute spiny tip .... 2. *P. echinata*

1. P. strobus L. White Pine. Woods; local in the n. one-fourth of Ill.

2. P. echinata Mill. Shortleaf Pine. Rocky soil, rare in the native condition, where it is known from Randolph and Union cos.; widely planted in plantations.

3. P. rigida Mill. Pitch Pine. Native to the e. of Ill.; occasionally encountered as an introduction in the state.

4. P. taeda L. Loblolly Pine. Native primarily s. of Ill.; frequently planted in Ill.

5. P. pungens Lamb. Table Mountain Pine. Native e. and s. of Ill.; rarely encountered as an introduction in Ill.

6. P. banksiana Lamb. Jack Pine. Sandy soil, rare; Cook, Lake, and Ogle cos.

7. P. sylvestris L. Scotch Pine. Native of Europe; occasionally encountered as an introduction in Ill.

8. P. resinosa Ait. Red Pine. Dry woods, rare; LaSalle Co.

2. *Larix* Mill.   Larch

1. Scales of cones glabrous, 10–20 in number; most of the leaves less than 2.5 cm long .... 1. *L. laricina*

1. Scales of cones pubescent, at least 30 in number; most of the leaves at least 2.5 cm long .... 2. *L. decidua*

1. L. laricina (DuRoi) K. Koch. American Larch. Tamarack. Bogs, rare; Cook, Lake, and McHenry cos.

2. L. decidua Mill. European Larch. Escaped from cultivation in a few ne. cos.

13. TAXODIACEAE

1. *Taxodium* Rich.   Bald Cypress

1. T. distichum (L.) Rich. Bald Cypress. Swamps; extreme s. Ill., n. along the Wabash River to Lawrence Co. (Including f. *confusum* Palmer & Steyerm.)

14. CUPRESSACEAE

1. Branches flattened; leaves all scale-like; fruit a woody cone ... 1. *Thu*
1. Branches not flattened; leaves all subulate or some scale-like, othe
   subulate; fruit a fleshy berry-like cone .... 2. *Juniper*

1. *Thuja* L.   Arbor Vitae

   1. T. occidentalis L. Arbor Vitae. White Cedar. Cliffs, bluffs, a
bogs, rare; Cook, Kane, Lake, and LaSalle cos.

2. *Juniperus* L.   Juniper

1. All leaves subulate, borne in whorls of three .... 1. *J. commun*
1. Some leaves subulate, others scale-like, borne opposite ....
2. Prostrate, trailing shrub; fruits more than 6 mm in diameter, 3- to
   seeded .... 2. *J. horizonta*
2. Upright tree; fruits 4–6 mm in diameter, 1- to 2-seeded . 3. *J. virginia*

   1. J. communis L. Common Juniper. Plants erect. Sandy soil, ra
Lake Co. *J. canadensis* Burgsdorf—J.
   var. depressa Pursh. Ground Juniper. Plants decumbent. Sa
dunes, rare; Cook and Lake cos. *J. canadensis* Burgsdorf—J.
   2. J. horizontalis Moench. Trailing Juniper. Sand dunes, rare; Co
and Lake cos.
   3. J. virginiana L. Red Cedar. Woods, cliffs, and fields; comm
throughout Ill. (Including var. *crebra* Fern. & Grisc.)

15. TYPHACEAE

1. *Typha* L.   Cat-tail

1. Pollen grains borne in groups of 4; stigma spatulate; leaves flat, usua
   8 or more per plant; staminate and pistillate portions of the inflore
   cence usually contiguous, but not always .... 1. *T. latifo*
1. Pollen grains borne singly; stigma linear; leaves somewhat conve
   usually less than 8 per plant; staminate and pistillate portions of th
   inflorescence separated .... 2. *T. angustifo*

   1. T. latifolia L. Common Cat-tail. June–Oct. Low areas, common;
every co.
   2. T. angustifolia L. Narrow-leaved Cat-tail. June–Oct. Wet situ
tions, occasional; scattered throughout the state.

16. SPARGANIACEAE

1. *Sparganium L. Bur-reed*

. Plants floating or suberect; stems to 30 cm long; staminate head 1; pistillate heads about 1 cm in diameter, the lowest short-pedunculate; beak of achene 0.5–1.5 mm long .... 1. *S. minimum*
. Plants erect; stems (5–) 30–120 cm tall; staminate heads 3–25; pistillate heads 1.5–3.5 cm in diameter, all sessile; beak of achene 2–6 mm long.
.... 2
. Central axis bearing 1–4 pistillate heads; achene stipitate, fusiform, tapering to the summit; stigma 1 3
. Central axis bearing only staminate heads; achene not stipitate, ob-pyramidal, truncate at the summit; stigmas 2 .... 5. *S. eurycarpum*
. At least one of the pistillate heads borne above the subtending bract (supra-axillary); inflorescence simple; achene usually greenish-brown, even at maturity .... 2. *S. chlorocarpum*
. All heads axillary in the subtending bract; inflorescence usually with 1 or more branches; achene pale or dark brown 4
. Leaves stiff, strongly keeled; branches of inflorescence without any pistillate heads; pistillate heads 2.5–3.5 cm in diameter; body of achene 5–7 mm long, shining, pale brown .... 3. *S. androcladum*
. Leaves soft, usually not strongly keeled; branches of inflorescence with 1–3 pistillate heads; pistillate heads 1.5–2.5 cm in diameter; body of achene 3–5 mm long, dull, dark brown .... 4. *S. americanum*

1. S. minimum (Hartm.) Fries. Least Bur-reed. July–Aug. Shallow water; McHenry Co.

2. S. chlorocarpum Rydb. Green-fruited Bur-reed. June–Aug. Shallow water; Cook, DeKalb, Lake, Lee, and Union cos.

3. S. androcladum (Engelm.) Morong. Bur-reed. June–July. Shallow water; scattered throughout the state, but not common.

4. S. americanum Nutt. Bur-reed. June–July. Shallow water; restricted to the northern one-fifth of the state.

5. S. eurycarpum Engelm. Bur-reed. June–Aug. Shallow water; occasional throughout the state, but less common in the extreme cos.

17. RUPPIACEAE

1. *Ruppia L. Ditch Grass*

1. R. maritima L. var. rostrata Agardh. Ditch Grass. July–Oct. Saline waters, rare; Henry, Lake, and Vermilion cos.

18. ZANNICHELLIACEAE

1. *Zannichellia* L. Horned Pondweed

1. Z. palustris L. Horned Pondweed. June–Oct. Ditches, spring-f
streams, and ponds; occasional in the n. one-half of the state; a
Jackson Co.

19. NAJADACEAE

1. *Najas* L. Naiad

1. Leaves seemingly entire or serrulate; plants monoecious; achene 1.5–3
mm long ....
1. Leaves coarsely toothed; plants dioecious; achene 4.0–7.5 mm long . .
.... 5. *N. mari*
2. Leaves linear, 0.5–1.4 mm wide, with slightly dilated bases; margir
spinules microscopic ....
2. Leaves filiform, 0.1–0.3 mm wide, with abruptly dilated bases; margir
spinules macroscopic, although minute ....
3. Achene lustrous, with 30–50 rows of obscure, minute, often square are
lae; style and 2 stigmas 0.8–1.6 mm long .... 1 *N. flex*
3. Achene dull, with 16–24 rows of distinct, large, hexagonal or rectangu
areolae; style and 2–3 stigmas less than 1 mm long   2. *N. guadalupens*
4. Achene with roughened appearance, with 22–40 rows of longer th
broad, rectangular areolae; style and 2–3 stigmas 0.8–1.2 mm long . . .
.... 3. *N. gracillin*
4. Achene with ribbed appearance, with 10–18 rows of broader than lor
rectangular areolae; style and 2 stigmas 1.0–1.4 mm long .. 4. *N. min*

1. N. flexilis (Willd.) Rostk. & Schmidt. Naiad. July–Sept. Shallo
water; occasional throughout the state.
2. N. guadalupensis (Spreng.) Magnus. Naiad. July–Oct. Alo
shores of ponds and shallow lakes; occasional to uncomm
throughout the state.
3. N. gracillima (A. Br.) Magnus. Naiad. July–Sept. Pools, pond
and lakes with sandy and muddy substrata, rare; Ford, Jackson, a
Williamson cos.
4. N. minor All. Naiad. July–Oct. Shallow waters along lake shore
occasional to common in the s. two-thirds of the state, absent els
where.
5. N. marina L. Naiad. July–Oct. In water, rare; Lake Co.

20. POTAMOGETONACEAE

1. *Potamogeton* L. Pondweed

1. Leaves septate, 0.3–0.5 (–2.0) mm broad; peduncles flexible ....
.... 1. *P. pectinat*

. Leaves not septate, some or all over 0.5 mm broad; peduncles rigid . 2
. Leaves uniform 3
. Leaves of two kinds, the floating usually shorter and broader than the submersed 20
. Stipules adnate to leaf-base; leaves auriculate at base .. 2. *P. robbinsii*
. Stipules free from leaf-base; leaves rounded, cordate, or tapering at base, never auriculate 4
. Leaves sharply serrulate; fruits 5–6 mm long .... 3. *P. crispus*
. Leaves entire or minutely denticulate; fruits 1.6–5.0 mm long 5
. Submersed leaves 5–75 mm broad 6
. Submersed leaves up to 5 mm broad 13
. Stems compressed; fruits 3-keeled .... 14. *P. epihydrus*
. Stems terete; fruits rounded on the sides, or with only one prominent keel 7
. Submersed leaves rounded or cordate at the base, some of them partly clasping the stem 8
. Submersed leaves tapering to the base, not clasping the stem 9
. Leaves entire, boat-shaped at the tip; fruits 4–5 mm long, dorsally keeled; stems zigzag .... 4. *P. praelongus*
. Leaves minutely denticulate, not boat-shaped at the tip; fruits 2–3.5 mm long, not keeled; stems not zigzag .... 5. *P. richardsonii*
. Leaves entire (sometimes crisped in *P. pulcher*) 10
. Leaves minutely denticulate 11
. Fruiting spike 4 cm long or longer; middle and upper submersed leaves broadly lanceolate to ovate, never crisped; fruit tapering to base .... 15. *P. amplifolius*
. Fruiting spike 2.0–3.5 cm long; middle and upper submersed leaves linear-lanceolate to lanceolate, sometimes crisped; fruit rounded at base .... 16. *P. pulcher*
. Fruits (including beak) 3.5–4.3 mm long .... 17 *P. nodosus*
. Fruits (including beak) 1.7–3.5 mm long 12
. Primary stems up to 1 mm thick; leaves more than 5 times longer than broad .... 20. *P. gramineus*
. Primary stems usually more than 1 mm thick; leaves less than 5 times longer than broad .... 18. *P. illinoensis*
. Submersed leaves with 15–35 nerves; fruits quadrate, 4–5 mm long .... 6. *P. zosteriformis*
. Submersed leaves with (1–) 3–7 (–11) nerves; fruits obovoid to ovoid to nearly orbicular, 1.5–3.5 mm long 14
. Spikes with 6 or more whorls of flowers 15
. Spikes with 1–5 whorls of flowers 17
. Leaves minutely denticulate .... 20. *P. gramineus*
. Leaves entire 16
. Spikes uniform; stipules free from base of leaf; fruits 3–5 mm long .... 12. *P. natans*
. Spikes of two or more kinds; stipules adnate to base of leaf; fruits 1.0–1.5 mm long .... 13. *P. diversifolius*
. Spikes subcapitate, 2–5 mm long; fruits keeled; sepaloid connectives 0.5–1.0 mm long; leaves glandless at base .... 7. *P. foliosus*
. Spikes elongate, more or less interrupted, 6–15 mm long; fruits rounded on back; sepaloid connectives 1.2–2.5 mm long; leaves biglandular at base (sometimes glands absent in *P. pusillus*) 18
. Leaves thin, translucent .... 8. *P. friesii*
. Leaves firm, opaque 19
. Stipules chartaceous, white, becoming fibrous .... 9. *P. strictifolius*

19. Stipules membranous, olive, not becoming fibrous ..... 10. *P. pusillu*
20. Leaves entire (submersed leaves sometimes crisped in *P. pulcher*) . *2*
20. Leaves minutely denticulate *2*
21. Submersed leaves up to 2 mm broad, 1- to 5-nerved *2*
21. Submersed leaves 5–75 mm broad, 7- to 21-nerved *2*
22. Spikes uniform; stipules free from base of leaf; fruits 1.6–5.0 mm lon
.... *2*
22. Spikes of two or more kinds, those in the axils of the submersed leave
subgloboid, those in the axils of the floating leaves elongate; stipule
adnate to base of leaf; fruits 1.0–1.5 mm long ...... 13. *P. diversifoliu*
23. Submersed leaves 0.1–0.5 mm broad, 1- to 3-nerved; fruits 1.6–2.2 m
long; floating leaves elliptic to narrowly obovate .... 11. *P. vase*
23. Submersed leaves 0.5–2.0 mm broad, 3- to 5-nerved; fruits 3–5 mm lon
floating leaves ovate .... 12. *P. natar*
24. Stems compressed; submersed leaves linear, 5–10 mm broad; fruit wi
3 sharp keels, capped with a minute, tooth-like beak .. 14. *P. epihydru*
24. Stems terete; submersed leaves lanceolate to ovate, 10–75 mm broad
fruit not sharply 3-keeled, capped with a prominent beak *2*
25. Fruiting spike 4 cm long or longer; stipules of submersed leaves subper
sistent, the middle and upper submersed leaves broadly lanceolate *t*
ovate, never crisped; fruit tapering to base; some nerves of floatin
leaves more prominent than other nerves .... 15. *P. amplifoliu*
25. Fruiting spike 2.0–3.5 cm long; stipules of submersed leaves falling awa
early; middle and upper submersed leaves lanceolate to linear-lan
ceolate, sometimes crisped; fruit rounded at base; all nerves of floatin
leaves uniform .... 16. *P. pulche*
26. Fruits (including beak) 3.5–4.3 mm long .... 17. *P. nodosu*
26. Fruits (including beak) 1.7–3.5 mm long *2*
27. Stems simple or once-branched; stipules strongly keeled *2*
27. Stems repeatedly branched; stipules faintly keeled *2*
28. Floating leaves 2.0–6.5 cm wide; fruits normally developed ....
.... 18. *P. illinoens*
28. Floating leaves 0.5–1.0 cm wide; fruits not maturing 19. *P.* X *hagstrom*
29. Floating leaves elliptic to ovate, the petioles usually as long as or longe
than the blades; stipules obtuse *3*
29. Floating leaves oblong- to linear-lanceolate, the petioles shorter than th
blades; stipules acuminate .... 22. *P.* X *rectifoliu*
30. Submersed leaves acute at apex; fruit strongly 3-keeled, well-develope
.... 20. *P. gramineu*
30. Submersed leaves cuspidate at apex; fruit obscurely 3-keeled, poorly de
veloped ....21. *P.* X *spathulaeform*

        1. P. pectinatus L. Fennel-leaved Pondweed. May–Sept. Calcareou
water; occasional throughout the state.
        2. P. robbinsii Oakes. Pondweed. June–Sept. Stagnant water, ver
rare; Lake Co.
        3. P. crispus L. Curly Pondweed. May–Sept. Native of Europe; natu
ralized in muddy to calcareous ponds and streams; occasional in th
state.
        4. P. praelongus Wulfen. Pondweed. June–Sept. Lakes and rivers
rare; Cook, Lake, and McHenry cos.
        5. P. richardsonii (Benn.) Rydb. Pondweed. June–Sept. Lakes an
rivers, rare; Cook, Kankakee, Lake, and McHenry cos.

6. P. zosteriformis Fern. Pondweed. June–Sept. Ponds, lakes, and streams, not common; known only from the n. half of the state.

7. P. foliosus Raf. Pondweed. June–Sept. Ponds, lakes, rivers, and streams; occasional throughout the state. (Including var. *macellus* Fern.)

8. P. friesii Rupr. Pondweed. June–Sept. Ponds and streams, rare; Cook and Lake cos.

9. P. strictifolius Benn. Pondweed. June–Sept. Lakes, very rare; Cook Co.

10. P. pusillus L. Pondweed. June–Sept. Ponds, lakes, and streams; occasional throughout the state, but rarer in the s. cos. (Including var. *minor* [Biv.] Fern. & Schub.; *P. berchtoldii* Fieber; *P. berchtoldii* Fieber var. *tenuissimus* [Mert. & Koch] Fern.)

11. P. vaseyi Robbins. Pondweed. June–Sept. Lakes and ponds, rare; Grundy and McHenry cos.

12. P. natans L. Pondweed. June–Sept. Lakes and streams; occasional throughout the state, except for the extreme s. cos.

13. P. diversifolius Raf. Pondweed. May–Oct. Quiet waters; occasional throughout the state.

14. P. epihydrus Raf. Pondweed. June–Sept. Quiet ponds and lakes, rare; Fulton, Hancock, and Lake cos.

15. P. amplifolius Tuckerm. Pondweed. June–Oct. Lakes and roadside ditches, rare; Cook, DuPage, and Lake cos.

16. P. pulcher Tuckerm. Pondweed. June–Sept. Shallow water, rare; Jackson, Mason, and Menard cos.

17. P. nodosus Poir. Pondweed. May–Oct. Ponds and streams; occasional throughout the state. *P. americanus* Cham. & Schlecht.—J.

18. P. illinoensis Morong. Pondweed. June–Oct. Ponds, lakes, and streams; occasional in the n. half of the state, rare in the s. half.

19. P. X hagstromii Benn. Pondweed. July–Sept. Lakes, very rare; Cook Co. Reputed to be a hybrid between *P. richardsonii* and *P. gramineus.*

20. P. gramineus L. Pondweed. June–Sept. Lakes, ponds, and streams, rare; Cook and Wabash cos.

21. P. X spathulaeformis (Robbins) Morong. Pondweed. June–Sept. Ponds and streams, very rare; Wabash Co. Reputed to be a hybrid between *P. illinoensis* and *P. gramineus.*

22. P. X rectifolius Benn. Pondweed. June–Sept. Ditches, in 1–3 feet of water; Cook Co. Reputed to be a hybrid between *P. nodosus* and *P. richardsonii.*

21. JUNCAGINACEAE

1. Inflorescence a spike-like raceme, without bracts; leaves all basal . . . . . .
. . . . 1. *Triglochin*
1. Inflorescence loosely racemose, bracteate; leaves basal and alternate . .
. . . . 2. *Scheuchzeria*

1. *Triglochin* L.   Arrow-grass

1. Carpels usually 6, the axis between them slender . . . . 1. *T. maritima*
1. Carpels 3, the axis between them broadly 3-winged . . . . . 2. *T. palustris*

1. T. maritima L. Arrow-grass. May–Aug. Sandy shores, swamps wet ditches; known only from the extreme ne. cos.; also Peoria Co.

2. T. palustris L. Arrow-grass. May–July. Low areas; known onl from the ne. cos.; also Peoria and Tazewell cos.

2. Scheuchzeria L. Arrow-grass

1. S. palustris L. var. americana Fern. Arrow-grass. May–July. Bogs Fulton, Lake, McHenry, and Menard cos. *S. palustris* L—G; *S ameri cana* (Fern.) G. N. Jones—J.

22. ALISMACEAE

1. Receptacle convex, bearing several rows of pistils; stamens 12 numerous (6–9 in *E. tenellus* var. *parvulus*); flowers unisexual or perfec
....
1. Receptacle flat, bearing a single row of pistils; stamens 6–9; flowers per fect .... 3. *Alism*
2. Achenes not winged; base of whorled inflorescence branches bearing bracts and several bracteoles; flowers perfect; stamens never more tha 21 .... 1. *Echinodoru*
2. Achenes winged; base of whorled inflorescence branches bearing bracts and no bracteoles, flowers mostly unisexual; stamens usuall more than 21 .... 2. *Sagittari*

1. Echinodorus Rich. Burhead

1. Plants erect, less than 10 cm tall; leaves linear to lanceolate; flowers a most only 6 mm broad; stamens 6–9; achenes 10–15, beakless or nearl so .... 1. *E. tenellus* var. *parvulu*
1. Plants erect, usually more than 10 cm tall, or plants creeping or arching leaves broadly ovate, rarely lanceolate; flowers at least 8 mm broad stamens 12–21; achenes more than 40, beaked ....
2. Scape erect; stamens 12; style longer than the ovary; beak of achen straight .... 2. *E. berteroi* var. *lanceolatu*
2. Scape creeping or arching; stamens 21; style shorter than the ovary beak of achene incurved .... 3. *E. cordifoliu*

1. E. tenellus (Mart.) Buchenau var. parvulus (Engelm) Fassett Small Burhead. Aug.–Sept. Wet ditches, rare; Cass and St. Clair cos *E. tenellus* (Mart.) Buchenau—F; *E. parvulus* Engelm.—G, J.

2. E. berteroi (Spreng.) Fassett var. lanceolatus (Wats. & Coult Fassett. Burhead. July–Sept. Wet ditches, edges of swamps; scattere in the s. three-fourths of the state, very rare elsewhere. *E. rostratu* (Nutt.) Engelm.—F, G, J.

3. E. cordifolius (L.) Griseb. Creeping Burhead. July–Sept. Wet di ches, edges of swamps; mostly restricted to the w. cos.; also Craw ford Co.

2. Sagittaria L. Arrowleaf

1. Filaments roughened with minute scales ....
1. Filaments glabrous ....
2. Peduncle and pedicel inflated; sepals tightly appressed to fruiting hea

. . . . 1. *S. calycina*

2. Peduncle and pedicel slender; sepals usually reflexed 3

3. Pistillate flowers and fruiting heads sessile or subsessile . . . 2. *S. rigida*

3. Pistillate flowers and fruiting heads long-pedunculate . 3. *S. graminea*

4. Mature achene with 1–3 facial wings; beak of achene inserted apically   5

4. Mature achene with marginal wings only; beak of achene inserted laterally . . . . 7. *S. latifolia*

5. Beak of achene less than 0.5 mm long, straight, with 1 small, narrow facial wing . . . . 4. *S. cuneata*

5. Beak of achene more than 0.5 mm long, usually reflexed at apex, usually with 2–3 facial wings with 1 well developed 6

6. Receptacle not echinate; flowers in whorls of 3 or 4 . . 5. *S. brevirostra*

6. Receptacle echinate; flowers in whorls of more than 4   6. *S. longirostra*

1. S. calycina Engelm. Arrowleaf. June–Sept. Marshes, pond margins, shorelines, shallow water; locally throughout cent. and s. Ill., apparently absent elsewhere. (Including var. *maxima* Engelm.) *S. montevidensis* Cham. & Schlecht.—G; *Lophotocarpus calycinus* (Engelm.) J. G. Sm.—F.

2. S. rigida Pursh. Arrowleaf. May–Oct. Shallow water, muddy sand, in swamps, margins of ponds, or along waterways; not common but scattered in Ill.

3. S. graminea Michx. Narrow-leaved Arrowleaf. May–Sept. Swamps, mud, sand, or shallow water; uncommon but scattered in the n. half of the state, rare in the s. half.

4. S. cuneata Sheld. Arrowleaf. June–Sept. Mud or water in sloughs, and along waterways; confined to the n. half of the state.

5. S. brevirostra Mack. & Bush. Arrowleaf. June–Sept. Sloughs, shorelines, and shallow water; occasional throughout the state. *S. engelmanniana* J. G. Sm. ssp. *brevirostra* (Mack. & Bush) Bogin— G.

6. S. longirostra (Micheli) J. G. Sm. Arrowleaf. July–Sept. Springy woods, rare; Pope Co. *S. engelmanniana* J. G. Sm. ssp. *longirostra* (J. G. Sm.) Bogin.—G; *S. australis* (J. G. Sm.) Small.

7. S. latifolia Willd. Common Arrowleaf. June–Oct. Swamps, sloughs, ponds, shorelines, shallow water, mud; occasional to common throughout the state. (Including f. *diversifolia* [Engelm.] B. L. Robins.)

3. *Alisma* L.   Water Plantain

1. Petals 3.5–6.0 mm long; flowers at least 7 mm broad; styles at anthesis as long as the ovaries; achenes at least 2.5 mm long . . . . 1. *A. plantago-aquatica* var. *americanum*

1. Petals 1.0–2.5 mm long; flowers at most 3.5 mm broad; styles at anthesis less than one-half as long as the ovaries; achenes less than 2.5 mm long . . . . 2. *A. subcordatum*

1. A. plantago-aquatica L. var. americanum Roem. & Schultes. Water Plantain. July–Aug. Shallow water, marshes, ditches not common; confined primarily to the n. half of the state; also Wabash Co. *A. triviale* Pursh—F, J.

2. A. subcordatum Raf. Small-flowered Water Plantain. July–Aug. Shallow water, marshes, ditches; common throughout the state.

23. BUTOMACEAE

1. *Butomus* L.   Flowering Rush

1. B. umbellatus L. Flowering Rush. July–Sept. Native of Europe rarely naturalized in Ill.; Cook Co.

24. HYDROCHARITACEAE

1. Leaves borne along the stem, the longest not more than 3 cm, the broadest not more than 5 mm .... 1. *Elode*
1. Leaves all basal, 2–200 cm long, 5–25 mm broad ....
2. Leaves narrow, elongate; fertile stamens 2, the filaments free ....
.... 2. *Vallisner*
2. Leaves ovate to orbicular; fertile stamens 6–12, the filaments united int a column .... 3. *Limnobiur*

1. *Elodea* Michx.   Waterweed

1. Leaves in whorls of 4 or 6, generally more than 2 cm long; petals 9–1 mm long .... 1. *E. dens*
1. Leaves opposite or in whorls of 3, generally less than 2 cm long; petal minute or up to 5 mm long ....
2. Staminate flowers pedicellate, not liberated at maturity, with sepal 3.5–5.0 mm long; leaves 1–5 mm broad .... 2. *E. canadens*
2. Staminate flowers sessile, liberated at maturity, with sepals 1.5–2.0 m long; leaves smaller, usually 0.5–1.5 mm broad .... 3. *E. nuttall*

1. E. densa (Planch.) Caspary. Elodea; Waterweed. July–Sept. Native to S. Am.; adventive in mine ponds; Franklin and Williamson cos *Anacharis densa* (Planch.) Vict.—G.

2. E. canadensis Michx. Elodea; Waterweed. July–Sept. Quiet water; occasional in the n. half of the state, rare in the s. half. *Anacharis canadensis* (Michx.) Rich.—G.

3. E. nuttallii (Planch.) St. John. Elodea; Waterweed. July–Sep Quiet water; occasional in the n. cos., not common in the s. cos *Anacharis nuttallii* Planch.—G; *E. occidentalis* (Pursh) St. John–J.

2. *Vallisneria* L.   Eelgrass

1. V. americana Michx. Eelgrass. June–Sept. Quiet water; restricte to the n. one-half of the state; also Alexander Co.

3. *Limnobium* Rich.   Frog's-bit

1. L. spongia (Bosc) Steud. Sponge Plant. July–Aug. Swamps, rare Alexander, Johnson, Massac, and Union cos.

25. POACEAE

1. Culms woody .... 80. *Arundinaria*
1. Culms herbaceous 2
2. Spikelets enclosed by a spiny bur .... 49. *Cenchrus*
2. Spikelets not enclosed by a spiny bur 3
3. Spikelets with one or more perfect florets 4
3. Spikelets unisexual (i.e., either all staminate or all pistillate) .. *Group G*
4. Inflorescence solitary, racemose, paniculate, or spicate, but not digitate
.... 5
4. Inflorescence digitate (the spikes and racemes radiating from near the
same point) .... *Group F*
5. Inflorescence spicate or spike-like, with one spike per culm .. *Group A*
5. Inflorescence solitary, racemose, or paniculate, but not composed of
single spikes 6
6. Each spikelet with 2 or more perfect florets 7
6. Each spikelet with one perfect floret (sterile or staminate lemmas may be
present, in addition) 8
7. Some part of the spikelet awned .... *Group B*
7. Spikelet without any awns .... *Group C*
8. Some part of the spikelet awned .... *Group D*
8. Spikelet without awns .... *Group E*

Group A

Inflorescence spicate or spike-like, with one spike per culm; spikelets with one or more perfect florets.

1. Spikelets cylindrical, borne at swollen rachis joints, the entire spikelet
falling at maturity; each glume with one awn and one tooth 32. *Triticum*
1. Spikelets not as above; rachis joints not swollen; glumes awned or awnless, but not with one awn and one tooth 2
2. Spikelets borne edgewise to the rachis; inner glume absent, except in
the terminal spikelet .... 4. *Lolium*
2. Spikelets borne flatwise to the rachis; glumes present on all spikelets   3
3. Each spikelet subtended and usually surpassed by one or more sterile
bristles (not be be confused with awns) .... 48. *Setaria*
3. Each spikelet not subtended by bristles 4
4. Each spikelet with two or more perfect florets (*Anthoxanthum* and *Phalaris* have three lemmas, but two of them are sterile) 5
4. Each spikelet with a single perfect floret (1–2 sterile lemmas present in
addition in *Anthoxanthum* and *Phalaris;* 1 staminate lemma present in
addition in *Holcus*) 19
5. At least some part of the spikelet awned 6
5. Spikelets awnless throughout 15
6. Upper spikelets paired, the lowermost solitary ..... 30. X *Agrohordeum*
6. Spikelets either all paired, all borne in threes, or all solitary 7
7. Spikelets either all paired or all borne in threes 8
7. Spikelets solitary 10
8. Spikelets in threes .... 29. *Hordeum*
8. Spikelets paired 9
9. Glumes to 4 cm long; axis of inflorescence rarely breaking apart at maturity .... 27. *Elymus*

9. Glumes 6–8 cm long; axis of inflorescence breaking apart at maturity
.... 28. *Sitanio*
10. Glumes awned 1
10. Glumes awnless 1
11. Glumes 3-nerved; annuals .... 32. *Triticur*
11. Glumes often 1-nerved; perennials 1
12. Spikelets more than 3 cm long; annuals .... 33. *Seca*
12. Spikelets up to 3 cm long; perennials .... 31. *Agropyro*
13. Blades 10–20 mm broad; annuals .... 32. *Triticur*
13. Blades 1–10 mm broad; perennials 1
14. Awn of lemma more than 1 mm long .... 31. *Agropyro*
14. Awn of lemma up to 1 mm long .... 9. *Koeler*
15. Annuals .... 32. *Triticu*
15. Perennials 1
16. Spikelets paired .... 27. *Elymu*
16. Spikelets solitary 1
17. Spikelets borne flatwise to the continuous rachis .... 31. *Agropyro*
17. Spikelets borne all around the articulated (jointed) rachis 1
18. Blades 3–10 mm broad; lemmas densely pubescent on the nerves....
.... 61. *Triden*
18. Blades 1–3 mm broad; lemmas merely scabrous .... 9. *Koeler*
19. Upper spikelets paired, the lowermost solitary ..... 30. X *Agrohordeu*
19. Spikelets either all paired, all borne in threes, or all solitary 2
20. Spikelets paired or in groups of three; glumes long-awned 2
20. Spikelets solitary 2
21. Spikelets paired .... 27. *Elymu*
21. Spikelets in threes .... 30. *Hordeu*
22. Some part of the spikelet awned 2
22. Spikelets not awned 2
23. Lemmas awned; glumes awned .... 24. *Phleu*
23. Lemmas awned; glumes awned or awnless 2
24. Lemma awned from the middle .... 16. *Calamagrost*
24. Lemma awned from the tip 2
25. Lemma I per spikelet, perfect 2
25. Lemmas 2–3 per spikelet, but only one perfect 2
26. Glumes united near the base .... 23. *Alopecur*
26. Glumes free at the base .... 65. *Muhlenberg*
27. Spikelets 5–10 mm long, each with one perfect floret and two emp
lemmas .... 20. *Anthoxanthu*
27. Spikelets 3.5–5.0 mm long, each with one perfect floret and one stami
ate floret .... 15. *Holc*
28. Glumes 9–15 mm long; lemma 7–14 mm long .... 17. *Ammophi*
28. Glumes to 7 (–10 in *Phalaris*) mm long; lemmas to 7 mm long 2
29. Each spikelet with one perfect floret and (1–)2 empty lemmas ....
.... 22. *Phalar*
29. Each spikelet with one perfect floret only 3
30. Lemma 3-nerved .... 65. *Muhlenberg*
30. Lemma 1-nerved 3
31. Spikes broad, one-fourth to nearly as broad as long .... 67. *Cryps*
31. Spikes slender, one-fifth or less as broad as long ...... 66. *Sporobol*

Group B

Inflorescence solitary, racemose, or paniculate, but not spicate of digitat
spikelets with 2 or more perfect flowers; some part of the spikelet awned

1. Lemmas 2-toothed at the apex 2
1. Lemmas acute or obtuse at the apex, not 2-toothed 9
2. Awn of lemma arising from between the teeth 3
2. Awn of lemma arising near the middle or base of the lemma 7
3. Lemmas 5- to 9-nerved 4
3. Lemmas 3-nerved 6
4. Callus of lemmas bearded . . . . 36. *Schizachne*
4. Callus of lemmas not bearded 5
5. Glumes much shorter than the entire spikelet . . . . 1. *Bromus*
5. Glumes equalling or longer than the uppermost floret . . 86. *Danthonia*
6. Panicles 3–5 (–8) cm long; spikelets 2- to 5-flowered . . . . . . 62. *Triplasis*
6. Panicles 10–20 cm long; spikelets 6- to 12-flowered . . . . 70. *Leptochloa*
7. Glumes 17–30 mm long; awn of lemmas up to 40 mm long . . 13. *Avena*
7. Glumes 2.5–5.0 mm long; awn of lemmas 2.5–6.0 mm long 8
8. Awn arising from near the middle of the lemma; lemmas 3-nerved . . . . .
. . . . 11. *Aira*
8. Awn arising from near base of lemma; lemmas 5-nerved . . . .
. . . . 12. *Deschampsia*
9. Lemmas 3-nerved . . . . 70. *Leptochloa*
9. Lemmas 5-nerved (all the nerves sometimes obscure in *Festuca*) . . . 10
10. Blades involute, about 1 mm in diameter, if flat, less than 2 mm broad
. . . . 11
10. Blades flat, 2–8 mm broad 12
11. Plants annual; stamen 1 . . . . 2. *Vulpia*
11. Plants perennial; stamens 3 . . . . 3. *Festuca*
12. Lemmas glabrous; spikelets not crowded in 1-sided panicles 3. *Festuca*
12. Lemmas ciliate along the keel; spikelets crowded in 1-sided panicles . . .
. . . . 8. *Dactylis*

Group C

Inflorescence solitary, racemose, or paniculate, but not spicate or digitate;
spikelets with 2 or more perfect flowers; spikelets awnless.

1. Lemmas distinctly 2-toothed at the apex 2
1. Lemmas acute to obtuse at the apex, not 2-toothed 3
2. Perennial; blades to 3 mm broad; panicle branches erect or spreading;
spikelets 5- to 12-flowered; glumes 1 cm long . . . . 1. *Bromus*
2. Annual; blades 5–15 mm broad; panicle branches lax; spikelets 2-
flowered; glumes 1.5–2.5 cm long . . . . 13. *Avena*
3. Glumes at least 15 mm long, as long as the spikelets . . . . 13. *Avena*
3. Glumes less than 10 mm long, shorter than the spikelets 4
4. Rachilla with long silky hairs, the hairs longer than the spikelets; culms
to 4 m tall . . . . 84. *Phragmites*
4. Rachilla without long silky hairs longer than the spikelets; culms to 1.5
m tall 5
5. Lemmas 3-nerved 6
5. Lemmas 5- to many-nerved, or apparently nerveless, with only the mid-
nerve conspicuous 13
6. Lemmas 6–10 mm long; grain beaked . . . . 40. *Diarrhena*
6. Lemmas 1.5–5.0 mm long; grain not beaked 7
7. Lemmas glabrous . . . . 60. *Eragrostis*
7. Lemmas pubescent 8
8. Lemmas densely hairy at base, frequently with a tuft of hairs 9
8. Lemmas pubescent only on the nerves 11

9. Lemmas villous at the base, but without a tuft of cobwebby hairs ... 1
9. Lemmas with a tuft of cobwebby hairs at the base, pubescent on th
     nerves .... 6. Po
10. Lemmas retuse or obtuse, 3.5–4.0 mm long .... 61. Trider
10. Lemmas acute and mucronate, 4.5 mm long .... 63. Redfield
11. Lemmas keeled .... 60. Eragrost
11. Lemmas rounded on the back 1
12. Lemmas 1.0–2.5 mm long; spikelets 1–5 mm long ...... 70. Leptochlo
12. Lemmas 4.0 mm long; spikelets 5–8 mm long .... 61. Trider
13. Lemmas apparently nerveless 1
13. Lemmas obviously nerved 1
14. Spikelets disarticulating above the glumes 1
14. Spikelets disarticulating below the glumes .... 10. Sphenophol
15. Glumes 2.0–4.5 mm long; plants of moist or dry woods .... 3. Festuc
15. Glumes up to 2 mm long; plants of waste ground ...... 5. Puccinell
16. Lemmas 4–10 mm long 1
16. Lemmas 1.5–5.0 mm long 2
17. Lemmas as broad as long; inflorescence with up to eight spikelets ...
     .... 7. Briz
17. Lemmas longer than broad; inflorescence with more than eight spikele
     1
18. Lemmas 4–10 mm long, with nine or more nerves 1
18. Lemmas to 7 (–8) mm long, 5- to 7-nerved 2
19. Spikelets compressed, 6- to 18-flowered .... 87. Chasmanthiu
19. Spikelets not compressed, 2- to 3-flowered .... 34. Melic
20. Lemmas obscurely 7-nerved; spikelets 10–20 mm long .... 35. Glycer
20. Lemmas 5-nerved; spikelets less than 10 mm long 2
21. Spikelets not crowded in 1-sided panicles, not compressed   3. Festuc
21. Spikelets crowded in 1-sided panicles, compressed .... 8. Dactyl
22. Lemmas distinctly keeled .... 6. Po
22. Lemmas rounded on the back 2
23. Nerves of lemma parallel to the summit 2
23. Nerves of lemma converging toward the summit 2
24. Sheaths closed; lodicules united .... 35. Glycer
24. Sheaths open; lodicules free from each other .... 5. Puccinell
25. Lemmas glabrous 2
25. Lemmas pubescent, at least on the nerves or the keel or at the base ..
     .... 6. Po
26. Plants annual; stamen 1 .... 2. Vulp
26. Plants perennial; stamens 3 .... 3. Festuc

Group D

Inflorescence solitary, racemose, or paniculate, but not composed of singl
spikes; each spikelet with one perfect floret (sterile or staminate lemma
may be present, in addition); some part of the spikelet awned.

1. Spikelets borne singly (i.e., not paired) ....
1. Spikelets borne in pairs 2
2. Lemmas 3-awned ....
2. Lemmas 1-awned or awnless ....
3. Spikelets borne on one side of a long, arching raceme; lemma rounde
     on the back; spikelets with one perfect lemma and 1–2 sterile ones ...
     .... 75. Bouteloua

3. Spikelets borne in a more or less erect inflorescence, not 1-sided; lemma inrolled around the palea; no sterile lemma present   79. *Aristida*

4. First glume reduced to a sheath, united with the lowest, swollen joint of the rachilla .... 44. *Eriochloa*

4. First glume not reduced to a sheath and not united with the rachillar joint 5

5. Lemma awnless; glumes awned 6

5. Lemma awned; glumes awnless or awned 7

6. Plants over 1 m tall; lemmas 7–10 mm long .... 77. *Spartina*

6. Plants less than 1 m tall; lemmas 2–4 mm long ...... 65. *Muhlenbergia*

7. Spikelets arranged in 4 or more crowded ranks, each spikelet composed of one fertile and one sterile floret .... 47. *Echinochloa*

7. Spikelets not arranged in 4 or more crowded ranks, each spikelet composed of one fertile floret (each spikelet composed of one fertile floret and one staminate floret in *Arrhenatherum* or one sterile floret sometimes in *Gymnopogon*) 8

8. Blades 1–3 mm broad 9

8. Blades 3 mm broad or broader 14

9. First glume less than 1 mm long .... 65. *Muhlenbergia*

9. First glume at least 1.5 mm long 10

0. At least one awn of lemma 2–4 cm long 11

0. Awn of lemma up to 2 cm long 12

1. Tufted annuals from a cluster of fibrous roots .... 79. *Aristida*

1. Cespitose or stout perennials .... 37. *Stipa*

2. Lemma 2.0–4.5 mm long 13

2. Lemma 0.5–1.6 mm long .... 18. *Agrostis*

3. Glumes 5-nerved; lemmas indurated .... 38. *Oryzopsis*

3. Glumes 1-nerved; lemmas not indurated .... 65. *Muhlenbergia*

4. Plants 2 m tall or taller; all leaves 2 cm broad or broader .. 85. *Arundo*

4. Plants usually less than 2 m tall; none or only a few of the leaves as much as 2 cm broad 15

5. Second glume 5- to 7-nerved 16

5. Second glume 1- to 3-nerved 17

6. Awns straight or curved, not twisted near base; second glume 7-nerved .... 38. *Oryzopsis*

6. Awns twisted near base; second glume 5-nerved .... 37. *Stipa*

7. Lemma (excluding awns) 5–10 mm long 18

7. Lemma (excluding awns) 1.5–5.0 mm long 20

8. First glume less than 1 mm long .... 39. *Brachyelytrum*

8. First glume, 2.5–8.0 mm long 19

9. Awn 10–20 mm long; spikelet (excluding awns) 7–10 mm long; lemma 5- to 7-nerved .... 14. *Arrhenatherum*

9. Awn to 1.5 mm long; spikelet (excluding awns) 2.5–6.5 mm long; lemma 3-nerved .... 19. *Cinna*

0. Spikelets remote along one side of a slender rachis, forming very slender unilateral spikes .... 71. *Gymnopogon*

0. Spikelets in contracted or open panicles 21

1. Lemma with a tuft of hairs at the base (on the callus), awned from near the middle .... 16. *Calamagrostis*

1. Lemma glabrous or pubescent, but without a large tuft of hairs on the callus, awned from the tip 22

2. Plants usually at least 1 m tall; spikelets disarticulating below the glumes .... 19. *Cinna*

2. Plants up to 1 m tall, usually smaller; spikelets disarticulating above the glumes .... 65. *Muhlenbergia*

23. Both spikelets pedicellate, the pedicels unequal in length ....
.... 50. *Miscanthus*
23. One spikelet sessile, the other pedicellate (or represented merely by the pedicel) 24
24. Pedicellate spikelet represented only by the pedicel . 53. *Sorghastrum*
24. Pedicellate spikelet present 25
25. Both spikelets of the pair with perfect florets .... 51. *Erianthus*
25. Only the sessile spikelet of the pair perfect 26
26. Inflorescence racemose or nearly spicate 27
26. Inflorescence paniculate .... 52. *Sorghum*
27. Flowering culms much branched into many short leafy branchlets terminated by 1–6 racemes 28
27. Flowering culms unbranched .... 56. *Bothriochloa*
28. Racemes 2 or more from the sheaths .... 54. *Andropogon*
28. Raceme solitary at the tip of the peduncle .... 57. *Schizachyrium*

Group E

Inflorescence solitary, racemose, or paniculate, but not spicate; each spikelet with one perfect floret (sterile or staminate lemmas may be present, in addition); no part of the spikelet awned.

1. Spikelets borne in pairs 2
1. Spikelets solitary (i.e., not borne in pairs) 4
2. One spikelet of the pair sessile, the other pedicellate 55. *Microstegium*
2. Both spikelets either sessile or pedicellate 3
3. First glume as long as or longer than the lemmas; plants 2.5–4.0 m tall
.... 50. *Miscanthus*
3. First glume absent or up to 0.5 mm long, much shorter than the lemmas; plants up to 1.5 m tall 4
4. Spikelets with long, tawny hairs longer than the spikelets ....
.... 42. *Trichachne*
4. Spikelets without long, tawny hairs exceeding the spikelets ....
.... 45. *Paspalum*
5. First glume reduced to a sheath and united with the lowest, swollen joint of the rachilla .... 44. *Eriochloa*
5. First glume absent, reduced, or normal, neither sheath-like nor united with a swollen rachillar joint 6
6. Both glumes absent .... 81. *Leersia*
6. Both glumes present, although the first often much reduced or, if absent, the plants not producing seeds 7
7. First glume absent; plants with creeping rhizomes, rarely producing seeds .... *Zoysia*
7. First glume present, although occasionally strongly reduced; rhizomes present or absent; plants producing seeds 8
8. First glume up to one-half (to ⅔ in a few species of *Panicum*) as long as second glume 9
8. First glume nearly as long as the second glume, not conspicuously different in size 12
9. Each floret subtended by one or more bristles .... 48. *Setaria*
9. Florets not subtended by bristles 10

* *Zoysia* is frequently planted in Illinois as a choice lawn grass, but no collections have ever been made of it as an adventive. Therefore, it is excluded from the text.

Group F

Inflorescence digitate (the spikes and racemes radiating from near the same point).

Group G

Spikelets unisexual (i.e., either all staminate or all pistillate).

1. Plants to 40 cm tall, dioecious; staminate spikelets 3- to 75-flowered   2
1. Plants 1–4 m tall, monoecious; staminate spikelets 1- to 2-flowered .. 5
2. Lemmas with a tuft of cobwebby hairs at base .... 6. *Poa*
2. Lemmas without a tuft of cobwebby hairs at base 3
3. Both staminate and pistillate spikelets 10- to 75-flowered 60. *Eragrostis*
3. Staminate spikelets (2-) 3- to 15-flowered, pistillate spikelets 1- to 9-flowered 4
4. Staminate spikelets (2-) 3-flowered; pistillate spikelets 1-flowered .....
.... 76. *Buchloë*
4. Staminate spikelets 8- to 15-flowered; pistillate spikelets 7- to 9-flowered
.... 78. *Distichlis*
5. Staminate spikelets 2-flowered; glumes membranous or coriaceous . 6
5. Staminate spikelets 1-flowered; glumes none 7
6. Annual; staminate and pistillate spikelets in different inflorescences; pistillate spikelets borne in pairs .... 59. *Zea*
6. Perennial; staminate and pistillate spikelets in the same inflorescence pistillate spikelets solitary .... 58. *Tripsacum*
7. Pistillate spikelets confined to the uppermost erect branches of the inflorescence, the staminate spikelets confined to the lower spreading branches; margin of leaf more or less smooth .... 82. *Zizania*
7. Pistillate and staminate spikelets on the same branches of the inflorescence; margin of leaf harsh and cutting .... 83. *Zizaniopsis*

1. Bromus L. Brome Grass

1. Some or all of the awns over 12 mm long; teeth of lemmas 2–5 mm long
.... 2
1. All or most of the awns less than 12 mm long; teeth of lemmas usually less than 2 mm long 3
2. Lemmas 16–21 mm long, scabrous or puberulent on the back; first glume 8–12 mm long; second glume 13–18 mm long; awns 20–30 mm long; blades and sheaths glabrous or short-pubescent .... 1. *B. sterilis*
2. Lemmas 10–12 mm long, villous throughout on the back, becoming hispidulous at the summit, rarely entirely glabrous; first glume 4–7 mm long; second glume 8–10 mm long; awns (10–) 12–15 mm long; blades and sheaths soft-pubescent .... 2. *B. tectorum*
3. First glume 3- to 5-nerved (1-nerved in *B. nottowayanus*); second glume 5- to 7-nerved; annuals or perennials 4
3. First glume 1-nerved; second glume 3- to 5-nerved; perennials. (*Bromus nottowayanus* has the first glume 1-nerved, the second glume 5- to 7-nerved.) 15
4. Perennials from rhizomes; blades 6–13 mm broad (occasionally 5 mm broad in *B. kalmii*) 5
4. Annuals from fibrous roots; blades 2–6 mm broad (to 8 mm broad only in *B. secalinus*) 7
5. Lemmas keeled; inflorescence erect .... 3. *B. marginatus*
5. Lemmas rounded on the back; inflorescence drooping 6
6. Awns 5–8 mm long; cauline leaves 6–8 .... 4. *B. nottowayanus*
6. Awns 1–3 mm long; cauline leaves 3–5 (–6) .... 5. *B. kalmii*
7. Lemmas strongly keeled on the back, 12–15 mm long .. 6 *B. willdenovii*

7. Lemmas rounded on the back, 5–11 (–12) mm long 8
8. Awns 0–6 mm long 9
8. Most or all the awns over 6 mm long 10
9. Blades softly villous; lemmas 9–11 (–12) mm long; awns 0–1 mm long ..
.... 8. *B. brizaeformis*
9. Blades harshly pubescent above, or glabrous; lemmas 5–8 mm long;
   awns 1–6 mm long, rarely absent .... 7. *B. secalinus*
0. Inflorescence erect or ascending 11
0. Inflorescence spreading or drooping 13
1. Lemmas plicate, conspicuously nerved; inflorescence compact ....
.... 9. *B. mollis*
1. Lemmas not plicate, faintly nerved; inflorescence open 12
2. Lower lemmas 7–9 mm long; branches of inflorescence solitary or
   paired, usually shorter than the spikelets; anthers 2.0–2.5 mm long .....
.... 10. *B. racemosus*
2. Lower lemmas 9–11 mm long; branches of inflorescence 2–6, usually
   much longer than the spikelets; anthers 1.5–2.0 mm long ....
.... 11. *B. commutatus*
3. Awn straight or nearly so; rachilla not exposed at maturity 14
3. Awn flexuous; rachilla exposed at maturity .... 13. *B. japonicus*
4. Lemmas all nearly the same length; anthers 4 mm long   12. *B. arvensis*
4. Lowest lemmas longer than the upper; anthers 2.0–2.5 mm long ....
.... 10. *B. racemosus*
5. Awns absent or up to 2 mm long; blades and sheaths glabrous ....
.... 14. *B. inermis*
5. Awns 2–8 mm long; blades and sheaths (particularly the lower) pubescent, rarely glabrous 16
6. Inflorescence narrow, erect; blades 2–3 mm broad ...... 15. *B. erectus*
6. Inflorescence spreading or drooping; blades (3–) 4–17 mm broad ... 17
7. Leaves 10–20 per culm, the blades auriculate at base .. 16. *B. purgans*
7. Leaves 5–8 per culm, the blades not auriculate 18
8. Lemmas pubescent throughout on the back, or glabrous ....
.... 17. *B. pubescens*
8. Lemmas pubescent only on the margins in the lower one-half to three-
   fourths of the lemma .... 18. *B. ciliatus*

1. B. sterilis L. Brome Grass. May–June. Native of Europe; adventive in waste ground; Champaign, Cook, and McDonough cos.

2. B. tectorum L. Downy Chess. April–July. Native of Europe; naturalized in waste areas, common; in every co.

3. marginatus Nees. Brome Grass. May–Aug. Native of the U.S.;
adventive in waste ground; Cook and Kane cos.

4. B. nottowayanus Fern. Brome Grass. June–Aug. Moist, wooded
ravines; Cook, Peoria, Stark, and Woodford cos.

5. B. kalmii Gray. Brome Grass. June–July. Dry woodlands, sandy
soil, not common; confined to the n. half of the state.

6. B. willdenovii Kunth. Rescue Grass. May–July. Native to tropical
America; introduced in Champaign Co. *B. catharticus* Vahl—F, G.

7. B. secalinus L. Chess. May–Aug. Native of tropical America;
waste ground, common; probably in every co.

8. B. brizaeformis Fisch. & Mey. Rattlesnake Chess. May–July. Native of Europe; waste ground, rare; Richland and Washington cos.

9. B. mollis L. Soft Chess. May–July. Native of Europe; waste
ground, not common; DuPage, Jackson, and Lawrence cos.

10. B. racemosus L. Chess. May–Aug. Native of Europe; waste ground; occasional throughout the state.

11. B. commutatus Schrad. Hairy Chess. May–Aug. Native of Europe. waste ground; common throughout the state. *B. racemosus* L.—G.

12. B. arvensis L. Chess. May–July. Native of Europe; waste ground. not common; scattered in the w. cent. cos.

13. B. japonicus Thunb. Japanese Chess. May–July. Native of Europe and Asia; waste ground, not common; scattered in the state.

14. B. inermis Leyss. Awnless Brome Grass. May–July. Native of Europe; roadsides, fields, waste ground, common; probably in every co. (Including f. *aristatus* [Schur] Fern.)

15. B. erectus Huds. Erect Brome Grass. July. Native of Europe waste ground, rare; St. Clair Co.

16. B. purgans L. Brome Grass. June–Sept. Moist, open woods; occasional in the n. half of the state, rare in the s. half. *B. latiglumis* (Shear) Hitchcock—F, G, J. (Including *B. latiglumis* f. *incanus* [Shear] Fern.)

17. B. pubescens Muhl. Canada Brome Grass. June–Aug. Moist open woods; common throughout the state. *B. purgans* L.—F, G, J. (Including *B. purgans* f. *laevivaginatus* Wieg.)

18. B. ciliatus L. Canada Brome Grass. June–Sept. Open woodlands common throughout the state. (Including var. *intonsus* Fern.)

2. *Vulpia* K. C. Gmel.

1. Awns of lemma, if present, up to 5.5 mm long .... 1. *V. octoflora*
1. Some or all the awns of the lemma 1 cm long or longer 2
2. Second glume at least twice as long as the first glume; first glume less than 3 mm long .... 2 *V myuros*
2. Second glume not twice as long as the first glume; first glume more than 3 mm long .... 3. *V. dertonensis*

1. V. octoflora (Walt.) Rydb. Six-weeks Fescue. May–July. Inflorescence appearing racemose; lower glume 3.5–4.5 mm long; awns 3.5–5.5 mm long. Dry soil; scattered throughout the state. *F. octoflora* Walt.—G, J.

var. tenella (Willd.) Fern. Six-weeks Fescue. May–July. Inflorescence loosely spictate; lower glume 2.5–4.0 mm long; awns 1–3 mm long. Dry soil; occasional throughout the state. *Festuca octoflora* Walt.—J.

var. glauca (Nutt.) Fern. Six-weeks Fescue. May–July. Inflorescence densely spicate; lower glume 1.5–3.0 mm long; awns absent or up to 2 mm long. Dry soil, not common; scattered in Ill.

2. V. myuros (L.) K. Gmel. Foxtail Fescue. June–July. Native of Europe; waste ground, rare; Johnson and Massac cos. *F. myuros* L.—G.

3. V. dertonensis All. June–July. Native of Europe; waste ground; Massac Co. *F. dertonensis* (All.) Aschers. & Graebn.—G.

3. *Festuca* L. Fescue

1. Leaf blades involute or plicate, 0.4–1.2 mm broad 2
1. Leaf blades flat, 3–11 mm broad 4

2. First glume 1–2 mm long; second glume 1.8–3.0 mm long; lemmas 2.5–3.5 mm long; awns absent, or to 0.5 mm long ...... 1. *F. capillata*

2. First glume 2.5–4.5 mm long; second glume 3.5–5.5 mm long; lemmas 4–7 mm long; awns 1.0–3.5 mm long 3

3. Lowest sheaths whitish, not becoming fibrous; lemmas essentially nerveless .... 2. *F. ovina*

3. Lowest sheaths brown or reddish, becoming fibrous; lemmas 3- to 5-nerved .... 3. *F. rubra*

4. Lemmas 5.5–10.0 mm long 5

4. Lemmas 3.3–5.2 mm long 6

5. Inflorescence 6- to 11-flowered; lemmas 5.5–8.0 mm long 4. *F. pratensis*

5. Inflorescence 4- to 5-flowered; lemmas 7–10 mm long 5. *F. arundinacea*

6. Inflorescence spreading at maturity; spikelets to 4 mm broad; lemmas acute or subacute .... 6. *F. obtusa*

6. Inflorescence ascending at maturity; spikelets about 5 mm broad; lemmas obtuse .... 7. *F. paradoxa*

1. F. capillata Lam. Slender Fescue. May–July. Native of Europe; waste ground, not common; scattered in metropolitan areas. *F. ovina* L. var. *capillata* (Lam.) Alef.—G.

2. F. ovina L. var. duriuscula (L.) Koch. Sheep Fescue. May–June. Native of Europe; waste ground, not common; scattered throughout the state. *F. ovina* L.—J.

3. F. rubra L. Red Fescue. June–July. Waste ground; infrequent throughout the state.

4. F. pratensis Huds. Meadow Fescue. May–Aug. Native of Europe; waste ground, common; in every co. *F. elatior* L.—F, G, J. (Including *F. elatior* f. *aristata* Holmb.)

5. F. arundinacea Schreb. Large Fescue. May–Aug. Native of Europe; waste ground, rare; scattered in Ill. *F. elatior* L. var. *arundinacea* (Schreb.) Wimmer—G, J.

6. F. obtusa Bieler. Nodding Fescue. May–July. Moist woodlands, occasional; scattered throughout the state.

7. F. paradoxa Desv. Fescue. May–July. Dry or moist woodlands; occasional in the s. three-fourths of the state, absent elsewhere.

4. Lolium L. Rye Grass

1. Glume as long as or longer than the spikelet, 15–20 mm long ....
.... 1. *L. temulentum*

1. Glume shorter than the spikelet, 4–12 mm long 2

2. Lemmas (or at least the uppermost) awned; spikelets 10- to 20-flowered; annual .... 2. *L. multiflorum*

2. Lemmas awnless; spikelets 6- to 10-flowered; perennial .. 3. *L. perenne*

1. L. temulentum L. Darnel. June–July. Native of Europe; waste ground, fields, not common; scattered in Ill.

2. L. multiflorum Lam. Italian Rye Grass. May–Sept. Native of Europe; waste ground, fields; occasional throughout the state.

3. L. perenne L. English Rye Grass. June–Sept. Native of Europe; waste ground, fields, lawns, common; probably in every co.

5. *Puccinellia* Parl.

1. Lemmas obscurely nerved .... 1. *P. distans*
1. Lemmas conspicuously nerved .... 2. *P. pallida*

    1. P. distans (L.) Parl. Alkali Grass. June–Oct. Native of Europe; waste ground; Cook Co.
    2. P. pallida (Torr.) Clausen. May–Aug. Swamps, very rare; Union Co. *Glyceria pallida* (Torr.) Trin.—F, G, J.

6. *Poa* L. *Bluegrass*

1. Plants dioecious; pistillate spikelets woolly; staminate spikelets glabrous or nearly so .... 4. *P. arachnifera*
1. Plants monoecious; spikelets perfect, variously pubescent or glabrous 2
2. Some spikelets transformed into bulblets .... 16. *P. bulbosa*
2. Spikelets normal 3
3. Lemmas without a tuft of cobwebby hairs at the base 4
3. Lemmas with a tuft of cobwebby hairs at the base 5
4. Tufted annual to about 30 cm tall, sometimes rooting at the lower nodes; lemmas elliptic to ovate .... 1. *P. annua*
4. Tufted perennial to 75 cm tall, not rooting at the lower nodes; lemmas oblong .... 3. *P. autumnalis*
5. Nerves and keel of the lemma glabrous (except for the cobwebby tuft) .... 8. *P. languida*
5. Keel and sometimes the nerves of the lemma pubescent 6
6. Keel of the lemma pubescent, the nerves glabrous 7
6. Keel and at least some of the nerves of the lemma pubescent 8
7. Culms beneath the panicle and the sheaths scabrous; lemmas sharply nerved; ligule of upper leaves 4–8 mm long .... 9. *P. trivialis*
7. Culms beneath the panicle and the sheaths usually glabrous; lemmas obscurely nerved; ligule of upper leaves about 1 mm long .... 10. *P. alsodes*
8. Marginal nerves of the lemma pubescent, the intermediate nerves glabrous 9
8. All nerves of the lemma pubescent 13
9. Plants with rhizomes; lemmas with 5 prominent nerves 10
9. Plants without rhizomes; lemmas with 3 prominent nerves and 2 obscure nerves 11
10. Basal leaves flat, at least as broad as the culm; culm compressed at base, 2–3 mm thick at base; glumes broadly lanceolate, straight .... 5. *P. pratensis*
10. Basal leaves involute or filiform, narrower than the culm; culm terete at base, 1–2 mm thick at base; glumes narrowly lanceolate, arching .... 6. *P. angustifolia*
11. Culms very weak, solitary or in small tufts; sheaths scabrous; lowest branches of the panicle mostly paired .... 11. *P. paludigena*
11. Culms more firm, usually densely tufted; sheaths usually glabrous; lowest branches of the panicle in clusters of 3–5 12
12. Ligule 0.5–1.0 mm long; anthers 1.2–1.6 mm long ..... 12. *P. nemoralis*
12. Ligule 2–5 mm long; anthers up to 1 mm long .... 13. *P. palustris*
13. First glume 2.5–3.5 mm long; lemma 3.5–4.5 mm long; blades 1–2 mm broad .... 14. *P. wolfii*

3. First glume 1.5–2.5 (–2.7) mm long; lemma 1.5–3.5 mm long; blades 2–5 mm broad 14

4. Tufted perennial; inflorescence reflexed or spreading, 10–20 cm long; lemmas with 5 distinct nerves .... 15. *P. sylvestris*

4. Tufted annual or rhizomatous perennial; inflorescence ascending; lemmas with 3 distinct nerves and 2 obscure nerves 15

5. Tufted annual to 30 cm tall; culms terete; anthers 0.1–0.2 mm long .... 2. *P. chapmaniana*

5. Rhizomatous perennial to 70 cm tall; culms compressed; anthers about 1 mm long .... 7. *P. compressa*

1. P. annua L. Annual Bluegrass. March–July. Native of Europe and Asia; mostly in moist waste ground; probably in every co.

2. P. chapmaniana Scribn. Annual Bluegrass. April–Sept. Fields and waste ground; occasional in the s. half of the state, less common in the n. half.

3. P. autumnalis Muhl. Bluegrass. March–June. Moist woodlands, very rare; Pope Co.

4. P. arachnifera Torr. Texas Bluegrass. Native to the sw. U.S.; roadsides and pastures; Winnebago Co.

5. P. pratensis L. Kentucky Bluegrass. April–July. Native of Europe and Asia; woods, fields, waste ground, common; in every co.

6. P. angustifolia L. Bluegrass. June–Sept. Woods and clearings, very rare; Union Co.

7. P. compressa L. Canadian Bluegrass. May–Aug. Native of Europe and Asia; dry soil, common; in every co.

8. P. languida Hitchc. Woodland Bluegrass. June–Sept. Moist woodlands, rare; Cook and JoDaviess cos.

9. P. trivialis L. Meadow Grass. May–Aug. Native of Europe; waste ground, rare; Cook and Stark cos.

10. P. alsodes Gray. Woodland Bluegrass. May–June. Moist woodlands, rare; Jackson and St. Clair cos.

11. P. paludigena Fern. & Wieg. Marsh Bluegrass. June–July. Bogs, rare; Kane Co.

12. P. nemoralis ·L. Woodland Bluegrass. June–Aug. Native of Europe; open woodlands, waste ground, rare; Champaign and Peoria cos.

13. P. palustris L. Fowl Bluegrass. June–Sept. Wet soil; occasional in the n. half of the state, rare in the s. half.

14. P. wolfii Scribn. Meadow Bluegrass. April–June. Meadows and woodlands, rare; Fulton, Henderson, and Peoria cos.

15. P. sylvestris Gray. Woodland Bluegrass. May–July. Moist woodlands; occasional throughout the state.

16. P. bulbosa L. Bulbous Blue Grass. Native of Europe; adventive in Ill.; DuPage Co.

7. *Briza* L. *Quaking Grass*

1. B. maxima L. Big Quaking Grass. June–July. Native of Europe; waste ground, rare; St. Clair Co.

8. *Dactylis* L. *Orchard Grass*

1. D. glomerata L. Orchard Grass. May–July. Native of Europe; waste ground, common; in every co.

9. *Koeleria* Pers. June Grass

1. K. macrantha (Ledeb.) Spreng. June Grass. June–Sept. Prairie sandy black-oak woods; occasional throughout the state. *K. cristai* Pers.—F, G, J.

10. *Sphenopholis* Scribn. Wedge Grass

1. First glume subulate, less than 0.5 mm broad; lemmas smooth to sc. brous; anthers less than 1 mm long .... 1. *S. obtusai*
1. First glume narrowly oblong, at least 0.5 mm broad; second lemma sc. brous; anthers 1.0–1.5 mm long .... 2. *S. nitic*

1. S. obtusata (Michx.) Scribn. Wedge Grass. May–July. Panic dense and spike-like; second glume firm, rounded or truncate at th apex. Woods and prairies, occasional; throughout the state.
var. major (Torr.) Erdman. Wedge Grass. May–July. Panicle loose second glume scarious, acute to apiculate at the apex. Moist wood moist prairies, occasional; throughout the state. *S. intermedia* (Rydb Rydb.—F, G, J.
2. S. nitida (Biehler) Scribn. Shining Wedge Grass. May–July. D woods, prairies, not common; scattered in Ill.

11. *Aira* L. Hairgrass

1. A. caryophyllaea L. Slender Hairgrass. May–June. Native c Europe; waste ground, rare; Piatt Co.

12. *Deschampsia* Beauv. Hairgrass

1. D. cespitosa (L.) Beauv. var. glauca (Hartm.) Lindm. Tufted Hai grass. June–July. Along creeks and in swamps, rare; confined to ex treme ne. Ill. *D. cespitosa* (L.) Beauv.—J.

13. *Avena* L. Oats

1. Spikelet 3-flowered; lemmas pubescent, with a bent awn .... 1. *A. fatu*
1. Spikelet 2-flowered; lemmas glabrous, awnless or with a straight awn
.... 2. *A. sativ*

1. A. fatua L. Wild Oats. May–Sept. Native of Europe; occasional escaped into waste ground.
2. A. sativa L. Oats. May–Aug. Native of Europe and Asia; oc casionally escaped into waste ground.

14. *Arrhenatherum* Beauv. Oat Grass

1. A. elatius (L.) Presl. Tall Oat Grass. July–Sept. Native of Europe occasionally escaped into waste ground; widely scattered in Ill.

15. *Holcus* L. Velvet Grass

1. H. lanatus L. Velvet Grass. June–July. Native of Europe; adver tive in waste ground; not common but scattered in Ill.

16. Calamagrostis Adans. Reed Grass

1. Callus hairs of lemma shorter than or equalling the lemma; spikelets up
to 4.5 mm long; awn of lemma straight 2
1. Callus hairs of lemma exceeding the lemma; spikelets 5–8 mm long; awn
of lemma slightly bent .... 3. *C. epigeios*
2. Blades flat (at least when fresh), 4–8 mm broad; panicle open, more or
less nodding; glumes spreading in fruit; lemmas translucent at tip .....
....1. *C. canadensis*
2. Blades involute, 2–4 mm broad when unrolled (rarely broader); panicle
contracted, spike-like, erect; glumes connivent at tip in fruit; lemmas
firm throughout ....2. *C. inexpansa*

1. C. canadensis (Michx.) Beauv. Bluejoint Grass. June–July. Panicle loosely flowered; spikelets 2.8–3.8 mm long; glumes distinctly exceeding the lemma, acute to acuminate; moist soil; occasional in the
n. half of the state, rare in the s. half, absent in the extreme s.

var. macouniana (Vasey) Stebbins. Bluejoint Grass. Aug. Panicle
densely flowered; spikelets 2.2–2.8 mm long; glumes nearly equalled
by the lemma, obtuse to acute; wet ditch, very rare; Henry Co.

2. C. inexpansa Gray var. brevior (Vasey) Stebbins. Northern Reed
Grass. June–July. Wet ground, rare; Cook, JoDaviess, Lake, and Winnebago cos. *C. inexpansa* Gray—J.

3. C. epigeios (L.) Roth. Feathertop. June–July. Native of Europe;
adventive in strip mine, rare; Randolph Co.

17. Ammophila Host. Beach Grass

1. A. breviligulata Fern. Beach Grass. July–Sept. Sand dunes; Cook
and Lake cos.

18. Agrostis L. Bent Grass

1. Annuals; lemma sharply nerved, with a flexuous awn to 10 mm long ...
....1. *A. elliottiana*
1. Perennials; lemma obscurely nerved, awnless (rarely a very short awn
present in a variety of *A. scabra*) 2
2. Tufted perennials without rhizomes; palea absent, or minute and nerveless 3
2. Rhizomatous or stoloniferous perennials; palea at least one-half as long
as the lemma, 2-nerved 5
3. Flat blades 1–2 mm broad; spikelets 1.2–2.0 mm long; lemma 0.5–1.0 mm
long ....2. *A. hyemalis*
3. Flat blades 2–6 mm broad; spikelets 2–3 mm long; lemma 1.3–2.0 mm
long 4
4. Panicle branches harshly scabrous, bearing florets only near the tip ...
....3. *A. scabra*
4. Panicle branches glabrous or nearly so, bearing florets from near the
middle to the tip .... 4. *A. perennans*
5. Some of the panicle branches bearing florets to base; ligule 2–6 mm
long ....5. *A. alba*
5. None of the panicle branches bearing florets to the base; ligule less than
2 mm long ....6. *A. tenuis*

1. A. elliottiana Schult. Awned Bent Grass. May–July. Dry soil, par
ticularly on bluffs; restricted to the s. one-half of Ill.

2. A. hyemalis (Walt.) BSP. Tickle Grass. March–June. Woods and
fields, common; probably in every co.

3. A. scabra Willd. Tickle Grass. June–Sept. Lemmas awnless. Moist
or dry, usually open, soil; occasional in the n. cos., rare elsewhere.

f. tuckermanii Fern. June–Sept. Lemmas awned. Moist ground
rare; Cook Co.

4. A. perennans (Walt.) Tuckerm. Upland Bent Grass. June–Sept.
Dry woodlands, common in the s. half of the state, becoming increas
ingly less common northward. (Including var. aestivalis Vasey.)

5. A. alba L. Red Top. June–Sept. Blades 5–10 mm broad; rhizomes
present; culms erect; panicle purple, the branches spreading. Fields
common; in every co. A. stolonifera L. var. major (Gaud.) Farw.—G.

var. palustris (Huds.) Pers. Creeping Bent Grass. June–Sept. Blades
1–5 mm broad; stolons present; culms decumbent; panicle straw
colored, the branches ascending. Wet ground; occasional throughout
the state. A. stolonifera L. var. major (Gaud.) Farw.—G; A. palustris
Huds.—J.

6. A. tenuis Sibth. Rhode Island Bent. June–Sept. Native of Europe
escaped from lawns into waste ground; scattered in some n. cos.

19. Cinna L. Wood Reed

1. Panicle gray-green, the branches mostly ascending; spikelets 4.0–6.5
mm long; second glume 3-nerved; awn less than 0.5 mm long ....
.... 1. C. arundinacea
1. Panicle green, the branches mostly spreading; spikelets 2.5–4.0 mm
long; second glume 1-nerved; awn 0.5–1.5 mm long ..... 2. C. latifolia

1. C. arundinacea L. Stout Wood Reed. July–Sept. Moist wood
lands, damp soil; occasional throughout the state.

2. C. latifolia (Trev.) Griseb. Drooping Wood Reed. July–Sept. Moist
woods, rare; DeKalb and Winnebago cos.

20. Anthoxanthum L. Vernal Grass

1. Spikelets brownish-green, 8–10 mm long; glumes pubescent; awns of
empty lemmas included or barely exserted; perennials to nearly 1 m tall
.... 1. A. odoratum
1. Spikelets whitish-green, 5–7 mm long; glumes glabrous; awns of empty
lemmas long-exserted; annuals to 35 cm tall .... 2. A. aristatum

1. A. odoratum L. Sweet Vernal Grass. June. Native of Europe
fields, rare; Cook, DuPage, and Lake cos.

2. A. aristatum Boiss. Annual Sweet Grass. Aug. Native of Europe
waste ground; Rock Island Co. A. puellii—F.

21. Hierochloë R. Br. Sweet Grass

1. H. odorata (L.) Beauv. Sweet Grass. May–June. Meadows; oc
casional in the n. one-fifth of the state.

22. Phalaris L. Canary Grass

1. Keel of glumes wingless, the glumes, 4.5–6.5 mm long; sterile florets 1–2 mm long; perennial from creeping rhizomes .... 1. *P. arundinacea*
1. Keel of glumes broadly winged, the glumes 7–10 mm long; sterile florets 2.5–4.5 mm long; annual .... 2. *P. canariensis*

    1. P. arundinacea L. Reed Canary Grass. May–July. Meadows and similar moist situations; occasional throughout the state. (Including f. *picta* [L.] Asch. & Graebn.)
    2. P. canariensis L. Canary Grass. June–July. Native of Europe; waste ground; occasional throughout the state.

23. Alopecurus L. Foxtail

1. Spike-like panicle 6–10 mm thick; spikelets 4.0–5.5 mm long; glumes acute, 4.0–5.5 mm long; lemma 4.0–5.5 mm long, the awn exserted 4–5 mm .... 1. *A. pratensis*
1. Spike-like panicle to 5.5 mm thick; spikelets 2.0–2.7 mm long; glumes obtuse, 2.0–2.7 mm long; lemma 2.0–2.7 mm long, the awn exserted only up to 4 mm 2
2. Perennial from slender rhizomes; awn exserted up to 1 mm .... 2. *A. aequalis*
2. Tufted annual; awn exserted from 1.5–4.0 mm .... 3. *A. carolinianus*

    1. A. pratensis L. Meadow Foxtail. June. Native of Eurasia; along railroads; known from a few cos. in the n. half of the state.
    2. A. aequalis Sobol. Foxtail. May–July. Wet ground, occasionally in shallow water, not common; scattered throughout the state.
    3. A. carolinianus Walt. Common Foxtail. April–July. Moist ground; occasional throughout the state.

24. Phleum L. Timothy

    1. P. pratense L. Timothy. June–Aug. Native of Europe; waste ground, fields, common; in every co.

25. Milium L. Millet Grass

    1. M. effusum L. Millet Grass. June–Aug. Moist woodlands, rare; Kane and Tazewell cos.

26. Beckmannia Host Slough Grass

    1. B. syzigachne (Steud.) Fern. American Slough Grass, July–Sept. Wet ground, very rare; Cook and Lake cos.

27. Elymus L. Wild Rye

1. Rhizomes present; glumes 3–4 mm broad; lemmas awnless .... 1. *E. arenarius*
1. Rhizomes absent; glumes up to 2.5 mm broad; lemmas awned (awnless in one variety of *E. virginicus*) 2

2. Glumes reduced to unequal filiform bristles, or absent; spikelets widely spreading .... 2. *E. hystrix*
2. Glumes subequal in length; spikelets ascending 3
3. Base of glumes indurated; paleas up to 8.5 mm long 4
3. Base of glumes thin or indurated for only 1 mm; paleas 8.5–12.0 (–14.0) mm long .... 6. *E. canadensis*
4. Glumes (0.8–) 1.0–2.0 mm wide, swollen on at least half or all of the adaxial surface .... 3. *E. virginicus*
4. Glumes 0.2–1.0 mm wide, indurated for 1–3 mm adaxially 5
5. Paleas 7.0–8.5 mm long, the apices bidentate .... 4. *E. riparius*
5. Paleas 5.5–6.5 mm long, the apices obtuse .... 5. *E. villosus*

1. E. arenarius L. Wild Rye. July–Oct. Sandy shores of Lake Michigan, rare; Cook and Lake cos. *E. arenarius* L. var. *villosus* Mey.—F; *E. mollis* Trin.—G.

2. E. hystrix L. Bottlebrush Grass. June–Aug. Lemmas glabrous Woodlands, common; in every co. *Hystrix patula* Moench—F, G, J.

var. bigeloviana (Fern.) Mohlenbrock. Bottlebrush Grass. June–Aug. Lemmas pubescent. Woodlands; occasional throughout the state. *Hystrix patula* Moench var. *bigeloviana* (Fern.) Deam—F *H. patula* Moench f. *bigeloviana* (Fern.) Gl.—G.

3. E. virginicus L. Virginia Wild Rye. June–Sept. Lemmas to 30 mm long, glumes to 25 mm long, both glabrous or scabrous. Fields and woodlands, low ground; common throughout the state. (Including var. *jejunus* [Ramaley] Bush.) Forma *hirsutiglumis* (Scribn.) Fern. has villous glumes and lemmas.

var. submuticus Hook. June–Sept. Glumes and lemmas acuminate or subulate-tipped. Fields and woodlands, not common.

var. glabriflorus (Vasey) Bush. June–Sept. Lemmas 35 mm long or longer, glumes 27 mm long or longer, both glabrous or scabrous Fields and woodlands, occasional throughout the state. Forma *aus tralis* (Scribn. & Ball) Fern. has hirsute glumes and lemmas.

4. E. riparius Wiegand. Wild Rye. July–Sept. Woodlands; not common but scattered.

5. E. villosus Muhl. Slender Wild Rye. June–Sept. Glumes and lemmas hispid to hirsute. Woodlands, common; in every co.

f. arkansanus (Scribn. & Ball) Fern. June–Sept. Glumes scabrous lemmas glabrous or scabrous. Woodlands, rare; DuPage, Henry Jackson, and Stark cos.

6. E. canadensis L. Nodding Wild Rye. June–Sept. Woods, road sides, dry prairies, common; in every co. (Including var. *glaucifolius* [Muhl.] Torr.; var. *brachystachys* [Scribn. & Ball] Farw.; var. *cres cendus* [Ramaley] Bush.)

28. Sitanion Raf.

1. S. hystrix (Nutt.) J. G. Sm. Squirrel-tail. July–Sept. Adventive in Il along a railroad; Mason Co.

29. Hordeum L.   Barley

1. Lateral spikelets of each group of three pedicellate and sterile ......
1. Lateral spikelets of each group of three sessile and perfect ....
2. Fertile spikelet subtended by awns less than 2 cm long; spikes less tha 2 cm broad ....

2. Fertile spikelet subtended by awns exceeding 2 cm long; spikes 2–8 cm
broad .... 3. *H. jubatum*
3. Four glumes of each group of 3 spikelets setiform, the other two glumes
dilated at the base; awn of lemma of central spikelet 8–15 mm long ....
.... 1. *H. pusillum*
3. All glumes setiform; awn of lemma of central spikelet 5–6 mm long ....
.... 2. *H. brachyantherum*
4. Annual; spikes erect; glumes flat, about 1 mm broad; awns of lemmas
60–150 mm long, or absent and replaced by a three-lobed appendage ..
.... 4. *H. vulgare*
4. Perennial; spikes nodding; glumes setiform; awns of lemmas 15–35 mm
long .... 5. *H. X montanense*

1. H. pusillum Nutt. Little Barley. May–July. Fields, waste ground;
common in the s. three-fourths of the state, rare in the n. one-fourth.
2. H. brachyantherum Nevski. Meadow Barley. May–July. Native to
N. Am.; adventive in a lawn; Jackson Co.
3. H. jubatum L. Squirrel-tail Grass. June–Sept. Fields; common in
the n. one-half of the state, occasional elsewhere.
4. H. vulgare L. Common Barley. June–Sept. Lemmas awned. Native
of Asia; occasional along country roads.
var. trifurcatum (Schlecht.) Alefeld. Pearl Barley. June–Sept.
Lemmas three-lobed at the apex, awnless. Native of Asia; occasional
along county roads.
5. H. X montanense Scribn. Barley. June–July. Prairies and roadsides, not common; Marshall, Peoria, and Stark cos. Reputed to be a
hybrid between *H. jubatum* and *Elymus virginicus*.

30. X Agrohordeum G. Camus

1. A. X macounii (Vasey) Lepage. Macoun's Wild Rye. June–July.
Native to w. N. Am.; adventive in Ill.; Cook Co. *Elymus macounii*
Vasey—F, G. Reputed to be a hybrid between *Agropyron trachycau-
lum* (Link) Malte and *Hordeum jubatum* L.

31. Agropyron Gaertn. Wheat Grass

1. Lemmas 5–7 mm long; glumes 2–5 mm long; spikelets pectinately
arranged; tufted plants 2
1. Lemmas 8–25 mm long; glumes 8–18 mm long; spikelets not pectinately
arranged; tufted or rhizomatous plants 3
2. Blades flat; spikelets 8–12 mm long; glumes abruptly tapering to the 2–3
mm long awn .... 1. *A. desertorum*
2. Blades involute (at least when dry); spikelets 5–7 mm long; glumes gradually tapering to the 2–5 mm long awn .... 2. *A. cristatum*
3. Tufted plants; spikelets at maturity (and on herbarium specimens) readily breaking up into individual florets when touched 4
3. Rhizomatous plants; spikelets at maturity falling in their entirety 5
4. Awn of lemmas 10–30 mm long; spikes dense; glumes 12–18 mm long
.... 3. *A. subsecundum*
4. Awn of lemmas absent or up to 2 mm long; spikes more slender; glumes
8–12 mm long .... 4. *A. trachycaulum*
5. Blades flat, 5–10 mm broad; sheaths pubescent (in Ill.) .... 5. *A. repens*
5. Blades involute when dry, 2–5 mm broad; sheaths glabrous (in Ill.) .....
.... 6. *A. smithii*

1. A. desertorum (Fisch.) Schult. Wheat Grass. June–Aug. Native of Russia; introduced in Ill.; Cook and JoDaviess cos.

2. A. cristatum (L.) Gaertn. Crested Wheat Grass. June–Aug. Native of Russia; adventive in Ill.; Fulton and JoDaviess cos.

3. A. subsecundum (Link) Hitchcock. Bearded Wheat Grass. July–Aug. Woodlands, fields, rare; confined to the extreme n. cos. *A. tra-chycaulum* (Link) Malte var. *unilaterale* (Cassidy) Malte—F.

4. A. trachycaulum (Link) Malte. Slender Wheat Grass. June–Aug. Fields and waste ground and along railroads, rare; known from a few n. cos.

5. A. repens (L.) Beauv. Quack Grass. June–July. Lemmas awnless. Native of Europe and Asia; adventive in fields and waste ground; common in the n. three-fourths of the state, rare in the s. one-fourth.

f. aristatum (Schum.) Holmb. Quack Grass. June–July. Lemmas with an awn up to 10 mm long. Native of Europe and Asia; adventive in waste ground; Cook Co.

6. A. smithii Rydb. Western Wheat Grass. June–Aug. Lemmas gla-brous, scabrous, or pubescent near base. Native of the w. U.S.; ad-ventive along railroads; occasional in the n. two-thirds of the state, rare in the s. one-third.

var. molle (Scribn. & Smith) Jones. Western Wheat Grass. June–Aug. Lemmas short-pilose throughout. Native of the w. U.S.; adven-tive along railroads; Cook, DuPage, and Will cos. *A. dasystachyum* (Hook.) Scribn.—F; *A. molle* (Scribn. & Smith) Rydb.—J.

32. Triticum L. Wheat

1. Spikelets compressed, disarticulating above the glumes; joints of rachis not swollen; blades 10–20 mm broad .... 1. *T. aestivum*
1. Spikelets cylindrical, falling in their entirety; joints of rachis swollen; blades 2–3 mm broad .... 2. *T. cylindricum*

1. T. aestivum L. Wheat. June–Aug. Sporadic in fields and along roadsides; several cultivated varieties may be found, including Bearded Wheat, with long awns on the lemmas.

2. T. cylindricum (Host) Ces. Jointed Goat Grass. June–Sept. Native of Europe; adventive in waste ground; scattered throughout the state. *Aegilops cylindrica* Host—F, G, J.

33. Secale L. Rye

1. S. cereale L. Rye. May–July. Occasionally adventive in waste ground; frequently planted by the highway department along new road right-of-ways.

34. Melica L. Melic Grass

1. Cauline leaves 3–4, 2–5 mm broad; sheaths scabrous; glumes nearly equal in length; first glume oblong, at least twice as long as broad; fer-tile lemmas usually 2 .... 1. *M. mutica*
1. Cauline leaves 5–8, 5–12 mm broad; sheaths glabrous; glumes unequal in length; first glume ovate, less than twice as long as broad; fertile lemmas usually 3 .... 2. *M. nitens*

1. M. mutica Walt. Two-flowered Melic Grass. May–June. Rocky woodlands, occasional; scattered throughout the state, but rare in the ne. cos.

2. M. nitens (Scribn.) Nutt. Three-flowered Melic Grass. May–July. Rocky woods, prairies; occasional throughout the state, but rare in the ne. cos.

35. Glyceria R. Br. Manna Grass

1. Spikelets at least 10 mm long; sheaths compressed 2
1. Spikelets 2–8 mm long; sheaths terete or subterete 4
2. Principal leaves 2–5 mm broad; lemmas shining, scabrous only on the nerves; pedicels very slender, all one-fourth to two-thirds the length of the spikelets .... 1. G. borealis
2. Principal leaves 6–18 mm broad; lemmas dull, scabrous between the nerves; pedicels thickened upward, less than one-fourth the length of the spikelets (except for the terminal ones) 3
3. Principal blades 6–12 mm broad; lemmas obscurely nerved, scabrous, 3.5–5.5 mm long; anthers over 1 mm long .... 2. G. septentrionalis
3. Principal blades 10–18 mm broad; lemmas sharply nerved, hirtellous, 2.5–3.0 mm long; anthers less than 1 mm long .... 3. G. arkansana
4. Lemmas 3–4 mm long, obscurely nerved; spikelets 3–4 mm broad ......
.... 4. G. canadensis
4. Lemmas 1.5–2.7 mm long, sharply nerved; spikelets 2.0–2.5 mm broad 5
5. Inflorescence 5–20 cm long; spikelets 2.0–4.5 mm long; first glume 0.5–1.0 mm long; second glume 0.8–1.3 mm long; lemmas 1.5–2.0 mm long .... 5. G. striata
5. Inflorescence 20–40 cm long; spikelets 5–6 mm long; first glume 1.2–2.0 mm long; second glume 1.5–2.5 mm long; lemmas 2.0–2.7 mm long ....
.... 6. G. grandis

1. G. borealis (Nash) Batchelder. Northern Manna Grass. June–July. Shallow water, rare; JoDaviess, Lake, and Stephenson cos.

2. G. septentrionalis Hitchc. Manna Grass. May–Aug. Shallow water, marshy soil, swampy meadows, occasional; scattered through-out the state.

3. G. arkansana Fern. Manna Grass. May–June. Shallow water of swamps, very rare; Union Co. G. septentrionalis Hitchc.—G, J.

4. G. canadensis (Michx.) Trin. Rattlesnake Manna Grass. June–Sept. Wet ground, rare; Cook and Peoria cos.

5. G. striata (Lam.) Hitchcock. Fowl Manna Grass. May–Aug. Spikelets green; uppermost branches of the panicle more or less nodding; lemmas with a minutely scarious apex. Moist soil, common; in every co.

var. stricta (Scribn.) Fern. Fowl Manna Grass. May–Aug. Spikelets purple; uppermost branches of the panicle ascending; lemmas with a broadly scarious apex. Wet ground; occasional in the n. one-third of the state, absent elsewhere.

6. G. grandis S. Wats. American Manna Grass. June–Aug. Wet ground, not common; restricted to the extreme n. cos.

36. Schizachne Hack. False Melic Grass

1. S. purpurascens (Torr.) Swallen. False Melic Grass. June. Moist wooded slope, very rare; JoDaviess Co.

114   POACEA

37. *Stipa* L. *Needle Grass*

1. Sheaths villous on the margins and at the summit; ligule less than 1 mm
   long; glumes 5–11 mm long; lemma 4.5–6.0 mm long, pubescen
   throughout, the awn 2–4 cm long .... 1. *S. viridu*
1. Sheaths more or less glabrous; ligule (at least of the upper leaves) 3–
   mm long; glumes 15–40 mm long; lemma 9–25 mm long, pubescent
   base, becoming glabrate above, the awn 10–20 cm long ....
2. Glumes 15–28 mm long; lemma 9–13 mm long, the flexuous but ob
   scurely geniculate awn 10–15 cm long; ligule of upper leaves 3–4 mm
   long .... 2. *S. coma*
2. Glumes 28–42 mm long; lemma 16–25 mm long, the twice genicula
   awn 12–20 cm long; ligule of upper leaves 4–6 mm long . 3. *S. sparte*

  1. S. virdula Trin. Feather Grass. June–July. Edge of woods an
along railroad track near pond, rare; Kane and McHenry cos.
  2. S. comata Trin. & Rupr. Needle Grass. July–Aug. Dry or loam
soil, usually in prairies, rare; Cook, Kane, and Winnebago cos.
  3. S. spartea Trin. Porcupine Grass. May–June. Sandy soil, partic
larly in prairies; occasional in the n. two-thirds of the state, nearly a
sent in the s. third.

38. *Oryzopsis* Michx. *Rice Grass*

1. Blades flat, 5–15 mm broad; spikelets (excluding the awn) 6–9 mm lon
   glumes acute to acuminate, 7–9 mm long, conspicuously 7-nerve
   lemma 5.5–8.5 mm long, with the awn 5–25 mm long ....
1. Blades involute, 1–2 mm broad; spikelets (excluding the awn) 3–4 m
   long; glumes obtuse, 3.5–4.0 mm long, obscurely 5-nerved; lemm
   3.5–4.0 mm long, with the awn 1–2 mm long .... 3. *O. punger*
2. Upper leaves longer than lower leaves; lemma dark brown to blackis
   the awn 12–25 mm long .... 1. *O. racemos*
2. Upper leaves shorter than lower leaves; lemma pale green to yellowis
   the awn 5–10 mm long .... 2. *O. asperifol*

  1. O. racemosa (J. E. Smith) Ricker. Rice Grass. July–Sept. Ric
rocky woodlands, rare; Grundy, LaSalle, Mason, Peoria, Vermilio
and Winnebago cos.
  2. O. asperifolia Michx. Rice Grass. April–July. Dry woodlands, ve
rare; Cook Co.
  3. O. pungens (Torr.) Hitchcock. Rice Grass. April–June. Dry so
very rare; Menard Co.

39. *Brachyelytrum* Beauv.

  1. B. erectum (Schreb.) Beauv. May–Aug. Moist or occasionally d
woodlands; occasional throughout the state.

40. *Diarrhena* Beauv.

  1. D. americana Beauv. var. obovata Gleason. June–Sept. Lo
shaded woods, moist ledges, base of limestone cliffs; throughout th
state, except for the ne. cos. *D. americana* Beauv.—F, J.

41. Digitaria Heist. Finger Grass

1. Culms rooting at the lower nodes, decumbent at the base; rachis broadly winged, about 1 mm broad 2
1. Culms erect, not rooting at the lower nodes; rachis narrowly winged, less than 1 mm broad 3
2. Sheaths (at least the lower) papillose-pilose; blades pilose to scabrous; spikelets 2.5–3.5 mm long; second glume about half as long as spikelet, usually 1.2–1.6 mm long; fertile lemma greenish-brown 1. *D. sanguinalis*
2. Sheaths glabrous; blades glabrous; spikelets 1.7–2.2 mm long; second glume about as long as spikelet, 1.7–2.2 mm long; fertile lemma dark brown to blackish .... 2. *D. ischaemum*
3. Racemes less than 10 cm long; spikelets 1.5–1.7 (–2.0) mm long; second glume and sterile lemma more or less glabrous to short-pubescent, 1.5–1.7 (–2.0) mm long .... 3. *D. filiformis*
3. Racemes over 10 cm long; spikelets 2.0–2.5 mm long; second glume and sterile lemma long-pubescent, 2.0–2.5 mm long .... 4. *D. villosa*

1. D. sanguinalis (L.) Scop. Crab Grass. June–Oct. Spikelets 2.5–3.0 mm long; sterile lemma appressed-pubescent. Native of Europe and Asia; adventive in waste ground and lawns; very common in every co.
   var. ciliaris (Retz.) Parl. Crab Grass. June–Oct. Native of Europe and Asia; adventive in waste ground; Perry Co.
2. D. ischaemum (Schreb.) Muhl. Smooth Crab Grass. July–Oct. Native of Europe; adventive in waste ground; throughout the state.
3. D. filiformis (L.) Koel. Slender Crab Grass. Aug.–Sept. Sandy soil; occasional in cent. Ill., absent in the extreme n., rare in the extreme s.
4. D. villosa (Walt.) Pers. Hairy Finger Grass. July–Sept. Sandy soil, rare; Jackson Co. *D. filiformis* (L.) Koel. var. *villosa* (Walt.) Fern.—F, G; *D. filiformis* (L.) Koel.—J.

42. Trichachne Nees Sour Grass

1. T. insularis (L.) Nees. Sour Grass. July–Sept. Native of the se. U.S., the W.I., and S.Am.; adventive in a roadside ditch; Williamson Co.

43. Leptoloma Chase Fall Witch Grass

1. L. cognatum (Schult.) Chase. Fall Witch Grass. July–Sept. Sandy soil; occasional throughout the state.

44. Eriochloa HBK. Cup Grass

1. Pedicels and rachis villous; spikelets about 5 mm long ... 1. *E. villosa*
1. Pedicels and rachis short-pilose; spikelets 3.5–5.0 mm long 2
2. Grain 2.0–2.5 mm long, with an awn to 1 mm long; blades pubescent, 3–7 mm broad .... 2. *E. contracta*
2. Grain 3 mm long, apiculate; blades glabrous, 5–10 mm broad .... 3. *E. gracilis*

1. E. villosa (Thunb.) Kunth. Cup Grass. Aug. Native of Asia; adventive in waste ground; Cook, DuPage, and Livingston cos.

2. E. contracta Hitchcock. Prairie Cup Grass. June–Sept. Moist so
native of the w. U.S.; Jackson and Union cos.

3. E. gracilis (Fourn.) Hitchcock. Cup Grass. July–Oct. Field borde
native of the w. U.S.; Union Co.

45. Paspalum L. Bead Grass

1. Rachis foliaceous, the margins folded over and clasping the spikelets
their bases ....
1. Rachis firm, narrow or broad, but the margins not folded over the rov
of spikelets ....
2. Racemes of each inflorescence 1–5; rachis shorter than the rows
spikelets .... 1. *P. dissectu*
2. Racemes of each inflorescence 5–50, usually more than 10; rach
longer than the rows of spikelets .... 2. *P. fluitar*
3. Rachis broad, over 1.5 mm wide; spikelets arranged in 4 rows ....
.... 3. *P. pubiflorum* var. *glabru*
3. Rachis narrower, less than 1.5 mm wide (about 1.5 mm wide in *P. le*
tiferum); spikelets in 2 rows (4 in some racemes of *P. floridanum*) ...
4. Spikelets 3.6 mm long or longer; culms robust, 1–2 m tall ....
.... 4. *P. floridanu*
4. Spikelets less than 3.2 mm long; culms slender, usually less than 1 m ta
(occasionally to 1.5 m in *P. lentiferum*) ....
5. Spikelets 2.5–3.2 (–3.4) mm long; sterile lemma 5-nerved, with later
nerves approximate at the margins ....
5. Spikelets 1.8–2.4 mm long; sterile lemma 3-nerved, the marginal nerve
obscure at maturity ....
6. Spikelets solitary; leaves glabrous to sparsely pilose .... 5. *P. lae*
6. Spikelets paired, or paired and solitary on the same raceme; leaves p
lose, becoming villous at base .... 6. *P. lentiferu*
7. Spikelets glabrous; nodes of culms glabrous; leaves glabrous
variously pubescent, but not velvety on both surfaces 7. *P. ciliatifoliu*
7. Spikelets pubescent, often densely so; nodes of culms pubescen
leaves velvety on both surfaces .... 8. *P. busl*

1. P. dissectum (L.) L. Bead Grass. June–Sept. Moist soil, edges
shallow swamps, very rare; Perry, Pulaski, and St. Clair cos.

2. P. fluitans (Ell.) Kunth. Swamp Bead Grass. June–Sept. Floatir
in shallow standing water; occasional in the s. two-thirds of the state

3. P. pubiflorum Rupr. var. glabrum (Vasey) Vasey. Bead Gras
June–Sept. In moist soil in ditches, along roadsides, and alor
streams; occasional in the s. one-half of the state, absent else
where. *P. geminum* Nash—J.

4. P. floridanum Michx. Giant Bead Grass. June–Sept. Low, moi
sandy soil; not common in the s. one-third of the state, absent else
where. *P. glabratum* (Engelm.) Mohr.—J.

5. P. laeve Michx. Bead Grass. June–Sept. Moist soil of roadsic
ditches, meadows, and stream borders; common in the s. one-half
the state, rare elsewhere. (Including var. *circulare* [Nash] Stone
P. circulare Nash—J.

6. P. lentiferum Lam. Bead Grass. June–Sept. Wet, roadsic
ditches, low post-oak flats, rare; Massac and Pulaski cos.

7. P. ciliatifolium Michx. Bead Grass. June–Sept. Dry or moi
sandy soils, open woods; common in the s. two-thirds of the state

rare elsewhere. (Including var. *muhlenbergii* [Nash] Fern. var. *stra-mineum* [Nash] Fern., *P. pubescens* Muhl., and *P. stramineum* Nash.)

8. P. bushii Nash. Hairy Bead Grass. June–Sept. Fields, along edge of woods; restricted to a few sw. cos. *P. ciliatifolium* Michx.—G.

46. Panicum L. Panic Grass

Group I

ikelets 1.0–1.9 mm long, glabrous.

Group II

kelets 1.0–1.9 mm long, pubescent.

2. Culms pilose with horizontally spreading hairs ...... 25. *P. praecociu*
2. Culms variously pubescent, but the hairs not horizontally spreading .
3. Upper surface of blades glabrous except for long hairs at the base; a tumnal form matted .... 24. *P. lanuginosu*
3. Upper surface of blades pilose or appressed-pubescent; autumnal for erect or spreading (becoming matted in a variety of *P. meridionale*) .
4. Spikelets 1.3–1.5 mm long; grain 1.2–1.3 mm long ....
4. Spikelets 1.6–1.9 mm long; grain 1.5–1.6 mm long ....
5. Panicle branches (at least the lower) drooping, the axes long-pilose ..
.... 24. *P. lanuginosu*
5. Panicle branches ascending, the axes glabrous or puberulent ....
.... 23. *P. meridiona*
6. Upper surface of leaves pilose, the hairs 3–5 mm long; first glume acu or acuminate, about ½ as long as the spikelet ...... 26. *P. subvillosu*
6. Uppersurface of leaves short-pubescent, the hairs less than 3 mm lon first glume obtuse and truncate, ¼–⅓ as long as the spikelet ....
.... 24. *P. lanuginosu*
7. Ligule 3–5 mm long .... 24. *P. lanuginosu*
7. Ligule less than 2 mm long ....
8. Grain 1.7–1.8 mm long ....
8. Grain 1.3–1.5 mm long ....
9. Panicle branches ascending, viscid; sheaths viscid; autumnal culr leafy to the base .... 18. *P. nitidu*
9. Panicle branches spreading, not viscid; sheaths not viscid; autumr culms essentially leafless below the middle .... 20. *P. dichotome*
10. Sheaths retrorsely pilose .... 16. *P. laxiflorю*
10. Sheaths glabrous or ciliate or ascending-pilose ....
11. Sheaths ascending-pilose, with long, soft hairs intermingled with sho crisp ones; blades up to 7 mm broad .... 29. *P. columbianu*
11. Sheaths glabrous or ciliate, without two types of hairs; blades 7–25 m broad ....
12. Spikelets ellipsoid, about 0.7 mm broad; nodes with reflexed hairs glabrous ....
12. Spikelets obovoid-spherical, 1.0–1.3 mm broad; nodes appresse puberulent or glabrous ....
13. Spikelets 1.5–1.6 mm long; nodes with reflexed hairs 17. *P. microcarp*
13. Spikelets 1.8 mm long or longer; nodes glabrous ...... 8. *P. rigidulu*
14. Panicle nearly as broad as long; culms spreading; nodes appresse puberulent .... 30. *P. sphaerocarp*
14. Panicle ¼–½ as broad as long; culms erect; nodes more or less glabro
.... 31. *P. polyanth*

Group III

Spikelets 2.0–2.9 mm long, glabrous.

1. Spikelets 1.2–1.7 mm broad ....
1. Spikelets 0.7–1.0 mm broad ....
2. Blades 3–6 mm broad; spikelets 2.1–2.4 mm long, 1.2–1.3 mm bro
.... 15. *P. linearifoli*
2. Blades 6–12 mm broad; spikelets 2.9–3.0 mm long, 1.7 mm bro
.... 34. *P. oligosanth*
3. At least some of the sheaths papillose-hispid ....
3. None of the sheaths papillose-hispid ....

Spikelets 2.5–4.0 mm long, acuminate .... 5. *P. capillare*
Spikelets up to 2.5 mm long, acute 5
Panicle at least half the entire length of the plant .... 5. *P. capillare*
Panicle up to ⅓ the entire length of the plant 6
Fruits stramineous; blades 6–10 mm broad; spikelets 0.9–1.0 mm broad; culms stout, to 100 cm tall .... 3. *P. gattingeri*
Fruits nigrescent; blades 2–8 mm broad; spikelets 0.7 mm broad; culms slender, to 50 cm tall .... 4. *P. philadelphicum*
At least the lower nodes bearded; sheaths softly pubescent .... 20. *P. dichotomum*
Nodes glabrous or sparsely pilose; sheaths glabrous, ciliate, pilose, or appressed-pubescent, but not softly pubescent 8
First glume ⅕–¼ as long as the spikelet .... 1. *P. dichotomiflorum*
First glume ⅓–½ as long as the spikelet 9
Ligule ciliate, 2–3 mm long .... 10. *P. longifolium*
Ligule 1 mm long or less 10
Spikelets 2.5–2.8 mm long; grain short-stipitate .... 9. *P. stipitatum*
Spikelets up to 2.5 mm long (rarely 2.5 mm long in *P. yadkinense*); grain not stipitate 11
Blades pilose on the upper surface near base 12
Blades glabrous, except sometimes for marginal cilia 13
Blades 1–5 mm broad; palea indurate at maturity, enlarging and dilating the spikelet .... 12. *P. hians*
Blades 5–10 mm broad; palea not indurate or enlarged at maturity .... 8. *P. rigidulum*
Spikelets 2.2–2.5 mm long 14
Spikelets 2.0–2.2 mm long 15
Sheaths with pale glandular spots; spikelets acute; first glume about ⅓ as long as spikelet .... 22. *P. yadkinense*
Sheaths without pale grandular spots; spikelets acuminate; first glume about ½ as long as spikelet .... 8. *P. rigidulum*
First glume about ½ as long as the spikelet; second glume and sterile lemma longer than the grain; grain 1.3 mm long, 0.6 mm broad .... 8. *P. rigidulum*
First glume about ⅓ as long as the spikelet; second glume and sterile lemma shorter than the grain; grain 1.8 mm long, 0.9 mm broad .... 20. *P. dichotomum*

Group IV

Spikelets 2.0–2.9 mm long, pubescent.
Nodes bearded, but with a sticky ring immediately beneath them .... 37. *P. scoparium*
Nodes beardless or, if with a beard, then without a sticky ring immediately beneath the nodes 2
Spikelets up to 2.5 mm long 3
Spikelets 2.5–2.9 mm long 13
Ligule 2–5 mm long; sheaths papillose-pubescent 4
Ligule 1 mm long or less; sheaths not papillose-pubescent (except *P. linearifolium*) 6
Pubescence of culms horizontally spreading; autumnal form freely branched; ligule 4–5 mm long .... 27. *P. villosissimum*

4. Pubescence of culms appressed or ascending; autumnal form sparse branched; ligule 2–3 mm long ....

5. Pubescence of blades and sheaths silky; upper surface of blades pube cent along both margins .... 27. *P. villosissimu.*

5. Pubescence of blades and sheaths short and stiff; upper surface blades more or less glabrous .... 28. *P. scoparioide*

6. Spikelets 1.2–1.5 mm broad (occasionally narrower in *P. mattamuske tense*); grain 2.0–2.1 mm long, 1.1–1.2 mm broad ....

6. Spikelets 0.9–1.1 mm broad; grain 1.5–1.9 mm long, 0.9–1.0 mm broa ....

7. Sheaths papillose-pilose; spikelets 1.3–1.5 mm broad .... 15. *P. linearifoliu*

7. Sheaths puberulent, velvety, or glabrous; spikelets 1.2–1.3 mm broad

8. Lowermost nodes bearded; lower sheaths velvety-pubescent .... 21. *P. mattamuskeeten.*

8. Nodes without a beard; lower sheaths glabrous or puberulent, but rare velvety ....

9. Sheaths puberulent; culms crisp-puberulent; spikelets at least 2.4 m long .... 38. *P. commutatu*

9. Sheaths glabrous; culms sparsely pilose on the nodes; spikelets 2.1–2 mm long .... 15. *P. linearifoliu*

10. Sheaths retrorsely pilose; spikelets papillose-pilose; grain 1.5 mm lor .... 16. *P. laxifloru.*

10. Sheaths puberulent or merely ciliate; spikelets not papillose; gra 1.7–1.9 mm long ....

11. Upper sheaths viscid-spotted; nodes with reflexed hairs   18. *P. nitidu*

11. Upper sheaths not viscid; nodes glabrous or sparsely pilose .... ΐ

12. Second glume and sterile lemma as long as the grain; plants green .. .. 19. *P. borea*

12. Second glume and sterile lemma shorter than the grain; plants ofte purplish .... 20. *P. dichotomu*

13. Sheaths with papillose hairs ....

13. Sheaths without papillose hairs .... ΐ

14. Blades 12–30 mm broad .... 40. *P. clandestinu.*

14. Blades 2–12 mm broad .... ΐ

15. Blades glabrous or scabrous on the upper surface .... ΐ

15. Blades hirsute or velvety on the upper surface .... ΐ

16. Blades 2–5 mm broad .... ΐ

16. Blades 6–12 mm broad .... 34. *P. oligosanthe*

17. Spikelets 1.6–1.7 mm broad; grain 2.4 mm long, 1.5–1.6 mm broad ... . 14. *P. perlongu*

17. Spikelets 1.3–1.5 mm broad; grain 2.0–2.1 mm long, 1.2 mm broad ... . 15. *P. linearifoliu*

18. Blades long-hirsute on both surfaces; grain 2.4–2.5 mm long .... 32. *P. wilcoxianu*

18. Blades velvety on both surfaces; grain 2.2 mm long .... 33. *P. malacophyllu*

19. Lowermost nodes bearded; lower sheaths velvety-pubescent .... 21. *P. mattamuskeeten.*

19. None of the nodes bearded; sheaths glabrous or puberulent, but not ve vety .... ΐ

20. Spikelets 2.6–2.8 mm long, obtuse to subacute; blades firm, more or le: cordate at base .... 38. *P. commutatu*

20. Spikelets more than 2.8 mm long, abruptly short-pointed; blades thi narrowed or slightly rounded at base .... 39. *P. joc*

Group V

Spikelets 3.0 mm long or longer, glabrous.

1. Spikelets obtuse .... 34. *P. oligosanthes*
1. Spikelets acute to acuminate 2
2. First glume over ½ as long as the spikelet; grain 2.4–3.0 mm long 3
2. First glume up to ½ as long as the spikelet; grain 1.7–2.3 mm long .....
.... 4
3. Sheaths papillose-hispid; panicle more or less nodding 6. *P. miliaceum*
3. Sheaths ciliate or villous at the throat; panicle ascending or spreading
.... 7. *P. virgatum*
4. Grain 1.4–1.5 mm broad; spikelets 1.5–1.7 mm broad ....
.... 13. *P. depauperatum*
4. Grain 0.8–1.0 mm broad; spikelets 0.9–1.2 mm broad 5
5. Sheaths glabrous, except for the ciliate margins; first glume ⅕–¼ as
long as the spikelet .... 1. *P. dichotomiflorum*
5. Sheaths papillose-pubescent (occasionally glabrous in *P. anceps*); first
glume ⅓–½ as long as the spikelet 6
6. Ligule less than 1 mm long; second glume and sterile lemma beaked-
acuminate; culms glabrous .... 11. *P. anceps*
6. Ligule 1–3 mm long; second glume and sterile lemma acuminate, but
not conspicuously beaked; culms usually pubescent, at least on the
nodes 7
7. Axis of panicle glabrous; grain 2.0 mm long .... 2. *P. flexile*
7. Axis of panicle short-pilose; grain 1.7–1.8 mm long ..... 5. *P. capillare*

Group VI

Spikelets 3.0 mm long or longer, pubescent.

1. Grain 2.1–2.5 mm long 2
1. Grain 2.8–3.5 mm long 8
2. Some or all the blades at least 12 mm broad 3
2. Blades 2–12 mm broad 4
3. Culms and usually the sheaths papillose-hispid; spikelet subacute to
acute .... 40. *P. clandestinum*
3. Culms and sheaths glabrous; spikelet abruptly short-pointed 39. *P. joori*
4. Blades glabrous above, glabrous or sparsely pilose beneath 5
4. Blades long-hirsute or velvety above and beneath 7
5. Blades 6–12 mm broad .... 34. *P. oligosanthes*
5. Blades 2–5 mm broad 6
6. Spikelets acute, the second glume and sterile lemma beaked ....
.... 13. *P. depauperatum*
6. Spikelets obtuse, the second glume and sterile lemma not beaked .....
.... 14. *P. perlongum*
7. Blades long-hirsute on both surfaces; grain 2.4–2.5 mm long ....
.... 32. *P. wilcoxianum*
7. Blades velvety on both surfaces; grain 2.2 mm long ....
.... 33. *P. malacophyllum*
8. Blades papillose-pubescent on both surfaces; first glume a little more
than ½ as long as the spikelet .... 36. *P. leibergii*
8. Blades not papillose-pubescent; first glume up to ½ as long as the
spikelet 9
9. Pubescence of sheath not papillose 10

1. *P. dichotomiflorum* Michx. Fall Panicum. June–Oct. Spikelet
2.4–3.5 mm long; at least some of the leaves over 5 mm broad; culm
mostly upright, not geniculate, the nodes not swollen; sheaths not
inflated. Fields, waste ground; common throughout the state.

var. *geniculatum* (Muhl.) Fern. Fall Panicum. June–Oct. Spikelet
2.4–3.5 mm long; at least some of the blades over 5 mm broad; culm
mostly spreading, geniculate, some of the nodes swollen; sheath
inflated. Low waste areas, fields; scattered throughout the state.

var. *puritanorum* Svenson. Panic Grass. June–Sept. Spikelet
1.7–2.3 mm long; blades up to 5 (–8) mm broad. Along streams, very
rare; Cook Co.

2. P. *flexile* (Gattinger) Scribn. Slender Panic Grass. July–Oct.
Moist, sandy soil; occasional in the s. one-third of the state, rare elsewhere.

3. P. *gattingeri* Nash. Panic Grass. July–Oct. Dry or moist open
ground, waste places; scattered throughout the state, except for the
n. three tiers of cos. *P. capillare* L. var. *campestre* Gatt.—G.

4. P. *philadelphicum* Bernh. Panic Grass. July–Oct. Dry, usually
sandy soil; locally scattered in the s. three-fifths of the state; also
DeKalb Co.

5. P. *capillare* L. Witch Grass. July–Oct. Lowest branches of panicle
included at the base; spikelets mostly 2.0–2.5 mm long; grain about
1.5 mm long. Fields, waste ground; common throughout the state.
P. capillare L. var. *agreste* Gatt.—G.

var. *occidentale* Rydb. Witch Grass. July–Oct. Open ground, very
rare; Henderson Co.

6. P. *miliaceum* L. Broomcorn Millet. June–Oct. Native of Europe
and Asia; escaped from cultivation into waste ground; scattered in Il

7. P. *virgatum* L. Switch Grass. June–Oct. Fields, waste ground,
prairies, rocky stream beds, woods; rather common throughout the
state.

8. P. *rigidulum* Bosc. Munro Grass. June–Oct. Panicle branches
spreading to ascending; spikelets 1.8–2.2 mm long. Moist soil,
prairies, along ponds or creeks, low woodlands; occasional in the
one-third of the state, less common elsewhere. *P. agrostoide*
Spreng.—F, G, J.

var. *condensum* (Nash) Mohlenbrock. Panic Grass. June–Oct. Panicle branches erect; spikelets 2.2–2.5 mm long. Moist soil, rare; Johnson and Massac cos. *P. agrostoides* Spreng, var. *condensum* (Nash)
Fern—F; *P. agrostoides* Spreng.—G.

9. P. *stipitatum* Nash. Panic Grass. June–Oct. Moist soil, rare; Johnson Co. *P. agrostoides* Spreng. var. *elongatum* Scribn.—G.

10. P. *longifolium* Torr. Panic Grass. June–Oct. Rocky ledges,
wooded ravines, very rare; Monroe Co.

11. P. *anceps* Michx. Panic Grass. June–Oct. Moist soil, woodland

prairies, roadside ditches, stream banks; common in the s. one-third of the state, rare to absent elsewhere.

12. P. hians Ell. Panic Grass. June–Oct. Low roadside ditch, rare; Alexander Co.

13. P. depauperatum Muhl. Panic Grass. May–Sept. Dry, open woodlands, prairies, waste ground; occasional throughout the state.

14. P. perlongum Nash. Panic Grass. May–Sept. Dry soil, particularly in prairies or upland woods; occasional in the n. half of the state, rare in the s. half.

15. P. linearifolium Scribn. Panic Grass. May–Sept. Sheaths papillose-pilose. Dry woods; occasional in the s. one-fourth of the state, rare elsewhere.

var. werneri (Scribn.) Fern. Panic Grass. May–Sept. Sheaths glabrous. Wooded slopes, very rare; LaSalle Co.

16. P. laxiflorum Lam. Panic Grass. May–Sept. Woodlands; common in the s. tip of the state, rare or absent elsewhere. P. xalapense HBK.—J.

17. P. microcarpon Muhl. Panic Grass. May–Oct. Wet ground, often in woods; occasional in the s. one-third of the state, absent elsewhere except Peoria Co. P. nitidum Lam. var. ramulosum Torr.—G.

18. P. nitidum Lam. Panic Grass. June–Sept. Xeric limestone blufftop, very rare; Jackson Co.

19. P. boreale Nash. Northern Panic Grass. June–Oct. Moist sand, very rare; Lake Co.

20. P. dichotomum L. Panic Grass. May–Oct. Nodes glabrous or sparsely pubescent, not bearded; grain slightly exserted. Dry soil, usually in woodlands; rather common in the s. one-fourth of the state, occasional to rare elsewhere.

var. barbulatum (Michx.) Wood. Panic Grass. May–Oct. At least the lowermost nodes bearded; grain included. Dry, usually rocky woods; occasional in the s. one-third of the state. P. dichotomum L.—G, J.

21. P. mattamuskeetense Ashe. Panic Grass. May–Sept. Along a levee, rare; Massac Co.

22. P. yadkinense Ashe. Panic Grass. May–Sept. Moist ground, rare; restricted to a few s. cos.

23. P. meridionale Ashe. Panic Grass. May–Sept. Vernal blades 1.5–4.0 cm long; plants greenish-yellow; autumnal form nearly erect. Sandy woodlands, rare; known from the extreme n. cos. and two s. cos.

var. albemarlense (Ashe) Fern. Panic Grass. May–Sept. Vernal blades 4.5–7.0 cm long; plants grayish; autumnal form spreading to ascending, but eventually forming mats. Margin of wet, peaty meadow, very rare; Kankakee Co. P. meridionale Ashe—G.

24. P. lanuginosum Ell. Panic Grass. May–Sept. Axis of panicle branches with some long-spreading hairs; upper surface of blades glabrous except for long hairs at the base; spikelets 1.6–1.9 mm long. Usually low, moist, open situations; common throughout the state. P. tennesseense Ashe—J.

var. implicatum (Scribn.) Fern. Panic Grass. May–Sept. Axis of panicle branches with some long-spreading hairs; upper surface of blades pubescent; spikelets 1.3–1.4 mm long. Swampy soil or in moist depressions of sandstone cliffs; occasional throughout the state. P. implicatum Scribn.—J.

var. lindheimeri (Nash) Fern. Panic Grass. May–Sept. Axis of panicle branches either glabrous or with appressed hairs; spikelets 1.4–1.8 mm long. Sandy soil, usually in woodlands; occasional throughout the state. *P. lindheimeri* Nash—J.

var. septentrionale (Fern.) Fern. Panic Grass. May–Sept. Axis of panicle branches either glabrous or with appressed hairs; spikelet 1.6–1.9 mm long. Moist soil; occasional in the s. cos.

25. P. praecocius Hitchc. & Chase. Panic Grass. May–Sept. Dry soil, often in prairies; occasional throughout the state, becoming rare in the extreme s. cos.

26. P. subvillosum Ashe. Panic Grass. June–Oct. Sandy soil in an open field, very rare; Lake Co.

27. P. villosissimum Nash. Hairy Panic Grass. May–Sept. Culms and sheaths spreading papillose-pilose; ligule 4–5 mm long; upper surface of blades appressed long-pilose. Sandy soil, often in woodlands; occasional throughout the state.

var. pseudopubescens (Nash) Fern. Panic Grass. May–Sept. Culm and sheaths appressed papillose-pilose; ligule 2–3 mm long; upper surface of blades glabrous or short-pilose. Sandy soil; occasional throughout the state, becoming rare in the s. cos. *P. pseudopubescens* Nash—J.

28. P. scoparioides Ashe. Panic Grass. June–Oct. Dry field, very rare; Lake Co. *P. villosissimum* Nash var. *scoparioides* (Ashe) Fern.—F., G.

29. P. columbianum Scribn. Panic Grass. June–Oct. Sandy woodlands; rare in the n. one-fourth of the state.

30. P. sphaerocarpon Ell. Panic Grass. June–Oct. Sandy soil; occasional in the s. two-fifths of the state, absent elsewhere, except for Christian Co.

31. P. polyanthes Schult. Panic Grass. June–Oct. Low ground, chiefly in woodlands; occasional in the s. one-third of the state; also Peoria Co.

32. P. wilcoxianum Vasey. Panic Grass. June–Sept. Dry soil, mostly in prairies; confined to the extreme nw. cos. and Pope Co.

33. P. malacophyllum Nash. Panic Grass. June–Oct. Limestone bluffs, dry woods, rare; Jackson and Pope cos.

34. P. oligosanthes Schult. Panic Grass. June–Oct. Spikelets 3.2–4.0 mm long; culms appressed-pubescent; first glume sparsely hirsute; grains 2.8–3.0 mm long. Sandy soil, mostly in woodlands; occasional throughout the state.

var. scribnerianum (Nash) Fern. Panic Grass. June–Oct. Spikelets 3.2–3.3 mm long; culms glabrous or spreading-pubescent; first glume glabrous; grains 2.8–2.9 mm long. Dry, sandy soil, often in prairies; occasional throughout the state. *P. scribnerianum* Nash—G, J.

var. helleri (Nash) Fern. Panic Grass. May–Sept. Spikelets 2.9–3.0 mm long; culms appressed-pubescent below; first glume glabrous; grains 2.4–2.5 mm long. Limestone ledge, rare; Randolph Co. *P. helleri* Nash—G.

35. P. ravenelii Scribn. & Merr. Ravenel's Panic Grass. June–Oct. Cherty ravines, sandstone ledges, rare; Hardin and Union cos.

36. P. leibergii (Vasey) Scribn. Leiberg's Panic Grass. June–Oct. Dry soil, mostly in sandstone woodlands; occasional in the n. half of the state, rare in the s. half.

37. P. scoparium Lam. Panic Grass. June–Oct. Moist fields, roadsides, rare; Pope Co.

38. P. commutatum Schult. Panic Grass. June–Oct. Leaves to 25 mm

broad, heart-shaped at base; spikelets 2.6–2.8 mm long. Mostly dry woodlands; occasional in the s. one-third of the state, absent elsewhere.

var. ashei Fern. Panic Grass. June–Oct. Leaves to 10 mm broad, narrowed at base; spikelets 2.4–2.7 mm long. Dry woodlands, rare; Jackson Co.

39. P. joori Vasey. Panic Grass. June–Oct. Low swampy woods, rare; Johnson Co. *P. commutatum* Schult. var. *joorii* (Vasey) Fern.—F; *P. commutatum* Schult.—G.

40. P. clandestinum L. Broad-leaved Panic Grass. June–Oct. Moist, sandy soil; common in the s. half of the state, becoming less frequent northward.

41. P. latifolium L. Broad-leaved Panic Grass. June–Oct. Dry, rocky woodlands; occasional throughout the state.

42. P. boscii Poir. Large-fruited Panic Grass. June–Oct. Leaves, sheaths, and culms glabrous or sparsely pubescent. Woodlands; common in the s. half of the state, rare or absent elsewhere.

var. molle (Vasey) Hitchc. & Chase. Large-fruited Panic Grass. June–Oct. Leaves, sheaths, and culms softly pubescent. Woodlands; occasional in the s. one-fourth of the state.

47. *Echinochloa* Beauv. *Barnyard Grass*

1. Second glume with an awn 2–10 mm long; sheaths papillose-hirsute (glabrous in f. *laevigata*); grain about three times longer than broad .... 1. *E. walteri*
1. Second glume awnless or with an awn less than 2 mm long; sheaths glabrous or scabrous; grain at most about twice as long as broad, usually shorter 2
2. Racemes of panicle slender, distant; grain 2.0–2.5 mm long; blades 2.–6 mm broad .... 2. *E. colonum*
2. Racemes of panicle broader, more crowded; grain 2.5–3.5 mm long; blades 5–25 mm broad 3
3. Fertile lemma with a weak, easily broken tip, with a ring of minute setae just below the tip; rachis and axis of panicle setose 4
3. Fertile lemma without a ring of setae just below the tip; rachis and axil of panicle not setose .... 5. *E. pungens*
4. Panicle green or purple; sterile lemma short-awned; pubescence of second glume and sterile lemma hispidulous or setose ..... 3. *E. crus-galli*
4. Panicle dark brownish-purple; sterile lemma awnless; pubescence of second glume and sterile lemma appressed .... 4. *E. frumentacea*

1. E. walteri (Pursh) Heller. Aug.–Oct. Sheaths (at least the lower) papillose-hirsute. Low ground, swamps, rarely in standing water; not common, but scattered throughout the state.

f. laevigata Wieg. Aug.–Oct. Sheaths glabrous. Swamps, rare; Union Co.

2. E. colonum (L.) Link. Jungle Rice. Aug.–Oct. Native of Europe and Asia; adventive in waste ground; Cook Co.

3. E. crus-galli (L.) Beauv. Barnyard Grass. Aug.–Oct. Native of Europe and Asia; adventive in waste ground; occasional throughout the state.

4. E. frumentacea (Roxb.) Link. Billion Dollar Grass. Aug.–Oct. Native of Asia; occasionally escaped from cultivation throughout the

state. *E. crusgalli* (L.) Beauv. var. *frumentacea* (Roxb.) W. Wight–
G, J.

5. E. pungens (Poir.) Rydb. Barnyard Grass. Aug.–Oct. Spikelet
and grain 3.5–4.5 mm long; awn of sterile lemma more than 3 mm
long. Low ground; occasional throughout the state. *E. crusgalli* (L.
Beauv.—J.

var. microstachya (Wieg.) Mohl. Barnyard Grass. Aug.–Oct. Spike
lets and grain less than 3.5 mm long; awn of sterile lemma less than
mm long or absent; panicle purple; second glume and sterile lemm
papillose-hispid. Low ground; Lake Co. *E. microstachya* (Wieg
Rydb.—G.

var. wiegandii Fassett. Barnyard Grass. Aug.–Oct. Spikelets an
grain less than 3 mm long; awn of sterile lemma less than 3 mm lon
or absent; panicle green; second glume and sterile lemma minutel
pubescent. Low ground; occasional throughout the state.

48. *Setaria* Beauv. Foxtail Grass

1. Perennial from short, knotty rhizomes; second glume 1.0–1.5 mm lon
   .... 1. *S. geniculat*
1. Annuals from fibrous roots; second glume 1.8–2.7 mm long ....
2. Spike with retrorsely scabrous bristles .... 2. *S. verticillat*
2. Spike with antrorsely scabrous bristles. ....
3. Each spikelet subtended by 5–20 bristles; blades more or less twisted .
   .... 3. *S. lutescen*
3. Each spikelet subtended by 1–3 bristles; blades usually not twisted ..
4. Spike strongly arching; blades strigose above, puberulent below ......
   .... 4. *S. fabe*
4. Spike erect or scarcely arching; blades glabrous (sometimes ciliate) .
5. Spike usually not green, often lobed or interrupted; first glume shor
   acuminate, 1.2–1.5 mm long; fertile lemma smooth .... 5. *S. italic*
5. Spike usually green, unlobed (except var. *major*); first glume acut
   0.7–1.0 mm long; fertile lemma rugose or rugulose .... 6. *S. virid*

1. S. geniculata (Lam.) Beauv. Perennial Foxtail. July–Sept. Field
and roadsides, not common; apparently absent from the n. one-thir
of the state.

2. S. verticillata (L.) Beauv. Bristly Foxtail. July–Sept. Native o
Europe and Asia; adventive in waste ground; occasional in the n. ha
of the state, rare in the s. half.

3. S. lutescens (Weigel) Hubb. Yellow Foxtail. June–Sept. Was
ground, common; in every co. *S. glauca* (L.) Beauv.—F, G.

4. S. faberi Herrm. Giant Foxtail. July–Oct. Native of Asia; adventiv
in waste ground, very common; in every co.

5. S. italica (L.) Beauv. Italian Millet. July–Sept. Native of Europ
and Asia; adventive in waste ground, not common.

6. S virdis (L.) Beauv. Green Foxtail. June–Sept. Culms to 75 c
tall; blades to 12 mm broad; panicle unlobed. Native of Europe ar
Asia; adventive in waste ground, common; in every co.

var. major (Gaudin) Pospichal. Giant Green Foxtail. June–Sep
Culms usually at least 1.5 m tall; blades to 25 mm broad; panicle a
pearing lobed near the base. Native of Europe; adventive in cultivate
fields; Champaign and LaSalle cos.

49. Cenchrus L. Sand Bur

1. C. longispinus (Hack.) Fern. Sand Bur. July–Sept. Sandy soil; occasional throughout the state. *C. pauciflorus* Benth.—G, J.

50. Miscanthus Anderss. Eulalia

1. Fertile lemma awned; blades up to 10 mm broad .... 1. *M. sinensis*
1. Fertile lemma awnless; blades 10–18 mm broad .. 2. *M. sacchariflorus*

    1. M. sinensis Anderss. Eulalia. Sept.–Oct. Native of Asia; escaped along roads; Cass and Jackson cos.
    2. M. sacchariflorus (Maxim.) Hack. Plume Grass. Aug.–Oct. Native of Asia; adventive in waste ground; known from four n. cos.

51. Erianthus Michx. Plume Grass

1. Awn of fertile lemma up to 6 mm long; blades usually not more than 10 mm broad .... 1. *E. ravennae*
1. Awn of fertile lemma 10–16 mm long; blades usually 10–25 mm broad   2
2. Panicle silvery, the axis villous; culms sericeous below the panicle; hairs subtending the spikelets much exceeding the spikelets; awn of fertile lemma twisted below .... 2. *E. alopecuroides*
2. Panicle tawny brown, the axis more or less glabrous; culms glabrous; hairs subtending the spikelets not exceeding the spikelets; awn of fertile lemma straight .... 3. *E. brevibarbis*

    1. E. ravennae (L.) Beauv. Ravenna Grass. Aug.–Oct. Native of Europe; escaped along roads; Jackson, Monroe, and Randolph cos.
    2. E. alopecuroides (L.) Ell. Plume Grass. Sept.–Oct. Open woods; occasional in the s. tip of the state.
    3. E. brevibarbis Michx. Brown Plume Grass. Aug.–Oct. Dry hills, very rare; sw. Ill.

52. Sorghum Moench Sorghum

1. Perennials from rhizomes; awn of fertile lemma twisted near base ... 2
1. Annuals; awn of fertile lemma usually not twisted near base 3
2. Culms up to 2 m tall; blades 10–20 mm broad; awn of fertile lemma 10–15 mm long .... 1. *S. halepense*
2. Culms more than 2 m tall; some or all of the blades 25–50 cm broad; awn of fertile lemma usually less than 10 mm long ...... 2. *S. almum*
3. Sessile spikelet 4.5–5.5 mm long, shorter than the pedicellate spikelet; awn of fertile lemma falling away early .... 3. *S. bicolor*
3. Sessile spikelet 6–7 mm long, as long as the pedicellate spikelet; awn of fertile lemma persistent .... 4. *S. sudanense*

    1. S. halepense (L.) Pers. Johnson Grass. June–Oct. Native of Europe and Asia; adventive in waste ground; occasional to common in the s. half of the state, becoming less common northward.
    2. S. almum Parodi. Sorghum Grass. July–Oct. Cultivated species of Argentina; adventive along roads and edges of fields; Jackson Co.
    3. S. bicolor (L.) Moench. Sorghum. Aug.–Oct. Native of Asia; occasionally escaped from cultivation; occasional throughout the state.

S. vulgare Pers.—F, G, J. Several cultivated varieties occasionally ar adventive—var. *drummondii* (Nees) Mohl., var. *caffrorum* (Retz Mohl., and var. *saccharatum* (L.) Mohl.

4. S. sudanense (Piper) Stapf. Sudan Grass. Sept.–Oct. Native c Africa; adventive along a highway; Jackson Co.

53. *Sorghastrum Nash   Indian Grass*

1. S. nutans (L.) Nash. Indian Grass. Aug.–Oct. Prairies, fields, ope woodlands; occasional to common throughout the state.

54. *Andropogon L.   Beardgrass*

1. Pedicellate spikelet staminate, 3–10 mm long; upper sheaths not infla ted; sessile spikelet 4.5–10.0 mm long .... 1. *A. gerard.*
1. Pedicellate spikelet undeveloped, with only the long-villous pedice present; upper sheaths inflated; sessile spikelet 2.5–5.0 mm long ....
2. Upper sheaths somewhat inflated, 2–6 cm long; culms more or less gla brous; racemes enclosed only at their base, the peduncles 2–10 mm long; awn straight or nearly so .... 2. *A. virginicu.*
2. Upper sheaths greatly inflated, 6–12 mm long; culms villous at the uppe nodes; racemes nearly entirely enclosed by the sheath, the peduncle more than 10 mm long; awn twisted or curved near base   . 3. *A. elliott.*

1. A. gererdii Vitman. Big Bluestem. July–Sept. Prairies; occasiona to common throughout the state. *A. furcatus* Muhl.—J.

2. A. virginicus L. Broom Sedge. Aug.–Oct. Fields and open woods occasional or common in the s. half of the state, absent in the n. half except for Cook, DuPage, Grundy, Kankakee, and Will cos.

3. A. elliottii Chapm. Elliott's Broom Sedge. Sept.–Oct. Fields, ope woodlands; restricted to the s. tip of the state.

55. *Microstegium Nees*

1. M. vimineum (Trin.) A. Camus. Eulalia. Sept.–Oct. Native of Asia adventive in waste ground; Massac and Pope cos. *Eulalia viminea* (Trin.) Kuntze—F, G.

56. *Bothriochloa Kuntze*

1. B. saccharoides (Swartz) Rydb. Silver Beardgrass. Native to the s. and w. of Ill.; adventive in waste ground; Alexander, Johnson, anc Sangamon cos. *Andropogon saccharoides* Sw.—F, G.

57. *Schizachyrium Nees*

1. S. scoparium (Michx.) Nash. Little Bluestem. Aug.–Oct. Prairies fields, open woodlands; occasional to common throughout the state *Andropogon scoparius* Michx.—F, G, J.

58. *Tripsacum L.   Gama Grass*

1. T. dactyloides (L.) L. Gama Grass. May–Sept. Low ground; oc casional to rare in the s. two-thirds of the state.

59. *Zea* L. Corn

1. Z. mays L. Corn. June–Sept. Escaped from cultivation along roads.

60. *Eragrostis* Beauv. Love Grass

1. E. hypnoides (Lam.) BSP. Pony Grass. July–Oct. Wet ground, usually in sandy or muddy areas; throughout the state.

2. E. reptans (Michx.) Nees. Pony Grass. July–Oct. Sandy soil; scattered throughout the state.

3. E. trichodes (Nutt.) Wood. Thread Love Grass. July–Oct. Inflorescence purplish; spikelets 3- to 6-flowered, 4–7 mm long; glumes and lemmas to 3 mm long. Open, sandy areas; restricted to the w. half of the state.

var. pilifera (Scheele) Fern. Thread Love Grass. July–Oct. Inflorescence yellowish; spikelets 8- to 15-flowered, 8–12 mm long; glumes and lemmas 3.0–3.5 mm long. Sand prairies; not common and confined to the Illinois River banks. *E. trichodes* (Nutt.) Wood—G, J.

4. E. cilianensis (All.) Mosher. Stinking Love Grass. May–Oct. Native of Europe and Asia; introduced in fields and waste ground; in every co. *E. megastachya* (Koel.) Link—F.

5. E. poaeoides Beauv. Love Grass. June–Oct. Native of Europe; adventive in fields, waste ground; probably in every co.

6. E. spectabilis (Pursh) Steud. Tumble-grass. June–Oct. Sandy soil; probably in every co. (Including var. *sparsihirsuta* Farw.)

7. E. curvula (Schrad.) Nees. Weeping Love Grass. Aug.–Oct. Native of Africa; introduced into waste ground; Morgan Co.

8. E. neomexicana Vasey. Love Grass. July–Oct. Native to the sw. U.S.; adventive along a railroad; St. Clair Co. *E. mexicana* (Hornem.) Link—G.

9. E. hirsuta (Michx.) Nees. Aug.–Sept. Dry soil, very rare; Massac Co.

10. E. pectinacea (Michx.) Nees. Love Grass. July–Oct. Waste ground, fields; in every co.

11. E. diffusa Buckl. Western Love Grass. July–Oct. Native of the w. U.S.; adventive along a road; Menard Co. *E. pectinacea* (Michx.) Nees—G.

12. E. capillaris (L.) Nees. Lace Grass. July–Sept. Dry, rocky soil, particularly in woodlands; occasional throughout the state.

13. E. pilosa (L.) Beauv. Love Grass. July–Sept. Native of Europe; adventive in waste ground; restricted to a few s. cos. *E. pectinacea* (Michx.) Nees—J.

14. E. frankii C. A. Meyer. Love Grass. July–Oct. Spikelets 2- to 5-flowered, 2–3 mm long. Sandy soil, occasional; scattered throughout the state.

var. brevipes Fassett. Love Grass. July–Oct. Spikelets 6- to 7-flowered, 3–4 mm long. Sandy soil, rare; Henderson Co.

61. Tridens Roem. & Schultes

1. Panicle loose, open; glumes oblong to ovate; lemmas 3.5–4.0 mm long
.... 1. *T. flavus*
1. Panicle contracted, spike-like; glumes linear-lanceolate; lemmas 2.5–3.0 mm long .... 2. *T. strictus*

1. T. flavus (L.) Hitchcock. Purple-top. June–Sept. Fields and edge of woodlands; common, except in the n. one-fifth of the state where it is apparently very rare. (Including f. *cuprea* Fosb.) *Triodia flava* (L.) Smyth—F, G.

2. T. strictus (Nutt.) Nash. July–Oct. Waste ground and fields; not common and scattered. *Triodia stricta* (Nutt.) Benth.—F, G.

62. Triplasis Beauv. Sand Grass

1. T. purpurea (Walt.) Chapm. Sand Grass. Aug.–Oct. Sandy soil, occasional in the n. three-fourths of the state, absent in the s. one-fourth.

63. Redfieldia Vasey Blowout Grass

1. R. flexuosa (Thurb.) Vasey. Blowout Grass. July–Oct. Native of the w. U.S.; adventive in Ill.; Hancock Co.

64. Calamovilfa (Gray) Hack. Sand Reed

1. C. longifolia (Hook.) Scribn. Sand Reed. July–Sept. Sandy areas; scattered in the n. half of the state; also St. Clair Co.

65. Muhlenbergia Schreb. Muhly

1. Panicle diffuse, open, at least 5 cm across; spikelets on pedicels longer than the lemmas 2
1. Panicle narrower, contracted to less than 2 cm thick 3
2. Rhizomes present; panicle 5–20 cm long; spikelets about 1.5 mm long; lemma glabrous, 1.3–1.7 mm long, awnless .... 1. *M. asperifolia*
2. Rhizome lacking; panicle 20–45 cm long; spikelets (excluding awns) 3.0–4.5 mm long; lemma scabrous, 3.0–4.5 mm long, with an awn 5–20 mm long .... 2. *M. capillaris*
3. Plants densely tufted from firm bases or declined and rooting at the lower nodes, without rhizomes 4
3. Plants rhizomatous, not rooting at the lower nodes 6
4. Lower portion of culms decumbent, rooting at the nodes; blades (1–) 2–4 mm broad, flat; glumes 0.1–1.5 (–1.7) mm long; lemma awned 5
4. Culms stiffly erect, tufted from firm bases; blades 1–2 mm broad, flat or involute; glumes 1.7–2.8 mm long; lemma acuminate, awnless ....
.... 3. *M. cuspidata*
5. Glumes obtuse, the first 0.1–0.2 mm long, the second 0.1–0.3 mm long; awn of lemma 1.5–4.0 mm long .... 4. *M. schreberi*
5. Glumes acute to aristate, the first 0.5 mm long, the second 1.0–1.5 (–1.7) mm long; awn of lemma 0.5–1.5 mm long .... 5. *M.* X *curtisetosa*
6. Internodes glabrous or minutely scabrous, but not puberulent or pilose (*M. racemosa* rarely puberulent at summit of internodes) 7
6. Internodes puberulent or pilose 10
7. Panicle to 4 mm broad; glumes ovate-lanceolate, 1.3–2.0 (–2.5) mm long
.... 8
7. Panicle 5–15 mm broad; glumes lance-subulate, 1.6–8.0 mm long ... 9
8. Body of lemma 1.7–2.3 mm long, awnless or with an awn 1–2 (–4) mm long .... 6. *M. sobolifera*
8. Body of lemma 2.5–3.3 mm long, with an awn 2–7 mm long 7. *M. bushii*
9. Spikelets 2–4 mm long; glumes 2–3 mm long .... 8. *M. frondosa*
9. Spikelets 4–8 mm long; glumes 4–8 mm long. .... 9. *M. racemosa*
10. Lemma glabrous at base .... 10. *M. glabrifloris*
10. Lemma pilose at base 11
11. Glumes (including awns) (3.2–) 4.5–8.0 mm long ..... 11. *M. glomerata*
11. Glumes 1.5–3.5 mm long 12

12. Glumes ovate-laneolate; anthers 1.0–1.5 mm long; mature grain 2.0–2.
mm long .... 12. *M. tenuiflora*
12. Glumes linear-lanceolate; anthers 0.3–0.8 mm long; mature grain 1.3–1.
mm long 1
13. Glumes silvery or whitish; ligules 1.0–2.5 mm long; anthers 0.5–0.8 mr
long .... 13. *M. sylvatica*
13. Glumes green or purplish; ligules 0.5–1.0 mm long; anthers 0.3–0.5 mr
long .... 14. *M. mexicana*

1. M. asperifolia (Nees & Meyen) Parodi. Scratch Grass. June–Sept
Sandy soil; scattered in the n. two thirds of the state, apparently ab
sent elsewhere.

2. M. capillaris (Lam.) Trin. Hair Grass. Sept.–Oct. Sandy woods
confined to the s. cos.

3. M. cuspidata (Torr.) Rydb. Muhly. Aug.–Oct. Gravelly soil, rare
Kankakee and Will cos.

4. M. schreberi J. F. Gmel. Nimble Will. July–Oct. Waste ground
woodlands; in every co.

5. M. X curtisetosa (Scribn.) Pohl. Muhly. July–Oct. Woods, rare
Champaign and Fulton cos. The Fulton Co. specimen is the type.

6. M. sobolifera (Muhl.) Trin. Muhly. July–Oct. Lemma awnless. Dry
rocky woodlands; occasional throughout the state.

f. setigera (Scribn.) Deam. Muhly. July–Oct. Lemma with an awn t
3 mm long. Dry woodlands; rare; Wabash Co.

7. M. bushii Pohl. Muhly. Aug.–Oct. Low woods, not common
M. brachyphylla Bush—F, G, J.

8. M. frondosa (Poir.) Fern. Muhly. Aug.–Sept. Lemma awnless
Moist woodlands, roadsides, fields; common throughout the state.

f. commutata (Scribn.) Fern. Muhly. Aug.–Sept. Lemma with an aw
4–11 mm long. Moist woodlands; occasional throughout the state.

9. M. racemosa (Michx.) BSP. Marsh Muhly. Aug.–Sept. Moist soi
occasional; not common in the e. cos.

10. M. glabrifloris Scribn. Muhly. July–Oct. Moist woodlands; oc
casional in the cent. and s. cos. absent from the n. cos. *M. glabriflora*
Scribn.—F, G, J.

11. M. glomerata (Willd.) Trin. Muhly. Aug.–Sept. Dry or wet ground
rare; Cook, DuPage, Kane, Kendall, Lake, McHenry, and St. Clair cos

12. M. tenuiflora (Willd.) BSP. Slender Muhly. July–Oct. Rock
woods, moist bluffs; occasional throughout the state.

13. M. sylvatica (Torr.) Torr. Muhly. Aug.–Oct. Rich, often rock
woods; occasional throughout the state.

14. M. mexicana (L.) Trin. Muhly. Aug.–Oct. Lemma awnless. Mois
soil, usually in woods; occasional throughout the state.

f. ambigua (Torr.) Fern. Muhly. Aug.–Oct. Lemma with an awn to
mm long. Moist soil, usually in woods, rare; Lake Co.

66. Sporobolus R. Br.   Dropseed

1. Panicle more or less open, spreading, over 10 cm long; perennials ..
1. Panicle contracted, spike-like, less than 10 cm long (except in *S. asper*
perennials or annuals ....
2. Sheaths densely villous, at least near the top; spikelets 1.8–2.5 long
glumes acute, the first 1.0–1.5 mm long, the second 2.0–2.5 long
lemma 2.0–2.5 mm long .... 1. *S. cryptandrus*

2. Sheaths glabrous or nearly so; spikelets 4–6 mm long; glumes subulate or acuminate, the first 2–4 mm long, the second 4–6 mm long; lemma 3.5–5.5 mm long .... 2. *S. heterolepis*

3. Perennials; blades (before drying) 2.5 mm broad; panicle 5–25 cm long; only the upper sheaths swollen 4

3. Annuals; blades 1–2 mm broad; panicle 1–5 cm long; all or nearly all sheaths swollen 5

4. Lemma glabrous; blade pilose on upper surface near base . 3. *S. asper*

4. Lemma sparsely villous; blade scabrous on margin and tip, but not pilose .... 4. *S. clandestinus*

5. Spikelets 3.5–6.5 mm long; first glume 2.8–4.0 mm long; second glume 3.0–4.5 mm long; lemma 3–5 mm long, minutely villous .... 5. *S. vaginiflorus*

5. Spikelets 2–3 mm long; first glume 1.5–2.5 mm long; second glume 1.7–2.7 mm long; lemma 2–3 mm long, mostly glabrous   6. *S. neglectus*

1. S. cryptandrus (Torr.) Gray. Sand Dropseed. Aug.–Oct. Sandy soil; occasional in the n. third of the state and in the w. cos.

2. S. heterolepis (Gray) Gray. Prairie Dropseed. Aug.–Sept. Dry soil, often prairies; occasional in the n. half of the state, rare in the s. half.

3. S. asper (Michx.) Kunth. Dropseed. Sept.–Oct. Dry, often sandy soil; occasional throughout the state.

4. S. clandestinus (Biehler) Hitchc. Dropseed. Aug.–Oct. Sandy soil, not common; scattered in Ill.

5. S. vaginiflorus (Torr.) Wood. Poverty Grass. Sept.–Oct. Dry soil; occasional throughout the state. (Including var. *inaequalis* Fern.)

6. S. neglectus Nash. Sheathed Dropseed. Sept.–Oct. Dry soil; occasional throughout the state.

67. *Crypsis Ait.*

1. C. schoenoides (L.) Lam. Aug.–Sept. Native of Europe; introduced into waste ground; Cook, Grundy, and St. Clair cos. *Heleochloa schoenoides* (L.) Host—F, G, J.

68. *Eleusine Gaertn.   Goose Grass*

1. E. indica (L.) Gaertn. Goose Grass. June–Oct. Native of Eurasia; adventive in lawns and waste ground; probably in every co.

69. *Dactyloctenium Willd.   Crowfoot Grass*

1. D. aegyptium (L.) Beauv. Crowfoot Grass. Aug. Native of the Old World; adventive in waste ground; St. Clair Co.

70. *Leptochloa Beauv.   Sprangletop*

1. Spikelets 2- to 4-flowered; lemmas 0.7–1.5 mm long; sheaths papillose-pilose; ligules 1–2 mm long 2

1. Spikelets 5- to 10-flowered; lemmas 2–8 mm long; sheaths glabrous or nearly so; ligules 3–7 mm long 3

2. Lemmas 1.0–1.5 mm long; 0.7–0.9 mm long; glumes acute 1. *L. filiformis*

2. Lemmas 0.7–1.0 mm long; grain 0.4–0.5 mm long; glumes aristate .... 2. *L. attenuata*

3. Blades becoming involute, 1–5 mm broad; lemma with an awn 0.5–1.
mm long ....
3. Blades flat, 5–10 mm broad; lemma apiculate, awnless   5. *L. panicoide*
4. Lemmas up to 4 mm long, pale green .... 3. *L. fasciculari*
4. Lemmas 5–8 mm long, purplish to lead-colored .... 4. *L. acuminat*

1. **L. filiformis** (Lam.) Beauv. Red Sprangletop. July–Sept. Lov
sandy soil, particularly along rivers; occasional in the s. two-fifths o
the state.

2. **L. attenuata** (Nutt.) Steud. Sprangletop. July–Sept. Sandy shore
occasional in the s. tip of the state. *L. filiformis* (Lam.) Beauv.—J.

3. **L. fascicularis** (Lam.) Gray. Salt Meadow Grass. July–Sept. We
soil; scattered in the s. three-fifths of the state; also Cook Co. *D
plachne fascicularis* (Lam.) Beauv.—F.

4. **L. acuminata** (Nash) Mohlenbr. Salt Meadow Grass. July–Sep
Native of the w. U.S.; adventive along a railroad and in saline wast
ground; Cook, DuPage, Grundy, Kane, Kankakee, Kendall, Lake, an
Will cos. *L. fascicularis* (Lam.) Gray var. *acuminata* (Nash) Gl.—G
Diplachne acuminata Nash—F.

5. **L. panicoides** (Presl) Hitchc. Salt Meadow Grass. July–Sept. Lov
areas, rare; Calhoun and Pike cos. *Diplachne halei* Nash—F.

71. *Gymnopogon* Beauv.   Beardgrass

1. **G. ambiguus** (Michx.) BSP. Beardgrass. July–Oct. Sandy o
gravelly soil in open areas, very rare; Pope Co.

72. *Schedonnardus* Steud.   Tumble Grass

1. **S. paniculatus** (Nutt.) Trel. Tumble Grass. July–Sept. Salt lick
very rare; Hancock Co.

73. *Cynodon* Rich.   Bermuda Grass

1. **C. dactylon** (L.) Pers. Bermuda Grass. July–Oct. Native of Europe
adventive in lawns and waste ground; occasional in the s. half of th
state, much less common in the n. half.

74. *Chloris* Swartz   Finger Grass

1. Blades 1–3 mm broad, obtuse; awn of fertile lemma 5–9 mm long; empt
lemma 1, with an awn 3.5–5.0 mm long; plants to 40 cm tall ....
.... 1. *C. verticilla*
1. Blades 3–5 mm broad, tapering to a long, fine point; awn of fertile lemm
1–5 mm long; empty lemmas 2, the upper awnless; plants usually over
m tall .... 2. *C. gayan*

1. **C. verticillata** Nutt. Windmill Grass. June–Sept. Native of the w
U.S.; adventive in lawns; scattered in Ill.

2. **C. gayana** Kunth. Finger Grass. June–Sept. Native of Africa; ad
ventive in a strip mine; Perry Co.

75. *Bouteloua Lag. Grama Grass*

1. Spikes 10–50, spreading or nodding, to 2 cm long, falling entire at maturity; spikelets 7–10 mm long .... 1. *B. curtipendula*

1. Spikes 1–3 (–6), straight or curved backward, 2–5 cm long, the florets falling from the glumes; spikelets 5–6 mm long 2

2. Rachis of inflorescence projecting 2–5 mm beyond last spike; second glume papillose-hirsute on keel; empty lemma glabrous at base; blades papillose-pubescent .... 2. *B. hirsuta*

2. Rachis of inflorescence not projecting; second glume sparsely pilose-papillose on keel; empty lemma long-villous at base; blades glabrous or scabrous .... 3. *B. gracilis*

1. B. curtipendula (Michx.) Torr. Sideoats Grama. July–Sept. Prairies, dry hills; occasional in the n. half of the state, rather common on the bluffs bordering the Mississippi River from JoDaviess to Alexander cos.

2. B. hirsuta Lag. Grama Grass. July–Sept. Prairies; confined to the nw. one-fourth of the state.

3. B. gracilis (HBK.) Lag. Blue Grama. July–Sept. Sand flats; JoDaviess Co.; also adventive at a strip mine in Williamson Co.

76. *Buchloë Engelm. Buffalo Grass*

1. B. dactyloides (Nutt.) Engelm. Buffalo Grass. Aug.–Sept. Prairies, rare; Peoria and Winnebago cos.

77. *Spartina Schreb. Cord Grass*

1. S. pectinata Lind. Cord Grass. June–Sept. Wet prairies, marshes; occasional throughout the state, although more common in the n. cos.

78. *Distichlis Raf. Salt Grass*

1. D. stricta (Torr.) Rydb. Salt Grass. July–Oct. Native of w. N. Am.; adventive along railroads and in saline waste areas; Champaign, Cook, DeKalb, Kane, and Kendall cos.

79. *Aristida R. Br. Three Awn*

1. Awns of lemma twisted and united into a basal column 8–15 mm long; glumes 20–30 mm long .... 1. *A. tuberculosa*

1. Awns of lemma free to base or united into a basal column up to 3 mm long; glumes up to 20 mm long (except *A. oligantha* and *A. ramosissima*) 2

2. Central awn of most lemmas at least 20 mm long 3

2. Central awn of lemma up to 20 mm long, usually considerably shorter   8

3. Lateral awns of lemma at least 12 mm long 4

3. Lateral awns of lemma up to 6 mm long, or even absent .... 7. *A. ramosissima*

4. Awns of lemma united at base into a column 1–3 mm long, the awns articulated with the summit of the lemma; sheaths villous on the margins .... 2. *A. desmantha*

4. Awns of lemma free to base, the awns not articulated with the summit of the lemma; sheaths glabrous or nearly so 5

5. All awns of lemma 35–70 mm long; glumes 12–32 mm long; lemma 12–20 mm long .... 3. *A. oligantha*

5. Central awn of lemma 15–33 mm long, the lateral awns 12–24 mm long; glumes 6–14 mm long; lemma 5.5–8.5 mm long 6

6. Perennial; first glume longer than second .... 4. *A. purpurascens*

6. Annual; first glume equal to second or shorter 7

7. Glumes essentially equal in length; central awn of lemma longer than lateral awns .... 5. *A. intermedia*

7. Second glume longer than first glume; awns of lemma equal in length .... 6. *A. necopina*

8. Lateral awns 12–20 mm long; central awn curved or divergent at base, not coiled 9

8. Lateral awns up to 12 mm long; central awn coiled at base (or bent conspicuously near base in *A. longespica*) 11

9. Central awn of lemma conspicuously bent near base; glumes obscurely 3-nerved .... 8. *A. longespica*

9. Awns of lemma not conspicuously bent near base; glumes sharply 1-nerved 10

10. Glumes nearly equal in length; central awn of lemma longer than the lateral ones .... 5. *A. intermedia*

10. Second glume longer than first glume; awns of lemma equal .... 6. *A. necopina*

11. Glumes essentially equal, the second 4–10 mm long; lemma 4.0–8.5 mm long; inflorescence usually reduced to a raceme 12

11. Second glume longer than first, usually by at least 2 mm, the second 7–15 mm long; lemma 7.5–10.5 mm long; inflorescence a slender panicle 13

12. Central awn conspicuously bent at base, not coiled, 6.5–20.0 mm long; lateral awns 3–15 mm long; glumes obscurely 3-nerved   8. *A. longespica*

12. Central awn loosely coiled at base, 3–10 mm long; lateral awns 0.7–3.3 mm long; glumes 1-nerved .... 9. *A. dichotoma*

13. Lateral awns of lemma 5–12 mm long; central awn 10–19 mm long .... 10. *A. basiramea*

13. Lateral awns of lemma 2–4 mm long; central awn 7.0–12.5 mm long ... . 11. *A. curtissi*

1. A. tuberculosa Nutt. Needle Grass. Aug.–Oct. Sandy soil; occasional in the n. half of the state; also Union Co.

2. A. desmantha Trin. & Rupr. Three Awn. Aug.–Sept. Sandy soil; very rare; Cass, Mason, and Morgan cos.

3. A. oligantha Michx. Three Awn. Aug.–Oct. Dry soil of fields and woods; along railroads; common throughout the state.

4. A. purpurascens Poir. Arrowfeather. Aug.–Sept. Sandy soil; occasional throughout the state.

5. A. intermedia Scribn. & Ball. Three Awn. Sept.–Oct. Sandy soil; rare; Grundy, Henry, Lake, Lee, McHenry cos.

6. A. necopina Shinners. Three Awn. Sept.–Oct. Sandy soil, rare; Lee Co.

7. A. ramosissima Engelm. Slender Three Awn. Aug.–Sept. Lateral awns 0.5–0.6 mm long. Dry soil of fields; occasional in the s. half of the state, absent in the n. half.

f. uniaristata (Gray) Mohlenbr. Slender Three Awn. Aug.–Sept. Lateral awns absent, or less than 0.5 mm long. Dry soil of fields, rare; Marion and St. Clair cos.

8. A. longespica Poir. Three Awn. Aug.–Oct. Lateral awns 3–4 mm long; central awn 6.5–13.0 mm long; glumes 4–6 mm long. Sandy soil, particularly in fields and along highways; occasional throughout the state, although rare or absent in the n. cent. cos.

var. geniculata (Raf.) Fern. Three Awn. Aug.–Sept. Lateral awns 4–15 mm long; central awn 10–20 mm long; glumes 5–9 mm long. Sandy soil, rare; Henry and Jackson cos.

9. A. dichotoma Michx. Three Awn. Aug.–Oct. Dry fields and along highways; occasional in the s. half of the state, less common in the n. half.

10. A. basiramea Engelm. Three Awn. Aug.–Oct. Dry, sandy soil, not common; confined to the n. four-fifths of the state; also Gallatin and Jackson cos.

11. A. curtissii (Gray) Nash. Three Awn. Aug.–Oct. Dry soil, rare; Massac and Ogle cos. A. dichotoma Michx. var. curtissii Gray—F.

80. Arundinaria Michx.   Cane

1. A. gigantea (Walt.) Chapm. Giant Cane. April–May. Low ground; occasional in the s. one-third of the state, absent elsewhere.

81. Leersia Sw.   Cut Grass

1. Spikelets broadly rounded, 3–4 mm broad, over half as wide as long ... . 1. L. lenticularis
1. Spikelets oblongoid, 1–2 mm broad, less than half as wide as long .. 2
2. Sheaths conspicuously retrorse-scabrous; blades spinulose on the margins; lowest panicle branches whorled; stamens 3 . 2. L. oryzoides
2. Sheaths glabrous or scaberulous; blades scaberulous; lowest panicle branches solitary; stamens 1–2 .... 3. L. virginica

1. L. lenticularis Michx. Catchfly Grass. Aug.–Oct. Low woods, swamps, marshes; occasional throughout the state, except for the extreme n. cos.

2. L. oryzoides (L.) Swartz. Rice Cutgrass. Aug.–Oct. Low, moist soil; occasional throughout the state.

3. L. virginica Willd. White Grass. July–Sept. Moist woodlands; common throughout the state. (Including var. ovata [Poir.] Fern.)

82. Zizania L.   Wild Rice

1. Z. aquatica L. Wild Rice. June–Sept. Pistillate lemma scabrous, slenderly nerved; aborted spikelets up to 1 mm broad, subulate, tapering into the awn. Shallow water, not common; scattered in Ill., except for the s. cos.

var. interior Fassett. Wild Rice. June–Sept. Pistillate lemma glabrous, broadly nerved; aborted spikelets 1.5–2.0 mm broad, linear, abruptly tapering into the awn. Shllow water, rare; Cook, Lake, and Union cos. Z. aquatica L.— J.

83. Zizaniopsis Doell & Aschers

1. Z. miliacea (Michx.) Doell & Aschers. Southern Wild Rice. Jul Oct. Edge of lake, rare; Montgomery Co.

84. Phragmites Trin. Reed

1. P. australis Trin. Reed. July–Sept. Moist soil; occasional in the half of the state, uncommon in the s. half. *P. communis* Trin.—F, G,

85. Arundo L.

1. A. donax L. Giant Reed. Sept.–Oct. Native along the Mediterr nean Sea; adventive along a gravel road in Pope Co.

86. Danthonia Lam. & DC. Wild Oat Grass

1. D. spicata (L) Beauv. Poverty Oat Grass; Curly Grass. May–Au Dry woods and bluffs; occasional to common throughout the state.

87. Chasmanthium Link Sea Oats

1. C. latifolium (Michx.) Yates. Sea Oats. July–Oct. Moist soil; cor mon in the s. half of the state, rare in the n. half. *Uniola latifo* Michx.—F, G, J. The type is from Ill.

26. CYPERACEAE

1. Flowers bisexual; achenes not enclosed in a sac-like structure (perigy ium) nor white in color (except for *Cyperus erythrorhizos*) . . . .
1. Flowers unisexual; achenes either enclosed in a sac-like structu (perigynium) or white or buff in color . . . .
2. Spikelets 3- or more-flowered (1-flowered in *Cyperus densicaespitos* which has a strong sweet odor) . . . .
2. Spikelets 1- to 2-flowered . . . .
3. Spikelets flattened; scales 2-ranked . . . .
3. Spikelets not flattened; scales spirally arranged . . . .
4. Inflorescence terminal; bristles absent; achene without a tubercle . . . .
. . . . 1. *Cyper*
4. Inflorescence axillary; bristles present; achene with a tubercle . . . .
. . . . 2. *Dulichiu*
5. Achene crowned by a tubercle . . . .
5. Achene without a tubercle . . . .
6. Spikelet one; leaves absent . . . . 3. *Eleocha*
6. Spikelets 2-many; leaves present . . . . 4. *Bulbosty*
7. Flowers not subtended by bristles or dilated scales (rudimentary scal only in *Lipocarpha*) . . . .
7. Flowers subtended by bristles or dilated scales although, in some sp cies of *Scirpus*, the scales sometimes falling away early . . . .
8. Bract subtending the inflorescence 1 . . . . 8. *Scirp*
8. Bracts subtending the inflorescence 2 or more . . . .

9. Styles the same thickness throughout .... 8. *Scirpus*
9. Styles dilated toward base .... 5. *Fimbristylis*
10. Flowers subtended by minute, dilated rudimentary perianth scales .....
.... 6. *Lipocarpha*
10. Flowers subtended by ovate-oblong scales or by bristles 11
11. Flowers subtended by ovate-oblong scales .... 7. *Fuirena*
11. Flowers subtended by bristles 12
12. Bristles pale or dark brown, reddish, or blackish at maturity ....
.... 8. *Scirpus*
12. Bristles white at maturity, forming "cottony" heads .... 9. *Eriophorum*
13. Achene crowned with a tubercle; bristles subtending the flower........
.... 10. *Rhynchospora*
13. Achene without a tubercle; bristles absent .... 11. *Cladium*
14. Achene white or buff, not enclosed in a perigynium .... 12. *Scleria*
14. Achene not white, enclosed in a perigynium .... 13. *Carex*

1. *Cyperus* L. Galingale

1. Stigmas 2; achenes lenticular 2
1. Stigmas 3; achenes trigonous 6
2. Spikelets 1-flowered; inflorescence of 1–3 sessile heads; plants with a strong sweet odor when fresh .... 1. *C. densicaespitosus*
2. Spikelets 5- to 35-flowered; inflorescence of 1-several radiating sessile spikelets and usually 1-several rays; plants without a distinctive sweet odor when fresh 3
3. Achenes black, nearly as broad as long, with horizontal wrinkles; scales straw-colored .... 2. *C. flavescens*
3. Achenes drab or gray, longer than broad, without horizontal wrinkles; scales usually suffused with purple 4
4. Styles divided nearly to base, persistent and conspicuously exserted to 4 mm from scales .... 3. *C. diandrus*
4. Styles divided to about the middle, falling away early, hidden by the scales or exserted to 2 mm from scales 5
5. Scales closely appressed, strongly suffused with purple (straw-colored in *C. rivularis* f. *elutus*) .... 4. *C. rivularis*
5. Scales with tips somewhat spreading, the spikelets appearing serrate, straw-colored or purple on the margins .... 5. *C. filicinus*
6. Scales with strongly recurved tips .... 6. *C. aristatus*
6. Scales with tips either appressed or slightly spreading 7
7. Cluster of spikelets spherical or globose, with spikelets radiating in all directions 8
7. Cluster of spikelets hemispherical, cylindrical, ellipsoid or lanceoloid, but not spherical or globose 11
8. Scales appressed, but with the tips shortly recurved; spikes to 8 mm across 9
8. Scales appressed to spreading, their tips straight; some of the heads over 1 cm across 10
9. Achenes linear, 1.0–1.4 mm long; perennial plants at least 35 cm tall at maturity .... 7. *C. pseudovegetus*
9. Achenes ellipsoid to oblongoid, 0.5–1.0 mm long; annual plants never more than 35 cm tall .... 8. *C. acuminatus*
10. Scales appressed; spikelets 2- to 3-flowered .... 9. *C. ovularis*
10. Scales spreading; spikelets 5- to several-flowered ... 10. *C. grayioides*
11. Spikelets arising from a common point on the axis 12

11. Spikelets arising from either side of an elongated axis .... 1!
12. Annual with soft bases; achenes tan, broadest above the middle ......
.... 11. *C. compressus*
12. Perennial with short, firm rhizomes; achenes light brown, dark brown, o
black, broadest at the middle 1:
13. Scales terminated by a mucro up to 1.5 mm long, 9- to 15-nerved
achenes light or dark brown .... 1.
13. Scales obtuse, or the uppermost acute, without a mucro, 5- to 9-nerved
achene black .... 14. *C. filiculmis*
14. Achenes 2.5–3.3 mm long, light brown; culms and margins of leave:
harshly scabrous; scales 2.5–4.0 mm long .... 12. *C. schweinitzi*
14. Achenes 1.2–1.8 mm long, dark brown; culms and margins of leave:
smooth; scales 2.0–2.5 mm long .... 13. *C. houghtoni*
15. Scales 1.0–1.5 mm long; achenes 0.8–1.0 mm long .... 1(
15. Scales 1.5–4.5 mm long; achenes 1.0–2.8 mm long 1
16. Achenes white; spikelets about 1 mm broad; scales closely arranged ..
.... 15. *C. erythrorhizo:*
16. Achenes brown or black; spikelet about 1.5 mm broad; scales somewha
remote from each other .... 16. *C. iri*
17. Scales remote, the tip of one just reaching the base of the one above i
giving the spikelet a zigzag appearance .... 17. *C. engelmann*
17. Scales approximate and overlapping 1:
18. Some or all of the mature spikelets reflexed; spikelets subterete ......
.... 18. *C. lancastriensi:*
18. None of the spikelets (except sometimes the lowest pair) reflexed; spike
lets flattened .... 1!
19. Rhizomes scaly and usually ending in a tuber; scales at the tips of th
spikelets slightly spreading, giving the spikelet an obtuse appearance .
.... 19. *C. esculentu*
19. Rhizomes absent or merely becoming hard and corm-like; scales at th
tips of the spikelet appressed, giving the spikelet a pointed appearanc
.... 2
20. Plants annual, without rhizomes; scales ferruginous or golden-browr
1.7–3.0 mm long; achenes obovoid-oblongoid ... 20. *C. ferruginescen*
20. Plants perennial, with hardened bases; scales straw-colored, 3.5–5.0 mr
long; achenes linear .... 21. *C. strigosu*

1. *C. densicaespitosus* Mattf. & Kükenth. July–Sept. Moist, ope
soil; restricted to the one-half of the state. *C. tenuifolius* (Steud
Dandy.—F.

2. *C. flavescens* L. July–Oct. Wet open soil, rare in the s. one-sixt
of the state; also Coles, McDonough, and Peoria cos. *C. flavescens* L
var. *poaeformis* (Pursh) Fern.—F.

3. *C. diandrus* Torr. June–Oct. Wet ground along banks and shore:
not common; scattered in the w. half of Ill.

4. *C. rivularis* Kunth. July–Oct. Scales strongly suffused with rec
brown or purple. Wet ground along banks and shores; throughout th
state, but rare in the s. one-third. (Including f. *elongatus* Boeckl.)

f. *elutus* (C. B. Clarke) Kükenth. July–Oct. Scales stramineou
throughout, except for the greenish midrib.

5. *C. filicinus* Vahl. July–Sept. Native to the s. U.S.; adventive in we
roadside ditch; Jackson Co.

6. *C. aristatus* Rottb. May–Oct. Moist soil; scattered throughout th
state. *C. inflexus* Muhl.—F, J.

7. C. pseudovegetus Steud. June–Oct. Moist soil; restricted to the s. one-fourth of the state. *C. virens* Gray—F.

8. C. acuminatus Torr. & Hook. June–Oct. Wet ground; scattered throughout the state except for very northernmost cos.

9. C. ovularis (Michx.) Torr. Round-headed Sedge. June–Sept. Dry sandy woods, old fields; common in the s. one-third of the state; also Peoria Co. (Including var. *robustus* Britt., var. *sphaericus* Boeckl., and var. *wolfii* [Wood] Kükenth.)

10. C. grayioides Mohlenbr. Aug. Sand prairies and blowouts, rare; Mason and Whiteside cos.

11. C. compressus L. Sept. Native of the s. U.S. and trop. Am.; adventive in roadside ditch; Lawrence Co.

12. C. schweinitzii Torr. June–Sept. Sandy soil; occasional in the n. half of the state; adventive in Jackson Co.

13. C. houghtonii Torr. June–Sept. Sandy soil, rare; Boone, Cook, Grundy, Kankakee Lake, and Will cos.

14. C. filiculmis Vahl. May–Oct. Plants with a central sessile glomerule and 3–6 (–8) well-developed rays; achenes (1.7–) 1.8–2.2 mm long. Rocky or sandy soil; common in the n. half of the state, s. to Madison Co. (Including *C.* X *mesochorus* Geise.)

var. macilentus Fern. May–Oct. Plants with only a central sessile glomerule or occasionally with 1–2 rays; achenes 1.2–1.8 (–2.0) mm long. Sandy soil; common in the n. half of the state, rare southward.

15. C. erythrorhizos Muhl. Aug.–Oct. Moist, often sandy soil; occasional throughout the state.

16. C. iria L. July–Oct. Native of Europe and Asia; adventive in wet meadows; Alexander Co.

17. C. engelmannii Steud. July–Oct. Wet ground; rare but scattered in Ill.

18. C. lancastriensis Porter. Aug.–Oct. Moist, sandy woods, rare; Massac and Pulaski cos.

19. C. esculentus L. Nut-grass. June–Oct. Spikelets to 15 mm long, 1.5–3.0 mm broad. Moist, frequently cultivated soil, commom; in every co.

var. leptostachyus Boeckl. June–Oct. Moist, often cultivated soil; scattered in Ill.

20. C. ferruginescens Boeckl. Aug.–Oct. Rich moist soil; scattered throughout the state.

21. C. strigosus L. June–Oct. Wet gorund, common; in every co. (Including f. *robustior* Kunth and var. *stenolepis* [Torr.] Kükenth.)

2. *Dulichium Rich. ex Pers. Three-way Sedge*

1. D. arundinaceum (L.) Britt. Three-way Sedge. July–Oct. Swamps and low ground in woods; throughout the state, but not common.

3. *Eleocharis R. Br. Spike Rush*

1. Spikelet about the same thickness as the culm; plants robust, often 1 m tall or taller 2
1. Spikelet usually conspicuously thicker than the culm; plants usually less than 1 m tall (except *E. rostellata*) 3
2. Culms round, septate by cross partitions; bristles shorter than the achene, or lacking .... 1. *E. equisetoides*

2. Culms 4-sided, sharply angled, not septate; bristles slightly longer than
the achene .... 2. *E. quadrangulata*
3. Style 2-cleft; achenes obovoid-lenticular (rarely sometimes tufted in *E.
olivacea*) 4
3. Style 3-cleft; achenes usually trigonous 9
4. Plants perennial, with rhizomes 5
4. Plants annual, with fibrous roots 8
5. Achenes at most 1.0 mm long, the small, conical tubercle $^1/_6$–$^1/_8$ as long
.... 3. *E. olivacea*
5. Achenes averaging somewhat longer than 1.0 mm, the conical tubercle
$^1/_2$–$^1/_4$ as long 6
6. Basal scales 1–3, never completely encircling the culm; culms averaging
1.3 mm wide 7
6. Basal scale 1, suborbicular, completely encircling the culm; culm
averaging 0.6 mm wide .... 6. *E. erythropoda*
7. Basal sheaths of mature culms with prominent V-shaped sinuses;
culms often soft and inflated .... 4. *E. smallii*
7. Basal sheaths truncate to slightly oblique at apex; culms usually rigid .
.... 5. *E. macrostachya*
8. Achenes lustrous black or purple; plants usually less than 15 cm tall ..
.... 7. *E. caribaea*
8. Achenes yellow to deep brown; plants usually 15–50 cm tall 8. *E. obtusa*
9. Tubercle conspicuously differentiated from achene; achenes up to 2 mm
long 10
9. Tubercle confluent with top of achene, long-conical; achenes 2–3 mm
long 14
10. Achenes with 8–19 longitudinal ribs ...... 9. *E. acicularis*
10. Achenes pitted or with papillae 11
11. Scales 1.5–2.2 mm long; culms 0.1–0.5 mm wide, capillary ...... 9. *E. acicularis*
11. Scales 2.5–3.0 mm long; culms about 1 mm wide, often inrolled ...... 10. *E. wolfii*
12. Tubercle not subulate; achene 1.0–1.2 mm long (including the tubercle);
perennials 13
12. Tubercle subulate; achene about 1.5 mm long (including the tubercle);
annual .... 13. *E. intermedia*
13. Culm with 5 vascular bundles and appearing (4-) 5-angled; mature
achene usually olivaceous and warty .... 11. *E. tenuis*
13. Culm with 6–14 vascular bundles and appearing low-angular or compressed; mature achene usually yellow, brown, or orange-brown, reticulate or slightly warty .... 12. *E. elliptica*
14. Stems flat, to 2 mm wide; scales elliptic, obtuse; plants tufted ....
.... 14. *E. rostellata*
14. Stems usually 3-angled, less than 1 mm wide; scales lanceolate, acute;
plants with stoloniferous rhizomes .... 15. *E. pauciflora*

1. E. equisetoides (Ell.) Torr. Horsetail Spike Rush. July–Oct. Wet
ground or standing water, very rare; Cook Co.
2. E. quadrangulata (Michx.) Roem. & Schultes. Square-stemmed
Spike Rush. June–Oct. Shallow water in ponds and lakes, rare; primarily in the s. one-fourth of the state. (Including var. *crassior* Fern.
3. E. olivacea Torr. July–Sept. Wet sands, very rare; Cook Co.
4. E. smallii Britt. June–Sept. Edge of swamps, sloughs, ponds,
streams; occasional throughout the state. *E. palustris* (L.) Roem. &
Schultes—G.

5. E. macrostachya Britt. June–Aug. Edge of swamps, sloughs, ponds, streams; occasional throughout the state. *E. palustris* (L.) Roem. & Schultes—G.

6. E. erythropoda Steud. June–Sept. Wet soil; common in the n. half of the state, rare elsewhere. *E. calva* (Gray) Torr.—F, G, J.

7. E. caribaea (Rottb.) Blake. June–Sept. Wet sand, very rare; Cook Co. (Including var. *dispar* [Hill] Blake.) *E. geniculata* (L.) Roem. & Schultes—F, G, J.

8. E. obtusa (Willd.) Schult. May–Oct. Tubercle more than ⅔ the width of the achene and ¼–½ the height of the achene. Wet ground; common throughout the state.

var. detonsa (Gray) Drap. & Mohlenbr. May–Oct. Tubercle more than ⅔ the width of the achene and up to ¼ the height of the achene. Wet ground; occasional throughout the state. *E. engelmannii* Steud.—F, G, J.

var. ovata (Roth) Drap. & Mohlenbr. May–Oct. Tubercle ½–⅔ the width of the achene. Wet ground; scattered but rare throughout the state. *E. ovata* (Roth) Roem. & Shultes—F, G, J.

9. E. acicularis (L.) Roem. & Schultes. July–Oct. Low wet ground; occasional throughout the state. (Including var. *gracilescens* Svenson and f. *inundata* Svenson.)

10. E. wolfii (Gray) Patterson. May–July. Wet ground; rare and scattered in Ill.

11. E. tenuis (Willd.) Schult. var. verrucosa (Svenson) Svenson. May–Sept. Low wet ground; moist crevices on dry bluffs; occasional throughout the state. *E. tenuis* (Willd.) Schult.—J.

12. E. elliptica Kunth. May–July. Culm with 6–8 vascular bundles and appearing with 6–8 low angles. Low areas; occasional throughout the state.

var. compressa (Sull.) Drap. & Mohlenbr. May–July. Culm with 9–14 vascular bundles and appearing compressed. Low areas; scattered throughout the state. *E. compressa* Sull.—F, G, J.

13. E. intermedia (Muhl.) Schult. July–Oct. River banks and swampy areas; fairly common along the Illinois River, rare or absent elsewhere.

14. E. rostellata (Torr.) Torr. July–Sept. Marshy, calcareous soils, rare; Lake and McHenry cos.

15. E. pauciflora (Lightf.) Link. July–Oct. Wet areas; restricted to the extreme ne. cos.

4. Bulbostylis (Kunth) C. B. Clarke

1. B. capillaris (L.) C. B. Clarke. July–Oct. Moist, sandy soil; occasional throughout the state. (Including var. *crebra* Fern. and var. *isopoda* Fern.)

5. Fimbristylis Vahl

1. Style branches 2; achene lenticular 2
1. Style branches 3; achene trigonous .... 4. *F. autumnalis*
2. Spikelets in a close head; achene 0.4–0.5 mm long; cespitose annuals .... 1. *F. vahlii*
2. Spikelets on elongated rays, cymose; achene 1.0–1.8 mm long; tufted perennials (or annuals) with hardened bases 3

3. Outer scales puberulent; achene 1.3–1.7 mm long, reticulate . . . .
. . . . 2. *F. puberula* var. *drummondii*
3. All scales glabrous; achene 1.0–1.3 mm long, with horizontal and transverse ribs, or sometimes irregularly tuberculate . . . . 3. *F. baldwiniana*

1. F. vahlii (Lam.) Link. July–Oct. Moist soil, rare; Cass Co.
2. F. puberula (Michx.) Vahl var. drummondii (Boeckl.) Ward. May–July. Moist or dry soil, rare; restricted to the n. half of the state; also St. Clair Co. *F. drummondii* Boeckl.—F, J; *F. caroliniana* (Lam.) Fern.—G.
3. F. baldwiniana (Schult.) Torr. June–Oct. Moist soil, rare; Alexander, Johnson, and Massac cos. *F. annua* (All.) Roem. & Schultes—G.
4. F. autumnalis (L.) Roem. & Schultes. June–Oct. Moist or rather dry, sandy soil; occasional throughout the state. (Including var. *mucronulata* [Michx.] Fern.)

6. *Lipocarpha* R. Br.

1. L. maculata (Michx.) Torr. June–Sept. Around a pond, very rare; Cass Co.

7. *Fuirena* Rottb. Umbrella Grass

1. F. scirpoides Michx. July. Along edge of lake, very rare; Hamilton Co.

8. *Scirpus* L. Bulrush

1. Inflorescence appearing lateral; lowest bract culm-like 2
1. Inflorescence appearing terminal; bracts leaf-like or scale-like 11
2. Achenes 1.5–4.0 mm long 3
2. Achenes 1.0–1.5 mm long . . . . 10. *S. koilolepis*
3. Spikelets sessile or subsessile; culm 3-angled 4
3. Spikelets (or most of them) on elongated pedicels; culm terete 9
4. Achene black (immature achenes whitish in *S. hallii*); plants with fibrous roots 5
4. Achene white, stramineous, or pale brown (rarely nearly black in *S. americanus*); plants with rhizomes 7
5. Achene smooth or pitted; scales obtuse or mucronulate 6
5. Achene transversely corrugated; scales awn-tipped . . . . 3. *S. hallii*
6. Achene smooth; culm scarcely angled . . . . 1. *S. smithii*
6. Achene pitted; culm 3-angled . . . . 2. *S. purshianus*
7. Scales entire; spikelets greenish or stramineous; culms slender, at most 2 mm thick; rhizome slender and weak 8
7. Scales ciliate, 2-cleft; spikelets reddish-brown; culms rather stout, mostly over 2 mm thick; rhizome stout and strong . . . 6. *S. americanus*
8. Culms 3-angled; leaves 2–3 . . . . 4. *S. torreyi*
8. Culms terete; leaves 4 or more . . . . 5. *S. subterminalis*
9. Style 2-cleft; spikelets ovoid, reddish-brown; achene plano-convex . 10
9. Style 3-cleft; spikelets usually ellipsoid, pale brown; achene trigonous . . . . 9. *S. heterochaetus*
10. Scales viscid, red-dotted, mucronulate . . . . 7. *S. acutus*
10. Scales glabrous, not red-dotted, awned . . . . 8. *S. validus*

1. Involucral leaves 2 or more 12
1. A true involucre absent, but represented by an elongated outer scale of the spikelet 23
2. Bristles firm or none, straight, shorter than to twice as long as the achene 13
2. Bristles slender, curled, much longer than the achene 15
3. Spikelets 1–2 mm broad .... 11. *S. micranthus*
3. Spikelets 5–10 mm broad 14
4. Achene sharply trigonous, 4–5 mm long; bristles stout, persistent .... 12. *S. fluviatilis*
4. Achene obtusely, if at all, trigonous, 2.5–4.0 mm long; bristles weak, deciduous .... 13. *S. paludosus*
5. Stolons present (sometimes apparently absent in *S. polyphyllus*); bristles concealed by the scales at maturity, retrorsely barbed 16
5. Stolens absent; bristles usually exceeding the scales at maturity, smooth or sparingly hairy 20
6. Culms with 10–20 leaves; scales reddish-brown ...... 14. *S. polyphyllus*
6. Culms with 2–10 leaves; scales brown or black 17
7. All sheaths greenish; bristles (if present) glabrous near base 18
7. Lowest sheaths reddish; bristles hairy throughout . 18. *S. microcarpus*
8. Bristles 0–3, shorter than the achenes .... 15. *S. georgianus*
8. Bristles usually 5 or 6, shorter than to slightly longer than the achenes 19
9. Lower leaf blades and sheaths usually nodulose-septate; scales mostly brownish; longer bristles frequently exceeding the achenes .... 16. *S. atrovirens*
9. Lower leaf blades and sheaths nearly smooth; scales mostly blackish; longer bristles usually shorter than or about equalling the achenes .... 17. *S. hattorianus*
10. Bristles at maturity equalling or shorter than the scales; scales usually with prominent green midribs .... 19. *S. pendulus*
10. Bristles at maturity much longer than the scales; scales usually with inconspicuous midribs 21
11. Spikelets and involucels reddish-brown .... 20. *S. cyperinus*
11. Spikelets and involucels whitish-brown or blackish 22
12. Spikelets and involucels whitish-brown .... 21. *S. pedicellatus*
12. Spikelets and involucels blackish .... 22. *S. atrocinctus*
13. Culms smooth, round; bristles longer than the achenes .... 23. *S. cespitosus*
13. Culms rough, triangular; bristles about as long as the achenes .... 24. *S. verecundus*

1. S. smithii Gray. July–Oct. Wet shores, very rare; Cass Co.
2. S. purshianus Fern. Aug.–Oct. Shores of lakes; rare in the s. two-thirds of the state. (Including f. *williamsii* [Fern.] Fern.) *S. smithii* Gray var. *williamsii* (Fern.) Beetle—G.
3. S. hallii Gray. Aug.–Oct. Shores of ponds, rare; Cass, Mason, and Menard cos.
4. S. torreyi Olney. July–Sept. Shores of ponds, very rare; St. Clair Co.
5. S. subterminalis Torr. July–Oct. Shallow water, very rare; Cook and Lake cos.
6. S. americanus Pers. June–Sept. Along shores, in marshes; scattered throughout the state.

7. S. acutus Muhl. Aug.–Sept. Spikelets on divergent rays to 5 c long. Shallow water; occasional throughout the state.

f. congestus (Farw.) Fern. Aug.–Sept. Spikelets all sessile, formin a dense glomerule. Shallow water, rare; Lake Co.

8. S. validus Vahl. June–Sept. Swamps, marshes, shallow wate common in the n. half of the state, less common elsewhere. S. valid Vahl var. creber Fern.—F. (Including var. condensatus [Peck] Beetle

9. S. heterochaetus Chase. July–Sept. Shores and swamps; sca tered throughout the state, but not common.

10. S. koilolepis (Steud.) Gl. May–June. Pastures, fields, rare; co fined to the s. tip of the state.

11. S. micranthus Vahl. Aug.–Oct. Scales narrowly obovate, sprea ing. Sandy banks; scattered throughout the state. Hemicarpha m crantha (Vahl) Pax—F, G, J. (Including var. minor [Schrad.] Frie land.)

var. drummondii (Nees) Mohlenbr. Aug.–Oct. Scales broadly ob vate to suborbicular, appressed. Wet, sandy soil; Cook and Lake co Hemicarpha drummondii Nees—F, G, J.

12. S. fluviatilis (Torr.) Gray. July–Sept. Margins of streams ar lakes; occasional throughout the state.

13. S. paludosus A. Nels. Bayonet-grass. July–Sept. Shores and ma shes, rare; Cook Co. S. maritimus var. paludosus (A. Nels.) Gl.—G.

14. S. polyphyllus Vahl. July–Sept. Low woodlands; rare and sca tered in Ill.

15. S. georgianus Harper. June–Sept. Moist soil; occasional throug out the state. S. atrovirens Willd. var. georgianus (Harper) Fern.— G, J.

16. S. atrovirens Willd. June–Aug. Wet soil, common; throughout t state.

17. S. hattorianus Mak. June–Aug. Ditches, very rare; Kankakee C

18. S. microcarpus Presl. Aug.–Sept. Marshes and swamps, ra Lake co. S. rubrotinctus Fern.—F, G, J.

19. S. pendulus Muhl. June–Aug. Low woods and along stream common; probably in every co. S. lineatus Michx.—F, G, J.

20. S. cyperinus (L.) Kunth. Aug.–Oct. Swamps and marshes; co mon throughout the state. (Including var. pelius Fern., var. rubricos [Fern.] Gilly, S. eriophorum Michx.)

21. S. pedicellatus Fern. July–Aug. Wet soil, rare; Cook and Po cos. S. cyperinus (L.) Kunth—J.

22. S. atrocinctus Fern. June–July. Swamps and boggy areas, ve rare; Lake Co.

23. S. cespitosus L. var. callosus Bigel. June–Aug. Boggy situatior rare; Lake and McHenry cos. S. cespitosus L.—J.

24. S. verecundus Fern. April. Cherty slopes, very rare; Union Co.

9. Eriophorum L. Cotton Sedge

Stamens 3; scales one-nerved 4
Stamen 1; scales several-nerved .... 5. *E. virginicum*
Peduncles glabrous; anthers 2.5–5.0 mm long .... 3. *E. angustifolium*
Peduncles puberulent; anthers 1.0–1.3 mm long ... 4. *E. viridi-carinatum*

1. E. gracile Koch. May–July. Bogs, rare; Peoria Co. *E. angustifolium* Honck.—J.

2. E. tenellum Nutt. June–Sept. Bogs, rare; Cass Co. *E. angustifolium* Honck.—J.

3. E. angustifolium Honck. June–Aug. Calcareous fens; scattered in the n. one-fourth of Ill. (Including var. *majus* Schultz.)

4. E. viridi-carinatum (Engelm.) Fern. May–Aug. Swamps and bogs, rare; Lake and Rock Island cos.

5. E. virginicum L. Aug.–Oct. Bogs, rare; Lake Co. (Including f. *album* [Gray] Wieg.)

10. *Rhynchospora* Vahl   Beaked Rush

Leaves 6–20 mm broad; achene 4.7–5.5 mm long, 2.5–3.0 mm broad; beak 12–20 mm long 2
Leaves 0.3–6.0 mm broad (7.0 rarely in *R. glomerata*); achene 1.2–2.6 mm long, 0.8–1.6 mm broad; beak 0.3–1.6 mm broad; beak 0.3–1.8 mm long 3
At least one of the usually 6 bristles longer than the achene ....
.... 1. *R. macrostachya*
None of the usually 3–5 bristles longer than the achene 2. *R. corniculata*
Bristles antrorsely hairy, rarely smooth or entirely lacking; tubercle 0.3–0.6 mm high, broadly deltoid .... 3. *R. globularis*
Bristles retrorsely hairy, rarely smooth; tubercle 0.6–1.8 mm long, conical 4
Spikelets lanceoloid; bristles 3.6–4.3 mm long; leaves capillary, at most 0.4 mm broad .... 4. *R. capillacea*
Spikelets ovoid to lance-ovoid; bristles 2.0–3.5 mm long (occasionally to 4.0 mm long in *R. glomerata*); leaves 0.5–7.0 mm broad 5
Achene with a conspicuous margin, smooth 6
Achene emarginate, or nearly so, slightly rugulose .... 7. *R. alba*
Achene 1.2–1.4 mm wide, with broad shoulders; bristles 3–4 mm long ..
.... 5. *R. glomerata*
Achene 0.8–1.2 mm wide, pyriform; bristles 2.0–2.8 mm long ....
.... 6. *R. capitellata*

1. R. macrostachya Torr. July–Oct. Swampy woods, very rare; Pulaski Co.

2. R. corniculata (Lam.) Gray. June–Sept. Swamps and low roadside ditches; restricted to the s. tip of Ill. (Including var. *interior* Fern.)

3. R. globularis (Chapm.) Small. June–July. Moist, sandy soil, very rare; Cook and Kankakee cos. (Including var. *recognita* Gale.)

4. R. capillacea Torr. July–Sept. Bogs and marshes; rare in the n. tip of the state; also St. Clair Co. (Including f. *leviseta* [E. J. Hill] Fern.)

5. R. glomerata (L.) Vahl. July–Oct. Moist, often sandy soil, rare; Kankakee and Pope cos.

6. R. capitellata (Michx.) Vahl. July–Oct. Moist soil; not common in the n. one-third of the state; absent elsewhere except for Pope Co.

7. R. alba (L.) Vahl. July–Sept. Sphagnum bogs, rare; Cook, Lak
and Peoria cos.

11. Cladium P. Browne Twig Rush

1. C. mariscoides (Muhl.) Torr. Aug.–Oct. Swamps and bogs; r
common and restricted to the extreme ne. corner of Ill.

12. Scleria Berg. Nut Rush

1. Achene smooth; tubercles of hypogynium none; coarse perennial . . . .
. . . . 1. *S. triglomera*
1. Achene papillose, pitted, or wrinkled; tubercles of hypogynium 0, 6, or
slender perennials or annuals . . . .
2. Achene papillose; tubercles of hypogynium 6 or 9; perennial . . . .
. . . . 2. *S. pauciflo*
2. Achene pitted or wrinkled; tubercles none; annual . . . .
3. Achene pitted; hypogynium 3-lobed; inflorescence closely panicula
achene 1.8–2.2 mm long . . . . 3. *S. reticula*
3. Achene wrinkled; hypogynium absent; inflorescence verticillate; ache
1.5 mm long . . . . 4. *S. verticilla*

1. S. triglomerata Michx. June–Sept. Moist or dry woods, fields; c
casional in the n. half of the state, rare elsewhere.
2. S. pauciflora Muhl. June–Sept. Culms and leaves glabrous
sparsely hirtellous. Dry soil, in woods or on bluffs; occasional in the
tip of the state, rare elsewhere.
var. caroliniana (Willd.) Wood. June–Sept. Culms and leav
densely pilose. Dry soil, rare; Lee Co.
3. S. reticularis Michx. Aug.–Oct. Sandy shore of pond, very ra
Cass Co.
4. S. verticillata Muhl. July–Sept. Bogs and marshes, rare; scatter
in the n. half of the state.

13. Carex L. Sedge

1. Spikes not all alike; stigmas 2 or 3; achenes lenticular, suborbicular,
trigonous . . . .
1. Spikes essentially alike; stigmas 2; achenes lenticular or plano-conv
. . . .
2. Stigmas 2; achenes lenticular or suborbicular . . . . Group
2. Stigmas 3; achenes trigonous . . . .
3. Perigynia pubescent (sometimes scarcely so) . . . . Group
3. Perigynia glabrous . . . .
4. Perigynia few; scales of staminate spike united at the base . . Group
4. Perigynia many; scales of staminate spike free at the base . . . .
5. Foliage pubescent (leaves only rough or scabrous keyed elsewhere)
. . . . Group
5. Foliage glabrous . . . .
6. Beak of perigynia not bidentate, at most emarginate, or absent . . . . .
. . . . Group
6. Beak of perigynia bidentate . . . . Group
7. Staminate flowers (or remnants) at the tips of some spikes (andro
nous) . . . . Group

. Staminate flowers (or remnants) at the base of some spikes (gynecandrous) .... Group 8

Group 1

ɔikes not all alike; styles 2; achenes lenticular or subglobose.

. Plants slender; culms rarely over 3.5 dm tall, but not more than 1 mm thick; lowest leaf-like bract long-sheathing; perigynia beakless, white-powdery or orange (drying brown), subglobose to plump-obovoid. .. 2
. Plants stout; culms more than 3.0 dm tall, more than 1 mm thick; lowest leaf-like bract sheathless or short-sheathing; perigynia short-beaked, brown, tawny, or green, compressed-lenticular 3
. Terminal spike gynecandrous; pistillate scales appressed; perigynia white-powdery .... 69. *C. garberi*
'. Terminal spike staminate; pistillate scales spreading; perigynia orange (drying brown) .... 70. *C. aurea*
. Staminate spikes long-pedunculate (especially the lower ones), arching-pendulous; pistillate scales with long, rough awns, the scales 2–4 times longer than the perigynia with the margins hyaline; achene wrinkled or constricted on one side .... 71. *C. crinita*
. Pistillate spikes sessile to short-pedunculate, erect to spreading; pistillate scales obtuse to long-acuminate, shorter than to slightly longer than the perigynia, with the margins brown or reddish; achene with a regular shape 4
. Lower pistillate spikes spreading or nodding; perigynia tapering to a prominent beak, the beak bending or twisting with drying .. 76. *C. torta*
. Pistillate spikes erect; perigynia beakless or straight, short-beaked .. 5
. Pistillate scales long-acuminate, much exceeding the perigynia; perigynia compressed-obovate, widest above the middle 75. *C. haydenii*
. Pistillate scales obtuse to acuminate, shorter than to slightly exceeding the perigynia; perigynia compressed-elliptic, widest at or below the middle 6
. Lower leaf sheaths easily tearing ventrally and with conspicuous cross-connecting filaments, the upper sheath subscabrous ventrally .... 73. *C. stricta*
. Lower leaf sheaths usually intact and without filaments, the upper ones smooth ventrally 7
'. Culms phyllopodic, with old leaves persisting; pistillate scales acuminate; perigynia 2.20–2.75 mm long, 1.50–1.75 mm wide .... 74. *C. emoryi*
'. Culms aphyllopodic, with the old leaves absent; pistillate scales obtuse to acute; perigynia 2.7–3.2 mm long, 1.50–2.75 mm wide .... 72. *C. substricta*

Group 2

ɔikes not all alike; stigmas 3; achenes trigonous; perigynia pubescent ɔmetimes scarcely so).

. Leaf blades or sheaths pubescent (only scabrous-puberulent in *C. debilis*) 2
. Leaf blades and sheaths glabrous 5
'. Spikes of two kinds, the topmost usually composed of staminate flowers

but occasionally with 1–3 perigynia; perigynia with a conspicuous bea
....

2. Spikes of one kind, the staminate scales conspicuous at the base and i
conspicuous or absent at the apex; perigynia without a conspicuou
beak ....

3. Leaf blades and sheaths scabrous-puberulent; staminate spike elongat
long-pedunculate, occasionally with 1–3 perigynia; perigynia stalked, th
beak long-tapering, scarcely pubesent or glabrous ...... 92. *C. debil*

3. Leaf blades and sheaths conspicuously pubescent; staminate spik
short, sessile, without any perigynia; perigynia sessile, the beak abrupt
contracted, conspicuously pubescent .... 68. *C. hirtifol*

4. Leaves usually shorter than the culms; pistillate spike linear, gradua
tapering to the base, 15–40 mm long .... 86. *C. virescer*

4. Leaves usually exceeding the culms; pistillate spikes cylindric to oblon
abrupt or rounded at the base, 5–20 mm long .... 87. *C. swar*

5. Perigynia beak less than 0.4 mm long or absent ....

5. Perigynia beak greater than 0.4 mm long ....

6. Pistillate scales obtuse, awnless, concealing the perigynia; bracts of th
inflorescence scale-like .... 66. *C. richardsor*

6. Pistillate scales acute to acuminate or awned, the perigynia easily see
bracts of the inflorescence leaf-like ....

7. Perigynia 2.5–3.5 mm long, ellipsoid, short-stipitate; pistillate scales ½–
the length of the perigynia, acute to acuminate .... 109. *C. digital*

7. Perigynia 3.5–4.5 mm long, obovoid, long-stipitate; pistillate scal
equalling or exceeding the perigynia, awned ...... 65. *C. peduncula*

8. Perigynia in a head, radiating in all directions, 12–18 mm long; leav
5–14 mm wide .... 132. *C. gra*

8. Perigynia in a spike, spreading to ascending; perigynia less than 10 m
long; leaves seldom more than 6 mm wide ....

9. Most or all culms shorter than or hidden by the leaves or leaf base
remains of old fruiting culms often persistent among the leaves ....

9. Most culms as long as or longer than the leaves, not hidden among t
leaf bases; older fruiting culms not persistent ....

10. Peduncle one per culm, all similar, elongated to 20 cm long, producir
both staminate and pistillate flowers; body of perigynia ellipso
oblongoid, ½–⅔ longer than wide .... 61. *C. nigromargina*

10. Peduncles 2–4 per culm, not all alike, one of them elongated and wi
staminate and pistillate flowers, the other 1–3 mostly short (up to 5 c
long) and with only pistillate flowers hidden among the tufted leaf base
body of perigynia broadly ellipsoid or ellipsoid-ovoid to subglobose,
to as long as wide ....

11. Leaf blades thin, soft, erect to ascending, 1.5–3.0 mm wide; perigyr
pubescent in the upper half, 2.25–4.00 mm long, membranaceous, w
the beak less than 2 mm long ....

11. Leaf blades thick, rigid, spreading at maturity, 2.0–4.5 mm wic
perigynia glabrous to sparsely pubescent in the upper half, 3.50–4.
mm long, subcoriaceous, with the beak at least 2 mm long 64. *C. ton*

12. Beak of perigynium 0.9–1.7 mm long, about ¾ as long as the perigyniu
body; pistillate scales long-acuminate .... 62. *C. umbella*

12. Beak of perigynium 0.5–1.0 mm long, about ½ as long as the perigyniu
body; pistillate scales acute .... 63. *C. abd*

13. Culms usually 4–12 dm tall, not densely tufted; pistillate spikes 10–
mm long, 3–6 times longer than thick; staminate spike usually lor
pedunculate, 2–5 cm long; plants of wet places ....

3. Culms usually less than 3 dm tall (sometimes taller in *C. pensylvanica*), densely tufted; pistillate spikes 5–10 mm long, less than 2 times longer than wide; staminate spike sessile to short-pedunculate, usually less than 2 cm long; plants of dry places 17

4. Perigynia 6–10 mm long, lanceoloid, evenly tapering to the beak, prominently nerved, the teeth 1.5–2.0 mm long .... 127. *C. trichocarpa*

4. Perigynia 2.5–5.5 mm long, ovoid to lance-ovoid, more abruptly tapering into the beak, obscurely nerved (especially in the upper ⅓), the beak about ¼ the total length of the perigynia, the teeth not over 1 mm long 15

5. Perigynia ovoid-lanceoloid, sparsely pubescent, 5.00–5.25 mm long, the beak 1.50–1.75 mm long ....128. *C. X subimpressa*

5. Perigynia ovoid to suborbicular, densely pubescent, 2.5–4.0 mm long, the beak 1.00–1.25 mm long 16

6. Leaf blades flat, some with revolute margins, 2–5 mm wide; achene obovoid to ovoid-ellipsoid; culms sharply triangular ..... 81. *C. lanuginosa*

6. Leaf blades involute-filiform, up to 2 mm wide; achene ellipsoid; culms obtusely triangular .... 80. *C. lasiocarpa*

7. Perigynium body suborbicular to ovoid, about as wide as long, slightly 3-angled 18

7. Perigynium body ellipsoid, longer than wide, noticeably 3-angled ... 19

8. Plants with long scaly rhizomes, the base coarsely fibrillose; ligule longer than wide; leaf blades 1–3 mm wide ... 56. *C. pensylvanica*

8. Plants without rhizomes, slightly or not at all fibrillose; ligule wider than long; leaf blades 3–7 mm wide .... 57. *C. communis*

9. Plants with elongated scaly rhizomes, slightly cespitose; perigynia whitish-green; staminate spike short-stalked ..... 60. *C. physorhyncha*

9. Plants with short stout rhizomes, densely cespitose; perigynia olive-green; staminate spike sessile 20

10. Culms weak, loosely spreading to reclining; all but the lowest spikes aggregated into an ovoid-ellipsoid head 1–2 cm long . 58. *C. emmonsii*

10. Culms stiffer, ascending; spikes interrupted or in a linear-cylindric head 1–5 cm long .... 59. *C. artitecta*

Group 3

pikes not all alike; stigmas 3; achenes trigonous; perigynia glabrous; cales of staminate flowers united at the base; spikes androgynous.

1. Culms equalling or longer than the leaves; perigynia beakless, the body ellipsoid; lowest pistillate scale of normal size .... 54. *C. leptalea*

1. Culms shorter than the leaves; perigynia long-beaked, the body subglobose; lowest pistillate scale prolonged and leaf-like ..... 55. *C. jamesii*

Group 4

pikes 2 or more; staminate scales free at the base; pubescence present on me of the foliage (scabrous leaves not included).

1. Staminate flowers (or remnants) and perigynia in different spikes, with the staminate spike overtopping the pistillate spike 2

1. Staminate flowers (or remnants) and perigynia in the same spike, with the staminate flowers conspicuous at the base 4

Perigynia 2.5–3.0 mm long, ellipsoid, beakless, entire   82. *C. pallescens*

2. Perigynia 4.5–12.0 mm long, ovoid to lanceoloid, beaked, deeply bide〉 tate or nearly emarginate ....

3. Pistillate spikes (as well as some androgynous ones) 5–12 cm lon〉 perigynia 8–12 mm long, the beak deeply bidentate .. 124. *C. atherod*〉

3. Pistillate spikes 1.0–2.5 cm long; perigynia 4.5–5.0 mm long, the be〉 merely emarginate .... 102. *C. hitchcockia*〉

4. Pistillate spikes over 2 cm long, narrowly cylindric; perigynia 3.5–6.0 m〉 long ....

4. Pistillate spikes less than 2 cm long, ovoid, short-cylindric; perigyr〉 2.00–3.75 mm long ....

5. Pistillate scales with an awn ¾ to as long as the main body of the sca〉 spreading; perigynia 4–6 mm long .... 91. *C. dav*〉

5. Pistillate scales long-acuminate or awn-tipped, appressed; perigyr〉 3.5–4.0 mm long .... 90. *C. oxyle*〉

6. Leaf blades and sheaths soft-pubescent ....

6. Leaf blades and sheaths sparsely hairy to nearly glabrous .... 84. *C. carolinia*〉

7. Perigynia 2.50–3.75 mm long, shorter than the prominently awned scal〉 .... 85. *C. bus*〉

7. Perigynia 2.0–2.5 mm long, longer than the obtuse to short-point〉 scales .... 83. *C. hirsute*〉

Group 5

Spikes 2 or more; plants glabrous; beak of perigynia not bidentate, at m〉 emarginate, or absent.

1. Culms without well-developed leaf blades .... 106. *C. plantagin*〉

1. Culms with well-developed leaf blades ....

2. Basal leaves, especially those of the sterile shoots, 10–30 mm wide ..

2. Basal leaves less then 10 mm wide ....

3. Blade of uppermost leaf never more than 3 times as long as the shea〉 ....

3. Blade of uppermost leaf many times longer than the sheath ....

4. Perigynia 5–7 mm long; staminate scales red-tinged . 107. *C. careya*

4. Perigynia 2.0–4.5 mm long; staminate scales pale or brown .... 108. *C. platyphy*〉

5. Perigynia sharply triangular, with flattish sides, symmetrical .... 110. *C. laxicul*〉

5. Perigynia rounded-triangular (at least below), with swollen sides, asy〉 metrical ....

6. Bracteal sheaths with the edges smooth or slightly rough; perigy〉 tapering to a straight or only slightly curved beak .... 112. *C. laxifl*〉

6. Bracteal sheaths with the edges strongly serrulate; perigynia mo〉 abruptly contracted into a strongly bent beak ....

7. Staminate spike sessile or subsessile ....

7. Staminate spike long-pedunculate, overtopping the pistillate spikes .. .. 113. *C. striat*〉

8. Some or all leaves over 14 mm wide; pistillate scales obtuse; ster〉 shoots culmless .... 111. *C. albursi*〉

8. None of the leaves over 12 mm wide; pistillate scales acute to arista〉 sterile shoots developing culms .... 114. *C. blan*〉

9. Beak of the perigynium bent or with the straight beak 0.6–1.0 mm long 10
9. Perigynia beakless, or with the nearly straight beak less than 0.5 mm long 16
0. Beak of the perigynium straight or only slightly bent 11
0. Beak of the perigynium abruptly bent 13
1. Perigynia with several elevated nerves; pistillate scales acute to smooth-awned 12
1. Perigynia with many impressed nerves; pistillate scales rough-awned .... 101. *C. oligocarpa*
2. Perigynia abruptly contracted into a 0.5 mm long beak, obtusely triangular above and below .... 112. *C. laxiflora*
2. Perigynia evenly tapered into a 0.6–1.0 mm long beak, sharply triangular (at least above) .... 116. *C. styloflexa*
3. Staminate spike sessile or subsessile 14
3. Staminate spike pedunculate, overtopping the pistillate spikes 15
4. Pistillate scales acute to aristate; sterile shoots developing culms .... 114. *C. blanda*
4. Pistillate scales obtuse; sterile shoots culmless ...... 111. *C. albursima*
5. Staminate spike brown or purplish; culms reddish at the base .... 115. *C. gracilescens*
5. Staminate spike green or whitish; culms green or brown at the base ... . 113. *C. striatula*
6. Terminal spike pistillate toward the apex, staminate at the base 17
6. Terminal (subterminal in *C. eburnea*) spike entirely staminate or at most with a few perigynia 20
7. Pistillate spikes drooping on long, slender peduncles . 89. *C. gracillima*
7. Pistillate spikes erect to ascending, sessile to short-pedunculate ... 18
8. Leaves 2–4 at the base of the culm, pale green, glaucous; perigynia whitish .... 77. *C. buxbaumii*
8. Leaves numerous at the base of the culm, dark green, not glaucous; perigynia dark green to brown 19
9. Perigynia nerveless, the tips spreading outward; spikes 4–8 times longer than thick .... 79. *C. shortiana*
9. Perigynia nerved, the tips ascending; spikes 1½–3 times longer than thick .... 84. *C. caroliniana*
0. Pistillate spikes drooping on long, weak peduncles 21
0. Pistillate spikes erect, ascending, or somewhat spreading, sessile to pedunculate 23
1. Pistillate spikes 2–6, 1.5–6.0 cm long; pistillate scales up to ½ as long as the perigynia and narrower than the perigynia, hyaline with a green center, lanceolate, acute to long-awned 22
1. Pistillate spikes 1–3, 1–2 cm long; pistillate scales as long as and as wide as the perigynia, brown or reddish, ovate, obtuse to cuspidate .... 78. *C. limosa*
2. Pistillate spikes 3–6 mm wide; perigynia lanceoloid, beaked, the tips spreading to recurving .... 88. *C. prasina*
2. Pistillate spikes 2–3 mm wide; perigynia ellipsoid, beakless, the tips ascending .... 89. *C. gracillima*
. Leaves capillary, 0.5 mm wide; perigynia 2 mm long ... 67. *C. eburnea*
. Leaves flattened, more than 1 mm wide; perigynia more than 2 mm long 24
. Staminate spike conspicuous, mostly long-pedunculate and overtopping the pistillate spikes 25

24. Staminate spike inconspicuous, mostly sessile or short-pedunculate, shorter than or equalling the terminal pistillate spike (sometimes over-topping in *C. glaucodea*) 3

25. Perigynia pale to dark green or glaucous, with many coarse or fine elevated nerves, tapering to the base, triangular in cross-section, closely enveloping the achene 2

25. Perigynia dark green or brown, impressed-nerved to nearly nerveless, rounded at the base, terete in cross-section, loosely enveloping the achene 2

26. Pistillate spikes loosely flowered; perigynia in 2 or 3 rows, the lower ones separated from one another 2

26. Pistillate spikes closely flowered; perigynia in 3–6 rows, mostly overlapping 2

27. Pistillate scales obtuse; most basal sheaths with poorly developed leaf blades; perigynia with a slight beak; achenes triangular-obovoid, stipitate .... 105. *C. woo*

27. Pistillate scales acute to acuminate; most basal sheaths with well developed leaf blades; perigynia barely short-beaked; achene triangular-ellipsoid, not stipitate .... 109. *C. digital*

28. Pistillate spikes linear-cylindric, 3.0–4.5 mm thick; perigynia appressed or ascending, 2.5–3.5 mm long .... 104. *C. tetanic*

28. Pistillate spikes oblongoid, 5–8 mm thick; perigynia spreading, 3–5 mm long .... 103. *C. mead*

29. Culms cespitose from fibrous or short root-stocks; leaves deep green; perigynia with many impressed nerves .... 96. *C. conoide*

29. Culm solitary from a creeping root-stock; leaves blue-green, glaucous; perigynia with a few elevated nerves .... 95. *C. craw*

30. Pistillate scales not more than ½ the perigynia length, acuminate with short, smooth awn; leaves blue-green, often glaucous 3

30. Pistillate scales equalling or exceeding the perigynia, with a long, scabrous awn; leaves dark green, rarely glaucous 3

31. Perigynia beaked, spreading to recurving, becoming squarrose, 2–4 mm long, the body subglobose to ovoid .... 94. *C. granular*

31. Perigynia beakless or slightly emarginate, ascending, 3–6 mm long, the body ellipsoid 3

32. Leaves somewhat glaucous; perigynia 4–6 mm long; staminate spike always shorter than the terminal pistillate spike; plants of wet places . ... 99. *C. flaccosperm*

32. Leaves conspicuously glaucous; perigynia 3–5 mm long; staminate spike often overtopping the terminal pistillate spike; plants of dry places ... . 100. *C. glaucode*

33. Base of plant brown or green; leaves 4–10 mm wide; mature perigynia inflated; pistillate spikes close together, none near the base of the culm .... 98. *C. grise*

33. Base of plant reddish or purplish; leaves 2–4 mm wide; mature perigynia tightly enveloping the achene; pistillate spikes widely separated, some near the base of the culm .... 97. *C. amphibo*

Group 6

Spikes 2 or more; plants glabrous; beak of perigynium bidentate.

1. Mature perigynia (9–) 12–18 mm long; staminate spike solitary .....

1. Mature perigynia 2–10 mm long; staminate spike none to 2 or more solitary, then the perigynia less than 10 mm long) ....

2. Developed pistillate spikes globose to hemispherical, about as wide as long; style straightish to coiled near the middle; perigynia beak much shorter than the body 3

2. Developed pistillate spikes oblong to cylindric, longer than wide; style bent or contorted near the base; perigynia beak equalling or longer than the body 4

3. Mature perigynia radiating in all directions, cuneate at the base, dull ... . 132. *C. grayi*

3. Mature perigynia wide-spreading to ascending, rounded at the base, often lustrous .... 133. *C. intumescens*

4. Mature perigynia almost horizontally spreading; body of achene mostly concave-truncate at the summit, as wide as long to wider than long, the sides concave more than ¾ their width, the projecting angles flattened and winglike .... 137. *C. gigantea*

4. Mature perigynia ascending to wide-spreading (but not horizontally); body of the achene subtruncate to acute, as wide as long to much longer than wide, the sides concave for less than ¾ their width, the projecting angles plumpish 5

5. Pistillate spikes more than 2 times longer than thick, long-cylindric; mature perigynia wide-spreading .... 136. *C. lupuliformis*

5. Pistillate spikes not more than 2 times longer than thick, short-cylindric to oblong; mature perigynia ascending to spreading 6

6. Leaf blades 2.5–7.0 mm wide; pistillate spikes loosely flowered; perigynia spreading, 9–14 mm long; achenes widest at the middle; plants usually solitary .... 134. *C. louisianica*

6. Leaf blades 5–11 mm wide; pistillate spikes compactly flowered; perigynia appressed-ascending, 10–17 mm long; achenes widest at or below the middle; plants most often cespitose .... 135. *C. lupulina*

7. Flowering culms usually 1–10 cm tall, mostly shorter than the leaves; pistillate spikes hidden among the leaf bases .... 64. *C. tonsa*

7. Flowering culms usually greater than 20 cm tall, mostly as long as or longer than the leaves; pistillate spikes conspicuous near the summit, or overtopping the leaves 8

8. Beak of the perigynium 0.2–0.4 mm long; foliage strongly glaucous; achenes poorly developed; rhizomes with conspicuous long, narrow, overlapping, fibrillose scales; culm base strongly fibrillose .... 125. *C.* X *fulleri*

8. Beak of the perigynium mostly greater than 2 mm long; foliage not glaucous (somewhat blue-green in *C. hyalinolepis*); achenes well-developed; rhizomes without conspicuous fibrillose scales; culm base non-fibrillose 9

9. Terminal spikes with perigynia above and staminate flowers (or remnants) below 10

9. Terminal spike staminate (occasionally with 1–2 perigynia, or absent or aborted in *C. frankii*) 11

10. Perigynia horizontally spreading; pistillate spikes globose to short-cylindric; pistillate scales acute to short-awned; styles bent or contorted .... 120. *C. squarrosa*

10. Perigynia appressed-ascending; pistillate spikes long-cylindric; pistillate scales mostly obtuse; styles straight or slightly curved . 121. *C. typhina*

11. Staminate spike 1 (occasionally absent or aborted in *C. frankii*) 12

11. Staminate spikes 2 or more 20

12. Leaves inrolled, becoming filiform and subterete   141. *C. oligosperma*

12. Leaves flattened 13

13. Pistillate spikes linear-cylindric, 10–15 times longer than thick, often

drooping on long weak peduncles; perigynia more than 4 times longe
than thick .... 92. *C. debilis*

13. Pistillate spikes subglobose, thick-cylindric or oblong, 2–5 times longe
than thick, ascending to spreading from stout peduncles; perigynia less
than 4 times longer than thick 1ₑ

14. Body of perigynium obovoid; pistillate scales narrowly linear, much ex
ceeding the perigynia; staminate spike occasionally absent or aborted
.... 119. *C. franki*

14. Body of perigynium lanceoloid, ellipsoid, or ovoid; pistillate scales ovate
to lanceolate, much shorter than to barely exceeding the perigynia
staminate spike conspicuous 1ₑ

15. Pistillate scales with a serrulate awn 1ₑ

15. Pistillate scales entire (ciliate in some), with or without an awn, if awned
the awn not serrulate 1ₑ

16. Perigynia (at least the lowest) reflexed, the teeth 1.0–2.2 mm long and
widely divergent; plants greenish to brown at the base . 129. *C. comosa*

16. Perigynia widely spreading to ascending, the teeth less than 0.75 mm
long and appressed to slightly spreading; plants reddish at the base 1

17. Perigynia 5–7 mm long, 1.5–2.0 mm thick, 12- to 20-nerved, slightly in
flated .... 130. *C. hystricina*

17. Perigynia 6–9 mm long, 2–4 mm thick, 7- to 10-nerved, strongly inflated
.... 131. *C. lurida*

18. Perigynia 2–3 mm long, the beak less than 1 mm long . 118. *C. viridula*

18. Perigynia 3.5–6.0 mm long, the beak 2.0–2.5 mm long 1ₑ

19. Pistillate spikes subglobose, erect, sessile; perigynia 3.5–4.5 mm long
the lower ones reflexed, about 10-ribbed .... 117. *C. cryptolepis*

19. Pistillate spikes elongate, drooping from long, weak peduncles
perigynia 5–6 mm long, ascending to spreading, 2-ribbed ....
.... 93. *C. sprengelii*

20. Perigynia firm, light to dark green, dull 2ₑ

20. Perigynia papery, stramineous to brown, shiny 2ₑ

21. Teeth of the perigynia 1–3 mm long .... 126. *C. laeviconica*

21. Teeth of the perigynia about 0.5 mm long 2ₑ

22. Perigynia with many elevated nerves; lower sheaths filamentous ven
trally; stems aphyllopodic, purplish at the base .... 122. *C. lacustris*

22. Perigynia with finely impressed nerves, or with a few elevated nerves
lower sheaths not filamentous ventrally; stems phyllopodic, green o
brown at the base .... 123. *C. hyalinolepis*

23. Perigynia broadly ovoid, 5–6 mm wide; achenes notched on one angle
.... 140. *C. tuckermani*

23. Perigynia narrowly ovoid, 2.5–3.5 mm wide; achenes with a regula
shape 2ₑ

24. Perigynia horizontally spreading to reflexed, the beak 2–3 mm long ...
.... 138. *C. retrorsa*

24. Perigynia ascending to slightly spreading, the beak 1–2 mm long ... 2ₑ

25. Teeth of the perigynia 1–3 mm long; styles straight   126. *C. laeviconica*

25. Teeth of the perigynia 0.5–1.0 mm long; styles curved to abruptly bent
.... 2ₑ

26. Perigynia widely spreading, with the lower ones sometimes reflexed
culms obtusely angled and smooth below the spikes; ligule slightly if a
all longer than wide; lower sheaths not filamentous ventrally ....
.... 139. *C. rostrata*

26. Perigynia ascending; culms sharply angled and rough below the spikes
ligule much longer than wide; lower sheaths somewhat filamentous ven
trally .... 142. *C. vesicaria*

Group 7

Spikes all essentially alike; stigmas 2; achenes lenticular; staminate flowers (or remnants) found at the tips of some of the spikes (androgynous).

1. Culms mostly solitary, from slender, elongated rhizomes or decumbent older culms 2
1. Culms several, cespitose, from fibrous roots or short, stout rhizomes 5
2. Inflorescence head-like, 8–17 mm long 3
2. Inflorescence elongate, 20–65 mm long 4
3. Culms erect, simple; apex of leaf sharply triangular; perigynia obscurely striate dorsally and ventrally, thin, with the margins serrate above ......
.... 1. *C. stenophylla*
3. Culms decumbent, often branching from the nodes; leaf flattened throughout; perigynia sharply nerved dorsally and ventrally, thick-spongy, with the margins entire throughout .... 3. *C. chordorrhiza*
4. Inflorescence 3–5 mm wide, green; staminate flowers inconspicuous; perigynia obscurely nerved dorsally, nerveless ventrally ....
.... 2. *C. praegracilis*
4. Inflorescence 5–12 mm wide, reddish-brown; staminate flowers conspicuous; perigynia distinctly nerved dorsally and ventrally . 4. *C. sartwellii*
5. Inflorescence branched near the base (some obscurely so), with several small spikelets on each branch 6
5. Inflorescence unbranched; spikes sessile on the main axis 15
6. Perigynia 4–7 mm long, attenuate, with the beak longer than the body, evenly tapering and coarsely serrate 7
6. Perigynia 2–4 (–5) mm long, ovoid to lanceoloid, with the beak shorter than the body, more or less abruptly contracted and finely serrate ... 9
7. Perigynia 6–7 mm long, obscurely nerved ventrally except at the base, the base abruptly enlarged and disklike, the beak 2–3 times the length of the body .... 29. *C. crus-corvi*
7. Perigynia 4–5 mm long, strongly nerved ventrally, rounded or slightly contracted at the base, the beak 1–2 times longer than the body 8
8. Sheaths with a thickened band at the orifice, smooth ventrally ....
.... 28. *C. laevivaginata*
8. Sheaths smooth at the orifice, cross-puckered ventrally .. 27. *C. stipata*
9. Flowering culms spongy, flattened when pressed; perigynia 3–4 (–5) mm long 10
9. Flowering culms hard, remaining angular when pressed; perigynia 2.0–3.5 mm long 11
10. Sheaths cross-puckered ventrally; spikes green; perigynia distinctly nerved dorsally, with the beak half as long as the body . 26. *C. conjuncta*
10. Sheaths smooth ventrally; spikes tawny; perigynia obscurely nerved dorsally, with the beak the length of the body .... 25. *C. alopecoidea*
11. Leaf sheaths cross-puckered ventrally; bracts filiform, conspicuous; pistillate scales awned 12
11. Leaf sheaths smooth ventrally; bracts inconspicuous or absent; pistillate scales acute to obtuse 13
12. Beak of the perigynium equalling the main body; leaves often longer than the culms .... 20. *C. vulpinoidea*
12. Beak of the perigynium shorter than the main body; leaves shorter than the culms .... 21. *C. annectens*
13. Leaves 3–8 mm wide; perigynia abruptly contracted into a 0.5 mm long beak, purple-black when mature, spongy at the base 22. *C. decomposita*

13. Leaves 1–3 mm wide; perigynia only slightly contracted or evenly tape
ing into a 1 mm long beak, tawny-brown when mature, solid at the bas
.... 1

14. Leaf sheaths striate-hyaline at the mouth; perigynia 2.0–2.5 mm lon(
convex ventrally, shiny brown .... 23. *C. diandr*

14. Leaf sheaths copper-tinged at the mouth; perigynia 2.5–3.5 mm long, fla
to concave ventrally, dull tawny-brown .... 25. *C. praire*

15. Spikelets with 1–3 perigynia and with 1–2 staminate flowers; perigyni
ellipsoid, abruptly contracted into a beak 0.25 mm long   30. *C. disperm*

15. Spikelets usually with 5 or more perigynia and with at least four stam
nate flowers; perigynia linear-lanceoloid to ovoid-lanceoloid, abruptl
contracted to a distinct beak at least 0.5 mm long 1

16. Perigynia corky-thickened or shrunken in the lower half, wide-spreadin
at maturity 1

16. Perigynia solid at the base, ascending at maturity 2

17. Beak of the perigynium smooth; pistillate scales deciduous, acumina
.... 1

17. Beak of the perigynium serrulate; pistillate scales persistent, acute . 1

18. Perigynia reflexed at maturity, broadly ovoid, gibbous and distinctl
nerved in the lower ventral half, 1.50–1.75 mm wide .... 6. *C retroflex*

18. Perigynia widely spreading at maturity, ovoid-lanceoloid, spongy an
wrinkled in the lower ventral half, 1.00–1.25 mm wide ... 7. *C. texens*

19. Perigynia gradually tapering into the beak; stigmas long, slender, fle:
uous, not twisted, light reddish; leaves 1–2 mm wide 2

19. Perigyia abruptly contracted into the beak; stigmas stout, tightly twiste(
dark red; leaves 2–3 mm wide .... 8. *C. convolut*

20. Culms arising from a long-creeping, fibrillose rhizome; perigynia linea
lanceoloid, 3.6–4.0 mm long, 0.7–0.8 mm wide; pistillate scales acum
nate to mucronate, ½ the length of the perigynia .... 10. *C. social*

20. Culms cespitose; perigynia lanceoloid, 3.0–3.5 mm long, 1.5 mm wide
pistillate scales obtuse to acute, ¾ the length of the perigynia ....
.... 9. *C. rose*

21. Leaf sheaths loose, easily friable ventrally, with cross-veins conspicuou
dorsally 2

21. Leaf sheaths tight, mostly intact ventrally, with the veins all parall(
dorsally 2

22. Inflorescence 1–3 cm long, ovoid to short-cylindric, with only the lowe
most spikes slightly separated; leaf blades 3–7 mm wide; perigynia fla
ventrally 2

22. Inflorescence 3–10 cm long, elongate, with most of the spikes remot
and the lowermost widely separated; leaf blades 5–10 mm wide
perigynia concave ventrally .... 19. *C. sparganioide*

23. Perigynia light green to buff at maturity, the beak ¼–⅓ the length of th
ovoid or suborbicular body .... 16. *C. gravid*

23. Perigynia deep green at maturity, the beak ⅓–½ the length of the ovoi
body 2

24. Scales ½ the length of the perigynia, acute to obtuse; leaf sheaths tru
cate at the mouth, the lowermost cross-puckered ... 18. *C. cephaloide*

24. Scales about the length of the perigynia, acuminate to aristate; lea
sheath concave at the mouth, none of them cross-puckered ....
.... 17. *C. aggrega*

25. Inflorescence ovoid to oblongoid; spikes in a head 2

25. Inflorescence linear, elongate; spikes interrupted 2

26. Perigyia more than 2 mm wide; scales about as long as or longer tha the perigynia; at least the lowest spike subtended by a conspicuous filiform bract . . . . 14. *C. muhlenbergii*

6. Perigynia less than 2 mm wide; scales much shorter than the perigynia; bracts subtending the spikes inconspicuous or absent 27

7. Perigynia broadest at the truncate-cordate base, flat ventrally, with the beak entire or nearly so; leaf sheaths thin at the mouth . . . . 12. *C. leavenworthii*

7. Perigynia broadest below the middle, round-tapering to the base, the margins raised ventrally, with the beak serrulate; leaf sheaths slightly thickened at the mouth . . . . 11. *C. cephalophora*

8. Perigynia 2 times longer than wide, 4.5–5.0 mm long, about 2.25 mm wide, with the beak at least 1.5 mm long; ligule much longer than wide . . . . 13. *C. spicata*

8. Perigynia less than 2 times longer than wide, 3–4 mm long, 2.25–0 mm wide, with the beak 1.00–1.25 mm long; ligule about as long as wide 29

9. Perigynia sessile, 3.0–3.5 mm long, 2.25–2.50 mm wide, wider than the scales; bracts slender based; achenes about 2 mm long . . . . 14. *C. muhlenbergii*

9. Perigynia stipitate, 3.5–4.0 mm long, 2.5–3.0 mm broad, mostly covered by the scales; bracts dilated at the base; achenes 2.2–2.5 mm long . . . . 15. *C. austrina*

Group 8

pikes essentially alike; stigmas 2; achenes lenticular or plano-convex; taminate flowers (or remnants) at the base of some spikes (gynecandrous).

1. Perigynia wing-margined (at least above), the lower part firm and without spongy thickenings 2

1. Perigynia without wings, at most thin margined, the lower part spongy-thickened 22

2. Plants mostly solitary, from well-developed rhizomes; lower 1–3 spikes usually pistillate, the middle spikes staminate, the terminal spikes gynecandrous . . . . 5. *C. foenea*

2. Plants mostly cespitose, without well-developed rhizomes; most spikes gynecandrous 3

3. Spikelets over 15 mm long, pointed at both ends; perigynia 6.5–10.0 mm long . . . . 36. *C. muskingumensis*

3. Spikelets less than 15 mm long, rounded at the apex or the base or both; perigynia mostly less than 6 mm long 4

4. Mature perigynia ovoid-lanceoloid to lanceoloid (2–) 3–5 times longer than wide 5

4. Mature perigynia ovoid to suborbicular, not more than 2½ times longer than wide, with some as wide as long 8

5. Sterile culms with spreading leaves common; mature perigynia 1.0–1.5 mm wide, the wing narrowed at or below the middle of the body and/or wingless at the base 6

5. Sterile culms uncommon, if present, the leaves ascending or clustered at the apex of the culm; mature perigynia 1.25–2.25 mm wide, the wing not distinctly narrowed at or below the middle, winged to the base . . 7

6. Spikes most often aggregated in an elongated head, ovoid-turbinoid; leaves 3–7 mm wide, stiff and often ascending . . . . . . 38. *C. tribuloides*

6. Spikes, or at least the lower 3–5, separated from each other, subglobose; leaves 1–3 mm wide, flaccid and spreading . . . . . . 39. *C. projecta*

7. Spikelets clavate at the base; perigynia 3–4 times longer than wide ..
.... 37. *C. scopari*

7. Spikelets abruptly contracted at the base; perigynia at most 2 time
longer than wide .... 42. *C. bebb*

8. Spikes globose; beaks of mature perigynia spreading to recurved, th
wing abruptly narrowed below the middle and wingless at the base
sterile culms with spreading, non-clustered leaves common ....
.... 41. *C. cristatel*

8. Spikes ovoid to lanceoloid-ovoid; beaks of mature perigynia ascendin
to appressed (somewhat spreading in *C. normalis*), the wing continuin
to the base; sterile leafy culms uncommon (except *C. normalis*) ....

9. Mature perigynia not more than 2 mm wide, mostly less than 4 mm lor
....1

9. Mature perigynia more than 2 mm wide, usually more than 4 mm lor
....1

10. Body of perigynium orbicular to broadly obovoid, widest near the sum
mit, abruptly beaked .... 46. *C. albolutescer*

10. Body of perigynium narrowly to broadly ovoid or suborbicular, wide
near the middle or at the base, tapering to the beak or abruptly beake
....1

11. Perigynia narrowly or broadly ovoid, 2–4 mm long, gradually taperin
into the beak 1

11. Body of perigynium suborbicular, 3.5–4.0 mm long, abruptly contracte
into the beak .... 45. *C. festucace*

12. Spikelets (especially the lower ones) well separated; leaves 0.5–2.5 m
wide .... 44. *C. tene*

12. Spikelets mostly overlapping; leaves 2.0–6.5 mm wide 1

13. Spikelets crowded into a head, abruptly contracted at the base, acute
the apex .... 42. *C. bebk*

13. Spikelets singly overlapping, clavate at the base, somewhat obtuse
the apex .... 43. *C. normal*

14. Perigynia widest above the middle, the body broadly obovoid to o
bicular 1

14. Perigynia widest at or below the middle, the body broadly ovoid to sul
orbicular 1

15. Perigynia 2.75–3.75 mm wide, the beaks ascending to appressed; pi
tillate scales long-acuminate to aristate; spikelets pointed at the apex
.... 53. *C. ala*

15. Perigynia 2.5–4.0 mm wide, the beaks spreading; pistillate scales obtus
to acuminate; spikelets rounded at the apex ....

16. Leaf sheaths green-striate ventrally; spikelets densely aggregate
perigynium 3–4 mm long, 2.5 mm wide .... 47. *C. cumula*

16. Leaf sheaths white-hyaline ventrally; spikelets slightly overlappin
perigynia 4–5 mm long, 3–4 mm wide .... 49 *C. brevi*

17. Leaf sheaths green-striate ventrally to the hyaline band at the mout
perigynia widest at the middle ....

17. Leaf sheaths narrowly or broadly white-hyaline ventrally to the mout
body of perigynium usually widest below the middle (except sometim
widest at the middle in *C. reniformis*) ....

18. Spikelets pointed; pistillate scales obtuse to cuspidate; perigynia nerv
less or obscurely 1- to 2-nerved ventrally, the body tapering from th
middle to the base .... 48. *C. suberec*

18. Spikelets obtuse; pistillate scales long-acuminate to aristate; perigyn
nerved ventrally, the body curved from the beak to the base   40. *C. rick*

19. Perigynia 5.25–6.00 mm long, 3.25–4.25 mm wide 20
19. Perigynia 3–5 mm long, 2.00–3.25 mm wide 21
20. Perigynia nerved dorsally and ventrally, hyaline and smooth, the body orbicular .... 52. *C. bicknellii*
20. Perigynia nerved ventrally, brown with wrinkles, the body broader than long .... 51. *C. reniformis*
21. Spikelets mostly aggregated, the tip broadly obtuse, the base abruptly contracted; perigynia finely nerved dorsally and ventrally, gradually tapering into the beak .... 50. *C. molesta*
21. Spikelets mostly separated, the tip usually pointed, the base gradually tapering; perigynia nerveless or nearly so ventrally, the body abruptly contracted into the beak .... 49. *C. brevior*
22. Perigynia 4–5 mm long, narrowly lanceoloid, appressed at maturity ....
.... 32. *C. bromoides*
22. Perigynia 2–4 mm long, ovoid or ellipsoid, spreading (except *C. trisperma*) at maturity 23
23. Spikes 2- to 4-flowered; perigynia ellipsoid, the beak 0.5 mm long, entire and emarginate, appressed at maturity .... 31. *C. trisperma*
23. Spikes (2-) 4- to 15-flowered; perigynia ovoid, the beak 0.5–1.0 mm long, serrulate, distinctly bidentate, spreading at maturity 24
24. Pistillate scales about ¾ the length of the body of the perigynia; perigynia ellipsoid, nerveless ventrally or only short-nerved at the base, with the beak about ¼ the total length .... 33. *C. interior*
24. Pistillate scales as long as or longer than the length of the body of the perigynia; perigynia ovoid-deltoid, obscurely to strongly nerved ventrally, with the beak about ⅓–½ the total length 25
25. Terminal spike sometimes totally staminate; pistillate scales lanceolate, acuminate, hyaline, with a wide brown center; perigynia deeply bidentate, the teeth usually bent or twisted, the beak about ½ the length of the body, setose-serrate .... 34. *C. sterilis*
25. Terminal spike always gynecandrous; pistillate scales ovate, acute, hyaline, with a narrow green center; perigynia shallowly bidentate, the teeth straight, the beak about ⅓ the length of the body, serrate ....
.... 35. *C. incomperta*

1. C. stenophylla Wahlenb. var. enervis (C. A. Mey.) Kükenth. April–May. Gravel-bluff prairie, very rare; Winnebago Co. *C. eleocharis* Bailey—G.

2. C. praegracilis W. Boott. April–Aug. Low prairies, roadsides, dry sterile soil, rare; DeKalb, Kane, and Winnebago cos.

3. C. chordorrhiza L. f. May–June. Sphagnous swamps, very rare; Lake and McHenry cos.

4. C. sartwellii Dew. April–July. Low wet prairies and meadows, marshes, peaty swamps, open cold bogs; occasional in the n. one-half of the state.

5. C. foenea Willd. May–June. Perigynia nerved on both faces, tapering gradually to the beak. Sandy, usually prairie, soil, not common; confined to the n. one-sixth of the state; also Menard Co. *C. siccata* Dewey—G, J.

var. enervis Evans & Mohlenbr. Perigynia nerveless on the ventral face, tapering abruptly to the beak. "Sandy plains," according to the collector; exact locality unknown.

6. C. retroflexa Muhl. April–June. Dry woods; occasional in the s. two-fifths of the state, rare or absent elsewhere.

7. C. texensis (Torr.) Bailey. April–June. Dry woods, old lawns; oc
casional in the s. one-half of Ill. *C. retroflexa* Muhl. var. *texensis*
(Torr.) Fern.—G.

8. C. convoluta Mack. April–June. Dry or moist woods, not com
mon; scattered throughout the state. *C. rosea* Schk.—G.

9. C. rosea Schk. April–June. Dry or moist woods; common
throughout the state.

10. C. socialis Mohlenbr. & Schwegm. April–June. Low woods, rare
Johnson, Massac, Pulaski, and Union cos.

11. C. cephalophora Muhl. April–July. Dry woods, lawns, and fields
common throughout the state.

12. C. leavenworthii Dewey. April–June. Dry open woods; not com
mon but scattered in Ill.

13. C. spicata Huds. June. Native of Europe; adventive along a
railroad; Champaign Co.

14. C. muhlenbergii Schk. May–July. Perigynia nerved on both faces
Dry woods and fields; occasional throughout the state.

var. enervis Boott. May–July. Perigynia nerveless on the upper face
Dry woods and fields; occasional throughout the state.

15. C. austrina (Small) Mack. May–July. Prairies, rare; Jackson and
Perry cos. *C. muhlenbergii* Schk.—J; *C. muhlenbergii* Schk. var. *aus
tralis* Olney—G.

16. C. gravida Bailey. May–July. Perigynia half as wide as long, ob
scurely nerved on the lower convex face. Prairies and fields; commo
in the n. three-fourths of the state, rare elsewhere.

var. lunelliana (Mack.) F. J. Herm. May–July. Perigynia $2/5$–$3/4$ as wid
as long, conspicuously nerved on the lower convex face. Prairies an
fields; scattered in the n. half of the state. *C. gravida* Bailey—J.

17. C. aggregata Mack. May–July. Woods, rare; Adams and Madiso
cos. *C. sparganioides* Muhl. var. *aggregata* (Mack.) Gl.—G.

18. C. cephaloidea Dewey. May–July. Woods; occasional in the n
half of the state, rare in the s. half. *C. sparganioides* Muhl. var. *cepha
loidea* (Dewey) Carey—G.

19. C. sparganioides Muhl. May–July. Dry or moist woods; oc
casional in the n. half of the state, rare in the s. half.

20. C. vulpinoidea Michx. May–Aug. Wet ground, common; probabl
in every co. (Including *C. setacea* Dewey.)

21. C. annectens Bickn. May–July. Perigynia yellow-brown, promir
ently nerved on the convex face. Fields, not common; widely sca
tered in Ill.

var. xanthocarpa (Bickn.) Wieg. May–July. Perigynia dark browr
obscurely nerved on the convex face. Fields, occasional; scattere
throughout the state. *C. brachyglossa* Mack.—J.

22. C. decomposita Muhl. June–Aug. Swamps, very rare; Pulaski an
Union cos.

23. C. diandra Schrank. May–Aug. Wet meadows, rare; confined t
the n. half of the state.

24. C. prairea Dewey. May–Aug. Wet meadows, rare; DeKalb, Star
and Winnebago cos.

25. C. alopecoidea Tuckerm. June–July. Wet meadows; occasional i
the n. half of the state; also Pope Co.

26. C. conjuncta Boott. May–June. Moist woods; occasion
throughout the state.

27. C. stipata Muhl. May–Aug. Leaves up to 8 mm. broad. Swamp
and moist woods; occasional throughout the state.

var. maxima Chapm. May–Aug. Leaves 8 mm broad or broader. Swamps and moist woods; occasional in the s. one-sixth of the state.

28. C. laevivaginata (Kükenth.) Mack. May–Aug. Swamps and wet woods, rare; scattered in Ill.

29. C. crus-corvi Shuttlew. June–Aug. Swamps and wet woods; occasional in the s. half of the state, becoming less common in the n. half.

30. C. disperma Dewey. May–Aug. Wet meadows; occasional in the n. one-third of the state.

31. C. trisperma Dewey. June–July. Bog, very rare; Lake Co.

32. C. bromoides Schk. May–July. Swamps and moist woods, not common; confined to the ne. corner of Ill.; also Jackson and Pope cos.

33. C. interior Bailey. May–Aug. Swamps and wet woods; occasional in the n. half of the state; also Christian and Pope cos.

34. C. sterilis Willd. May–Aug. Swampy meadows; occasional in the n. half of the state. C. muricata L. var. sterilis (Carey) Gl.—G.

35. C. incomperta Bickn. May–Aug. Swampy woods, rare; Pope Co.

36. C. muskingumensis Schwein. July–Aug. Swamps and wet woods; occasional throughout the state.

37. C. scoparia Schk. May–Aug. Marshes and wet prairies; occasional throughout the state.

38. C. tribuloides Wahlenb. June–Sept. Wet woods, marshes, wet prairies; occasional to common throughout the state.

39. C. projecta Mack. June–Aug. Wet woods; not common, but scattered in Ill.

40. C. richii (Fern.) Mack. June–Aug. Along a railroad; Menard Co. C. straminea Willd.—F, G.

41. C. cristatella Britt. June–Aug. Wet ground; occasional throughout the state. (Including f. catelliformis [Farw.] Fern.)

42. C. bebbii Olney. June–Aug. Wet ground; occasional in the n. half of the state, absent elsewhere.

43. C. normalis Mack. May–Aug. Moist or dry woods, prairies; occasional to common throughout the state.

44. C. tenera Dewey. May–Aug. Wet ground; occasional in the n. half of the state, less common in the s. half.

45. C. festucacea Schk. May–July. Wet woods and wet prairies; occasional throughout the state.

46. C. albolutescens Schw. May–Aug. Wet ground; occasional throughout the state.

47. C. cumulata (Bailey) Mack. June–Sept. Moist soil, very rare; Kankakee Co.

48. C. suberecta (Olney) Britt. May–July. Wet meadows and wet prairies; occasional in the n. one-fourth of the state, rare elsewhere.

49. C. brevior (Dewey) Mack. May–July. Prairies and meadows; occasional to common throughout the state.

50. C. molesta Mack. May–July. Moist soil, rare; Jackson and St. Clair cos. C. brevior (Dewey) Mack—J.

51. C. reniformis (Bailey) Small. May–June. Wet ground, very rare; Massac Co.

52. C. bicknellii Britt. May–July. Dry or moist prairies and meadows; occasional or common in the n. three-fourths of the state.

53. C. alata Torr. & Gray. May–July. Wet ground, very rare; Massac Co.

54. C. leptalea Wahlenb. June–Aug. Bogs and wet meadows; know from the extreme ne. cos. and Peoria, Tazewell, and Woodford cos.

55. C. jamesii Schwein. May–June. Moist or dry woods; occasion throughout the state.

56. C. pensylvanica Lam. April–June. Perigynia less than 3 mm lon the beak about one-fourth as long as the body. Dry woods; o casional to common in the n. half of Ill., much rarer southward.

var. distans Peck. Perigynia 3–4 mm long, the beak at least one-ha as long as the body. Lake Co.

57. C. communis Bailey. May–July. Woods, rare; Peoria Co.

58. C. emmonsii Dewey. April–June. Dry woods, rare; Union Co.

59. C. artitecta Mack. April–June. Dry woods; common in the s. ha of the state, becoming less common northward. C. nigromargina Schwein. var. muhlenbergii (Gray) Gl.—G.

60. C. physorhyncha Liebm. April–June. Dry, rocky woods, rare; Ra dolph and Union cos. C. nigromarginata Schwein. var. muhlenberg (Gray) Gl.—G.

61. C. nigromarginata Schwein. April–June. Woods, rare; Pope an Wabash cos.

62. C. umbellata Schk. March–June. Dry woods and on sandstor rocks; occasional in the n. and s. cos., rare in the cent. cos.

63. C. abdita Bickn. April–June. Dry woods and prairies, rare; Jac son and Pope cos.

64. C. tonsa (Fern.) Bickn. April–June. Woods, very rare; Pope Co.

65. C. pedunculata Muhl. April–May. Woods, rare; JoDaviess, Kan and Winnebago cos.

66. C. richardsonii R. Br. May–June. Dry, rocky woods, rare; Ha cock, Lake, Menard, Peoria, and Winnebago cos.

67. C. eburnea Boott. May–Aug. Rocky soil, not common; scattere in the n. cos. and the cos. along the Mississippi River.

68. C. hirtifolia Mack. May–June. Woods; occasional throughout th state.

69. C. garberi Fern. June–July. Sandy soil along Lake Michiga Cook and Lake cos.

70. C. aurea Nutt. June–July. Moist soil, very rare; Lake Co.

71. C. crinita Lam. May–Aug. Awns of lower pistillate scales at lea twice as long as the perigynia; perigynia 1–2 mm thick. Low wood swamps; not common but scattered throughout the state.

var. brevicrinis Fern. May–Aug. Awns of lower pistillate scales up t twice as long as the perigynia; perigynia 2–3 mm thick. Swamp woods, rare; Pope Co.

72. C. substricta (Kükenth.) Mack. May–Aug. Wet ground; occasion in the n. one-sixth of the state, rare or absent elsewhere. C. aquatili Wahl.—G; C. aquatilis Wahl. var. altior (Rydb.) Fern.—F.

73. C. stricta Lam. May–Aug. Swamps; occasional in the n. three fourths of the state. (Including var. strictior [Dewey] Carey.)

74. C. emoryi Dewey. May–June. Swamps and low woods; not con mon but scattered in the n. one-third of the state; also Fayette Co. (stricta Lam. var. elongata (Boeckl.) Gl.—G.

75. C. haydenii Dewey. May–Aug. Wet ground; occasional in the half of the state.

76. C. torta Boott. May–July. In shallow, rocky streams; confined the s. one-sixth of the state.

77. C. buxbaumii Wahlenb. May–Aug. Wet meadows and bogs; o

casional in the n. half of the state; also Clay Co. (Including f. *hetero-stachya* Anderss.)

78. C. limosa L. May–Aug. Bogs, rare; Cook, McHenry, Peoria, Taze-well, and Will cos.

79. C. shortiana Dewey. May–June. Moist woods, wet meadows, along streams; occasional to common throughout the state.

80. C. lasiocarpa Ehrh. May–Aug. Bogs, swamps, wet roadside ditches; occasional in the n. half of the state. *C. lasiocarpa* Ehrh. var. *americana* Fern.—F, G.

81. C. lanuginosa Michx. May–Aug. Wet woods, meadows, and road-side ditches; occasional throughout the state. *C. lasiocarpa* Ehrh. var. *latifolia* (Boeckl.) Gl.—G.

82. C. pallescens L. May–Aug. Wet ground, very rare; "N. Ill.," accord-ing to the collector. *C. pallescens* L. var. *neogaea* Fern.—F.

83. C. hirsutella Mack. May–Aug. Dry woods and fields; common in the s. half of the state, becoming less common northward. *C. com-planata* Torr. & Hook. var. *hirsuta* (Bailey) Gl.—G.

84. C. caroliniana Schwein. May–July. Wet ground; Alexander, Jack-son, McDonough, and Pope cos.

85. C. bushii Mack. May–July. Moist or dry woods and fields; oc-casional to common in the s. half of the state; also McDonough Co.

86. C. virescens Muhl. May–Aug. Moist or dry woods; not common but scattered in the ne. and s. cos.; also Edgar Co.

87. C. swanii (Fern.) Mack. May–July. Dry woods and fields; not com-mon but scattered throughout the state.

88. C. prasina Wahlenb. May–June. Wet ground, very rare; Adams Co.

89. C. gracillima Schwein. May–Aug. Moist ground; occasional in the n. one-third of the state.

90. C. oxylepis Torr. & Hook. April–June. Swampy woods, rare; Har-din, Johnson, and Union cos.

91. C. davisii Schwein. & Torr. May–June. Moist ground; occasional to common throughout the state.

92. C. debilis Michx. var. rudgei Bailey. May–Aug. Dry woods, very rare; Hardin Co. *C. debilis* Michx.—J.

93. C. sprengelii Dewey. May–Aug. Moist woods and fields; oc-casional in the n. half of the state.

94. C. granularis Muhl. May–July. Perigynia ovoid to subglobose, 1.5–2.5 mm thick. Wet ground; occasional to common throughout the state.

   var. haleana (Olney) Porter. May–July. Perigynia oblongoid-ovoid to ellipsoid, up to 1.5 mm thick. Wet ground; rare but scattered in Ill. *C. granularis* Muhl.—G, J.

95. C. crawei Dewey. June–July. Moist soil, rare; Boone, Cook, Grundy, and Lake cos.

96. C. conoidea Schk. May–Aug. Moist soil; occasional in the n. half of the state; also Massac Co.

97. C. amphibola Steud. May–June. Moist woods, not common; Alex-ander, Cook, DuPage, Jackson, Johnson, St. Clair, and Stark cos. (Including var. *rigida* [Bailey] Fern.)

98. C. grisea Wahlenb. May–July. Dry or moist woods; common throughout the state. *C. amphibola* Steud. var. *turgida* Fern.—F.

99. C. flaccosperma Dewey. April–June. Rich woods; occasional in the s. one-fifth of the state; also McDonough Co.

100. C. glaucodea Tuckerm. May–July. Woods; occasional in the s.

half of the state. *C. flaccosperma* Dewey var. *glaucodea* (Tuckerm.)
Kükenth.—F; *C. flaccosperma* Dewey—J.

101. C. oligocarpa Schk. May–July. Moist woods; occasional throughout the state.

102. C. hitchcockiana Dewey. May–July. Moist woods, not common;
apparently confined to the ne. one-fourth of the state.

103. C. meadii Dewey. May–June. Prairies; occasional in the n. half of
the state, becoming much less common southward.

104. C. tetanica Schk. May–July. Bogs, wet prairies, low woods; occasional in the n. half of the state, rare in the s. half.

105. C. woodii Dewey. May–June. Rich woods, rare; Cook, Kankakee
and Will cos. *C. tetanica* Schk. var. *woodii* (Dewey) Wood—G.

106. C. plantaginea Lam. April–June. Rich woods, very rare; Cook Co.

107. C. careyana Torr. May–June. Rich woods; rare and scattered in
the s. half of the state.

108. C. platyphylla Carey. April–June. Rich woods, very rare; McHenry
Co.

109. C. digitalis Willd. May–July. Dry woods, rare; Hardin, Massac, and
Pope cos.

110. C. laxiculmis Schwein. May–July. Rich woods, rare; Fulton, Kane,
Peoria, and Will cos.

111. C. albursina Sheldon. April–June. Rich woods; occasional
throughout the state. *C. laxiflora* Lam. var. *latifolia* Boott—G.

112. C. laxiflora Lam. April–June. Rich woods, very rare; Pope Co.

113. C. striatula Michx. May–June. Rich woods, very rare; Pope and
Union cos. *C. laxiflora* Lam. var. *angustifolia* Dewey—G.

114. C. blanda Dewey. April–June. Rich woods; occasional to common throughout the state. *C. laxiflora* Lam. var. *blanda* (Dewey)
Boott—G.

115. C. gracilescens Steud. April–June. Rich woods; occasional
throughout the state. *C. laxiflora* Lam. var. *gracillima* (Boott) Robins.
& Fern.—G.

116. C. styloflexa Buckl. April–June. Rich woods, very rare; Jackson
Co.

117. C. cryptolepis Mack. June–Aug. Moist ground, very rare; Cook
and Lake cos. *C. flava* L. var. *fertilis* Peck—F.

118. C. viridula Michx. June–Sept. Wet ground, rare; Cook, Lake, and
McHenry cos.

119. C. frankii Kunth. June–Sept. Rich woods, wet ground; occasional
in the s. three-fourths of the state; also Kendall and Will cos.

120. C. squarrosa L. June–Sept. Wet woods and ditches; common in
the s. half of the state, becoming less common northward.

121. C. typhina Michx. June–Sept. Wet woods and ditches; occasional
in the s. half of the state, less common in the n. half.

122. C. lacustris Willd. May–Aug. Swamps and wet ground; occasional throughout the state.

123. C. hyalinolepis Steud. May–July. Wet ground; occasional to common in the s. half of the state, absent elsewhere. *C. lacustris* Willd.
var. *laxiflora* Dewey—G.

124. C. atherodes Spreng. June–Aug. Wet ground, very rare; Winnebago Co.

125. C. X fulleri Ahles. June–Aug. Wet ground, very rare; Winnebago
Co.

126. C. laeviconica Dewey. June–Aug. Wet ground; occasional in the
n. half of the state; also Lawrence and St. Clair cos.

127. C. trichocarpa Muhl. June–Aug. Wet ground; occasional in the n. half of the state.

128. C. X subimpressa Clokey. June–July. Wet ground, rare; Macon, Perry, and St. Clair cos. Reputed to be a hybrid between *C. hyalinolepis* Steud. and *C. lanuginosa* Michx.

129. C. comosa Boott. June–Aug. Swamps and wet ground; occasional in the n. half of the state; also Gallatin and Union cos.

130. C. hystricina Muhl. June–Aug. Swamps and wet ground; occasional in the n. half of the state, rare elsewhere.

131. C. lurida Wahlenb. June–Oct. Swamps and wet ground; occasional throughout the state.

132. C. grayi Carey. June–Oct. Low woods and wet ground; occasional throughout the state. (Including var. *hispidula* Gray.)

133. C. intumescens Rudge. May–Sept. Low woods and wet ground, not common; Cook, Johnson, and Peoria cos.

134. C. louisianica Bailey. June–Sept. Achene ellipsoid-ovoid, widest near the middle, not conspicuously knobbed. Swamps and low woods, not common; confined to the s. one-fourth of the state.

135. C. lupulina Muhl. June–Oct. Wet ground; occasional throughout the state. (Including var. *pedunculata* Gray.)

136. C. lupuliformis Sartwell. June–Oct. Wet ground; not common but scattered throughout the state.

137. C. gigantea Rudge. May–Sept. Wet ground, rare; Johnson and Union cos.

138. C. retrorsa Schw. July–Oct. Wet ground; rare and scattered in Ill.

139. C. rostrata Stokes var. utriculata (Boott) Bailey. June–Oct. Wet ground, very rare; Henderson and Winnebago cos. *C. rostrata* Stokes—J.

140. C. tuckermanii Boott. June–Sept. Low woods and wet ground, rare; Cook, Hancock, Lake, and Winnebago cos.

141. C. oligosperma Michx. June–Aug. Bogs, very rare; Kane and Lake cos.

142. C. vesicaria L. June–Aug. Swamps and wet ground; occasional in the n. half of the state, much less common in the s. half. (Including var. *monile* [Tuckerm.] Fern.)

27. ARACEAE

1. True spathe absent; ovary 2- to 3-celled; fruit dry; leaves linear, the mid-vein not central . . . . 1. *Acorus*
1. Spathe present; ovary 1-celled; fruit fleshy; leaves broad, the midvein central 2
2. Flowers perfect; spadix nearly globoid; perianth parts 4; fruit 8–12 cm in diameter; leaves simple, ovate; plants with rhizomes . 2. *Symplocarpus*
2. Flowers unisexual; spadix elongate; perianth absent; fruit to 6 cm long, narrower; leaves simple, arrowhead-shaped, or compound; plants with corms or fleshy roots 3
3. Staminodia present; berries brownish or greenish; leaves simple, arrowhead-shaped 4
3. Staminodia absent; berries red; leaves compound . . . . 5. *Arisaema*
4. Plants rhizomatous; spathe tubular only at base; flowers confined to the lower half of the spadix . . . . 3. *Arum*

4. Plants with fleshy roots; spathe tubular at both ends, opening at th middle; flowers covering all or most of the spadix .... 4. *Peltandr*

1. *Acorus* L.   Sweet Flag

1. A. calamus L. Sweet Flag. May–Aug. Marshes and other lo areas; occasional throughout the state.

2. *Symplocarpus* Salisb.   Skunk Cabbage

1. S. foetidus (L.) Nutt. Skunk Cabbage. Feb.–March. Swamps an other low areas; occasional in the n. three-fifths of the state.

3. *Arum* L.   Arum

1. A. italicum Mill. Arum. May. Native of w. Europe and N. Af.; rare escaped in Ill.; Jackson Co.

4. *Peltandra* Raf.   Arrow Arum

1. P. virginica (L.) Kunth. Arrow Arum. May–June. Swamps, shallo water, wet ditches; occasional in the s. four-fifths of the state; als Cook Co.

5. *Arisaema* Mart.   Jack-in-the-Pulpit

1. Leaflets (3–) 5–13; spathe convolute below and above, open near th middle; spadix slender-tapering, exserted beyond the spathe ....
.... 1. *A. dracontiur*
1. Leaflets 3; spathe convolute below, open and arching above; spad club-shaped, obtuse, covered by the arching spathe ... 2. *A. triphyllur*

1. A. dracontium (L.) Schott. Green Dragon. May–June. Rich wooc lands, common; probably in every co.
2. A. triphyllum (L.) Schott. Jack-in-the-Pulpit. April–May. Latera leaflets strongly asymmetrical; extended portion of spathe 4–7 c wide; fruiting head 3–6 cm long. Rich woods, common; in every c A. atrorubens (Ait.) Blume—F.
   var. pusillum Peck. Small Jack-in-the-Pulpit. May. Lateral leaflet scarcely asymmetrical; extended portion of spathe to 3 cm wide; frui ing head to 2 cm long. Rich wooldlands; known only from Alexande Co. A. triphyllum (c.) Schoot.—F.

28. LEMNACEAE

1. Rootlets 1-several per plant; reproductive pouches lateral, 2 per frond
1. Rootlets none; reproductive pouch basal, 1 per plant ....
2. Rootlets (1–) 2 or more per plant .... 1. *Spirodel*
2. Rootlet 1 per plant .... 2. *Lemn*
3. Frond thin, linear .... 3. *Wolffiel*
3. Frond thick, globose or ellipsoid .... 4. *Wolffi*

1. *Spirodela Schleiden*

1. Fronds orbicular, 5- to 8-nerved; rootlets 2–10 .... 1. *S. polyrhiza*
1. Fronds obovate to reniform, obscurely 3- (5-) nerved; rootlets (1–) 2–5
.... 2. *S. oligorhiza*

    1. S. polyrhiza (L.) Schleiden. Duckweed. Standing water; throughout the state; probably in every co.
    2. S. oligorhiza (Kurtz) Hegelm. Duckweed. Standing water; Alexander, Johnson, and Union cos.

2. *Lemna L. Duckweed*

1. Fronds spatulate with long, persistent stipes, often submerged in compact masses; rootlet frequently absent .... 1. *L. trisulca*
1. Fronds orbicular to elliptic, floating; rootlet present 2
2. Fronds 3- to 5-nerved 3
2. Fronds 1-nerved or obscurely nerved 6
3. Root sheaths cylindrical 4
3. Root sheaths winged 5
4. Lower surface of frond flattened or only weakly convex, apex rounded to acute, frond usually 3-nerved .... 2. *L. minor*
4. Lower surface of frond moderately to strongly convex, apex broadly rounded or often obtuse, frond 3- to 5-nerved .... 3. *L. gibba*
5. Fronds weakly to strongly asymmetrical, thick, the nerves often inconspicuous .... 4. *L. perpusilla*
5. Fronds symmetrical or nearly so, membranous, the 3 nerves prominent
.... 5. *L. trinervis*
6. Fronds 1-nerved, the lower suface green 7
6. Fronds obscurely nerved, the lower surface frequently reddish-purple ..
.... 8. *L. obscura*
7. Fronds ovate-elliptic to elliptic, weakly to moderately asymmetrical, often floating in compact masses .... 6. *L. valdiviana*
7. Fronds obovate to orbicular, symmetrical, usually solitary   7. *L. minima*

    1. L. trisulca L. Duckweed. Standing water; local throughout the state, but rare in the s. cos.
    2. L. minor L. Duckweed. Standing water; rather common throughout the state.
    3. L. gibba L. Duckweed. Standing water; Knox, Mason, and Will cos.
    4. L. perpusilla Torr. Duckweed. Standing water; scattered throughout the state, but not common.
    5. L. trinervis (Austin) Small. Duckweed. Standing water; throughout the state, but not common.
    6. L. valdiviana Phil. Duckweed. Standing water; in the s. one-third of the state; also Lake Co.
    7. L. minima Phil. Duckweed. Standing water; Carroll, Madison, and Will cos.
    8. L. obscura (Austin) Daubs. Duckweed. Standing water; scattered throughout the state.

3. Wolffiella Hegelm.

1. W. floridana (J. D. Smith) Thompson. Duckweed. Standing water; scattered throughout the state, but not common.

4. Wolffia Horkel

1. Plants punctate, flattened above, with a single papule, or with papule absent ....
1. Plants epunctate, convex above, frequently with a median row of small papules .... 3. W. columbiana
2. Upper surface with a single acuminate papule, ovate-elliptic .... 1. W. papulifera
2. Upper surface flattened, without a papule, elliptic ...... 2. W. punctata

1. W. papulifera Thompson. Water Meal. Standing water; scattered throughout the state, but apparently not common.
2. W. punctata Griseb. Water Meal. Standing water; scattered in Ill. but not common.
3. W. columbiana Karst. Water Meal. Standing water; throughout the state

29. XYRIDACEAE

1. Xyris L.   Yellow-eyed Grass

1. Plants twisted, from a bulbous base; lateral sepals 3–4 mm long, the keel ciliate .... 1. X.torta
1. Plants not twisted, from a non-bulbous base; lateral sepals 4–6 mm long, the keel toothed but not ciliate .... .2. X. jupicai

1. X. torta Sm. Twisted Yellow-eyed Grass. July–Aug. Low, sandy areas; restricted in the n. half of the state.
2. X. jupicai L. Rich. Yellow-eyed Grass. July–Aug. Low, sandy areas, very rare. X. caroliniana Walt.—F, G.

30. COMMELINACEAE

1. Calyx radially symmetrical; corolla radially symmetrical; fertile stamens 6, the filaments villous; ovary 3-celled, with all cells fertile; inflorescence appearing umbellate, from an elongated, leaf-like bract 1. Tradescantia
1. Calyx bilaterally symmetrical, with two of the sepals connate at the base; corolla bilaterally symmetrical; fertile and sterile stamens each 3 (rarely 3 plus 2), the filaments glabrous; ovary 3-celled, with one cell aborted; inflorescence cymose, from a spathe .... 2. Commelina

1. Tradescantia L. Spiderwort

1. Stems flexuous; leaves usually 2–4 cm broad; cymes several, terminal and lateral; sepals 5–10 mm long .... 1. *T. subaspera*
1. Stems straight; leaves usually 0.5–1.5 cm broad; cyme usually solitary, terminal; sepals 8–15 mm long 2
2. Stems generally 4–10 dm tall; leaves and stems glabrous and glaucous; sepals glabrous or with tuft of eglandular hairs at the tip   2. *T. ohiensis*
2. Stems generally 0.5–4.0 dm tall; leaves and stems glabrous or pubescent, but not glaucous; sepals pubescent throughout with glandular or eglandular hairs 3
3. Pubescence of sepals eglandular .... 3. *T. virginiana*
3. Pubescence of sepals glandular-viscid .... 4. *T. bracteata*

1. T. subaspera Ker. Spiderwort. May–Aug. Woodlands; occasional in the s. two-thirds of the state, rare in the n. one-third.

2. T. ohiensis Raf. Spiderwort. April–Aug. Edges of woodlands, prairies, common; probably in every co.

3. T. virginiana L. Spiderwort. April–June. Woodlands and prairies; common in the s. two-thirds of the state, rare in the n. one-third.

4. T. bracteata Small. Prairie Spiderwort. May–Aug. Prairies, very rare; Jersey, Mason, and Morgan cos.

2. Commelina L. Day Flower

1. Plants rooting at the nodes; margins of spathes free; seeds black, rugose or reticulate 2
1. Plants not rooting at the nodes; margins of spathes united for about one-third the way up from the base; seeds reddish or brown, smooth or puberulent, not rugose or reticulate 3
2. Plants annual; leaf-sheaths glabrous at summit; anthers 6; seeds 4; petals 10–15 mm long, the lower petal white; seeds 3.5–4.0 mm long, rugose .... 1. *C. communis*
2. Plants perennial; leaf-sheaths ciliate at summit; anthers 5; seeds 5; petals 6–8 mm long, the lower petal blue; seeds 2.0–2.5 mm long, reticulate .... 2. *C. diffusa*
3. Plants with rhizomes; lowest petal blue, only slightly smaller than the other two; seeds about 6 mm long, reddish .... 3. *C. virginica*
3. Plants with thick, fleshy roots; lowest petal white, much smaller than the other two; seeds about 3 mm long, brown .... 4. *C. erecta*

1. C. communis L. Common Day Flower. June–Oct. Native of Asia; moist waste ground; occasional throughout the state.

2. C. diffusa Burm. f. Day Flower. July–Oct. Moist woodlands and moist waste ground; occasional in the s. one-third of the state, absent in the n. two-thirds, except for Jersey and Mason cos.

3. C. virginica L. Day Flower. June–Sept. Low woodlands, not common; restricted to the s. one-fourth of the state; also McDonough Co.

4. C. erecta L. Day Flower. June–Sept. Moist or dry sandy soil; occasional in the w. cos.; also Cook, Grundy, and Kankakee cos. (Including var. *angustifolia* [Michx.] Fern., f. *crispa* [Woot.] Fern., and var. *deamiana* Fern.)

31. PONTEDERIACEAE

1. Perianth funnelform, bilaterally symmetrical; stamens 6; ovary 3-celled
with 2 cells abortive; fruit a utricle; inflorescence a spike-like panicle
with more than 8 flowers; perianth segments blue, with the uppermost
marked with yellow .... 1. *Pontederia*
1. Perianth salverform, radially symmetrical; stamens 3; ovary 1-celled or
partially 3-celled; fruit a capsule; inflorescence a spike of 2–8 flowers, or
flower solitary; perianth segments blue, white, or pale yellow, but never
blue marked with yellow ....
2. Flower pale yellow; stamens uniform; ovary 1-celled; capsule indehiscent, few-seeded; leaves linear, 2–6 mm broad .... 2. *Zosterella*
2. Flowers blue or white; two stamens shorter than the third; ovary partially
3-celled; capsule dehiscent, many-seeded; leaves broader, at least 1 cm
broad .... 3. *Heteranthera*

1. Pontederia L.   Pickerelweed

1. P. cordata L. Pickerelweed. May–Sept. Wet situations, sometimes
in standing water; occasional in most of Ill., except for the e. cen.
cos. (Including var. *angustifolia* [Pursh] Torr.)

2. Zosterella Small   Water Star Grass

1. Z. dubia (Jacq.) Small. Water Star Grass. July–Aug. Shallow
water, muddy shores, not common; confined to the n. one-half of the
state. *Heteranthera dubia* (Jacq.) MacM.—F, J.

3. Heteranthera R. & P.   Mud Plantain

1. Flower solitary; perianth tube 2–4 cm long; filaments glabrous; stigma
lobed; leaves lanceolate to ovate, tapering, rounded, or subcordate at
the base .... 1. *H. limosa*
1. Flowers 2–8 in a spike; perianth tube 5–10 cm long; filaments pilose;
stigma capitate; leaves orbicular, cordate .... 2. *H. reniformis*

1. H. limosa (Sw.) Willd. Mud Plantain. June–July. Muddy shores,
shallow water; Alexander, Hardin, and St. Clair cos.
2. H. reniformis R. & P. Mud Plantain. July–Aug. Muddy shores,
shallow water; restricted to the s. one-third of the state.

32. JUNCACEAE

1. Plants hairy; capsule 3-seeded .... 1. *Luzula*
1. Plants glabrous; capsule several-seeded .... 2. *Juncus*

1. Luzula DC.   Wood Rush

1. Flowers solitary (rarely paired) at the tips of the inflorescence rays;
seeds over 2 mm long, including the strongly curved caruncle ....
.... 1. *L. acuminata*

1. Flowers crowded in glomerulate spikes; seeds 1.2–2.0 mm long, including the conical caruncle .... 2. *L. multiflora*

> 1. *L. acuminata* Raf. Wood Rush. May. Open woods, rare; LaSalle and Ogle cos. *L. saltuensis* Fern.—J.
>
> 2. *L. multiflora* (Retz.) Lejeune. Wood Rush. April–May. Rays of umbel erect to ascending. Woods and shaded cliffs; common in the s. cos., less common northward. (Including *L. bulbosa* [Wood] Rydb.) *L. campestris* (L.) DC.—G.
>
> var. echinata (Small) Mohl. Wood Rush. April–May. Rays of umbel horizontally spreading to reflexed. Woods and cliffs; occasional to common in the s. one-fifth of the state, rare elsewhere. (Including var. *mesochorea* Herman.) *L. echinata* (Small) F. J. Herm.—F, J.

2. *Juncus* L. Rush

1. Leaf sheaths without blades, apiculate or mucronate; inflorescences appearing lateral on the stems 2
1. Leaf sheaths with definite blades; inflorescences terminal 3
2. Stems densely cespitose; stamens 3, anthers 0.5–0.8 mm long; seeds 0.5 mm long; capsules beakless .... 1. *J. effusus* var. *solutus*
2. Stems single at intervals from elongate rhizomes; stamens 6, anthers 1.5 to nearly 2.0 mm long; seeds 1 mm long; capsules with beaks 0.5–1.0 mm long .... 2. *J. balticus* var. *littoralis*
3. Flowers in heads, not prophyllate (i.e., not with bractlets) 4
3. Flowers borne singly on the inflorescence branches, not in heads, prophyllate 15
4. Leaves flat, not terete and cross-septate; anthers purplish-brown 5
4. Leaves terete, cross-septate; anthers yellow 6
5. Stems solitary, approximately 3.5 cm apart on conspicuous, scaly rhizomes, 4.9–8.8 (–10.2) dm tall; leaves 2.0–6.5 mm wide; heads (13–) 20–135 .... 3. *J. biflorus*
5. Stems cespitose, 0.45–5.90 dm tall; leaves 1.0–2.5 (–2.9) mm wide; heads 3–28 (–32) .... 4. *J. marginatus*
6. Seeds fusiform, 0.8–1.9 mm long, caudate 7
6. Seeds ellipsoid to oblongoid or ovoid, 0.4–0.6 mm long, apiculate ... 8
7. Seeds 1.2–1.9 mm long, the tails comprising (one-third) ½–⅝ the total length of the seeds; heads 5- to 50-flowered; stamens 3 5. *J. canadensis*
7. Seeds 0.8–1.2 mm long, the tails comprising ¼–²/₅ the total length of the seeds; heads 2- to 5- (10-) flowered; stamens 3 or 6 6. *J. brachycephalus*
8. Stamens 6 9
8. Stamens 3 11
9. Involucral leaves shorter than the inflorescences; heads hemispherical or ellipsoid, 2–7 mm wide, 2- to 9-flowered; sepals 1.9–3.0 mm long, acuminate to acute or obtuse; capsules oblongoid or ellipsoid, acute or obtuse .... 7. *J. alpinus*
9. Involucral leaves usually exceeding the inflorescences; heads hemispherical or spherical, 8–15 mm wide, 9- to 90-flowered; sepals 2.5–5.0 mm long, subulate; capsules lanceoloid, subulate 10
10. Sepals 2.5–4.0 mm long; petals equalling to exceeding the sepals by 0.8 mm; heads 8–11 (–12) mm wide; stems to 6 dm tall ...... 8. *J. nodosus*
10. Sepals 4–5 mm long; petals 1 mm shorter than to nearly equalling the sepals; heads 10–15 mm wide; stems to 10.7 dm tall .... 9. *J. torreyi*

11. Capsules linear-lanceoloid, acute, exceeding the sepals by at least 1.
mm (usually by 2.0 mm) .... 10. *J. diffusissimu*
11. Capsules ellipsoid or oblongoid, obtuse to subulate, shorter than to ex
ceeding the sepals by 1 mm 1
12. Capsules oblongoid, subulate, exceeding the sepals by 0.75–1.00 mm
perianth segments subulate .... 11. *J. scirpoide*
12. Capsules ellipsoid, acute or obtuse, shorter than to exceeding the sepal
by 0.75 mm; perianth segments subulate or acuminate 1
13. Perianth segments acuminate; heads 2- to 35-flowered, hemispherical t
spherical; petals 0.7 mm shorter than to equalling the sepals; capsule
slightly shorter than the petals to exceeding the sepals by 0.75 mm   1
13. Perianth segments subulate; heads densely 50- to 80-flowered, spher
cal; petals distinctly (approximately 1 mm) shorter than the sepals; cap
sules usually 0.5 mm shorter than the petals ...... 14. *J. brachycarpu*
14. Sepals 2.0–2.5 mm long; heads 150–280, 3–5 mm wide, 2- to 7- (8-
flowered; leaves 1.5–4.5 mm wide .... 12. *J. nodatu*
14. Sepals 3–4 mm long; heads (1–) 2–82, 5–10 mm wide, 5- to 35-flowerec
leaves 1–3 mm wide .... 13. *J. acuminatu*
15. Annual; auricles absent; sepals 4–7 mm long; inflorescences comprisin
¼–⁴/₅ of the total height of the plants .... 15. *J. bufoniu*
15. Perennial; auricles present; sepals 2.3–6.0 mm long; inflorescence
comprising less than one-half of the total height of the plants 1
16. Sepals obtuse; anthers 1 mm long, three times longer than the filaments
.... 16. *J. gerard*
16. Sepals acuminate, subulate, or aristate; anthers shorter than to as lon
as the filaments 1
17. Leaves terete, at least distally; capsules usually exceeding the perianths
seeds 0.5–1.3 mm long 1
17. Leaves flat or involute; capsules shorter than to exceeding the perianth
by 0.1 mm; seeds 0.3–0.5 mm long 1
18. Leaves involute near the summit of the sheath, becoming closed an
terete above; inflorescences 1.5–6.5 (–8.0) cm long; petals acute or ob
tuse; capsules exceeding the sepals by (0.75–) 1.0–1.6 mm; seeds 0.5–0
mm long, apiculate at both ends .... 17. *J. green*
18. Leaves terete throughout; inflorescences (1.0–) 2.0–3.5 cm long; peta
acuminate or aristate; capsules slightly shorter than to exceeding th
sepals by 1 mm; seeds 1.0–1.3 mm long, caudate at both ends ....
.... 18. *J. vase*
19. Petals 0.2 mm shorter than to exceeding the sepals by 0.5 mm; tips (
the inflorescence branches incurved; leaves to 13 cm long ....
.... 19. *J. secundu*
19. Petals 1 mm shorter than to equalling the sepals; tips of the inflores
cence branches not incurved; leaves to 30 cm long 2
20. Auricles friable, not firm or rigid, scarious, hyaline, lanceolate, white (
brownish, prolonged 1.0–4.5 (–5.0) mm beyond point of insertior
perianth segments spreading .... 20. *J. tenu*
20. Auricles firm at apex or rigid, cartilaginous or membranceous, o
casionally prolonged to 2 mm beyond point of insertion; perianth se
ments spreading or appressed 2
21. Auricles cartilaginous, opaque, rigid, often slightly flaring, obtuse, ye
low or orange-brown, less than 1 mm long and usually 0.75 mm long (
less; perianth segments spreading; prophylls obtuse to acute ....
.... 21. *J. dudle*
21. Auricles membranous, hyaline, not cartilaginous or rigid, usually firm

apex, pale or brown, very slightly prolonged to exserted 2 mm beyond point of insertion; perianth segments appressed; prophylls acuminate to aristate .... 22. *J. interior*

1. J. effusus L. var. solutus Fern. & Wieg. Soft Rush. June–Sept. Wet ground of shores, swamps, and ditches; common in the s. half of the state, occasional in the n. half. *J. effusus* L.—J.

2. J. balticus Willd. var. littoralis Engelm. Rush. June–Sept. Wet, sandy shores, swamps, not common; confined to the n. one-fourth of the state. *J. balticus* Willd.—J.

3. J. biflorus Ell. Rush. May–Sept. Heads predominantly 2- to 4- (5-) flowered. Wet ground; occasional to common in the s. half of the state; also Cook, Kankakee, and Winnebago cos.

f. andinus Fern. & Grisc. Rush. May–Sept. Heads predominantly 6- to 10-flowered. Wet ground; scattered in the s. half of the state.

4. J. marginatus Rostk. Rush. June–Sept. Wet, often sandy ground of pond borders, ditches, fields; occasional throughout the state.

5. J. canadensis J. Gay. Rush. July–Oct. Wet ground of bogs, swamps, meadows; occasional in the northern half of the state, becoming less common in the southern half. (Including f. *conglobatus* Fern.).

6. J. brachycephalus (Engelm.) Buch. Rush. July–Sept. Marshes, wet ground along bodies of water; local in the n. half of the state; also Johnson and Montgomery cos.

7. J. alpinus Vill. var. rariflorus Hartm. Rush. July–Aug. One or several of the flowers elevated above the others on slightly elongated pedicels. Wet sandy shores and marshes, rare; confined to extreme ne. Ill. *J. richardsonianus* Schult.—J.

var. fuscescens Fern. Rush. July–Aug. Flowers sessile or equally short-pedicellate. Wet sandy shores and marshes, rare; confined to extreme ne. Ill.

8. J. nodosus L. Rush. July–Aug. Wet, often sandy ground of marshes, swamps, fields, and shores; confined to the n. one-half of the state; also St. Clair and Washington cos.

9. J. torreyi Coville. Rush. July–Aug. Wet ground; common throughout the state. (Including var. *globularis* Farw.)

10. J. diffusissimus Buckl. Rush. July–Aug. Wet ground, ditches; local in the s. one-third of the state.

11. J. scirpoides Lam. Rush. July–Aug. Wet and often sandy ground, rare; Cass, Lawrence, and Menard cos.

12. J. nodatus Coville. Rush. July–Sept. Wet grounds, ditches, edges of water bodies; occasional in the s. one-half of the state; also Hancock Co.

13. J. acuminatus Michx. Rush. July–Sept. Wet, low ground of ditches, ponds; occasional to common throughout the state.

14. J. brachycarpus Engelm. Rush. July–Aug. Low, wet ground of fields, prairies, ditches, roadsides; occasional throughout the state, except for the nw. cos.

15. J. bufonius L. Toad Rush. June–Aug. Terminal flowers and flowers on the inflorescence branches borne singly, separated. Wet roadsides, fields; local and scattered in Ill.

var. congestus Wahlb. Rush. July–Aug. Many of the terminal flowers and flowers on the inflorescence branches closely aggregated into 2- and 4-flowered fascicles. Wet roadsides, rare.

16. J. gerardii Loisel. Black Grass. July–Aug. Adventive in disturbe(
marshy area; Cook Co.

17. J. greenei Oakes & Tuckerm. Rush. July–Sept. Wet, usually sand
ground of meadows, prairies, and fields; local in a few n. cos.

18. J. vaseyi Engelm. Rush. July–Sept. Wet meadows, bogs, shore:
Cook, McHenry, and Winnebago cos.

19. J. secundus Beauv. Rush. July–Sept. Open sandstone ledges an
outcroppings; confined to the s. one-sixth of the state.

20. J. tenuis Willd. Path Rush. June–Oct. Moist or dry ground, ofte
in disturbed areas; in every co.

21. J. dudleyi Wieg. Rush. June–Oct. Moist ground; occasion:
throughout the state.

22. J. interior Wieg. Rush. June–Oct. Wet ground of fields, prairie
ditches, roadsides; common throughout the state.

33. LILIACEAE

1. Ovary superior ....
1. Ovary inferior .... ⟨
2. Leaves evergreen, rigid; stems woody .... 25. Yucc
2. Leaves deciduous, mostly not rigid; stems herbaceous ....
3. Flowers borne in umbels; leaves and bulbs usually with a strong odor (
onion (except Nothoscordum and Medeola) ....
3. Flowers borne variously, but not in urnbels ....
4. Leaves in two whorls below the flowers .... 23. Medeo
4. Leaves never whorled, usually all basal ....
5. Bulb with a strong odor of garlic or onion; ovary 3-celled, with 1-
ovules per cell .... 21. Alliu
5. Bulb lacking an odor of garlic or onion; ovary 3-celled, with 6–10 ovul
per cell .... 22. Nothoscordu
6. Leaves in one or two or several whorls below the flowers ....
6. Leaves alternate, opposite, or basal, not whorled ....
7. Leaves net-veined; plants usually less than 50 cm tall; flower solitary .
.... 24. Trilliu
7. Leaves parallel-veined; plants usually more than 50 cm tall; flowe
usually several .... 7. Liliu
8. Perianth parts 4; stamens 4; style 2-lobed; leaves 1–3, ovate, cordate .
.... 20. Maianthemu
8. Perianth parts 6; stamens 6; style 1, simple or 3-cleft, or styles 3, d
tinct; leaves 1-several, narrower, if ovate, then not cordate ....
9. Flower solitary, one per plant, borne on a leafless scape; leaves 1-
basal .... 14. Erythroniu
9. Flowers numerous, or if borne singly, then the stem leafy ....
10. Flowers axillary from the cauline leaves ....
10. Flowers terminal ....
11. Flowers yellow, at first terminal, at length appearing axillary; style deep
3-cleft; fruit a capsule .... 15. Uvula
11. Flowers white or greenish, axillary from the first; style unbranched; fr
a berry ....
12. Flowers perfect, more than 7 mm long; leaves broad; fruit blue or bla
.... 16. Polygonatu
12. Flowers unisexual, less than 5 mm long; leaves reduced to minu

scales, in the axils of which are filiform branchlets (commonly mistaken for leaves); fruits red .... 19. *Asparagus*

3. Flower borne singly 14
3. Flowers in spikes, racemes, panicles, or irregular clusters 15
4. Style deeply 3-cleft; flowers at most 4 cm long, basically yellowish; plants rhizomatous, to 50 cm tall .... 15. *Uvularia*
4. Style undivided; flowers at least 5 cm long, basically orange; plants bulbous, almost always well over 50 cm tall .... 7. *Lilium*
5. Flowers at least 5 cm long, irregularly arranged (rarely only 4 cm long in *Hosta*) 16
5. Flowers less than 5 cm long, arranged in spikes, racemes, or panicles 18
6. Plants bulbous; stems leafy .... 7. *Lilium*
6. Plants with fleshy roots, tubers, or rhizomes; stems nearly leafless . 17
7. Flowers yellow or orange; leaves linear, to 2 cm broad 8. *Hemerocallis*
7. Flowers lilac or whitish; leaves lanceolate, at least 3 cm broad 9. *Hosta*
8. Leaves cauline, elliptic to ovate, not grass-like; style 1; fruit a berry 19
8. Leaves basal, if cauline, then long and narrow (grass-like); styles 1 or 3; fruit a capsule 20
9. Perianth parts united into a tube; raceme one-sided ... 18. *Convallaria*
9. Perianth parts free, except at base; raceme or panicle not one-sided ... . 17. *Smilacina*
0. Style 1; plants bulbous and with a racemose inflorescence 21
0. Styles 3; plants rhizomatous or, if bulbous, then with a paniculate inflorescence 25
1. Perianth parts basically white 22
1. Perianth parts lavender or purple or blue (occasionally white in *Camassia*) 23
2. Perianth parts free, with a green stripe down the back, without scales .. .. 10. *Ornithogalum*
2. Perianth parts united into a long tube, without a green stripe down the back, scaly .... 11. *Aletris*
3. Perianth parts essentially free; raceme less dense 24
3. Perianth parts united nearly to tip; raceme dense .... 12. *Muscari*
4. Each perianth segment with three veins .... 13. *Camassia*
4. Each perianth segment with one vein .... 31. *Scilla*
5. Axis of inflorescence glabrous; plants bulbous or rhizomatous 26
5. Axis of inflorescence with a sticky exudate (glutinous), or pubescent; plants rhizomatous 28
6. Plants dioecious, rhizomatous; inflorescence a spike-like raceme; capsule loculicidal .... 6. *Chamaelirium*
6. Plants monoecious or the flowers perfect; plants bulbous; inflorescence a panicle; capsule septicidal 27
7. Leaves principally basal; perianth parts with a large gland just below the middle .... 4. *Zigadenus*
7. Leaves principally cauline; perianth eglandular .... 5. *Stenanthium*
8. Nearly all leaves basal; inflorescence densely racemose; all flowers perfect .... 1. *Tofieldia*
8. Leaves cauline; inflorescence paniculate; some flowers unisexual .. 29
9. Perianth parts with 2 glands near the base, clawed; flowers yellow or greenish .... 3. *Melanthium*
9. Perianth parts eglandular and without claws; flowers maroon .... 2. *Veratrum*
0. Perianth scaly on the back .... 11. *Aletris*
0. Perianth not scaly 31

LILIACEAE

31. Flowers campanulate, with the white perianth segments tipped with green .... 26. *Leucojum*
31. Flowers not campanulate, the perianth parts not green-tipped 32
32. Flowers with a corona, often borne singly or, if in an umbel of 2–6, white 33
32. Flowers without a corona, borne in spikes or, if in an umbel of 2–6, yellow 34
33. Flowers recurved; perianth tube (not corona) up to 2 (–4) cm long; stamens included within the corona; ovules several per cell of ovary; bract one .... 27. *Narcissus*
33. Flowers ascending; perianth tube (not corona) 5–10 cm long; stamens exserted from the corona; ovules 2 per cell of ovary; bracts 2-several . ... 28. *Hymenocallis*
34. Perianth parts united to form a tube, greenish-yellow, leaves large, glabrous, fleshy .... 29. *Polianthes*
34. Perianth parts free, bright yellow; leaves small, pubescent, not fleshy . ... 30. *Hypoxis*

1. *Tofieldia* Huds.   False Asphodel

1. T. glutinosa (Michx.) Pers. False Asphodel. June–July. Bogs, rare; restricted to the extreme ne. cos.

2. *Veratrum* L.   False Hellebore

1. V. woodii Robbins. False Hellebore. July–Aug. Rich, moist woodlands, rare; apparently confined to a few cent. cos.

3. *Melanthium* L.   Bunch-flower

1. M. virginicum L. Bunch-flower. June–July. Meadows, wet prairies; occasional in the w. cent. cos., rare in the sw. cos.; absent elsewhere.

4. *Zigadenus* Michx.   Camass

1. Z. glaucus Nutt. White Camass. July–Sept. Limestone cliffs, low areas near rivers, rare; JoDaviess and Kane cos.

5. *Stenanthium* Gray   Grass-leaved Lily

1. S. gramineum (Ker) Morong. Grass-leaved Lily. June–Aug. Moist woods and along streams, not common; confined to the s. half of the state. (Including var. *robustum* [S. Wats.] Fern.)

6. *Chamaelirium* Willd.   Devil's Bit

1. C. luteum (L.) Gray. Fairy Wand. May–June. Low, wooded hillsides, rare; Hardin, Massac, and Pope cos.

7. *Lilium* L.   Lily

1. All the leaves alternate, the upper with bulblets in the axils; stems scabrous .... 1. *L. lancifolium*

1. Some of the leaves borne in whorls; bulblets absent; stems glabrous   2
2. Flowers erect; perianth parts with claws; only the uppermost group of
leaves in a whorl; plants less than 1 m tall . . . .
. . . . 2. *L. philadelphicum* var. *andinum*
2. Flowers nodding; perianth parts without claws; several whorls of leaves
borne on the stem; plants generally over 1 m tall 3
3. Bulbs yellow; margins and nerves of leaves roughened; midvein of outer
perianth parts rounded on the back; anthers 8–15 mm long; filament attached 1–2 mm from end of anther . . . . 3. *L. michiganense*
3. Bulbs white; margins and nerves of leaves smooth; midvein of outer
perianth parts sharply ridged on the back; anthers 17–25 mm long; filament attached 4–8 mm from end of anther . . . . 4. *L. superbum*

1. L. lancifolium Thunb. Tiger Lily. July–Aug. Native of e. Asia; escaped from cultivation into waste ground. *L. tigrinum*—F, G, J.

2. L. philadelphicum L. var. andinum (Nutt.) Ker. Wood Lily. June–
July. Dry woodlands, occasional; restricted to the upper half of the
state; also Macoupin Co. *L. umbellatum* Pursh—J.

3. L. michiganense Farw. Turk's-cap Lily. May–July. Moist woodlands and prairies; rather common throughout the state.

4. L. superbum L. Superb Lily. July. Low, moist woodlands, rare;
Jackson and Pope cos. *L. michiganense* Farw.—G.

8. Hemerocallis L.   Day Lily

1. Flowers orange . . . . 1. *H. fulva*
1. Flowers yellow . . . . 2. *H. lilio-asphodelus*

1. H. fulva L. Orange Day Lily. June–Aug. Introduced from Eurasia;
roadsides and waste ground, common; probably in every co.

2. H. lilio-asphodelus L. Yellow Day Lily. June–Aug. Introduced
from Asia; rarely escaped into waste ground. *H. flava* L.—F, G, J.

9. Hosta Tratt   Plantain Lily

1. H. lancifolia (Thunb.) Engl. Plantain Lily. Aug. Native of e. Asia;
rarely escaped into disturbed areas. *H. japonica* (Thunb.) Voss—F, G,
J.

10. Ornithogalum L.   Star-of-Bethlehem

1. O. umbellatum L. Star-of-Bethlehem. April–June. Native of
Europe; adventive in fields, open woods, and along roads; common
throughout the state.

11. Aletris L.   Colic Root

1. A. farinosa L. Colic Root. June–Aug. Moist, sandy prairies, sandy
flats; restricted to the n. one-third of the state.

12. Muscari Mill.   Grape Hyacinth

1. Leaves flat, most or all of them over 3 mm broad 2
1. Leaves terete, not more than 3 mm broad . . . . 4. *M. atlanticum*

2. Raceme up to 6 cm long; perianth to 6 mm long ....
2. Raceme at least 9 cm long; perianth 9–11 mm long .... 3. *M. comosu*
3. All flowers fertile, purple or blue; perianth 3.5–5.0 mm long ....
.... 1. *M. botryoide*
3. Fertile flowers deep violet with white teeth; sterile flowers pale blu
perianth 5–6 mm long .... 2. *M. armeniacu*

    1. M. botryoides (L.) Mill. Grape Hyacinth. April. Native of Europ
escaped to fields, particularly grassy places; occasional in the s. ha
of the state.
    2. M. armeniacum Leicht. Heavenly Blue. April–May. Native of sv
Asia; escaped from cultivation in Piatt Co.
    3. M. comosum (L.) Mill. Grape Hyacinth. April. Native of s. Euro
and Asia; adventive in Jackson Co.
    4. M. atlanticum Boiss. & Reut. Blue Bottle. April–May. Native
Europe; occasionally escaped into waste ground. *M. racemosum* (l
Mill.—F, G, J.

13. Camassia Lindl. Wild Hyacinth

1. Scape with 0–2 (–3) deciduous bracts; inflorescence at anthesis 3–5 c
broad; capsule as broad as long; perianth parts white to pale blue
pale lilac; plants beginning to flower in early April ..... 1. *C. scilloide*
1. Scape with 3–24 persistent bracts; inflorescence at anthesis 2–3 (–3.
cm broad; capsule longer than broad; perianth parts deep lavender
pale purple; plants beginning to flower in May .... 2. *C. angus*

    1. C. scilloides (Raf.) Cory. Wild Hyacinth. April–June. Moist prairi
and woodlands; occasional throughout the state.
    2. C. angusta (Engelm. & Gray) Blankinship. Wild Hyacinth. Ma
July. Prairies and woodlands, very rare; Macon Co.

14. Erythronium L. Dog-tooth Violet

1. Flowers yellow; stigmas united .... 1. *E. americanu*
1. Flowers white; stigmas free .... 2. *E. albidu*

    1. E. americanum Ker. Yellow Dog-tooth Violet; Yellow Trout Li
April–May. Moist woodlands, frequently on shaded bluffs; occasion
in the ne., e. cent., and s. cos.; also Calhoun Co.
    2. E. albidum Nutt. White Dog-tooth Violet; White Trout Lily. Apr
May. Woods and occasionally fields; common throughout the state

15. Uvularia L. Bellwort

1. Leaves perfoliate, puberulent beneath; flowers bright yellow, t
perianth parts nearly all over 25 mm long .... 1. *U. grandiflo*
1. Leaves sessile, not perfoliate, glabrous; flowers stramineous, t
perianth parts nearly all less than 25 mm long .... 2. *U. sessilifo*

    1. U. grandiflora Sm. Yellow Bellwort. April–May. Rich woodland
common throughout the state.
    2. U. sessilifolia L. Sessile-leaved Bellwort. April–May. Rich woo
lands, not common; confined to the s. one-third of the state.

16. Polygonatum Mill. Solomon's Seal

1. Leaves pilose on the nerves beneath .... 1. *P. pubescens*
1. Leaves glabrous beneath 2
2. Leaves more or less clasping or sheathing at the base, the largest ones with over 100 nerves; perianth 17–20 mm long, the lobes 5–7 mm long .... 2. *P. commutatum*
2. Leaves sessile, the largest ones with less than 100 nerves; perianth 10–17 mm long, the lobes 3–4 mm long .... 3. *P. biflorum*

    1. P. pubescens (Willd.) Pursh. Small Solomon's Seal. May–June. Moist, shaded woods; restricted to a few extreme n. cos.
    2. P. commutatum (Schult.) A. Dietr. Solomon's Seal. April–June. Woods, common; in every co.
    3. P. biflorum (Walt.) Ell. Small Solomon's Seal. April–June. Dry woods, sandstone cliffs; restricted to the s. tip of the state.

17. Smilacina Desf. False Solomon's Seal

1. Flowers paniculate; stamens longer than the perianth parts; perianth parts to 3 mm long; pedicels less than 4 m long; berry red ....
.... 1. *S. racemosa*
1. Flowers racemose; stamens shorter than or equalling the perianth parts; perianth parts 4–6 mm long; pedicels at least 4 mm long; berry black or greenish-black .... 2. *S. stellata*

    1. S. racemosa (L.) Desf. False Solomon's Seal; False Spikenard. April–June. Rich, moist woods, common; in every co.
    2. S. stellata (L.) Desf. Small False Solomon's Seal. May–June. Moist woodlands, prairies; occasional in the n. three-fifths of the state; also Crawford and Wabash cos.

18. Convallaria L. Lily-of-the-Valley

    1. C. majalis L. Lily-of-the-Valley. May–June. Native of Europe; occasionally escaped from cultivation.

19. Asparagus L. Asparagus

    1. A. officinalis L. Asparagus. May–June. Native of Europe; commonly escaped from cultivation; probably in every co.

20. Maianthemum Weber False Lily-of-the-Valley

    1. M. canadense Desf. False Lily-of-the-Valley. May–June. Leaves glabrous beneath; transverse veins of leaf conspicuous in transmitted light. Moist woodlands, rare; Cook Co.
    var. interius Fern. False Lily-of-the-Valley. May–June. Leaves pubescent beneath; transverse veins of leaf obscure in transmitted light. Moist woodlands, occasional; restricted to the n. one-fourth of the state.

21. *Allium* L. Onion

1. Leaves absent at flowering time, some or all of them, when present, ov
2.5 cm broad .... 1. *A. tricoccu*
1. Leaves present at flowering time, up to 1.5 cm broad ....
2. Leaves flat or channelled, not hollow ....
2. Leaves terete, hollow ....
3. Leaves extending about half-way up the stem; flowers whitish, greenis
or deep purple ....
3. Leaves basal or nearly so; some of the flowers usually pinkish ......
4. Umbel producing bulblets; flowers whitish or greenish ... 2. *A. sativu*
4. Umbel not producing bulblets; flowers deep purple ....
.... 3. *A. ampeloprasum* var. *atroviolaceu*
5. Umbel producing bulblets .... 4. *A. canaden.*
5. Umbel not producing bulblets ....
6. Stems solid .... 5. *A. porru*
6. Stems hollow ....
7. Ovary and capsule crested near apex; outer bulb scales membranou
perianth parts shorter than the stamens at maturity ....
7. Ovary and capsule without crests; outer bulb scales fibrous; perian
parts usually longer than or about equal to the stamens at maturity . .
.... 8. *A. mutab*
8. Umbel nodding; leaves soft .... 6. *A. cernuu*
8. Umbel erect; leaves stiff .... 7. *A. stellatu*
9. Leaves extending nearly to middle of stem; umbel bulblet-bearing ...
.... 9. *A. vinea*
9. Leaves basal or nearly so; umbel sometimes bulblet-bearing only
number 11 ....
10. Flowers pink or purplish, the pedicels shorter than or barely equalli
the flowers; umbel not bulblet-bearing; stem not strongly inflated bel
the middle .... 10. *A. schoenoprasu*
10. Flowers white or greenish, the pedicels much longer than the flowe
umbel sometimes bulblet-bearing; stem strongly inflated below the m
dle .... 11. *A. ce*

1. A. tricoccum Ait. Wild Leek; Ramp. June–July. Petioles and le
sheaths reddish; blades mostly 2.6–6.0 cm broad, elliptic. Moist, ri
woodlands; occasional in the n. half of the state, rare elsewhere.
var. burdickii Hanes. Wild Leek; Ramp. June–July. Petioles and le
sheaths greenish or whitish; blades mostly 0.8–2.0 cm broad, la
ceolate. Low, moist woods; occasional in the n. one-fourth of t
state.
2. A. sativum L. Garlic. June–Sept. Native of Europe and w. As
occasionally escaped from cultivation.
3. A. ampeloprasum L. var. atroviolaceum (Boiss.) Regel. Wild Lee
July–Sept. Native of se. Europe and Asia; rarely escaped along road
Pope and Union cos.
4. A. canadense L. Wild Onion. May–July. Dry woodlands, prairi
waste ground, common; in every co.
5. A. porrum L. Leek. May–July. Cultivated plant rarely escap
from cultivation; Johnson Co.
6. A. cernuum Roth. Nodding Onion. July–Sept. Wooded banks, r
common; restricted to ne. Ill.
7. A. stellatum Ker. Cliff Onion. July–Sept. Hill prairies in calcarec
areas; Jackson, McHenry, Monroe, Randolph, and Union cos.

8. A. mutabile Michx. Wild Onion. June–Aug. Dry areas, not common; DuPage, Jackson, Johnson, Kane, and Will cos.

9. A. vineale L. Field Garlic. May–Aug. Native of Europe; naturalized in waste ground, common; in every co.

10. A. schoenoprasum L. Chives. June–Aug. Bulbs several. Native of Europe; rarely escaped from cultivation; Jackson Co.

var. sibiricum (L.) Hartm. June–Aug. Bulbs 1–2. Native of Europe and Asia; rarely escaped from cultivation; Kankakee Co.

11. A. cepa L. Onion. May–Aug. Native of sw. Asia; rarely escaped from cultivation; Union Co.

22. *Nothoscordum Kunth   False Garlic*

1. N. bivalve (L.) Britt. False Garlic. April–June. Dry woods, bluffs, and prairies; rather common in the s. half of the state, apparently absent elsewhere.

23. *Medeola L.   Cucumber Root*

1. M. virginiana L. Indian Cucumber Root. May–June. Rich, moist woodlands, rare; Cook and LaSalle cos.

24. *Trillium L.   Wake Robin*

1. Flower sessile, basically maroon or green (rarely yellow). 2
1. Flower pedunculate, white or occasionally pink or purple, never green or yellow 5
2. Leaves petiolate; petals abruptly tapering to distinct claws; sepals reflexed . . . . 1. *T. recurvatum*
2. Leaves sessile; petals scarcely or gradually tapering, usually without distinct claws; sepals spreading or erect, not reflexed 3
3. Stamens one-half as long as petals; connectives 1–3 mm long 4
3. Stamens one-third as long as petals; connectives less than 1 mm long . . . . 4. *T. cuneatum*
4. Stems and veins on lower leaf surface glabrous; anthers 5–6 times longer than the filaments; pollen yellow . . . . 2. *T. sessile*
4. Summit of stems and veins on lower leaf surface minutely scabrous; anthers 4 times longer than the filaments; pollen olive to brownish . . . . 3. *T. viride*
5. Ovary and fruit 3-angled . . . . 5. *T. nivale*
5. Ovary and fruit 6-angled 6
6. Ovary purple (in Ill. specimens) . . . . 6. *T. erectum*
6. Ovary white 7
7. Petals strongly recurved, 1.5–2.5 cm long; peduncle 10–40 mm long; anthers 4–7 mm long . . . . 7. *T. cernuum* var. *macranthum*
7. Petals spreading, 2–6 cm long; peduncle 25–100 mm long; anthers (6–) 7–15 mm long 8
8. Stigmas erect; anthers slightly longer than the filaments . . . . 8. *T. grandiflorum*
8. Stigmas spreading; anthers about twice as long as the filaments . . . . 9. *T. flexipes*

1. T. recurvatum Beck. Red Trillium; Wake Robin. March–May. Rich, moist woodlands, common; in every co. (Including f. *luteum* Clute and f. *shayii* Palmer & Steyerm.)

2. T. sessile L. Sessile Trillium; Sessile Wake Robin. March–Apr
Rich woodlands, occasional; scattered throughout the state, exce
in the w. cos. (Including f. *viridiflorum* Beyer.)

3. T. viride Beck. Green Trillium. April–May. Rich woodland
prairies, not common; Jackson, Macoupin, Pike, and Union cos.

4. T. cuneatum Raf. Trillium. April. Rich, moist woodlands, ve
rare; Jackson Co.

5. T. nivale Riddell. Snow Trillium. March–April. Rich woodland
occasional; restricted to the n. three-fifths of the state.

6. T. erectum L. Purple Trillium. May. Rich woods, very rare; La
and McHenry cos.

7. T. cernuum L. var. macranthum Eames & Wieg. Nodding Trilliu
May–June. Moist woodlands, very rare; Cook and McHenry cos.

8. T. grandiflorum (Michx.) Salisb. Large White Trillium. April–Ma
Rich, moist woodlands; occasional in the n. half of the state, very ra
elsewhere.

9. T. flexipes Raf. White Trillium. April–May. Rich woodlands; o
casional throughout the state. *T. gleasonii* Fern.—F.

25. Yucca. L. Adam's Needle

1. Y. filamentosa L. var. smalliana (Fern.) Ahles. Yucca; Adam
Needle. May–July. Native of the east coast; escaped from cultivatic
along roads and in cemeteries. *Y. filamentosa* L.—F, G, J.

26. Leucojum L. Snowflake

1. L. aestivum L. Summer Snowflake. April–May. Native of Europ
rarely escaped to roadside ditches; Pope Co.

27. Narcissus L. Narcissus

1. Perianth yellow; corona yellow, tubular, about as long as the perianth
.... 1. *N. pseudo-narciss*
1. Perianth white; corona white, with red margin, cupular, less than on
fourth as long as the perianth .... 2. *N. poetic*

1. N. pseudo-narcissus L. Daffodil. April–May. Native of Europe; o
casionally escaped from cultivation; scattered in Ill.

2. N. poeticus L. Poet's Narcissus. April–May. Native of Europ
rarely escaped from cultivation along roads; scattered but not co
mon in Ill.

28. Hymenocallis Salisb. Spider Lily

1. H. occidentalis (LeConte) Kunth. Spider Lily. July–Sept. Lc
woods, swamps; restricted to the s. one-sixth of the state; also W
bash Co.

29. Polianthes L. Aloe

1. P. virginica (L.) Shinners. American Agave. May–July. Sandsto
outcroppings; dry woodlands, occasional; restricted to the s. on
fourth of the state; also Jersey Co. *Agave virginica* L.—F, G, J.

30. *Hypoxis* L. *Star Grass*

1. H. hirsuta (L.) Coville. Yellow Star Grass. April–June. Dry woods, prairies, fields, sandstone outcroppings, calcareous fens; occasional throughout the state.

31. *Scilla* L. *Squill*

1. S. sibirica Andr. April–May. Native of Eurasia; rarely spreading from cultivation; McDonough Co.

34. SMILACACEAE

1. Stems woody, with few to many prickles, rarely without prickles; ovule 1 per cell of ovary 2
1. Stems herbaceous, without prickles; ovules 2 per cell of ovary 5
2. Leaves pale beneath, usually glaucous .... 1. *S. glauca*
2. Leaves green on both sides 3
3. Stems flexuous, with stout spines, or spines absent; leaves subcoriaceous to coriaceous 4
3. Stems not flexuous, with weak spines, or spines absent; leaves membranaceous .... 4. *S. hispida*
4. Leaves thick-margined, coriaceous, often blotched with white, often panduriform; peduncles much longer than the subtending petioles; berries not glaucous .... 2. *S. bona-nox*
4. Leaves thin-margined, subcoriaceous, not blotched with white, never panduriform; peduncles about as long as the subtending petioles; berries glaucous .... 3. *S. rotundifolia*
5. Stems twining or climbing, with numerous tendrils all along the stem; peduncles borne from the axils of developed leaves 6
5. Stems erect, with few (or no) tendrils near the apex; peduncles borne from bladeless sheaths 8
6. Leaves glabrous and glaucous beneath .... 5. *S. herbacea*
6. Leaves pubescent beneath 7
7. Blades light green, not shiny beneath; petiole generally short; berries blue .... 6. *S. lasioneuron*
7. Blades dark green, shiny beneath; petiole long; berries black ....
.... 7. *S. pulverulenta*
8. Basal leaves narrowly ovate or elliptical, the base mostly truncate to subcordate; petioles usually equal to or longer than blade . 8. *S. illinoensis*
8. Basal leaves broadly ovate, the base cordate; petioles generally equal to or shorter than blade .... 9. *S. ecirrata*

1. S. glauca Walt. Catbrier. May–June. Leaves minutely pubescent beneath. Dry woods, edges of fields and bluffs; common in the s. one-third of Ill., absent elsewhere.
var. leurophylla Blake. Catbrier. May–June. Leaves glabrous beneath. Edges of swamps; Johnson, Pope, and Union cos.
2. S. bona-nox L. Catbrier. May–June. Leaves spinulose, with patches of white coloration. Dry woods and fields, on bluffs; confined to the s. one-fourth of the state.

var. hederaefolia (Beyrich) Fern. Catbrier. May–June. Leaves enti
or with very weak marginal spinules, green throughout. Woods ar
fields, not common; restricted to the s. one-sixth of the state.

3. S. rotundifolia L. Catbrier. April–May. Dry woods, edge of field
common in the s. one-third of the state, absent elsewhere.

4. S. hispida Muhl. Bristly Catbrier. May–June. Usually moi
woods; probably in every co. S. tamnoides L. var. hispida (Muh
Fern.—F.

5. S. herbacea L. Carrion Flower. April–June. Moist woods, ve
rare; Jackson Co.

6. S. lasioneuron Hook. Carrion Flower. April–June. Moist wood
edge of fields; in every co. S. herbacea L. var. lasioneura (Hook.)
DC.—F, G.

7. S. pulverulenta Michx. Carrion Flower. April–June. Moist or d
woods; edges of fields; occasional in the s. two-fifths of the stat
rare elsewhere. S. herbacea L. var. pulverulenta (Michx.) Gray—G.

8. S. illinoensis Mangaly. Carrion Flower. May–June. Thickets, ofte
near roads; occasional in the n. three-fourths of the state; abse
elsewhere.

9. S. ecirrata (Engelm.) S. Wats. Carrion Flower. May–June. Moi
woods; occasional in the n. one-half of Ill., becoming uncommon
the s. cos.

35. DIOSCOREACEAE

1. All leaves (except sometimes the lowermost) alternate; capsule 1.5–2
cm long; seeds (including wing) 7–14 mm broad; petiole essentially g
brous at point of attachment of blade .... 1. D. villo
1. Lowest leaves whorled, becoming opposite or alternate on the upp
part of the stem; capsule 2.5–3.0 cm long; seeds (including wing) 15–
mm broad; petiole puberulent at point of attachment of blade ....
.... 2. D. quaterna

1. D. villosa L. Wild Yam. June–July. Dry or moist woodlands, co
mon; probably in every co. (Including f. glabrifolia [Bartlett] Fern.)

2. D. quaternata [Walt.] J. F. Gmel. Wild Yam. June–July. Dry
moist woodlands, occasional; confined to the s. one-third of the stat
(Including var. glauca [Muhl.] Fern.)

36. IRIDACEAE

1. Sepals somewhat recurved; petals spreading or erect; styles petal-lik
stamens concealed by the arching styles .... 1. I
1. Sepals and petals similar, spreading or tubular; styles not petal-lik
stamens not concealed ....
2. Perianth tubular, forming a short funnel, scarlet; upper three perian
segments longer than the lower three .... 2. Gladiol

2. Perianth spreading and orange, or rotate and blue or white; all perianth segments essentially equal 3
3. Flowers orange, with purplish spots, at least 3 cm broad; filaments distinct; flowers bracteate, the bracts soon withering; capsule at least 2 cm long; seeds attached to a central column, resembling a blackberry; plants rhizomatous .... 3. *Belamcanda*
3. Flowers blue or white, less than 2 cm broad; filaments united to apex; flowers borne from 1 or 2 persistent spathes; capsule up to 6 mm long; seeds not attached to a central column; plants with fibrous roots ......
.... 4. *Sisyrinchium*

1. *Iris* L. Iris

1. Upper surface of sepals with a beard of hairs 2
1. Sepals without a beard of hairs 3
2. Stems very short or none, bearing one flower; leaves to 15 cm long, to 7 mm broad; sepals to 17 mm broad .... 1. *I. pumila*
2. Stems to nearly 1 m tall, bearing several flowers; leaves to 100 cm long, to 3 cm broad; sepals over 20 mm broad .... 2. *I. germanica*
3. Rhizomes stout, at least 1 cm thick; leaves 40–100 cm long; flowering stem usually branched, 20–100 cm long; spathes subtending flower unequal in length; flowers several per stem; sepals 1- to 2-ridged, or without ridges; capsule obtusely 3-angled or 6-angled, 3–9 cm long; plants often of more aquatic situations 4
3. Rhizomes slender, less than I cm thick; leaves (at flowering time) to 25 cm long, at maturity as long as 40 cm; flowering stem unbranched, to 4.5 cm long; spathes subtending flower nearly equal in length; flowers 1–2 per stem; sepals sharply 3-ridged above; capsule sharply 3-angled, about 1 cm long; plants of rich woodlands, usually near streams ....
.... 7. *I. cristata*
4. Ovary and capsule 6-angled; capsule indehiscent; leaves rather soft . 5
4. Ovary and capsule 3-angled (occasionally 6-angled in the yellow-flowered *I. pseudacorus*); capsule dehiscent; leaves firm 6
5. Flowering stem 20–40 cm tall; flowers dark blue; capsule 3–5 cm long; leaves 15–30 mm broad; seeds more or less globoid ... 3. *I. brevicaulis*
5. Flowering stem at least 50 cm tall; flowers coppery; capsule 4.5–7.5 cm long; leaves 10—15 mm broad; seeds flattened .... 4. *I. fulva*
6. Flowers usually some shade of blue or violet; sepals not 2-ridged on upper surface; perianth tube constricted above the ovary . 5. *I. shrevei*
6. Flowers basically yellow; sepals 2-ridged on upper surface; perianth not constricted above the ovary .... 6. *I. pseudacorus*

1. I. pumila L. Dwarf Iris. March–April. Native of Europe and Asia; escaped into waste ground; DuPage Co.

2. I. germanica L. Bearded Iris. April–June. Escaped from cultivation into waste land; McLean, Piatt, and Pope cos.

3. I. brevicaulis Raf. Blue Iris. May–July. Marshes and wet prairies; scattered in the s. two-thirds of the state.

4. I. fulva Ker. Swamp Red Iris. May–June. Swamps, usually in shallow water; Alexander, Hamilton, Johnson, Massac, Pulaski, and Union cos.

5. I. shrevei Small. Wild Blue Iris. May–June. Wet situations, common; probably in every co. *I. virginica* L. var. *shrevei* (Small) E. Anders—F.

6. I. pseudacorus L. Yellow Iris. June–Aug. Native of Europe; scattered throughout Ill.
7. I. cristata Ait. Dwarf Crested Iris. April–May. Lowland woods
usually along streams; restricted to the s. tip of Ill.

2. Gladiolus L.   Gladiolus

1. G. X colvillei Sweet. Scarlet Gladiolus. July–Aug. Escaped from
cultivation along roads; Johnson and Massac cos. Reputed to be a
hybrid between G. concolor Salisb. and G. cardinalis Curt.

3. Belamcanda Adans.   Blackberry Lily

1. B. chinensis (L.) DC. Blackberry Lily. July–Aug. Native of Asia
adventive along roads; scattered throughout the state.

4. Sisyrinchium L.   Blue-eyed Grass

1. Spathes long-pedunculate from the axils of leaf-like bracts; capsules
dark brown 2
1. Spathes sessile, terminal; capsules pale brown, stramineous, or greenish (occasionally becoming dark brown in S. montanum) 3
2. Leaves bright green; stems broadly winged, 3–5 mm wide; capsule 4–6
mm long; plants becoming black on drying .... 1. S. angustifolium
2. Leaves pale green or glaucous; stems narrowly winged, 1–3 mm wide
capsule 3.0–4.5 mm long; plants generally not becoming black on drying
.... 2. S. atlanticum
3. Spathes 2 .... 3. S. albidum
3. Spathe 1 4
4. Capsules 2–4 mm long; margins of outer bract free to base or united for
less than 2 mm 5
4. Capsules 4–6 mm long; margins of outer bract united for 2–6 mm above
base .... 6. S. montanum
5. Plants pale green or glaucous, drying pale; stems winged; margins of
outer bract free to base; flowers pale blue or white ... 4. S. campestre
5. Plants dark green, drying blackish; stems marginate; margins of outer
bract united for 1–2 mm above base; flowers bright violet ....
.... 5. S. mucronatum

1. S. angustifolium Mill. Blue-eyed Grass. May–June. Low woodlands, moist prairies, common; in nearly every co. S. graminoides
Bickn.—G; S. bermudiana L.—J.
2. S. atlanticum Bickn. Blue-eyed Grass. May–June. Wet prairies
very rare; Kankakee Co.
3. S. albidum Raf. Blue-eyed Grass. April–June. Open woodlands
fields, prairies, common; probably in every co.
4. S. campestre Bickn. Blue-eyed Grass. May–June. Prairies, particularly in sandy soil; generally restricted to nw. and w. cent. cos; Pope
Co.
5. S. mucronatum Michx. Blue-eyed Grass. May–June. Sandy
prairies, rare; Hancock and Mason cos. S. campestre Bickn.—J.
6. S. montanum Greene. Blue-eyed Grass. May–June. Plants pale
green or glaucous, drying pale; capsule whitish, greenish, or stramineous, even at maturity. Sandy prairies, rare; Cook, Lake, and Winnebago cos. S. campestre Bickn.—J.

var. crebrum Fern. Blue-eyed Grass. May–June. Plants bright green, drying blackish; capsule greenish or pale brown at first, becoming dark at maturity. Fields, very rare; Kankakee Co.

37. BURMANNIACEAE

1. *Thismia Griff.*

1. T. americana N. E. Pfeiffer. Aug. Wet prairies, undoubtedly extinct; Cook Co.

38. ORCHIDACEAE

1. Plants with green leaves at flowering time 2
1. Plants without green leaves, at least at flowering time 15
2. Leaves whorled .... 10. *Isotria*
2. Leaves alternate or basal 3
3. Lip inflated, sac-like, at least 18 mm long; anthers 2 ... 1. *Cypripedium*
3. Lip flat or, if inflated, less than 18 mm long; anther 1 4
4. Leaves 1 or 2, basal (leaf 1–2 and cauline in *Malaxis*) 5
4. Leaves more than 2, basal or cauline or, if only 1–2, then the leaf cauline 8
5. Flowers totally pink, resupinate; leaf generally 1, linear to linear-lanceolate .... 2. *Calopogon*
5. Flowers greenish-yellow, madder-purple, or white and pink, not resupinate; leaves generally 2, lanceolate to elliptic to ovate to orbicular 6
6. Flowers spurred, usually fragrant; anther persistent, difficult to detach 7
6. Flowers spurless, not fragrant; anther easily and soon detachable .... 5. *Liparis*
7. Flowers white and pink; sepals and petals united to form a hooded structure (galea) behind the column .... 3. *Orchis*
7. Flowers greenish-yellow; sepals and petals not forming a galea .... 4. *Habenaria*
8. Leaves 1–2, cauline .... 6. *Malaxis*
8. Leaves 3 or more, basal or cauline or both 9
9. All leaves basal 10
9. At least 1 or more cauline leaves present (basal leaves also may be present) 11
0. Leaves green throughout; lip not sac-like .... 7. *Spiranthes*
0. Leaves conspicuously marked with white; lip sac-like .... 8. *Goodyera*
1. Flowers spurred; anther persistent, not easily detached .. 4. *Habenaria*
1. Flowers not spurred; anther easily detached 12
2. Leaves both basal and cauline; flowers 1–2, terminal, or numerous, sessile and spicate 13
2. Leaves all cauline; flowers numerous, pedicellate, racemose 14
3. Cauline leaf 1; sepals and petals free; flowers 1–2, pink, at least 15 mm long .... 9. *Pogonia*
3. Cauline leaves 2 or more; upper sepal united with petals; flowers numerous, white, at most 10 mm long .... 7. *Spiranthes*

14. Flowers white (rarely pink), nodding, 1.0–1.5 cm long, the lip not sac
like; sepals and petals lanceolate; leaves weak, oval ...... 11. *Triphor*
14. Flowers green and purple, ascending, well over 1.5 cm long, the lip sac
like at the base; sepals and petals ovate; leaves firm, lanceolate to ovat
.... 12. *Epipact*
15. Green leaves never produced; stem brown or purple, with colore
sheaths; rhizomes coralline 1
15. Green leaves produced, but these usually absent at flowering time
plants with fleshy roots or with corms or tubers connected in a series 1
16. Lip without longitudinal ridges; flowers at most 1 cm long, brow
yellow-green, or white spotted with purple .... 13. *Corallorhiz*
16. Lip with 5–6 longitudinal ridges; flowers 1.7–2.3 cm long, madder-purp
.... 14. *Hexalectr*
17. Leaves more than 1; plants with fleshy roots, but without connecte
tubers; flowers whitish .... 7. *Spiranthe*
17. Leaf 1; plants with connected tubers; flowers yellow or brown o
madder-purple 1
18. Leaves green on both sides; flowers not spurred .... 15. *Aplectrur*
18. Leaves purple beneath; flowers spurred .... 16. *Tipular*

1. *Cypripedium* L. Lady's-slipper

1. Lip cleft down the middle, pink; leaves 2, basal .... 1. *C. acau*
1. Lip not cleft, yellow, white, or with pink or purple streaks; leaves mor
than 2, cauline ....
2. Sepals and petals acuminate, yellow or greenish-yellow, or streaked wit
purple or brown ....
2. Sepals and petals obtuse, white .... 6. *C. regina*
3. Lip yellow ....
3. Lip white, occasionally marked with purple streaks ....
4. Lateral petals purple-brown; staminodium ovate .... 2. *C. calceolu*
4. Lateral petals yellow-green striped with brown; staminodium oblong .
.... 3. *C. calceolus* var. *pubescens* X *C. candidu*
5. Lip white; staminodium narrowly triangular ....
.... 4. *C calceolus* var. *parviflorum* X *C. candidu*
5. Lip white, streaked with purple; staminodium oblong .. 5. *C. candidu*

1. C. acaule Ait. Lady's-slipper Orchid. May–July. Bogs and ac
woodlands, very rare; Cook, Lake, McHenry, and Ogle cos.

2. C. calceolus L. var. parviflorum (Salisb.) Fern. Small Yello
Lady's-slipper Orchid. May–June. Lateral petals 3.5–5.0 cm lon
sepals usually purplish. Bogs and marshes, rare; restricted to the
cos. C. parviflorum Salisb.—J.

var. pubescens (Willd.) Correll. Yellow Lady's-slipper Orchid. Apri
May. Lateral petals 5–8 cm long; sepals usually yellow-green. Moist o
dry woodlands, not common; scattered throughout the state. C. pa
viflorum Salisb.—J.

3. C. calceolus var. pubescens X C. candidum. June. Shade
ground, very rare; Cook, Henderson, and Woodford cos. Reputed t
be a hybrid between C. calceolus var. pubescens and c. candidum. C
X favillianum J. T. Curtis.

4. C. calceolus var. parviflorum X C. candidum. Lady's-slipper O
chid. June. Low, springy areas, very rare; McHenry and Cook co
Reputed to be a hybrid between C. calceolus var. parviflorum and C
candidum. C. X andrewsii A. M. Fuller.

5. C. candidum Muhl. White Lady's-slipper Orchid. May–June. Bogs, swamps, and wet prairies, very rare; restricted to the n. half of Ill.; also from Macoupin and St. Clair cos.

6. C. reginae Walt. Showy Lady's-slipper Orchid. June–July. Bogs; low, springy areas, rare; restricted to the n. one-half of the state.

2. Calopogon L. C. Rich. Grass Pink Orchid

1. C. tuberosus (L.) BSP. Grass Pink Orchid. May–July. Wet prairies, not common; confined to the n. half of the state; also Jackson and Macoupin cos. C. pulchellus (Salisb.) R. Br.—F, G, J.

3. Orchis L. Orchis

1. O. spectabilis L. Showy Orchis. April–June. Rich woodlands, not common; scattered throughout the state.

4. Habenaria Willd. Fringed Orchis

1. Leaves 2, basal, suborbicular; lip elongated, entire, 8–20 mm long, the spur 13–40 mm long 2
1. Leaves 1-several, cauline, linear to lanceolate to oval; lip lobed, toothed, erose, or fringed, if entire and elongated, then the lip 4–8 mm long and the spur 2–12 mm long 3
2. Lip linear, 12–20 mm long; flowers greenish-white, with pedicels 5–13 mm long; stem bracted .... 1. H. orbiculata
2. Lip triangular-elongated, 8–12 mm long; flowers yellow-green, sessile; stem without bracts .... 2. H. hookeri
3. Lip entire, erose, 2- to 3-toothed, or very shallowly 3-lobed, with the lobes entire; spur 2–12 mm long; flowers sessile 4
3. Lip fringed or deeply 3-lobed and toothed; spur 12 mm long or longer; flowers pedicellate 8
4. Leaf 1; inflorescence 2–6 cm long; all bracts shorter than the flowers; spur 8–12 mm long; roots slender .... 3. H. clavellata
4. Leaves 2-several; inflorescence 5–30 cm long; at least the lower bracts longer than the flowers; spur 2–8 mm long; rootstocks tuberous 5
5. Sepals at least twice as long as the spur, the spur 2–3 mm long; flowers greenish, tinged with purple; lip 2- to 3-toothed ....
   .... 4. H. viridis var. bracteata
5. Sepals shorter than to as long as the spur, the spur 3–8 mm long; flowers greenish-white, greenish-yellow, or white; lip entire or erose . 6
6. Lip erose, with a tubercle borne near the summit .... 5. H. flava
6. Lip entire, without a tubercle 7
7. Flowers greenish-white, faintly odorous; lip gradually broadened at the base .... 6. H. hyperborea
7. Flowers creamy-white, with a spicy fragrance; lip abruptly broadened at the base .... 7. H. dilatata
8. Lip simple, fringed 9
8. Lip deeply 3-lobed, the lobes fringed, long-toothed, or erose 10
9. Flowers orange; leaves lanceolate, some or all over 2 cm broad ....
   .... 8. H. ciliaris
9. Flowers white; leaves linear to linear-lanceolate, less than 2 cm broad
   .... 9. H. blephariglottis
10. Flowers yellow-green or white; lobes of the lip fringed 11
10. Flowers reddish-purple; lobes of the lip long-toothed or erose 12

11. Flowers yellow-green or greenish-white; sepals 4.5–7.0 mm long; spur 14–20 mm long .... 10. *H. lacera*
11. Flowers white; sepals 7–12 mm long; spur 20–48 mm long ....
.... 11. *H. leucophaea*
12. Lip shallowly erose, 13–22 mm long, the terminal lobe notched ....
.... 12. *H. peramoena*
12. Lip long-toothed, 9–13 mm long, the terminal lobe not notched ....
.... 13. *H. psycode*

1. H. orbiculata (Pursh) Torr. Round-leaved Orchid. July–Aug. Moist woodlands, very rare; Cook and Kane cos.

2. H. hookeri Torr. Hooker's Orchid. June–July. Moist woodlands, rare; Cook, Hancock, and Lake cos.

3. H. clavellata (Michx.) Spreng. Wood Orchid. July–Aug. Moist shaded areas, rare; Cass, Cook, JoDaviess, Kankakee, Lake, Pope, and Will cos.

4. H. viridis (L.) R. Br. var. bracteata (Muhl.) Gray. Bracted Green Orchid. May–June. Rich woodlands, rare; confined to the n. one-half of the state. *H. bracteata* (Muhl.) R. Br.—J.

5. H. flava (L.) R. Br. Tubercled Orchid. June–Aug. Bracts of lowest flowers equalling the flowers; inflorescence lax; lip suborbicular. Low, shaded areas, rare; Johnson, Massac, and Wabash cos.

var. herbiola (R. Br.) Ames & Correll. Tubercled Orchid. June–July. Moist woodlands, rare; restricted to the n. one-half of the state; also St. Clair and Wabash cos.

6. H. hyperborea (L.) R. Br. var. huronensis (Nutt.) Farw. Green Orchid. June–Aug. Swampy areas, rare; known only from the n. one-half of the state. *H. huronensis* (Nutt.) Spreng.—J; *H. hyperborea* (L.) R. Br.—G.

7. H. dilatata (Pursh) Hook. White Orchis. June–July. Springy situations, very rare; Carroll, Kankakee, and McHenry cos.

8. H. ciliaris (L.) R. Br. Yellow Fringed Orchid. June–Aug. Low ground, rare; Cook Co.; also reported from Kankakee and Union cos.

9. H. blephariglottis (Willd.) Hook. White Fringed Orchid. June–July. Low areas, very rare; Macon Co.

10. H. lacera (Michx.) Lodd. Green Fringed Orchid. June–July. Swamps; rare in the n. three-fourths of Ill.; also Massac and Pope cos.

11. H. leucophaea (Nutt.) Gray. White Fringed Orchid. June–July. Swampy areas, not common; scattered in the n. two-thirds of the state.

12. H. peramoena Gray. Purple Fringeless Orchid. June–Sept. Low woodlands, wet prairies; occasional in the s. two-fifths of the state; absent elsewhere.

13. H. psycodes (L.) Spreng. Purple Fringed Orchid. June–Aug. Low woodlands, bogs, rare; Cook, Lake, and Winnebago cos.

5. *Liparis* Rich. Twayblade Orchid

1. Pedicels 5–10 mm long, as long as the flowers; sepals greenish-white, 10–12 mm long; lateral petals 10–12 mm long, madder-purple; lip 8–10 mm long, madder-purple, flat .... 1. *L. liliifolia*
1. Pedicels 4–5 mm long, usually a little shorter than the flowers; sepals yellow-green, 4–6 mm long; lateral petals 4–6 mm long, yellowish; lip 4–

mm long, yellow-green, turned up slightly along the margins ....
.... 2. *L. loeselii*

1. L. liliifolia (L.) Rich. Twayblade Orchid. May–July. Rich woodlands, occasional; scattered throughout the state.
2. L. loeselii (L.) Rich. Lesser Twayblade Orchid. June–July. Low
woodlands, bogs, rare; restricted to the n. half of the state.

6. *Malaxis* Sw.   Adder's Mouth Orchid

1. Pedicels 4–9 mm long; lip bilobed, ascending; upper sepal 1.2–1.6 mm
long .... 1. *M. unifolia*
1. Pedicels 1–3 mm long; lip entire, deflexed; upper sepal 2.0–2.5 mm long
.... 2. *M. monophylla* var. *brachypoda*

1. M. unifolia Michx. Adder's Mouth Orchid. June–Aug. Dry or moist
woodlands, rare; Hancock, Henderson, and Menard cos.
2. M. monophylla (L.) Sw. var. brachypoda (Gray) F. Morris. Adder's
Mouth Orchid. July. Bogs, very rare; Kane and McHenry cos. *M. brachypoda* (Gray) Fern.—F; *M. monophylla* (L.) Sw.—G.

7. *Spiranthes* Rich.   Ladies' Tresses

1. Flowers produced in 2–3 rows, the spikes usually crowded; rachis pubescent (see also *S. vernalis* under second no. 1 of couplet) 2
1. Flowers produced in a single, twisted row, often appearing secund, the
spikes laxly-flowered; rachis glabrous (pubescent in *S. vernalis*) 5
2. Lip yellow; flowers produced from May to July .... 1. *S. lucida*
2. Lip white or cream; flowers produced from August to October 3
3. Sepals and petals 4–5 mm long; lip 4–5 mm long, with two basal, conspicuous, incurved callosities .... 2. *S. ovalis*
3. Sepals and petals 7–12 mm long; lip 7–12 mm long, with two basal,
rather obscure (sometimes prominent in *S. cernua*), rounded callosities
.... 4
4. Leaves usually persisting, at least through anthesis; flowers white, not
strongly scented .... 3. *S. cernua*
4. Leaves usually absent during anthesis; flowers cream, strongly scented
.... 4. *S. magnicamporum*
5. Rachis pubescent; sepals, petals, and the lip 6–10 mm long, the lip yellowish; lowest cauline leaves usually resembling the basal leaves,
present at flowering time .... 5. *S. vernalis*
5. Rachis glabrous; sepals, petals, and the lip 3–4 mm long, the lip white or
white with a green area; all leaves basal, withered or absent at flowering
time 6
6. Tuberous roots several; lip white with a green central area; largest
leaves usually 2–4 cm long 7
6. Tuberous root 1 or occasionally 2 or 3; lip white throughout; largest
leaves at most 2 cm long .... 8. *S. tuberosa*
7. Leaves present but withered at flowering time, thin; spike little spiraling;
summit of lip green with a broad white border; plants flowering from
mid-June to early September, averaging by August 5 ...... 6. *S. lacera*
7. Leaves absent at flowering time, thick; spike strongly spiraling; summit
of lip green with a narrow white border; plants flowering from late July
to early October, averaging by September 2 .... 7. *S. gracilis*

1. S. lucida (H. H. Eaton) Ames. Yellow-lipped Ladies' Tresses. May–June. Wet situations, rare; Cook, Hancock, Will, and Woodford cos.

2. S. ovalis Lindl. Ladies' Tresses. Sept.–Oct. Rich woodlands, not common; scattered in the s. three-fourths of the state.

3. S. Cernua (L.) Rich. Nodding Ladies' Tresses. Aug.–Oct. Moist or dry situations in usually open areas; occasional throughout the state.

4. S. magnicamporum Sheviak. Aug.–Oct. Dry prairies; scattered in most of the state, except the southernmost cos.

5. S. vernalis Engelm. & Gray. Ladies' Tresses. July–Aug. Rich woods, prairies, rare; Effingham, Massac, Pope, St. Clair, and Wabash cos.

6. S. lacera (Raf.) Raf. Slender Ladies' Tresses. June–Sept. Mostly dry woodlands, rare; restricted to the n. half of the state. S. gracilis (Bigel.) Beck—J.

7. S. gracilis (Bigel.) Beck. Slender Ladies' Tresses. July–Oct. Dry open woodlands, rare; restricted to the s. one-third of Ill.

8. S. tuberosa Raf. Little Ladies' Tresses. July–Oct. Dry, open wood lands; confined to the s. one-half of the state. S. grayi Ames—J.

8. Goodyera R. Br. Rattlesnake Plantain Orchid

1. G. pubescens (Willd.) R. Br. Rattlesnake Plantain Orchid. June–Aug. Rich, moist woodlands; scattered throughout the state.

9. Pogonia Juss. Pogonia

1. P. ophioglossoides (L.) Ker. Snake-mouth. June–July. Low ground, rare; restricted to the n. one-fourth of the state.

10. Isotria Raf. Whorled Pogonia

1. I. verticillata (Willd.) Raf. Whorled Pogonia. May–June. Lowland woods, very rare; Pope Co.

11. Triphora Nutt. Nodding Pogonia

1. T. trianthophora (Sw.) Rydb. Nodding Pogonia. Aug.–Oct. Rich woodlands, not common; scattered throughout the state.

12. Epipactis Sw. Helleborine

1. E. helleborine (L.) Crantz. Helleborine. July–Aug. Native of Europe; adventive in disturbed woodlands; Cook, DuPage, Lake, and McHenry cos. E. latifolia (L.) Crantz—G.

13. Corallorhiza Chat. Coral-root Orchid

1. Lip with two lateral lobes or teeth; plants flowering generally from mid May to mid-August. ....

1. Lip entire or erose, not lobed or toothed; plants flowering generally from March to mid-May or from mid-August to October ....

2. Stems yellowish; sepals and petals yellow-green, rarely spotted with purple, 4–5 mm long, 1-nerved; spur absent; lip 4–5 mm long, with 2 short

lateral lobes; capsule greenish, 6–10 mm long; rhizome white . . . .
. . . . 1. *C. trifida*

2. Stems purplish; sepals and petals white with purple spots, 6–8 mm long, 3-nerved; spur present; lip 6–8 mm long, with 2 well-developed lateral teeth or lobes; capsule brownish, 10–20 mm long; rhizome brown . . . . .
. . . . 2. *C. maculata*

3. Stems purple; sepals and petals greenish-yellow, spotted with purple, linear-lanceolate, 6–8 mm long; lip 5–6 mm long, notched at apex, long-clawed; capsule 8–12 mm long; flowering in early spring 3. *C. wisteriana*

3. Stems purple or brown below, greenish above; sepals and petals purplish-green to purple, oblong, 3–5 mm long; lip 3–4 mm long, undulate at apex but not notched, short-clawed or clawless; capsule 5–8 mm long; flowering in autumn . . . . 4. *C. odontorhiza*

    1. C. trifida Chat. Pale Coral-root Orchid. May. Moist woods, very rare; Cook Co.
    2. C. maculata Raf. Spotted Coral-root Orchid. June–Aug. Woods rich in humus, rare; confined to the n. two-fifths of the state.
    3. C. wisteriana Conrad. Wister's Coral-root Orchid. March–May. Rich woodlands; scattered in the s. half of the state; also LaSalle Co.
    4. C. odontorhiza (Willd.) Nutt. Fall Coral-root Orchid. Aug.–Oct. Woodlands; scattered throughout the state, but not common.

14. Hexalectris Raf. Crested Coral-root Orchid

    1. H. spicata (Walt.) Barnh. Crested Coral-root Orchid. June–Sept. Dry woodlands; limestone ledges; Jackson, Monroe, Pope, and Randolph cos.

15. Aplectrum (Nutt.) Torr. Putty-root Orchid

    1. A. hyemale (Muhl.) Torr. Putty-root Orchid. May–June. Rich woods, occasional; scattered throughout the state.

16. Tipularia Nutt. Crane-fly Orchid

    1. T. discolor (Pursh) Nutt. Crane-fly Orchid. July–Aug. Rich woods, rare; confined to the s. tip of Ill.

39. SAURURACEAE

1. Saururus L. Lizard's-tail

    1. S. cernuus L. Lizard's-tail. May–Sept. Swampy woods, sometimes in standing water; occasional to common in the two-thirds of Ill.; also Cook, DeKalb, Grundy, and Kankakee cos.

40. SALICACEAE

1. Leaves twice as long as broad or longer; bud scale one; catkins not drooping . . . . 1. *Salix*

1. Leaves never twice as long as broad; bud scales several; catkins droop
ing .... 2. *Populu*

1. Salix L. Willow

1. One or more glands present at upper end of petiole ....
1. Glands absent at upper end of petiole ....
2. Leaves whitish beneath; capsule 7–10 mm long .... 6. *S. serissim*
2. Leaves green beneath; capsule less than 7 mm long ....
3. Leaves with long-tapering, almost tail-like, tip .... 5. *S. lucic*
3. Leaves merely acute to short-acuminate at tip .... 4. *S. pentandr*
4. Leaves purplish, at least some of them opposite .... 24. *S. purpure*
4. Leaves green or whitish, alternate ....
5. Leaves glabrous on the lower surface ....
5. Leaves pubescent, at least on the lower surface **1**
6. Leaves remotely denticulate; petioles less than 3 mm long   10. *S. interic*
6. Leaves closely crenate or crenate-serrate; petioles 3 mm long or longe
....
7. Leaves entire, revolute; some of the stems creeping   17. *S. pedicellar*
7. Leaves crenate, serrate, or denticulate; all stems upright ....
8. Stipules absent or minute and falling away early on vegetative sprou
and young branchlets ....
8. Stipules persistent on vegetative sprouts and young branchlets **1**
9. Leaves green beneath, coarsely undulate-serrate .... 7. *S. fragil*
9. Leaves whitish beneath, finely serrulate **1**
10. Branchlets "weeping"; capsule 1.0–2.5 mm long ..... 8. *S. babylonic*
10. Branchlets not "weeping"; capsule 3–9 mm long **1**
11. Leaves tapering to a long-acuminate, almost tail-like, tip............
.... 3. *S. amygdaloide*
11. Leaves acute to acuminate, but not tail-like ....
12. Flowers appearing before the leaves; teeth along margin of blade not e
tending all the way to the base .... 20. *S. petiolar*
12. Flowers appearing with the leaves; teeth along margin of blade exten
ing all the way to the base .... 9. *S. alb*
13. Leaves green on both sides ....
13. Leaves pale on the lower surface ....
14. Leaves lanceolate, rarely as much as 2 cm broad (except sometim
those on the sprouts), tapering to base; each staminate flower with tw
glands at the base; stamens 3 .... 1. *S. nig*
14. Leaves oblong-lanceolate, some of them at least 2 cm broad, rounded
even subcordate at base; each staminate flower with one gland at th
base; stamens 2 .... 11. *S. rigid*
15. Leaves irregularly crenate-serrate; flowers appearing before leaves e
pand; capsule puberulent .... 18. *S. discol*
15. Leaves finely serrulate; flowers appearing with the leaves; capsule gl
brous or granular ....
16. Leaves lanceolate, often falcate; each staminate flower with two glan
at base; stamens 4–8; capsule 4–6 mm long, the pedicel at least half
long .... 2. *S. carolinian*
16. Leaves oblong to narrowly ovate, not falcate; each staminate flower wi
one gland at base; stamens 2; capsule 7–10 mm long, the pedicel mu
less than half as long .... 13. *S. glaucophylloides* var. *glaucophy*
17. Young branchlets and new leaves covered with white wool ....
.... 23. *S. candi*

7. Young branchlets and new leaves without white wool 18
8. Leaves entire or undulate 19
8. Leaves serrulate or crenate or denticulate 21
9. Catkins appearing before the leaves; young branchlets densely tomentose or pilose .... 19. *S. humilis*
9. Catkins appearing with the leaves; young branchlets sericeous, sparsely
pilose, or glabrate 20
0. Capsules up to 6 mm long; bracts brown, with a black tip ....
.... 25. *S.* X *subsericea*
0. Capsules 6–10 mm long; bracts yellowish .... 16. *S. bebbiana*
1. Leaves silvery-silky on lower surface 22
1. Leaves pubescent beneath, but not with silvery-silky hairs 25
2. Petiole up to 3 mm long; margin of blade remotely denticulate ....
.... 10. *S. interior*
2. Petiole 3 mm long or longer; margin of blade finely serrulate 23
3. Teeth along margin of blade not extending all the way to the base .....
.... 20. *S. petiolaris*
3. Teeth along margin of blade extending all the way to the base 24
4. Young branchlets silky; flowers appearing with the leaves .. 9. *S. alba*
4. Young branchlets glabrous or glabrate; flowers appearing before the
leaves .... 21. *S. sericea*
5. Leaves narrowed or rounded at base, not subcordate; stipules on
sprouts and young branchlets inconspicuous and falling away early   26
5. Leaves subcordate at base (tapering in *S. eriocephala*); stipules on
sprouts and young branchlets large and persistent 29
6. Young branchlets densely tomentose or pilose .... 19. *S. humilis*
6. Young branchlets sparsely pilose or glabrate 27
7. Upper surface of leaves lustrous; flowers appearing before the leaves 28
7. Upper surface of leaves dull; flowers appearing with the leaves ....
.... 16. *S. bebbiana*
8. Capsule 7–12 mm long, minutely pubescent .... 18. *S. discolor*
8. Capsule 6–7 mm long, densely gray-hairy .... 22. *S. caprea*
9. Branchlets permanently tomentulose 30
9. Branchlets glabrous or glabrate 31
0. Leaves lustrous above, acute to short-acuminate ...... 15. *S. syrticola*
0. Leaves dull above, tapering to a tail-like tip .... 14. *S. eriocephala*
1. Leaves glaucous beneath, tapering to a tail-like tip   12. *S.* X *myricoides*
1. Leaves green beneath, acute to short-acuminate .... 11. *S. rigida*

1. S. nigra Marsh. Black Willow. April–May. Along streams, common; in every co.

2. S. caroliniana Michx. Ward's Willow. May. Wet soil, not common;
Jackson, Madison, Perry, St. Clair and Williamson cos.

3. S. amygdaloides Anderss. Peach-leaved Willow. April–May. Along
streams; occasional throughout the state.

4. S. pentandra L. Bay-leaved Willow. May. Native of Europe; escaped along creek in Champaign Co.

5. S. lucida Muhl. Shining Willow. May. Bogs and wet sandy areas,
not common; confined to the n. one-fourth of the state.

6. S. serissima (Bailey) Fern. Autumn Willow. June. Marshes and
bogs, very rare; Lake Co.

7. S. fragilis L. Crack Willow. April–May. Native of Europe; occasionally escaped in all parts of the state.

8. S. babylonica L. Weeping Willow. April–May. Native of Europ and Asia; rarely escaped from cultivation.

9. S. alba L. White Willow. April–May. Leaves silky beneath. Nativ of Europe; occasionally escaped in most parts of Ill.

var. calva G. F. W. Mey. April. Leaves glabrate, branchlets browr Native of Europe; occasionally escaped in Ill.

var. vitellina (L.) Stokes. Golden Willow. April. Leaves glabrate branchlets yellow. Native of Europe; rarely escaped; Champaign, Kar kakee, and Vermilion cos. *S. vitellina* L.—J.

10. S. interior Rowlee. Sandbar Willow. April–May. Leaves glabrat beneath. Sandy shores of rivers and streams; common throughou the state.

f. wheeleri (Rowlee) Rouleau. April–May. Leaves silvery-silky be neath. Sandy shores; occasional in most parts of the state.

11. S. rigida Muhl. Heart-leaved Willow. April–May. Moist ground; o casional throughout the state. (Including var. *angustata* [Pursh] Fern

12. S. X myricoides (Muhl.) Carey. May. Moist soil, rare; Jackson Co Reputedly a hybrid between *S. rigida* Muhl. and *S. sericea* Marsh.

13. S. glaucophylloides Fern. var. glaucophylla (Bebb) Schneie Blue-leaf Willow. May. Open sand, calcareous pond borders, mars borders, not common; confined to the n. one-third of Ill. *S. glauco phylloides* Fern.—G, J.

14. S. eriocephala Michx. April. Low woods, rare; confined to th s. one-third of Ill.

15. S. syrticola Fern. Sand-dune Willow. June. Sand dunes, rar Cook and Lake cos. *S. cordata* Michx.—G, J.

16. S. bebbiana Sarg. Bebb Willow. May. Bogs, occasional; confine to the n. one-fourth of Ill.

17. S. pedicellaris Pursh var. hypoglauca Fern. April–May. Bog rare; confined to the n. one-third of the state. *S. pedicellar* Pursh—J.

18. S. discolor Muhl. Pussy Willow. March–May. Marshes ar swamps; occasional throughout the state, but less common in the cos. (Including var. *latifolia* Anderss.)

19. S. humilis Marsh. Prairie Willow. March–May. Branchlets 2 m or more thick, leaves pilose beneath, the petiole 3 mm long or longe Prairies; occasional throughout the state.

var. hyporhysa Fern. March–May. Branchlets 2 mm or more thic leaves glabrate beneath, the petiole 3 mm long or longer. Prairies; o casional in the s. cos.

var. microphylla (Anderss.) Fern. Sage Willow. March–Apr Branchlets less than 2 mm thick, leaves with petioles less than 3 m long. Prairies; Cook, Kankakee, and McDonough cos. *S. trist* Ait.—G.

20. S. petiolaris Sm. Petioled Willow. April–June. Low prairies, ri woods, marshes, bogs; confined to the n. one-fourth of the sta *S. gracilis* Anderss. var. *textoris* Fern.—F.

21. S. sericea Marsh. Silky Willow. March–May. Low ground, alo streams, in bogs; occasional and scattered throughout the state.

22. S. caprea L. Goat Willow. March–April. Native of Europ frequently planted but rarely escaped from cultivation; Jackson a Wabash cos.

23. S. candida Fluegge. Hoary Willow. April–May. Bogs; confined the n. one-third of the state.

24. S. purpurea L. Purple Osier. April–May. Native of Europe; occasionally cultivated but rarely escaped in Ill.; Cook and Lake cos.

25. S. X subsericea (Anderss.) Schneid. April–June. Low ground, rare; Cook and Kankakee cos. Reputed to be a hybrid between *S. petiolaris* Sm. and *S. sericea* Marsh.

2. *Populus* L. Poplar

1. Petioles flat 2
1. Petioles round 6
2. Leaves triangular-ovate to rhombic 3
2. Leaves ovate to suborbicular 4
3. Leaves triangular-ovate; tree with a broad crown, the branches spreading .... 1. *P. deltoides*
3. Leaves rhombic; tree with a narrow crown, the branches strongly ascending .... 2. *P. nigra* var. *italica*
4. Margin of leaf dentate, with 5–25 teeth (averaging 10–20); buds pubescent 5
4. Margin of leaf finely crenate, with 20 or more teeth (averaging 31); buds glabrous or nearly so .... 5. *P. tremuloides*
5. Margin of leaf with 5–15 teeth (averaging 10); petiole 5–10 cm long (averaging 7 cm) .... 3. *P. grandidentata*
5. Margin of leaf with 12–25 teeth (averaging 20); petiole 3–6 cm long (averaging 5.5 cm) .... 4. *P. X smithii*
6. Leaves covered with a white felt on the lower surface .... 6. *P. alba*
6. Leaves glabrous or variously pubescent beneath, but not covered with a white felt on the lower surface 7
7. Leaves rounded or truncate at base, if cordate, the buds heavily resinous 8
7. Leaves cordate at base; buds not heavily resinous . 10. *P. heterophylla*
8. Leaves sinuate-dentate; buds to 6 mm long, not resinous .... 7. *P. canescens*
8. Leaves serrulate or crenulate; buds to 25 mm long, resinous 9
9. Twigs usually glabrous; leaves glabrous, or with a few hairs on the midvein beneath .... 8. *P. balsamifera*
9. Twigs usually pubescent; leaves pubescent beneath .. 9. *P. gileadensis*

1. P. deltoides Marsh. Cottonwood. March–May. Along rivers and streams, common; probably in every co. (Including var. *missouriense* [Henry] Rehd.)

2. P. nigra L. var. italica Muenchh. Lombardy Poplar. March–May. First found in Europe; rarely escaped from cultivation.

3. P. grandidentata Michx. Large-toothed Aspen. April. Disturbed areas in and around woodlands; occasional to common in the n. half of the state, extending s. to Marion and Wabash cos.

4. P. X smithii Boivin. Barnes' Aspen. April–May. Low woods; LaSalle and Peoria cos. Reputed to be a hybrid between *P. grandidentata* Michx. and *P. tremuloides* Michx.

5. P. tremuloides Michx. Quaking Aspen. April–May. Low ground of woods, marshes, and bogs; common in the n. cos., absent in the extreme s. cos.

6. P. alba L. White Poplar. April–May. Native of Europe; frequently spreading from cultivation.

7. P. canescens (Ait.) Sm. Gray Poplar. April–May. Native of Europe;

200   SALICACEA▮

infrequently escaped from cultivation; Adams, Champaign, and Lak▮ cos.

8. P. balsamifera L. Balsam Poplar. April. Dunes near Lake M▮ chigan, rare; Cook and Lake cos. (Including *P.* X *jackii* Sarg.)

9. P. X gileadensis Rouleau. Balm-of-Gilead. April. Rarely escape▮ from cultivation; Lake Co. Reported to be a hybrid between *P. de▮ toides* Marsh. and *P. balsamifera* L.

10. P. heterophylla L. Swamp Cottonwood. April–May. Swamps ar▮ low woods, often in standing water; occasional in the s. one-third ▮ Ill.

41. MYRICACEAE

1. *Comptonia L'Hér.* Sweet-fern

1. C. peregrina (L.) Coult. Sweet-fern. April–May. Sand flats ar▮ barrens; confined to the n. one-fourth of Ill. *Myrica asplenifolia* L.—▮

42. JUGLANDACEAE

1. Husk of nuts not splitting; pith chambered .... 1. *Jugla▮*
1. Husk of nuts splitting (at least partially) at maturity; pith not chamber▮
.... 2. *Car▮*

1. *Juglans L. Walnut*

1. Branchlets and husk of nuts downy; fruit ellipsoid .... 1. *J. ciner▮*
1. Branchlets and husk of nuts not downy; fruit globose (very rarely ell▮ soid) .... 2. *J. nig▮*

1. J. cinerea L. Butternut. April–May. Rich woodlands; occasior▮ throughout the state.
2. J. nigra L. Black Walnut. April–May. Rich woodlands; comm▮ throughout the state. (Including f. *oblonga* [Marsh.] Fern.)

2. *Carya Nutt. Hickory*

1. At least some of the leaves with nine or more leaflets (if only 7 in *C. c▮ diformis,* then the buds mustard-yellow and very elongated) ....
1. Leaves with (3–) 5–7 leaflets (occasionally 9 in the very tomentose *tomentosa*) ....
2. Nut cylindric, sweet; buds with yellow hairs .... 1. *C. illinoen▮*
2. Nut usually compressed, bitter; buds with yellow glands or yellow sc▮ finess ....
3. Buds with yellow glands; leaflets strongly falcate; fruit with wings ▮ tending to the base .... 2. *C. aquat▮*
3. Buds with yellow scurfiness; leaflets not usually strongly falcate; fr▮ with wings extending only about half-way to the base 3. *C. cordiforn▮*
4. Buds yellow-lepidote ....

4. Buds not yellow-lepidote 6
5. Rachis (at least in spring) with reddish hairs .... 4. *C. texana*
5. Rachis without reddish hairs .... 5. *C. pallida*
6. Terminal bud up to 1.2 cm long; leaflets without cilia along the margin 7
6. Terminal bud over 1.2 cm long; leaflets with cilia along the margin .. 8
7. Husk of fruit readily splitting; bark becoming platy or even shaggy at maturity; leaflets mostly 7; outermost bud scales hairy throughout on the margins .... 6 *C. ovalis*
7. Husk of fruit tardily splitting or not splitting at all; bark tight at maturity; leaflets mostly 5; outermost bud scales hairy only at the tip 7. *C. glabra*
8. Branchlets densely pubescent; bark tight at maturity; kernel bitter .....
.... 8. *C. tomentosa*
8. Branchlets glabrous or sparsely pubescent; bark shaggy at maturity; kernel sweet 9
9. Leaflets mostly 5; twigs without conspicuous raised orange lenticels ...
.... 9. *C. ovata*
9. Leaflets mostly 7–9; twigs with conspicuous raised orange lenticels ....
.... 10. *C. laciniosa*

1. C. illinoensis (Wang.) K. Koch. Pecan. April–May. Bottomland woods; occasional throughout the state except for the ne. one-fourth. A hybrid with *C. laciniosa* Sarg. (*Carya* X *nussbaumeri* Sarg.) is known from Ill.

2. C. aquatica (Michx. f.) Nutt. Water Hickory. March–April. Swamps and wet woods, rare; confined to the s. one-eighth of the state.

3. C. cordiformis (Wang.) K. Koch. Bitternut Hickory. May–June. Moist or dry woods; common throughout the state; in every co.

4. C. texana Buckl. Black Hickory. April–May. Dry woods and bluffs; occasional to rare in the s. half of the state; apparently absent elsewhere. (Including var. *arkansana* [Sarg.] Sarg. and var. *villosa* [Sarg.] Little.)

5. C. pallida (Ashe) Engl. & Graebn. Pale Hickory. April–May. Wooded slopes, rare; Union Co.

6. C. ovalis (Wang.) Sarg. Sweet Pignut Hickory. April–June. Leaflets not glandular beneath; husk of fruit narrowly winged or wingless; fruit ellipsoid to nearly spherical. Woodlands; occasional to common in the s. half of the state. (Including var. *obcordata* [Muhl. & Willd.] Sarg.)

var. obovalis Sarg. April–May. Leaflets not glandular beneath; husk of fruit narrowly winged or wingless; fruit obovoid. Woodlands; occasional in the s. half of the state.

var. odorata (Marsh.) Sarg. April–May. Leaflets glandular-viscid beneath; husk of fruit strongly winged. Woodlands; rare in the s. tip of the state.

7. C. glabra (Mill.) Sweet. Pignut Hickory. April–May. Husk of fruit up to 2.5 mm thick. Woodlands; occasional to common in the s. half of the state, becoming rare elsewhere.

var. megacarpa Sarg. April–May. Husk of fruit about 3.5 mm thick. Woodlands; rare in the s. one-fourth of the state, apparently absent elsewhere.

8. C. tomentosa (Poir.) Nutt. Mockernut Hickory. May–June. Moist or dry woods; occasional in the s. three-fourths of the state, rare elsewhere. A hybrid with *C. illinoensis* (*C.* X *schneckii* Sarg.) is known from Ill.

9. C. ovata (Mill.) K. Koch. Shagbark Hickory. April–May. Fru
3.5–6.0 cm long; some or all the leaflets 5 cm wide or wider. Ric
woodlands, upland woodlands; occasional to common througho
the state. (Including var. *pubescens* Sarg.)

 var. nuttallii Sarg. Small Shagbark Hickory. April. Fruit 1.5–2.0 c
long; some or all the leaflets 5 cm wide or wider. Rich woodlanc
rare in the s. tip of Ill.

 var. fraxinifolia Sarg. Ash-leaved Shagbark Hickory. April. Fru
3.5–4.0 cm long; none of the leaflets up to 5 cm wide. Rich woc
lands, rare; apparently confined to the s. half of the state.

10. C. laciniosa (Michx.) Loud. Kingnut Hickory. April–May. Be
tomland woods; occasional to rare in the s. two-thirds of the state.

43. BETULACEAE

1. Nuts borne in elongated catkins or in woody "cones"; staminate flowe
 with a calyx ....
1. Nuts subtended or enclosed by large bracts; staminate flowers withou
 calyx ....
2. Nuts borne in elongated catkins; pistillate scales 3-lobed, herbaceous
 .... 1. *Bett*
2. Nuts borne in woody "cones"; pistillate scales 5-lobed, woody 2. *Alr*
3. Leaves broadly ovate, cordate, doubly serrate .... 3. *Cory*
3. Leaves narrowly ovate or oval, not cordate, once-serrate ....
4. Bracts inflated, enclosing the nut; bark rough; lateral nerves of lea
 usually branched near the margin .... 4. *Ostr*
4. Bracts foliaceous, subtending the nut; bark smooth; lateral nerves
 leaves unbranched .... 5. *Carpir*

1. Betula L. Birch

1. Leaves with 8 or more pairs of lateral veins ....
1. Leaves with up to 7 (–8) pairs of lateral veins ....
2. Bark yellowish- or silvery-gray, peeling off in thin layers; leaves th
 green beneath ....1. *B. lu*
2. Bark brownish to pinkish, peeling off in shaggy pieces; leaves firm, p
 beneath .... 2. *B. ni*
3. Bark peeling off in thin layers; small trees ....
3. Bark close, not peeling off in thin layers; shrubs ....
4. Leaves pubescent along the veins; most or all the fruiting spikes 3
 long or longer .... 3. *B. papyrif*
4. Leaves glabrous beneath; most or all the fruiting spikes less than 3
 long .... 4. *B. populife*
5. Leaves acute to acuminate, some or all over 3 cm long ....
5. Leaves obtuse to subacute, not more than 3 cm long ..... 7. *B. pun*
6. Bark dark brown; fruiting spikes less than 1 cm thick 5. *B.* X *sandbe*
6. Bark gray; fruiting spikes at least 1 cm thick .... 6. *B.* X *purp*

 1. B. lutea Michx. f. Yellow Birch. May–June. Boggy woods, ra
DuPage, Lake, Lee, Ogle, and Winnebago cos.

 2. B. nigra L. River Birch. April–May. Along rivers and streams;
casional in the s. half of the state, becoming less common northwa

3. B. papyrifera Marsh. Paper Birch. May–June. Wooded ravines, low dune ridges, rare; Carroll, Cook, JoDaviess, and Lake cos. (Including var. *cordifolia* [Regel] Fern.)

4. B. populifolia Marsh. Gray Birch. May–June. Thickets, very rare; Cook and Winnebago cos.

5. B. X sandbergii Britt. Sandberg's Birch. May–June. Boggy woods, rare; Kane and Lake cos. Reputed to be a hybrid between *B. lutea* and *B. pumila*.

6. B. X purpusii Schneid. Purpus' Birch. May–June. Boggy woods, rare; Lake Co. Reputed to be a hybrid between *B. lutea* and *B. pumila*.

7. B. pumila L. Dwarf Birch. May–June. Bogs; confined to Boone, Cook, Kane, Lake, McHenry, and Winnebago cos. (Including var. *glabra* Regel and var. *glandulifera* Regel.)

2. Alnus B. Ehrh.   Alder

1. Leaves dentate or denticulate, with up to 7 pairs of lateral veins . . . .
. . . . 1. *A. glutinosa*
1. Leaves serrate or serrulate, with 8 or more pairs of lateral veins 2
2. Leaves ovate to elliptic, whitened beneath . . . . 2. *A. rugosa*
2. Leaves obovate, green beneath . . . . 3. *A. serrulata*

1. A. glutinosa (L.) Gaertn. Black Alder. April–May. Native of Europe; becoming naturalized along rivers in the n. tip of the state; Cook, DuPage, and Will cos.

2. A. rugosa (DuRoi) Spreng. var. americana (Regel) Fern. Speckled Alder. May–June. Moist thickets, rare; Boone, Cook, Kane, Lake, McHenry, and Winnebago cos. *A. rugosa* (DuRoi) Spreng.—J.

3. A. serrulata (Ait.) Willd. Smooth Alder. April–May. Along rocky streams; occasional in the s. one-third of the state. (Including var. *noveboracensis* [Britt.] Fern.)

3. Corylus L.   Hazelnut

1. C. americana Walt. Hazelnut. March–April. Thickets, dry, disturbed woods, common; in every co. (Including f. *missouriensis* [A. DC.] Fern.)

4. Ostrya Scop.   Hop Hornbeam

1. O. virginiana (Mill.) K. Koch. April–May. Woodlands; in every co. (Including var. *lasia* Fern. and f. *glandulosa* [Spach] Macbr.)

5. Carpinus L.   Ironwood

1. C. caroliniana Walt. Blue Beech. April–May. Moist woods; probably in every co. (Including var. *virginiana* [Marsh.] Fern.)

44. FAGACEAE

1. Leaves unlobed 2
1. Leaves lobed . . . . 3. *Quercus*

2. Fruits with a prickly covering ....
2. Fruits acorns, not prickly .... 3. *Quercu*
3. Bark smooth; nuts triangular .... 1. *Fagu*
3. Bark rough; nuts flattened .... 2. *Castane*

1. *Fagus L.   Beech*

1. *F. grandifolia* Ehrh. Beech. April–May. Rich woods; occasional
the s. one-fourth of the state, extending along the extreme e. border
Ill. to Vermilion Co.; also Lake Co. (Including var. *caroliniana* [Loud
Fern. & Rehd. and f. *mollis* Fern. & Rehd.)

2. *Castanea Mill.   Chestnut*

1. *C. dentata* (Marsh.) Borkh. Chestnut. June–July. Rocky wood
originally in the s. tip of the state, particularly Pulaski Co., now near
extinct, except for a few sprouts here and there and a single tree
Williamson Co.

3. *Quercus L.   Oak*

1. Leaves entire ....
1. Leaves toothed or lobed ....
2. Leaves stellate-pubescent beneath; cup of acorn at least 1.5 cm broa
.... 1. *Q. imbricar*
2. Leaves glabrous or nearly so beneath; cup of acorn up to 1.2 cm. broa
.... 2. *Q. phello*
3. Leaves toothed ....
3. Leaves lobed ....
4. Fruits on stalks at least 2 cm long; veins of leaves not reaching the tip
the teeth .... 3. *Q. bicolo*
4. Fruits up to 1 cm long; veins of leaves reaching the tip of the teeth   .
5. Leaves velvety-tomentose beneath; cup of acorn at least 2.5 cm broa
.... 14. *Q. michaux*
5. Leaves glabrous or minutely pubescent, but not velvety ....
6. Leaves with pointed teeth; acorns up to 2 cm long   15. *Q. muhlenberg*
6. Leaves with rounded teeth; acorns 2 cm long or longer .. 16. *Q. prin*
7. Lobes of leaf with bristle-tips ....
7. Lobes of leaf without bristle-tips ....
8. Leaves permanently uniformly pubescent throughout on the lower su
face ....
8. Leaves glabrous beneath, or with tufts of axillary hairs, or irregular
pubescent ....
9. Leaves much broader in the upper half, with 3 (–5) broad, shallow lob
.... 3. *Q. marilandic*
9. Leaves broader at or below the middle with (3–) 5–11 narrower, deep
lobes ....
10. Lower leaf surface grayish or yellowish; scales of buds reddish-brow
glabrous or nearly so; terminal winter buds up to 6 mm long ....
10. Lower leaf surface brownish or reddish-brown; scales of buds grayis
tomentose; terminal winter buds about 6–12 mm long ... 6. *Q. velutir*
11. Terminal and usually one or two lateral lobes of leaf curved 4. *Q. falca*
11. All lobes of leaf straight or nearly so .... 5. *Q. pagodaefol*
12. Leaves much broader in the upper half, with 3 (–5) broad, shallow lob
.... 3. *Q. marilandic*

2. Leaves broader at or below the middle, with 5–11 narrower, deeper lobes
.... 13
3. Terminal winter buds 6–12 mm long, gray-tomentose; scales along edge
of acorn cup not appressed; upper leaf surface with pubescent midnerve
.... 6. *Q. velutina*
3. Terminal winter buds up to 4 mm long, not gray-tomentose; scales along
edge of acorn cup appressed; upper leaf surface with glabrous midnerve
.... 14
4. Leaves lobed less than half-way to middle; cup covering less than one-
fourth of acorn .... 7. *Q. ruba*
4. Leaves lobed at least half-way to middle; cup covering at least one-
fourth of acorn 15
5. Acorn cup covering much less than one-half of the acorn 16
5. Acorn cup covering about one-half of the acorn 18
6. Acorn up to 1.5 cm long; cup not exceeding 1.5 cm in diameter ....
.... 8. *Q. palustris*
6. Acorn more than 1.5 cm long; cup more than 1.5 cm in diameter ... 17
7. Buds glabrous; leaves not long-tapering at the tip; acorn cup mostly 2
cm broad or broader .... 9. *Q. shumardii*
7. Buds puberulent, or at least ciliate; leaves tending to be long-tapering at
the tip; acorn cup mostly 1.5–2.0 cm broad .... 10. *Q. nuttallii*
8. Acorn cup up to 1.5 cm in diameter, with pubescent scales ....
.... 11. *Q. ellipsoidalis*
8. Acorn cup at least 1.5 cm in diameter, with glabrous scales ....
.... 12. *Q. coccinea*
9. Leaves completely glabrous beneath .... 17. *Q. alba*
9. Leaves pubescent beneath 20
0. Leaves 3- to 5-lobed, the upper three with squarish tips, forming a cross;
acorn up to 1 cm across, the cup unfringed and covering less than half
of the nut .... 18. *Q. stellata*
0. Leaves 5- to 11-lobed, the upper three without squarish tips; acorn 1 cm
broad or broader, the cup either fringed or covering half to nearly all the
nut 21
1. Scales of cup forming a fringe around acorn; cup ½–⅘ covering the
acorn .... 19. *Q. macrocarpa*
1. Scales of cup not forming a fringe; cup nearly completely covering the
acorn .... 20. *Q. lyrata*

1. *Q. imbricaria* Michx. Shingle Oak. April–May. Moist or dry woods,
edges of fields; occasional to common throughout the state. Hybrids
between this species and *Q. falcata* (*Q.* X *anceps* E. J. Palmer), *Q.
marilandica* (Q. X *tridentata* [A. DC.] Engelm), *Q. palustris* (*Q.* X *exacta* Trel.), *Q. rubra* (Q X *runcinata* [A. DC.] Engelm.), and *Q. velutina*
(*Q.* X *leana* Nutt.) have been reported from the state.
2. *Q. phellos* L. Willow Oak. April–May. Swampy woods; Alexander,
Massac, Pulaski, and Union cos. Hybrids between this species and *Q.
palustris* (*Q.* X *schochiana* Dieck) and *Q. veluntina* (Q. X *filialis* Little)
have been reported from the state.
3. *Q. marilandica* Muenchh. Blackjack Oak. April–May. Upland
woods, bluff-tops; occasional to common in the s. three-fourths of
the state, absent elsewhere. Hybrids between this species and *Q. imbricaria* (Q. X *tridentata* [A. DC.] Engelm.) and Q. *velutina* (Q. X
bushii Sarg.) have been reported from the state.
4. *Q. falcata* Michx. Spanish Oak. April–May. Moist or dry woods;
occasional in the s. one-fifth of the state. Hybrids between this spe-

cies and *Q. imbricaria* (*Q.* X *anceps* E. J. Palmer) and *Q. phellos* (*Q.* X *ludoviciana* Sarg.) have been found in the state.

5. Q. pagodaefolia (Ell.) Ashe. Cherry-bark Oak. April–May. Rich woods; occasional in the s. one-fifth of the state. *Q. falcata* Michx. var. *pagodaefolia* Ell.—F, G; *Q. falcata* Michx.—J.

6. Q. velutina Lam. Black Oak. April–May. Young branchlets and lower surface of leaves glabrate at maturity, except in the leaf axils. Upland woods; common throughout the state; in every co. Hybrids between this species and *Q. imbricaria* (*Q.* X *leana* Nutt.), *Q. marilandica* (*Q.* X *bushii* Sarg.), and *Q. phellos* have been reported from the state.

f. missouriensis (Sarg.) Trel. April–May. Branchlets and lower surface of leaves permanently pubescent. Upland woods; Saline Co.

7. Q. rubra L. Red Oak. April–May. Upland woods; common throughout the state. Including var. *borealis* [Michx. f.] Farw.) Hybrids between this species and *Q. imbricaria* (*Q.* X *runcinata* [A. DC] Engelm.) have been reported from the state.

8. Q. palustris Muenchh. Pin Oak. April–May. Wet ground; occasional to common in all parts of Ill. except the extreme nw. co. Hybrids between this species and *Q. imbricaria* (*Q.* X *exacta* Trel.) and *Q. phellos* (*Q.* X *schochiana* Dieck) have been reported from the state.

9. Q. shumardii Buckley. Shumard's Oak. April–May. Low woods; occasional in the s. one-third of the state; also McLean Co. (Including *Q. shumardii* Buckley var. *schneckii* [Britt.] Sarg.)

10. Q. nuttallii E. J. Palmer. Nuttall's Oak. April–May. Wet woods; very rare; Alexander Co.

11. Q. ellipsoidalis E. J. Hill. Hill's Oak. April–May. Upland woods; occasional in the n. one-fourth of the state.

12. Q. coccinea Muenchh. Scarlet Oak. April–May. Upland woods; occasional in the s. one-third of the state.

13. Q. bicolor Willd. Swamp White Oak. April–May. Low woods and swamps; occasional throughout the state. Hybrids between this species and *Q. alba* (*Q.* X *jackiana* Schneider), *Q. lyrata* (*Q.* X *humidicola* E. J. Palmer), and *Q. macrocarpa* (*Q.* X *hillii* Trelease) have been reported from the state.

14. Q. michauxii Nutt. Basket Oak. April–May. Low woods and swamps; occasional in the s. one-third of the state.

15. Q. muhlenbergii Engelm. Yellow Chestnut Oak. April–May. Moist or dry woods; occasional throughout the state. (Including f. *alexanderi* [Britt.] Trel.) *Q. prinoides* Willd. var. *acuminata* (Michx.) Gl.—Hybrids between this species and *Q. alba* (*Q.* X *deamii* Trel.) and *Q. macrocarpa* (*Q.* X *fallax* E. J. Palmer) have been reported from the state.

16. Q. prinus L. Rock Chestnut Oak. April–May. Rocky woods; rare; Alexander, Hardin, Saline, and Union cos.

17. Q. alba L. White Oak. April–May. Upland woods; common throughout the state; in every co. (Including f. *repanda* [Michx.] Trel. and f. *latiloba* [Sarg.] Palmer & Steyerm.) Hybrids between this species and *Q. bicolor* (*Q.* X *jackiana* Schneider), *Q. macrocarpa* (*Q.* X *bebbiana* Schneider), *Q. muhlenbergii* (*Q.* X *deamii* Trel.), and *Q. stellata* (*Q.* X *fernowi* Trel.) have been reported from Ill.

18. Q. stellata Wangh. Post Oak. April–May. Upland woods, bluffs; occasional to common in the s. half of the state; also Grundy Co.

Hybrids between this species and *Q. alba* (*Q.* X *fernowi* Trel.) have been reported from Ill.

19. Q. macrocarpa Michx. Bur. Oak. April–May. Bottomland woods; occasional to common throughout the state; in every co. Hybrids between this species and *Q. alba* (*Q.* X *bebbiana* Schneider), *Q. bicolor* (*Q.* X *hillii* Trel.), and *Q. muhlenbergii* (*Q.* X *fallax* E. J. Palmer) have been reported from Ill. (Including f. *olivaeformis* [Michx. f.] Trel.)

20. Q. lyrata Walt. Overcup Oak. April–May. Bottomland woods; occasional in the s. half of the state.

45. ULMACEAE

1. Leaves with strong lateral veins arising from the main vein at a distance above the very base of the blade; pith of branches solid; sepals united; fruit a samara or nut 2
1. Leaves with a pair of strong lateral veins arising from the main vein at the very base of the blade; pith of branches chambered; sepals free; fruit a drupe .... 3. *Celtis*
2. Fruit a samara; all flowers perfect, appearing before the leaves; leaves mostly doubly toothed .... 1. *Ulmus*
2. Fruit a wingless nut; at least some of the flowers unisexual, appearing with the leaves; leaves mostly singly toothed .... 2. *Planera*

1. Ulmus L. Elm

1. Upper surface of leaves harshly pubescent; winter buds rusty-pubescent; samaras eciliate on the margins but pubescent at the center of both sides .... 1. *U. rubra*
1. Upper surface of leaves glabrous or pubescent, but never harshly so; winter buds glabrous or pubescent, but not with rusty pubescence; samaras ciliate or, if eciliate, the center of each side glabrous ...┊....... 2
2. None of the branches corky-winged 3
2. Some of the branches corky-winged 4
3. Leaves doubly serrate, usually strongly asymmetrical at the base; samaras ciliate; flowers pendulous from long pedicels .... 2. *U. americana*
3. Leaves mostly singly serrate, usually nearly symmetrical at the base; samaras eciliate; flowers sessile or nearly so .... 3. *U. pumila*
4. Buds glabrous or nearly so; leaves sessile or on petioles up to 3 mm long; samaras (excluding cilia) up to 5 mm broad .... 4. *U. alata*
4. Buds downy-pubescent; leaves on petioles 3 mm long or longer; samaras (excluding cilia) 1.0–1.5 cm broad 5
5. Most of the mature leaves over 8 cm long, glabrous above; flowers pendulous from long pedicels; samaras ciliate .... 5. *U. thomasii*
5. Most of the mature leaves less than 8 cm long, scabrous above; flowers sessile or nearly so; samaras eciliate .... 6. *U. procera*

1. U. rubra Muhl. Slippery Elm. Feb.–April. Woods, often in disturbed areas, common; in every co.

2. U. americana L. American Elm. Feb.–April. Rich woods, wooded flood plains, common; in every co.

3. U. pumila L. Siberian Elm. May. Native of Asia; escaped int
waste areas; occasional to common.

4. U. alata Michx. Winged Elm. Feb.–March. Wooded slopes, cliff
or low woodlands; common in the s. one-fourth of the state; ap
parently absent elsewhere.

5. U. thomasii Sarg. Rock Elm. March–May. Rich woods, not com
mon in the n. half of Ill. absent elsewhere.

6. U. procera Salisb. English Elm. April. Native of Europe; rarely e
caped in Ill.; Jackson Co.

2. *Planera* J. F. Gmel.   *Water Elm*

1. P. aquatica [Walt.] J. F. Gmel. Water Elm. March–April. Swamp
woods rare; Alexander, Jackson, Johnson, Massac, Pope, Pulask
Randolph, and Union cos.

3. *Celtis* L.   *Hackberry*

1. Leaves conspicuously several-toothed ....
1. Leaves entire or only sparingly toothed ....
2. Leaves harshly scabrous on the upper surface ...... 1. *C. occidental*
2. Leaves smooth or nearly so on the upper surface ....
3. Most of the leaves more than half as long as broad, very strongly asyr
   metrical at the base .... 1. *C. occidental*
3. Most of the leaves less than half as long as broad, only slightly asymme
   rical at the base ....
4. Drupe 8–11 mm long, dark purple or dark brown .... 1. *C. occidental*
4. Drupe 5–7 mm long, orange, pale brown, or red .... 2. *C. laeviga*
5. Leaves harshly scabrous on the upper surface ....
5. Leaves smooth or nearly so on the upper surface ....
6. Leaves mostly over half as broad as long .... 3. *C. tenuifol*
6. Leaves mostly less than half as broad as long .... 2. *C. laeviga*
7. Leaves more than half as broad as long, acute to short-acuminate at th
   apex .... 3. *C. tenuifol*
7. Leaves less than half as broad as long, long-tapering at the apex ....
   .... 2. *C. laeviga*

1. C. occidentalis L. Hackberry. April–May. Leaves harshly scabro
on the upper surface. Low woods; common throughout the state.

var. pumila (Pursh) Gray. Hackberry. April–May. Leaves more
less smooth on the upper surface, most of them more than half
broad as long. Low woods, bluffs; occasional throughout the state.

var. canina (Raf.) Sarg. Hackberry. April–May. Leaves more or le
smooth on the upper surface, most of them less than half as broad
long. Low woods; occasional throughout the state.

2. C. laevigata Willd. Sugarberry. April–May. Leaves more or le
smooth on the upper surface, entire or nearly so, the petioles esse
tially glabrous. Low woods; occasional to common in the s. half
the state.

var. smallii (Beadle) Sarg. Sugarberry. April–May. Leaves more
less smooth on the upper surface, regularly toothed, the petioles e
sentially glabrous. Low woods, occasional in the extreme s. cos.

var. texana Sarg. Dwarf Hackberry. April–May. Leaves harshly sc
brous on the upper surface, entire or nearly so, the petioles pube
cent. Dry woods and cliffs; occasional in the s. one-fifth of the stat

3. C. tenuifolia Nutt. Dwarf Hackberry. April–May. Leaves more or less smooth on the upper surface, membranaceous, the petioles glabrous or nearly so. Dry woods and cliffs; occasional, mostly in the s. one-fifth of the state. *C. pumila* Pursh—J.

var. georgiana (Small) Fern. & Schub. Dwarf Hackberry. April–May. Leaves harshly scabrous on the upper surface, leathery, the petioles densely hairy. Dry woods and cliffs; occasional, mostly in the s. one-fifth of the state. *C. pumila* Pursh—J.

46. MORACEAE

1. Trees or shrubs; latex present 2
1. Herbs or vines; latex absent 4
2. Leaves toothed, often lobed; branches without spines; fruit 1–2 cm in diameter, white, orange, or red 3
2. Leaves entire, unlobed; branches with short spines; fruit 10–20 cm in diameter, greenish-yellow .... 3. *Maclura*
3. Leaves glabrous or short-hairy on the lower surface, not velvety; most or all the petioles less than 3 cm long, glabrous or appressed-pubescent; bark roughened .... 1. *Morus*
3. Leaves velvety on the lower surface; most or all the petioles more than 3 cm long, pilose; bark smooth .... 2. *Broussonetia*
4. Stems climbing or trailing; leaves simple although often lobed .... 4. *Humulus*
4. Stems erect; leaves compound .... 5. *Cannabis*

1. Morus L.   Mulberry

1. Lower surface of leaves pubescent on the blade as well as on the veins .... 1. *M. rubra*
1. Lower surface of leaves glabrous, or pubescent only on the veins, or the hairs confined to axillary tufts .... 2. *M. alba*

1. M. rubra L. Red Mulberry. April–May. Lowland or upland woods, edges of fields; common throughout most of Ill.

2. M. alba L. White Mulberry. April–May. Fruits white or pale pink, rarely dark purple or black, usually over 1 cm long. Native of Asia; naturalized throughout the state. (Including f. *skeletoniana* [Schneider] Rehder.)

var. tatarica (L.) Loudon. Russian Mulberry. Fruits red, up to 1 cm long. Native of Europe and Asia; naturalized along the edges of woods; occasional throughout the state. *M. tatarica* L.—J.

2. Broussonetia L'Hér.   Paper Mulberry

1. B. papyrifera (L.) L'Hér. Paper Mulberry. April–May. Native of Asia; occasionally naturalized in the s. one-fourth of the state.

3. Maclura Nutt.   Osage Orange

1. M. pomifera (Raf.) Schneider. Osage Orange. May–June. Native of the s. cent. U.S.; apparently naturalized in Ill.; occasional throughout the state.

4. Humulus L. Hops

1. Most of the leaves 3-lobed, with resinous glands beneath   1. *H. lupulu*
1. Most of the leaves 5- to 7-lobed, without resinous glands beneath....
.... 2. *H. japonicu*

    1. H lupulus L. Common Hops. July–Aug. Fencerows, thickets, d
wooded slopes; occasional throughout the state. *H. american*
Nutt.—J.
    2. H. japonicus Sieb. & Zucc. Japanese Hops. July–Sept. Native
Asia; naturalized in waste areas; occasional and scattered in Ill.

5. Cannabis L. Hemp

    1. C. sativa L. Marijuana. June–Oct. Native of Asia; naturalized ar
escaped into waste areas; occasional throughout the state, althoug
apparently less common in the southernmost cos.

47. URTICACEAE

1. Leaves opposite (a few alternate leaves may occur in *Boehmeria*) ..
1. Leaves all alternate ....
2. Leaves and usually the stems pubescent ....
2. Leaves and stems glabrous .... 3. *Pi*
3. Pistillate flowers with free sepals; stinging hairs often present 1. *Urt*
3. Pistillate flowers with united sepals; stinging hairs never present....
.... 2. *Boehme*
4. Leaves serrate; stinging hairs present .... 4. *Lapor*
4. Leaves essentially entire; stinging hairs absent .... 5. *Parieta*

1. Urtica L. Nettle

1. Perennials often attaining a height of 1 m or more; inflorescences 2
long or longer .... 1. *U. dio*
1. Annuals never attaining a height of 1 m; inflorescences less than 2
long ....
2. Leaves dentate, tapering to the base .... 2. *U. ur*
2. Leaves crenate, some of them cordate at the base ....
.... 3. *U. chamaedryoi*

    1. U. dioica L. Stinging Nettle. June–Sept. Rich woods, moist wa
ground; occasional to common in the n. half of the state, rare in the
half. (Including *U. gracilis* Ait. and *U. procera* Muhl.)
    2. U. urens L. Burning Nettle. June–Sept. Native of Europe a
Asia; rarely escaped into waste areas.
    3. U. chamaedryoides Pursh. Nettle. April–June. Swampy woo
river banks; Alexander and Jackson cos.

2. Boehmeria Jacq. False Nettle

    1. B. cylindrica (L.) Sw. False Nettle. July–Oct. Leaves smooth
slightly scabrous above. Moist woods, common; throughout the sta

var. drummondiana Wedd. False Nettle. July–Oct. Leaves harshly scabrous above. Edge of bogs and marshes, rare; Alexander, Dewitt, Jackson, Lake, and Will cos. *B. drummondiana* Wedd.—J.

3. Pilea Lindl. Clearweed

1. Achenes green, averaging 1 mm wide .... 1. *P. pumila*
1. Achenes black, averaging 1.5 mm wide 2
2. Plants leafy above the middle; achenes averaging 1.5 mm long ....
.... 2. *P. fontana*
2. Plants leafy from near the base; achenes averaging 2.0 mm long ....
.... 3. *P. opaca*

1. P. pumila (P.) Gray. Clearweed. July–Sept. Leaves mostly tapering to the base, with 3–11 teeth on each margin. Moist, usually shaded, ground, common; probably in every co. (Including var. *deamii* [Lunell] Fern.)
2. P. fontana (Lunell) Rydb. July–Sept. Moist soil; occasional and scattered in Ill.
3. P. opaca (Lunell) Rydb. July–Sept. Moist soil; occasional and scattered in Ill.

4. Laportea Gaud. Wood Nettle

1. L. canadensis (L.) Wedd. Wood Nettle. June–Sept. Moist soil; common throughout the state.

5. Parietaria L. Pellitory

1. P. pensylvanica Muhl. Pellitory. May–Sept. In shade beneath cliffs and along buildings; common throughout the state.

48. SANTALACEAE

1. Comandra Nutt. False Toadflax

1. C. richardsiana Fern. False Toadflax. May–Aug. Woods, prairies, occasional; throughout the state, except for the very southernmost cos. *C. umbellata* (L.) Nutt.—J.

49. LORANTHACEAE

1. Phoradendron Nutt. Mistletoe

1. P. flavescens (Pursh) Nutt. Mistletoe. Sept.–Oct. On various deciduous trees, mostly in low areas; occasional in the s. one-sixth of the state and extending along the Wabash River to Clark Co.

50. ARISTOLOCHIACEAE

1. Leaves 2, basal; stamens 12; calyx regular .... 1. *Asar*
1. Leaves more than 2, cauline; stamens 6; calyx irregular   2. *Aristoloc*

1. Asarum L. Wild Ginger

1. A. canadense L. Wild Ginger. April–May. Tips of calyx lobes
tended into a point 5–20 mm long. Rich woodlands; restricted to the
half of the state. *A. canadense* var. *acuminatum* Ashe—F; *A. acumi*
tum (Ashe) Bickn.—J.

var. reflexum (Bickn.) Robins. Wild Ginger. April–May. Tips of ca
lobes merely short-pointed, the point not exceeding 5 mm in leng
Rich woodlands; common throughout Ill. *A. reflexum* Bickn.—J.

2. Aristolochia L. Birthwort

1. Erect herb; lower surface of leaves pubescent, but not white-woo
calyx dark purple, to 15 mm long .... 1. *A. serpenta*
1. High-climbing woody twiner; lower surface of leaves white-woolly; ca
yellow, with a purplish orifice, over 15 mm long ...... 2. *A. toment*

1. A. serpentaria L. Virginia Snakeroot. May–July. Some or all
leaves over 2 cm across midway from base to apex. Rich woodlan
local in the s. half of the state.

var. hastata (Nutt.) Duchartre. Narrow-leaved Snakeroot. June–J
Leaves never more than 2 cm across midway from base to apex. R
or swampy woods, rare; Alexander and Johnson cos. *A. na*
Kearney—J.

2. A. tomentosa Sims. Dutchman's Pipe-vine. May–June. Chi
calcareous woods, rare; confined to the s. half of Ill.

51. POLYGONACEAE

1. Climbing shrubs with tendrils; pedicels winged on one side by the ca
.... 1. *Brunnic*
1. Prostrate or ascending or erect herbs or, if climbing, then without
drils; pedicels not winged by the calyx ....
2. Leaves needle-like .... 2. *Polygon*
2. Leaves linear or broader, not needle-like ....
3. Sepals 6, the inner 3 longer than the outer in fruit .... 3. *Rur*
3. Sepals usually 5, the inner not longer than the outer in fruit ....
4. Stems retrorse-prickly .... 5. *Polygon*
4. Stems not retrorse-prickly ....
5. Leaves hastate-deltoid .... 4. *Fagopyr*
5. Leaves not hastate-deltoid .... 5. *Polygon*

1. Brunnichia Banks

1. B. cirrhosa Banks. Ladies' Eardrops. June–Aug. Swampy woo
confined to the s. one-sixth of the state.

2. Polygonella Michx. Jointweed

1. P. articulata (L.) Meisn. Jointweed. July–Oct. Sandy soil; restricted to the n. one-third of the state.

3. Rumex L. Dock

Some of the leaves hastate (rarely entire in one form of *R. acetosella*); plants dioecious 2
Leaves not hastate; plants monoecious 3
Plants with stolons or slender rhizomes; achenes exserted from calyx .. .. 1. *R. acetosella*
Plants with a taproot; achenes enclosed by the calyx . 2. *R. hastatulus*
Fruiting sepals with spinulose bristles 4
Fruiting sepals entire or only minutely toothed 5
Stems hollow; tubercle of fruit long and slender; bristles longer than width of fruiting sepals .... 3. *R. maritimus*
Stems firm; tubercle of fruit only slightly longer than broad; bristles not longer than width of fruiting sepals .... 4. *R. obtusifolius*
Only one of the 3 fruiting sepals with a tubercle 6
Each of the fruiting sepals with a tubercle 8
Some of the leaves over 10 cm broad; each fruiting sepal 8–10 mm broad .... 5. *R. patientia*
None of the leaves 10 cm broad; each fruiting sepal up to 6 mm broad 7
Leaves wavy along the margins .... 6. *R. crispus*
Leaves flat .... 7. *R. altissimus*
Leaves with conspicuous wavy margins .... 6. *R. crispus*
Leaves flat, entire or crenulate 9
Leaves crenulate; lateral veins of leaves forming right angles with vertical veins .... 8. *R. orbiculatus*
Leaves entire; lateral veins of leaves ascending 10
Fruiting pedicels two to five times as long as the calyx .... 9. *R. verticillatus*
Fruiting pedicels shorter than or barely longer than the calyx 11
Leaves narrowly lanceolate, never more than 3 cm broad .... 10. *R. mexicanus*
Leaves broadly lanceolate, at least some of them over 3 cm broad .... 7. *R. altissimus*

1. R. acetosella L. Sour Dock. May–Aug. Leaves hastate. Native of Europe; adventive in fields, roadsides, waste areas; in every co.

f. integrifolius (Wallr.) G. Beck. Leaves unlobed. Native of Europe; rarely adventive in Ill.; Kane Co.

2. R. hastatulus Baldw. Sour Dock. May–Aug. Sandy soil, rare; Madison and St. Clair cos.

3. R. maritimus L. var. fueginus (Phil.) Dusén. July–Oct. Sandy shores; scattered throughout the state. *R. fueginus* Phil.—J.

4. R. obtusifolius L. Bitter Dock. June–Sept. Native of Europe; adventive in waste ground; occasional throughout the state.

5. R. patientia L. Patience Dock. June–July. Native of Europe and Asia; adventive in waste ground; scattered throughout the state.

6. R. crispus L. Curly Dock. May–Aug. Native of Europe; adventive in waste ground; in every co. (Including f. *unicallosus* Peterm.)

7. R. altissimus Wood. Pale Dock. May–Aug. Moist soil; occasional to common throughout the state.

8. R. orbiculatus Gray. Water Dock. June–Sept. Moist soil; stricted to the n. two-thirds of the state.

9. R. verticillatus L. Swamp Dock. June–Sept. Wet ground; comm in the n. half of the state, occasional in the s. half.

10. R. mexicanus Meisn. Dock. July–Sept. Moist soil; scatter throughout the state. *R. triangulivalvis* (Danser) Rech. f.—J.

4. *Fagopyrum Mill. Buckwheat*

1. F. esculentum Moench. Buckwheat. June–Sept. Native of As adventive in waste ground; occasional in the n. three-fourths of state, apparently absent elsewhere. *F. sagittatum* Gilib.—F.

5. *Polygonum L. Smartweed*

1. Stems reflexed-prickly ....
1. Stems not reflexed-prickly ....
2. Achenes trigonous; leaves sagittate .... 1. *P. sagittat*
2. Achenes lenticular; leaves hastate .... 2. *P. arifoli*
3. Stems twining or trailing, neither erect nor prostrate ....
3. Stems erect or prostrate ....
4. Achenes dull, granular; outer sepals keeled .... 3. *P. convolvu*
4. Achenes shining, not granular; outer sepals winged ....
5. Calyx in fruit up to 6 mm long; achenes up to 3 mm long   4. *P. cristat*
5. Calyx in fruit more than 6 mm long; achenes more than 3 mm long .
    .... 5. *P. scande*
6. Flowers borne in small axillary clusters ....
6. Flowers borne in terminal and/or axillary spikes or racemes ....
7. Leaves plicate; stems angular .... 6. *P. ter*
7. Leaves flattened; stems terete or nearly so ....
8. Outer sepals flat or nearly so .... 7. *P. avicul*
8. Outer sepals cucullate ....
9. Achenes shiny ....
9. Achenes dull ....
10. Leaves blue-green; calyx up to 2 mm long .... 8. *P. prolific*
10. Leaves yellow-green; calyx 3 mm long or longer ....
11. Achenes green, mostly all 4 mm long or longer .... 9. *P. exsert*
11. Achenes brown, mostly less than 4 mm long .... 10. *P. ramosissim*
12. Leaves yellow-green; achenes brown; pedicels exserted from the ocr at maturity .... 11. *P. erect*
12. Leave blue-green; achenes olive; pedicels included within the ocreae maturity .... 12. *P. achore*
13. Outer sepals broadly winged ....
13. Outer sepals unwinged ....
14. Leaves cordate at base .... 13. *P. sachaline*
14. Leaves truncate at base .... 14. *P. cuspidat*
15. Styles persistent as beaks on the achenes; calyx usually 4-parted ...
    .... 15. *P. virginian*
15. Styles deciduous; calyx usually 5-parted ....
16. Ocreae with bristles ....
16. Ocreae without bristles ....
17. Spikes mostly 1–3 .... 25. *P. amphib*
17. Spikes usually more than 3 ....
18. Peduncles with stipitate glands .... 16. *P. car*

    1. P. sagittatum L. Tear Thumb. July–Oct. Swampy woods, wet ground; occasional throughout the state.

    2. P. arifolium L. var. pubescens (Keller) Fern. Tear Thumb. July–Oct. Wet ground, very rare; McHenry and Macon cos. *P. arifolium* L.—J.

    3. P. convolvulus L. Black Bindweed. May–Oct. Native of Europe; fields and edges of woods; occasional to common throughout the state.

    4. P. cristatum Engelm. & Gray. Crested Bindweed. Aug.–Sept. Woods; occasional in the s. four-fifths of the state, rare elsewhere. (Including *P. dumetorum* L.) *P. scandens* L. var. *cristatum* (Engelm. & Gray) Gl.—G.

    5. P. scandens L. False Buckwheat. Aug.–Oct. Woods and thickets; occasional to common throughout the state

    6. P. tenue Michx. Slender Knotweed. July–Sept. Dry, sandy soil; crevices of sandstone blufftops; occasional throughout the state.

    7. P. aviculare L. Knotweed. June–Oct. Leaves thin, linear to lanceolate, more or less acute. Native of Europe; waste ground; common throughout the state. (Including var. *vegetum* Ledeb.)

    var. littorale (Link) W. D. J. Koch. Knotweed. June–Oct. Leaves thick, oblong to elliptic, obtuse. Waste ground, rare; scattered in (Including *P. buxiforme* Small.)

8. P. prolificum (Small) Robins. Knotweed. July–Oct. Waste grour rare; Johnson Co.

9. P. exsertum Small. Long-fruited Knotweed. Aug.–Oct. Banks a shores, usually in sand; occasional throughout the state. *P. ramos simum* Michx.—G, J.

10. P. ramosissimum Michx. Knotweed. July–Oct. Sandy soil; c casional throughout the state.

11. P. erectum L. Knotweed. Aug.–Oct. Waste ground; occasional common throughout the state.

12. P. achoreum Blake. Knotweed. Aug.–Oct. Waste ground, ra Champaign and JoDaviess cos. *P. erectum* L.—J.

13. P. sachalinense F. Schmidt. Giant Knotweed. Aug.–Oct. Native e. Asia; rarely adventive in Ill.; Jackson Co.

14. P. cuspidatum Sieb. & Zucc. Japanese Knotweed. Aug.–Sept. N tive of e. Asia; occasionally spreading from cultivation in Ill.

15. P. virginianum L. Virginia Knotweed. July–Sept. Woods; comm throughout the state. *Tovara virginiana* (L.) Raf.—F.

16. P. careyi Olney. Carey's Smartweed. July–Sept. Sandy soil, ra Kankakee Co.

17. P. orientale L. Prince's Feather. June–Oct. Native of Eurasia; caped into waste ground; occasional and scattered throughout state.

18. P. punctatum Ell. Smartweed. July–Oct. Wet ground; comm throughout the state. (Including var. *leptostachyum* [Meisn.] Small.

19. P. hydropiper L. Smartweed. June–Oct. Native of Europe; adve tive in wet ground; occasional throughout the state.

20. P. persicaria L. Lady's Thumb. May–Sept. Native of Europe; a ventive in waste ground; common throughout the state.

21. P. cespitosum Blum var. longisetum (DeBruyn) Steward. Cree ing Smartweed. June–Oct. Native of se. Asia; adventive in shad ground and in lawns; scattered throughout the state. *P. longisetu* DeBruyn—J.

22. P. setaceum Baldw. var. interjectum Fern. Smartweed. July–O Wet ground, not common; confined to the s. one-third of the sta *P. hydropiperoides* Michx.—J; *P. hydropiperoides* Michx. var. *taceum* (Baldw.) Gl.—G.

23. P. hydropiperoides Michx. Mild Water Pepper. June–Oct. W ground, sometimes in shallow water; occasional to common throug out the state.

24. P. opelousanum Riddell. Water Pepper. July–Oct. Wet grou not common; confined to the s. one-fourth of the state. *P. hydro peroides* Michx.—J; *P. hydropiperoides* Michx. var. *opelousan* (Riddell) Stone—G.

25. P. amphibium L. var. stipulaceum (Coleman) Fern. Water Sma weed. June–Sept. Ocreae usually with bristles. Wet ground; casional in the n. half of the state, rare elsewhere.*P. fluitans* Eato hartwrightii (Gray) G. N. Jones—J; *P. natans* Eaton f. *hartwrig* (Gray) Stanford—G.

f. fluitans (Eat.) Fern. Water Smartweed. June–Sept. Ocreae with bristles. In water; occasional in the n. half of the state, rare elsewhe *P. fluitans* Eaton—J; *P. natans* Eaton—G.

26. P. coccineum Muhl. Water Smartweed. July–Oct. Wet grou

occasional throughout the state. (Including var. *pratincola* [Greene] Stanford.)

27. P. longistylum Small. Smartweed. July–Oct. Wet ground, rare; Alexander, St. Clair, and Union cos. *P. pennsylvanicum* L.—J.

28. P. scabrum Moench. Smartweed. June–Aug. Native of Europe; rarely adventive along railroads; Champaign and Lee cos. (Including *P. tomentosum* Schrank.) *P. lapathifolium* L.—G, J.

29. P. lapathifolium L. Pale Smartweed. July–Oct. Wet ground; common throughout the state.

30. P. pensylvanicum L. Common Smartweed. July–Oct. Leaves strigose; peduncles with spreading hairs. Wet ground; occasional throughout the state.

var. durum Stanford. July–Sept. Leaves strigose; peduncles strigose. Wet ground, rare; Pope Co.

var. laevigatum Fern. July–Oct. Leaves glabrous. Wet ground, common; in every co.

52. CHENOPODIACEAE

Leaves reduced to small, opposite scales; stems jointed, fleshy . . . . 1. *Salicornia*
Leaves linear-subulate or broader; stems not conspicuously jointed or fleshy (sometimes succulent in *Salsola*) 2
Leaves linear-subulate to linear, up to 3 mm broad 3
Leaves linear-lanceolate or broader, over 3 mm broad. 7
Leaves at most about 1 cm long . . . . 2. *Polycnemum*
Leaves 2 cm long or longer 4
Uppermost leaves spinose . . . . 3. *Salsola*
Uppermost leaves not spinose 5
Sepals 5 6
Sepal 1 . . . . 6. *Corispermum*
Stems villous above; leaves more or less pubescent; all sepals winged in fruit . . . . 4. *Kochia*
Stems glabrous; leaves glabrous; only 1 or 3 of the sepals winged in fruit . . . . 5. *Suaeda*
Fruiting calyx horizontally winged . . . . 7. *Cycloloma*
Fruiting calyx sometimes variously keeled or ridged, but not horizontally winged 8
Flowers unisexual; fruit enclosed by 2 bracteoles . . . . 8. *Atriplex*
Flowers perfect; fruit enclosed by the persistent calyx 9. *Chenopodium*

1. Salicornia L. Glasswort

1. S. europaea L. Glasswort. Aug.–Oct. Native of Europe, Asia, and Africa; rarely adventive in soils with a high mineral content; Cook Co.

2. Polycnemum L.

1. P. majus A. Br. July–Sept. Native of Europe; rarely adventive along a railroad; Monroe Co.

3. Salsola L.   Saltwort

1. S. kali L. var. tenuifolia Tausch. Russian Thistle. July–Sept. N
tive of Europe and Asia; adventive along railroads and on sa
beaches; occasional in the n. half of the state, rare in the s. ha
S. pestifer Nels.—J.

4. Kochia Roth   Burning-bush

1. K. scoparia (L.) Roth. Summer Cypress. July–Sept. Native
Europe and Asia; naturalized in waste areas; common in the n. half
the state, occasional to rare elsewhere. (Including var. culta Farw.)

5. Suaeda Forsk.   Sea-Blite

1. S. depressa (Pursh) S. Wats. Sea-Blite. Aug.–Oct. Native of the
US.; adventive along a highway, apparently in response to saltir
Cook Co.

6. Corispermum L.   Bugseed

1. C. hyssopifolium L. Common Bugseed. July–Sept. Sandy sc
rare; Cook, Lake, Mason, Menard, and Whiteside cos.

7. Cycloloma Moq.   Winged Pigweed

1. C. atriplicifolium (Spreng.) Coult. Winged Pigweed. July–Au
Sandy soil; occasional throughout the state.

8. Atriplex L.   Orach

1. Most or all the lowermost leaves opposite or subopposite; stems a
leaves not mealy or scurfy (rarely mealy in young specimens of A. h
tensis) ....
1. All the leaves alternate; stems and leaves mealy or scurfy ....
2. Bracts subtending fruits 10–15 mm broad; seeds 2–4 mm across ....
.... 1. A. horten
2. Bracts subtending fruits up to 5 mm broad; seeds 1–2 mm across ...
.... 2. A. pate
3. Stems and leaves mealy; bracts subtending fruits entire or dentate ..
.... 3. A. ros
3. Stems and leaves scurfy; bracts subtending fruits deeply cleft ......
.... 4. A. argent

1. A. hortensis L. Garden Orach. Aug.–Sept. Native of Asia; rar
escaped from cultivation; DuPage, Lake, and Vermilion cos.
2. A. patula L. Spear Scale. July–Sept. Waste ground, often in sal
areas; occasional in the n. half of the state, uncommon elsewhe
(Including A. hastata L. and A. patula L. var. hastata [L.] Gray.)
3. A. rosea L. Red Scale. Aug.–Oct. Native of Europe and As
rarely adventive in alkaline waste ground; Cook Co.
4. A. argentea Nutt. Silver Scale. July–Oct. Native of the w. U.
rarely adventive in Ill.; Cook, Hancock, and Menard cos.

9. *Chenopodium* L.   Goosefoot

1. Plants with glandular hairs or resinous sessile glands on the leaves and/or the stems; plants strongly aromatic 2
1. Plants without glandular-pubescence or without sessile glands; plants usually not aromatic 3
2. Calyx lobes glandular-pubescent; stems with spreading, glandular hairs .... 1. *C. botrys*
2. Calyx lobes not glandular-pubescent; stems glabrous or with appressed, glandless hairs .... 2. *C. ambrosioides*
3. Lower surface of leaves densely white-mealy 4
3. Lower surface of leaves not white-mealy 12
4. Leaves entire, mostly linear, oblong, or narrowly lanceolate 5
4. Leaves lobed or toothed, mostly lanceolate, ovate, deltoid, or rhombic 6
5. Leaves 1-nerved, linear; pericarp somewhat firmly attached to seed ....
.... 3. *C. pallescens*
5. Leaves 3-nerved, oblong or narrowly lanceolate; pericarp readily removed from seed .... 4. *C. desiccatum*
6. Most of the seeds vertical, about 0.5 mm broad, free from the pericarp; calyx lobes mostly 3–4 .... 5. *C. glaucum*
6. Most of the seeds horizontal, 0.8–2.0 mm broad, firmly attached to the pericarp (easily removed in *C. desiccatum*); calyx lobes 5 7
7. Seeds honeycombed on the surface 8
7. Seeds smooth, or at least not honeycombed on the surface 9
8. Calyx lobes conspicuously keeled; seeds 0.8–1.5 mm broad ....
.... 6. *C. berlandieri*
8. Calyx lobes slightly or not at all keeled; seeds 1.5–2.0 mm broad ....
.... 7. *C. bushianum*
9. Pericarp easily removed from the seed .... 4. *C. desiccatum*
9. Pericarp firmly attached to the seed 10
10. Seed 1.1–1.5 mm broad .... 8. *C. album*
10. Seeds 0.8–1.2 mm broad 11
11. Lowermost leaves serrate; middle leaves entire; calyx lobes not covering the fruit. .... 9. *C. strictum*
11. Lowermost and middle leaves coarsely toothed; calyx lobes covering the fruit .... 10. *C. missouriense*
12. Calyx mostly 3-lobed; seeds nearly all vertical 13
12. Calyx 5-lobed; seeds horizontal 15
13. Mature flower-heads 5–10 mm in diameter; fruit berry-like ....
.... 11. *C. capitatum*
13. Mature flower-heads less than 5 mm in diameter; fruit not berry-like ...
.... 14
14. Seeds 0.8–1.0 mm broad; annuals; calyx fleshy .... 12. *C. rubrum*
14. Seeds 1.5 mm broad; perennials; calyx herbaceous ....
.... 13. *C. bonus-henricus*
15. Seeds 1.5–2.5 mm broad 16
15. Seeds less than 1.5 mm broad 17
16. Seeds honeycombed, the pericarp closely adherent to the seeds; leaves less than 5 cm long .... 7. *C. bushianum*
16. Seeds smooth, the pericarp easily removed from the seeds; most or all the leaves over 5 cm long .... 14. *C. gigantospermum*
17. Seeds honeycombed .... 6. *C. berlandieri*
17. Seeds smooth 18
18. All leaves entire .... 15. *C. polyspermum*

18. Some or all of the leaves toothed **1**
19. Pericarp easy to remove from the seed .... 16. *C. standleyanu*
19. Pericarp closely adherent to the seed *2*
20. Seeds 1.2–1.5 mm broad, dull .... 17. *C. mura*
20. Seeds up to 1.2 mm broad, lustrous *2*
21. Leaves thin, lustrous on the upper surface .... 18. *C. urbicu*
21. Leaves thickish, not lustrous on the upper surface *2*
22. Lowermost leaves serrate; middle leaves entire; calyx lobes not coverir
   the fruit .... 9. *C. strictu*
22. Lowermost and middle leaves coarsely toothed; calyx lobes covering tl
   fruit .... 10. *C. missourien*

1. C. botrys L. Jerusalem Oak. July–Oct. Native of Europe and Asi
adventive in waste ground; occasional throughout the state.
2. C. ambrosioides L. Mexican Tea. Aug.–Nov. Inflorescence i
terspersed with leafy bracts; lobes of calyx minutely keeled. Native
trop. Am.; adventive in waste ground; common throughout the stat
except for the extreme n. cos.
   var. anthelminticum (L.) Gray. Mexican Tea. Aug.–Nov. Inflore
cence without leafy bracts; lobes of calyx not keeled. Native of tro
Am.; adventive in waste ground; occasional in the s. half of the state
3. C. pallescens Standl. Narrow-leaved Goosefoot. June–Sep
Rocky ground, waste areas; occasional in the n.-cent. counties. (I
cluded in *C. leptophyllum* Nutt. by F and G and in *C. pratericola* Ryo
by J.)
4. C. desiccatum A. Nels. var. leptophylloides (Murr.) Wahl. Narro
leaved Goosefoot. June–Oct. Waste ground, occasional; mostly in tl
n. half of the state. (Included in *C. leptophyllum* Nutt. by F and G a
in *C. pratericola* Rydb. by J.)
5. C. glaucum L. Oak-leaved Goosefoot. July–Sept. Native
Europe; adventive in waste ground; occasional throughout the stat
6. C. berlandieri Moq. var. zschackei (Murr.) June–Oct. Was
ground; occasional throughout the state. *C. berlandieri* Moq.—J;
album L.—G.
7. C. bushianum Aellen. Goosefoot. Aug.–Oct. Fields, woods, was
ground; occasional throughout the state. *C. paganum* Reich.—G, J
8. C. album L. Lamb's Quarters. May–Oct. Flower clusters contir
ous; margin of calyx yellow. Waste ground; common throughout t
state; in every co.
   var. lanceolatum (Muhl.) Coss. & Germ. Lamb's Quarters. May–O
Flower clusters interrupted; margin of calyx white. Waste ground; c
casional throughout the state. *C. lanceolatum* Muhl.—F.
9. C. strictum Roth var. glaucophyllum (Aellen) Wahl. Goosefo
July–Sept. Waste ground, rare.
10. C. missouriense Aellen. Goosefoot. Sept.–Oct. Fields, wa
ground; occasional throughout the state. *C. berlandieri* Moq.—F;
album L.—G.
11. C. capitatum (L.) Aschers. Strawberry Blite. May–Aug. Native
Europe and Asia; rarely escaped in Ill.; McHenry and Peoria cos.
12. C. rubrum L. Coast Blite. July–Oct. Native of the w. U.S.; rar
adventive in Ill.; Cook Co.
13. C. bonus-henricus L. Good King Henry. July–Oct. Native
Europe; rarely adventive in Ill.; Cook Co.
14. C. gigantospermum Aellen. Maple-leaved Goosefoot. June–O

Shaded ledges, rocky woods; occasional throughout the state. *C. hybridum* L.—G, J; *C. hybridum* var. *gigantospermum* (Aellen) Rousseau—F.

15. C. polyspermum L. Many-seeded Goosefoot. June–Oct. Native of Europe; rarely adventive in waste ground; Jackson Co.

16. C. standleyanum Aellen. Goosefoot. June–Oct. Woodlands; occasional throughout the state. C. boscianum Moq.—F, J.

17. C. murale L. Nettle-leaved Goosefoot. June–Nov. Native of Europe; adventive in waste areas; scattered throughout the state.

18. C. urbicum L. City Goosefoot. July–Oct. Native of Europe; adventive in waste areas; scattered throughout the state, except for the s. one-fifth.

53. AMARANTHACEAE

1. Leaves alternate .... 1. *Amaranthus*
1. Leaves opposite 2
2. Pubescence stellate; flowers axillary .... 2. *Tidestromia*
2. Pubescence simple; flowers terminal 3
3. Leaves ovate, glabrous or sparsely hairy beneath; inflorescence paniculate .... 3. *Iresine*
3. Leaves linear to elliptic, densely pubescent beneath; inflorescence spicate .... 4. *Froelichia*

1. Amaranthus L. Amaranth

. Leaves with a pair of spines in the axils .... 1. *A. spinosus*
. Leaves without axillary spines 2
. Plants monoecious, with both staminate and pistillate flowers on the same plant. 3
. Plants dioecious, with staminate and pistillate flowers on separate plants 9
. Flowers in short axillary clusters; plants mostly low and spreading (sometimes erect in *A. albus*) 4
. Flowers in elongated terminal and axillary spikes; plants erect 5
. Seeds less than 1 mm long; bracts subulate, much longer than the calyx .... 2. *A. albus*
. Seeds more than 1 mm long; bracts narrowly ovate, about equalling the calyx or a little longer .... 3. *A. graecizans*
. Calyx of pistillate flowers 3.0–3.3 mm long, up to half as long as the bracts 6
. Calyx of pistillate flowers 1.5–2.0 mm long, more than half as long as the bracts 7
. Sepals of pistillate flowers acute; stamens 3 .... 4. *A. powellii*
. Sepals of pistillate flower obtuse to emarginate ...... 5. *A. retroflexus*
. Sepals of pistillate flowers acute; calyx usually longer than the fruit .... 6. *A. hybridus*
. Sepals of pistillate flowers obtuse; calyx shorter than the fruit 8
. Sepals of pistillate flowers overlapping at their margins; terminal spike arching .... 7. *A. caudatus*
. Sepals of pistillate flowers not overlapping at their margins; terminal spike erect .... 8. *A. cruentus*

9. Pistillate flowers with 5 sepals 1
9. Pistillate flowers lacking a calyx, or sometimes with 1 or 2 sepals on
.... 1
10. Bracts of pistillate flowers longer than the calyx; leaves acute; seed
1.3–1.4 mm across .... 9. *A. palme.*
10. Bracts of pistillate flowers shorter than the calyx; leaves obtuse; seed
1.0–1.1 mm across 1
11. Fruits dehiscent .... 10. *A. torre*
11. Fruits indehiscent .... 11. *A. ambigen*
12. Fruit indehiscent or splitting irregularly; bracts without a conspicuou
midnerve; outer sepals as long as inner sepals .... 12. *A. tuberculatu*
12. Fruit circumscissile; bracts with a conspicuous midnerve; outer sepa
longer than inner sepals .... 13. *A. tamariscinu*

1. A. spinosus L. Spiny Pigweed. July–Oct. Native of trop. Am
waste ground; common in the s. two-thirds of the state, rare els
where.

2. A. albus L. Tumbleweed. July–Sept. Fields and waste areas; cor
mon throughout the state.

3. A. graecizans L. Prostrate Amaranth. July–Sept. Native of the *
U.S.; waste ground; common throughout the state.

4. A. powellii S. Wats. Smooth Pigweed. July–Oct. Native of the s
U.S.; occasional throughout the state, becoming very common nor*
ward.

5. A. retroflexus L. Rough Pigweed. Aug.–Oct. Fields and was
areas; in every co.

6. A. hybridus L. Green Amaranth. Aug.–Oct. Fields and was
ground; common throughout the state.

7. A. caudatus L. Purple Amaranth. July–Oct. Native of trop. Am
rarely escaped from cultivation; Jackson Co.

8. A. cruentus L. Purple Amaranth. July–Oct. Native of Asia; rare
escaped from cultivation; occasional in the ne. corner of the state.

9. A. palmeri S. Wats. Palmer's Amaranth. Aug.–Oct. Native of t*
sw. U.S.; rarely adventive in Ill.; Cook, DeKalb, and Kane cos.

10. A. torreyi (Gray) Benth. Torrey's Amaranth. July–Oct. Native
the w. U.S.; rarely adventive in Ill.; Crawford Co. *A. arenicola* I. *
Johnston—F, J.

11. A. ambigens Standl. Water Hemp. July–Oct. Moist soil, rare; Wi
nebago Co. The type was collected by M. S. Bebb from Winneba*
Co. *Acnida altissima* (Riddell) Riddell—J.

12. A. tuberculatus (Moq.) Sauer. Water Hemp. Aug.–Oct. W
ground, often on sand bars; scattered throughout the state. *Acni*
altissima Riddell—G, J; *Acnida altissima* var. *subnuda* (S. Wat
Fern.—F; *Acnida subnuda* (S. Wats.) Standl.—G, J; *Acnida altissir*
var. *prostrata* (Uline & Bray) Fern.—F.

13. A. tamariscinus Nutt. Water Hemp. July–Oct. Wet soil; scatter*
throughout the state. *Acnida tamariscina* (Nutt.) Wood—F, G, J.

2. *Tidestromia Standl.*

1. T. lanuginosa (Nutt.) Standl. July–Oct. Native of w. U.S.; rare
adventive along a railroad; Cook Co.

3. *Iresine* P. Br.   Bloodleaf

1. I. rhizomatosa Standl. Aug.–Oct. Wet woodlands, rare; confined to the s. one-fourth of the state.

4. *Froelichia Moench*   Cottonweed

1. Flowering spikes over 1 cm thick; leaves oblanceolate to elliptic; stem usually unbranched from near the base .... 1. *F. floridana*
1. Flowering spikes less than 1 cm thick; leaves linear to lanceolate; stems usually branched from near the base .... 2. *F. gracilis*

1. F. floridana (Nutt.) Moq. var. campestris (Small) Fern. Cottonweed. May–Sept. Sandy soil; occasional in the n. two-thirds of the state, apparently absent elsewhere. *F. campestris* Small—J.
2. F. gracilis (Hook.) Moq. Cottonweed. May–Sept. Native of w. U.S.; adventive, particularly along railroads; occasional throughout the state.

54. NYCTAGINACEAE

1. *Mirabilis* L.   Four-o'clock

1. Leaves linear, at least 10 times as long as broad, glaucous   1. *M. linearis*
1. Leaves linear-lanceolate to ovate, at most about 5 times as long as broad, not glaucous 2
2. Leaves more or less ovate, cordate or truncate at base; stems glabrous or nearly so .... 2. *M. nyctaginea*
2. Leaves linear-lanceolate to elliptic, tapering or somewhat rounded at the base; stems hirsute or with short, incurved hairs in two longitudinal lines
.... 3
3. Stems hirsute; fruit rugose .... 3. *M. hirsuta*
3. Stems with short, incurved hairs in two longitudinal lines; fruit tuberculate (at least on the angles) .... 4. *M. albida*

1. M. linearis (Pursh) Heimerl. Narrow-leaved Umbrella-wort. June–Aug. Native to the w. U.S.; rarely adventive in waste areas; Cook, St. Clair, and Will cos. *Oxybaphus linearis* (Pursh) Robins.—G.
2. M. nyctaginea (Michx.) MacM. Wild Four-o'clock. May–Aug. Native to the w. U.S.; naturalized in waste areas, particularly along railroads; common throughout the state. *Oxybaphus nyctagineus* (Michx.) Sweet—G.
3. M. hirsuta (Pursh) MacM. Hairy Umbrella-wort. July–Aug. Native to the w. U.S.; rarely adventive in waste areas; Cook, DuPage, and JoDaviess cos. *Oxybaphus hirsutus* (Pursh) Sweet—G.
4. M. albida (Walt.) Heimerl. Pale Umbrella-wort. July–Aug. Native to the s. cent. U.S.; rarely adventive in waste areas; Cook, Grundy, and Logan cos. *Oxybaphus albidus* (Walt.) Sweet—G.

224 PHYTOLACCACEA

55. PHYTOLACCACEAE

1. Phytolacca L. Pokeweed

1. P. americana L. Pokeweed. July–Oct. Fields, woods, waste areas occasional to common throughout the state.

56. AIZOACEAE

1. Mollugo L. Indian Chickweed

1. M. verticillatus L. Carpetweed. June–Oct. Native of trop. Am.; na uralized in waste areas; common throughout the state.

57. PORTULACACEAE

1. Flowers yellow .... 1. Portulac
1. Flowers not yellow ....
2. All leaves from base of plant .... 2. Talinu
2. Leaves cauline ....
3. Only one pair of flat leaves on each stem .... 3. Clayton
3. Several alternate terete leaves on each stem .... 1. Portulae

1. Portulaca L. Purslane

1. Leaves spatulate, flat; plants glabrous; flowers up to 1 cm across ....
.... 1. P. olerace
1. Leaves linear, terete; plants hairy at the nodes; flowers at least 2 c
across .... 2. P. grandiflo

1. P. oleracea L. Purslane. June–Oct. Native of Europe; naturalize in waste areas; occasional to common throughout the state.
2. P. grandiflora Hook. Rose Moss. June–Oct. Native of S. Am.; c casionally escaped from cultivation.

2. Talinum Adans. Fameflower

1. Flowers pale pink; stamens usually 5, never more than 8 ....
.... 1. T. parvifloru
1. Flowers rose; stamens 10 or more ....
2. Petals up to 8 mm long; stamens 10–25; capsule up to 5 mm lo
.... 2. T. rugospermi
2. Petals 10 mm long or longer; stamens usually more than 25; capsule 6
mm long .... 3. T. calcinc

1. T. parviflorum Nutt. Small Flower-of-an-hour. June–July. posed sandstone cliffs, rare; Johnson, Pope, and Union cos.

2. T. rugospermum Holz. Flower-of-an-hour. June–Aug. Sandy soil, cliffs; occasional to rare in the n. one-third of the state.

3. T. calycinum Engelm. Large Flower-of-an-hour. June–Sept. Edge of sandstone cliff, very rare; Randolph Co.

3. *Claytonia* L. *Spring Beauty*

1. C. virginica L. Spring Beauty. March–May. Woods, fields; common throughout the state. (Including f. *robusta* [Somes] Palmer & Steyerm.)

58. CARYOPHYLLACEAE

1. Leaves with scarious stipules at their base 2
1. Leaves without stipules 4
2. Petals absent; leaves elliptic to oblanceolate; styles 2-cleft ....
.... 1. *Paronychia*
2. Petals 5; leaves linear-filiform; styles 3 or 5 3
3. Leaves opposite; petals pink or white; styles 3 .... 2. *Spergularia*
3. Leaves whorled; petals white; styles 5 .... 3. *Spergula*
4. Petals absent; flowers green 5
4. Petals present; flowers variously colored, not green 6
5. Fruit with 1 seed; styles 2; sepals united into a cup .... 4. *Scleranthus*
5. Fruit with several seeds; styles 5; sepals free .... 5. *Sagina*
6. Sepals free, or united only at the very base 7
6. Sepals united into a tube 12
7. Petals emarginate, bifid, or jagged toothed along the margins 8
7. Petals entire 11
8. Petals jagged toothed along the margins .... 6. *Holosteum*
8. Petals emarginate or bifid 9
9. Styles 3 .... 7. *Stellaria*
9. Styles 5 10
10. Styles alternate with the sepals; capsule ovoid .... 8. *Myosoton*
10. Styles opposite the sepals; capsule cylindric .... 9. *Cerastium*
11. Stamens 5; styles 3 .... 5. *Sagina*
11. Stamens 10; styles 3 .... 10. *Arenaria*
12. Calyx subtended by 2–6 bracts .... 11. *Dianthus*
12. Calyx not subtended by bracts 13
13. Petals bifid .... 12. *Tunica*
13. Petals entire, or fringed at the tip 14
14. Styles 5 15
14. Styles 2 or 3 16
15. Calyx lobes longer than calyx tube .... 13. *Agrostemma*
15. Calyx lobes shorter than calyx tube .... 14. *Lychnis*
16. Styles 3; calyx obviously 10-nerved .... 15. *Silene*
16. Styles 2; calyx obviously 5-nerved, or the nerves obscure 17
17. Calyx up to 5 mm long, conspicuously nerved .... 16. *Gypsophila*
17. Calyx over 8 mm long, obscurely nerved .... 17. *Saponaria*

1. *Paronychia* Mill. *Forked Chickweed*

1. Stems glabrous; sepals obtuse .... 1. *P. canadensis*
1. Stems pubescent; sepals mucronulate .... 2. *P. fastigiata*

1. P. canadensis (L.) Wood. Forked Chickweed. June–Oct. Dry woods; occasional throughout the state.

2. P. fastigiata (Raf.) Fern. Forked Chickweed. June–Oct. Dry woods; occasional in all but the northernmost cos. (Including var. *paleacea* Fern.)

2. Spergularia J. & C. Presl

1. Capsule 5 mm long or longer; seeds 0.6 mm long or longer, narrowly winged .... 1. *S. media*
1. Capsule up to 5 mm long; seeds up to 0.6 mm long, wingless   2. *S. rubra*

1. S. media (L.) C. Presl. July–Sept. Native of Europe; rarely adventive along a highway, in response to salting; Cook, DuPage, and Kane cos.

2. S. rubra (L.) J. & C. Presl. July–Sept. Native of Europe and Asia; rarely adventive in Ill.; Cook Co.

3. Spergula L.

1. S. arvensis L. June–Sept. Native of Europe; rarely adventive in Ill.; Champaign, Cook, and Marion cos.

4. Scleranthus L.

1. S. annuus L. April–Aug. Native of Europe; rarely adventive in Ill.; Kankakee and Winnebago cos.

5. Sagina L.   Pearlwort

1. S. decumbens (Ell.) Torr. & Gray. Pearlwort. April–May. Moist open areas; occasional in the s. half of the state, rare in the n. half.

6. Holosteum L.   Jagged Chickweed

1. H. umbellatum L. Jagged Chickweed. April–May. Native of Europe; rapidly becoming common throughout the s. half of the state, rare elsewhere.

7. Stellaria L.   Chickweed

1. Median leaves ovate, petiolate; petals shorter than the sepals ....
.... 1. *S. media*
1. Median leaves linear or lanceolate or elliptic, sessile; petals longer than the sepals ....
2. Stems puberulent in lines, at least in the upper half of the stem; leaves elliptic; capsule shorter than the sepals .... 2. *S. pubera*
2. Stems glabrous or essentially so; leaves lanceolate or linear; capsule longer than the sepals ....
3. Leaves lanceolate; seeds roughened; sepals strongly nerved ....
.... 3. *S. graminea*
3. Leaves linear; seeds smooth; sepals nerveless or essentially so ......
.... 4. *S. longifolia*

1. S. media (L.) Cyrillo. Common Chickweed. Jan.–Dec. Native of Europe and Asia; in every co.

2. S. pubera Michx. Great Chickweed. March–May. Wooded bluffs, rare; Cook and Pope cos.

3. S. graminea L. Common Stitchwort. May–June. Native of Europe; adventive in moist areas; scattered throughout the state.

4. S. longifolia Muhl. Chickweed. May–July. Moist ground; scattered throughout the state.

8. *Myosoton Moench*

1. M. aquaticum (L.) Moench. Giant Chickweed. June–Sept. Native of Europe; naturalized in moist waste ground; occasional in the n. one-sixth of the state; also Clark, Cumberland, and Sangamon cos. *Stellaria aquatica* (L.) Scop.—G, J.

9. *Cerastium L.   Mouse-ear Chickweed*

1. Margins of bracts scarious 2
1. Bracts herbaceous throughout 4
2. Petals about as long as the sepals; stems more or less viscid 3
2. Petals at least twice as long as the sepals; stems not viscid ....
.... 3. *C. velutinum*
3. Stamens 10; plants perennial, with basal offshoots .... 1. *C. vulgatum*
3. Stamens 5; plants annual, without basal offshoots ...... 2. *C. pumilum*
4. Pedicels at least twice as long as the capsules 5
4. Pedicels up to twice as long as the capsules or shorter 7
5. Stamens 4 or 5 .... 4. *C. tetrandrum*
5. Stamens 10 6
6. Filaments glabrous; stems viscid throughout; sepals pubescent but not distinctly bearded at the tip; capsule curved .... 5. *C. nutans*
6. Filaments ciliate; stems viscid only near apex; sepals long-bearded at the tip; capsule straight .... 6. *C. brachypetalum*
7. Sepals acute; capsule about twice as long as the sepals 7. *C. viscorum*
7. Sepals obtuse to subacute; capsule about three times as long as the sepals .... 8. *C. brachypodum*

1. C. vulgatum L. Common Mouse-ear Chickweed. May–Sept. Native of Europe and Asia; naturalized in waste ground; in every co.

2. C. pumilum Curtis. Mouse-eared Chickweed. April–May. Native of Europe; rarely found in waste ground; Jackson Co.

3. C. velutinum Raf. Field Mouse-ear Chickweed. May–June. Rocky woods; scattered throughout the state. *C. arvense* L.—G; *C. arvense* L. var. *villosum* (Muhl.) Hollick & Britt.—F. (Including *C. arvense* L. f. *oblongifolium* [Torr.] Pennell.)

4. C. tetrandrum Curtis. March–April. Native of Europe; rarely adventive along roadsides; Champaign and Union cos.

5. C. nutans Raf. Nodding Mouse-ear Chickweed. March–June. Moist ground; common throughout the state, except for the extreme nw. cos.

6. C. brachypetalum Pers. April–May. Native of Europe; adventive along roadsides; Pulaski and Union cos.

7. C. viscosum L. Mouse-ear Chickweed. April–May. Native of

Europe; adventive in moist soils; occasional in the s. half of the state; absent elsewhere except for Boone, DeKalb, and Kane cos.

8. C. brachypodum (Engelm.) B. L. Robins. Mouse-ear Chickweed. April–May. Native of the w. U.S.; occasional in the s. half of the state; also McLean and Winnebago cos.

10. *Arenaria* L.   Sandwort

1. Leaves subulate, setaceous, filiform, or linear 2
1. Leaves oval, oblong, or ovate 3
2. Leaves often in fascicles; seeds about 1 mm long .... 1. *A. stricta*
2. Leaves opposite; seeds 0.5–0.6 mm long .... 2. *A. patula*
3. Perennial with slender rhizomes; leaves obtuse to subacute, most or all of them over 7 mm long; sepals obtuse, 2–3 mm long; seeds 1 mm long or longer .... 3. *A. lateriflora*
3. Annual; leaves acuminate, rarely over 7 mm long; sepals acuminate, 3–4 mm long; seeds 0.5–0.6 mm long .... 4. *A. serpyllifolia*

1. A. stricta Michx. Stiff Sandwort. May–July. Sand ridges, hill prairies, limestone glade; occasional or rare in the n. one-fourth of the state; also Tazewell Co.

2. A. patula Michx. Slender Sandwort. May–June. Wooded slopes; rare; Cook, Grundy, Kankakee, St. Clair, and Will cos.

3. A. lateriflora L. Sandwort. May–June. Woodlands; occasional in the n. half of the state; apparently absent elsewhere.

4. A. serpyllifolia L. Thyme-leaved Sandwort. April–June. Native of Europe; adventive in sandy waste ground; scattered throughtout the state.

11. *Dianthus* L.   Pink

1. Flowers solitary, on long pedicels; leaves less than 2 cm long; bracts about half as long as the calyx .... 1. *D. deltoides*
1. Flowers clustered, sessile or nearly so; leaves (or most of them) more than 2 cm long; bracts about as long as the calyx 2
2. Pubescent annual; leaves linear, or linear-lanceolate, to 4 mm broad; calyx pubescent .... 2. *D. armeria*
2. Glabrous perennial; leaves lanceolate to ovate-lanceolate, mostly at least 4 mm broad; calyx glabrous .... 3. *D. barbatus*

1. D. deltoides L. Maiden Pink. June–Aug. Native of Europe; rarely escaped from cultivation; Kane and Vermilion cos.

2. D. armeria L. Deptford Pink. May–Aug. Native of Europe; adventive in fields and along roads; occasional throughout the state.

3. D. barbatus L. Sweet William. June–Aug. Native of Europe; rarely adventive in Ill.; Hancock and Jackson cos.

12. *Tunica* Scop.

1. T. saxifraga (L.) Scop. Saxifrage Pink. June–July. Native of Europe; rarely adventive in Ill.; Champaign Co.

13. Agrostemma L.

1. A. githago L. Corn Cockle. May–July. Native of Europe; adventive in fields and waste ground; occasional throughout the state.

14. Lychnis L.   Campion

1. Plants viscid-pubescent; teeth of calyx not twisted; petals bifid 2
1. Plants white-woolly; teeth of calyx twisted; petals emarginate ....
.... 3. *L. coronaria*
2. Flowers white, opening at dusk; teeth of calyx narrowly lanceolate; capsule ovoid .... 1. *L. alba*
2. Flowers pink or red, opening in the morning; teeth of calyx deltoid; capsule globose .... 2. *L. dioica*

1. L. alba Mill. Evening Campion. May–Aug. Native of Europe; adventive in waste ground; common in the n. two-thirds of the state, rare elsewhere.
2. L. dioica L. Red Campion. June–Aug. Native of Europe; rarely adventive in Ill.; Winnebago Co.
3. L. coronaria (L.) Desr. Mullein Pink. June–Aug. Native of Europe; rarely escaped from cultivation; Cook Co.

15. Silene L.   Catchfly

1. Petals deep red, crimson, or scarlet 2
1. Petals white or pink or purplish 3
2. Stem with 15 or more pairs of leaves; petals entire or emarginate ......
.... 1. *S. regia*
2. Stem with up to 8 pairs of leaves; petals bifid .... 2. *S. virginica*
3. Leaves whorled; petals fringed .... 3. *S. stellata*
3. Leaves opposite; petals bifid 4
4. Flowers up to 4 mm across .... 4. *S. antirrhina*
4. Flowers 1 cm or more across 5
5. Stems glutinous below each node; flowers pink or purple 5. *S. armeria*
5. Stems not glutinous below each node; flowers white or only pink-based
.... 6
6. Stems viscid-pubescent or hirsute; calyx pubescent 7
6. Stems and calyx glabrous 8
7. Flowers nodding, opening at dusk; viscid-pubescent annual ....
.... 6. *S. noctiflora*
7. Flowers ascending, opening during the day; hirsute biennial ....
.... 7. *S. dichotoma*
8. Plants green; leaves acuminate; flowers solitary in the upper axils .....
.... 8. *S. nivea*
8. Plants glaucous; leaves acute; flowers in cymose panicles 9
9. Calyx up to 15 mm long in fruit, scarcely inflated; vertical nerves of calyx sparsely or not at all connected by cross-veins .... 9. *S. cserei*
9. Calyx up to 20 mm long in fruit, inflated; vertical nerves of calyx regularly connected by cross-veins .... 10. *S. cucubalus*

1. S. regia Sims. Royal Catchfly. July–Aug. Dry soil, rare; Clark, Cook, Lawrence, St. Clair, Wabash, Will, and Winnebago cos.

2. S. virginica L. Firepink. April–July. Dry woods; occasional in the ne. quarter of the state, rare or absent elsewhere.

3. S. stellata (L.) Ait. f. Starry Campion. June–Aug. Woods; occasional throughout the state.

4. S. antirrhina L. Sleepy Catchfly. May–July. Fields, waste ground; occasional to common throughout the state.

5. S. armeria L. Sweet William Catchfly. June–July. Native of Europe; rarely escaped from gardens; Cook, Kane, and Will cos.

6. S. noctiflora L. Night-flowering Catchfly. June–July. Native of Europe; adventive in waste ground; occasional in the n. three-fifths of the state, absent elsewhere.

7. S. dichotoma Ehrh. Forked Catchfly. June–July. Native of Europe; rarely adventive in waste ground; Champaign, Cook, Douglas, McLean, and Winnebago cos.

8. S. nivea (Nutt.) Otth. Snow Campion. June–July. Woods, calcareous fens; occasional in the n. two-thirds of the state, absent elsewhere.

9. S. cserei Baumg. Glaucous Campion. June–July. Native of Europe; adventive along railroads and in waste ground; occasional to common in the n. one-third of the state; also Coles, Lawrence, St Clair, and Vermilion cos.

10. S. cucubalus Wibel. Bladder Catchfly. May–July. Native of Europe; adventive in waste ground; occasional in the n. two-thirds of the state.

16. Gypsophila L.

1. Petals at least twice as long as the calyx; calyx 3–5 mm long ....
   .... 1. G. elegans
1. Petals about as long as the calyx; calyx up to 3 mm long 2
2. Leaves cuneate, 1-nerved; teeth of calyx suborbicular, with a narrow green center .... 2. G. paniculata
2. Leaves subcordate, 3- to 5-nerved; teeth of calyx ovate, with a broad green center .... 3. G. perfoliata

1. G. elegans Bieb. Baby's Breath. May–June. Native of Europe and Asia; rarely escaped from cultivation; Champaign Co.

2. G. paniculata L. Baby's Breath. June–Aug. Native of Europe and Asia; rarely escaped from cultivation; Cook, Mason, and Winnebago cos.

3. G. perfoliata L. Baby's Breath. June–Oct. Native of Europe and Asia; rarely escaped from cultivation; Cook and McHenry cos.

17. Saponaria L.

1. Each petal with a scale at the base; flowers 1.8–2.5 cm broad, usually pink or white .... 1. S. officinalis
1. Petals without scales at the base; flowers less than 1.5 cm broad, usually red .... 2. S. vaccaria

1. S. officinalis L. Bouncing Bet. June–Sept. Native of Europe; adventive in waste ground; common throughout the state.

2. S. vaccaria L. Cow Herb. June–Aug. Native of Europe; adventive in waste ground; scattered in the n. half of the state; also Jackson Co.

59. CERATOPHYLLACEAE

1. Ceratophyllum L. Hornwort

1. Achenes with 2 basal spines; ultimate leaf segments toothed on the margins .... 1. *C. demersum*
1. Achenes with several spines both lateral and basal; ultimate leaf segments not toothed on the margins .... 2. *C. echinatum*

    1. C. demersum L. Coontail. July–Sept. Quiet waters; common throughout Ill.
    2. C. echinatum Gray. Coontail. July–Sept. Quiet waters, not common; scattered in Ill.

60. NYMPHAEACEAE

1. Flowers yellow; sepals 5–6; leaves oval .... 1. *Nuphar*
1. Flowers white or pinkish; sepals 4; leaves orbicular ...... 2. *Nymphaea*

1. Nuphar Smith Pond Lily

    1. N. luteum L. ssp. variegatum (Engelm.) Beal. Bullhead Lily. May–Aug. Petiole conspicuously flattened; sepals usually red-tinged. Ponds, rare; Cook and Lake cos. *N. variegatum* Engelm.—F, G, J.
    ssp. macrophyllum (Small) Beal. Yellow Pond Lily. May–Aug. Petiole terete to more or less flattened; sepals green. Ponds; occasional throughout the state. *N. advena* (Ait.) Ait. f.—F, G, J.

2. Nymphaea L. Water Lily

1. Petals subacute at tip; flowers fragrant; seeds 2 mm long 1. *N. odorata*
1. Petals rounded at tip; flowers not fragrant; seeds 3–4 mm long ....
.... 2. *N. tuberosa*

    1. N. odorata Ait. Fragrant Water Lily. June–Sept. Lakes and ponds, rare; Cook, Franklin, Johnson, McHenry, Perry, and Union cos. (Including var. *gigantea* Tricker.)
    2. N. tuberosa Paine. White Water Lily. June–Aug. Ponds and streams; occasional in the n. two-thirds of Ill.

61. NELUMBONACEAE

1. Nelumbo Adans. Lotus

    1. N. lutea (Willd.) Pers. American Lotus. July–Aug. Lakes and ponds; occasional throughout Ill.

62. CABOMBACEAE

1. Leaves uniform; stamens 12–18; carpels 4 or more .... 1. *Brasenia*
1. Leaves dimorphic; stamens 3–6; carpels 2–4 .... 2. *Cabomba*

1. Brasenia Schreb. Watershield

1. B. schreberi Gmel. Watershield. June–Sept. Ponds and quie
streams, rare; scattered in Ill.

2. Cabomba Aubl.

1. C. caroliniana Gray. Carolina Watershield. May–Sept. Ponds
rare; Franklin, Massac, Union, and Wabash cos.

63. RANUNCULACEAE

1. Flowers yellow or white or occasionally pinkish ....
1. Flowers red, blue, purple, or green 1
2. Stems climbing; leaves opposite .... 17. *Clemati*
2. Stems erect or creeping or floating; leaves basal or alternate ....
3. Flowers yellow ....
3. Flowers white or occasionally pinkish ....
4. Sepals and petals present and differentiated; leaves variously toothed o
lobed, but not crenate; fruit an achene .... 1. *Ranunculu*
4. Sepals yellow, petal-like; petals absent; leaves crenate; fruit a follicle .
.... 2. *Caltha*
5. Plants aquatic; sepals and petals each 5, differentiated   1. *Ranunculu*
5. Plants not true aquatics; petals absent or, if present, either stamen-lik
or with a spur ....
6. Flowers spurred .... 3. *Delphiniur*
6. Flowers without a spur ....
7. Flowers numerous in racemes, panicles, or corymbs; sepals inconspic
uous, falling away as the flower opens ....
7. Flowers 1–4, never in a raceme; sepals showy, petal-like, persistent   1
8. Leaves simple, although deeply lobed .... 4. *Trautvetter*
8. Leaves variously compound ....
9. Flowers unisexual, arranged in much-branched panicles . 5. *Thalictrur*
9. Flowers perfect, arranged in racemes or corymbs 1
10. Fruit fleshy, berry-like; raceme simple .... 6. *Actae*
10. Fruit dry, follicular; raceme sparingly branched .... 7. *Cimicifug*
11. Leaves 3-lobed, all basal .... 8. *Hepatic*
11. Leaves either not 3-lobed or not all basal ....
12. Leaves on the stem alternate ....
12. Leaves on the stem opposite or whorled ....
13. Leaves ternately compound; sepals 5; pedicels glabrous . 10. *Isopyru*
13. Leaves simple, palmately lobed; sepals 3; pedicels pubescent........
.... 9. *Hydrast*
14. Leaves ternately compound; roots tuberous-thickened   11. *Anemonel*

14. Leaves deeply to shallowly palmately lobed, but not ternately divided; plants rhizomatous, or with a woody caudex .... 12. *Anemone*
15. Leaves all basal 16
15. At least some of the leaves cauline 17
16. Leaves linear, entire; sepals spurred; flowers greenish; receptacle elongated .... 13. *Myosurus*
16. Leaves palmately lobed; sepals not spurred; flowers pinkish or light purplish; receptacle not elongated .... 8. *Hepatica*
17. One or more petals or sepals prolonged backward into a spur 18
17. None of the perianth parts spurred 19
18. Flowers red and yellow, with each of the five petals prolonged backward into a long spur .... 14. *Aquilegia*
18. Flowers purple, blue, or green; one of the petal-like sepals prolonged backward into a spur .... 3. *Delphinium*
19. Cauline leaves opposite or whorled 20
19. Cauline leaves alternate 21
20. Sepals 4, thick and fleshy; leaves opposite .... 17. *Clematis*
20. Sepals usually 5–20, thin; leaves whorled .... 12. *Anemone*
21. Flowers green .... 15. *Helleborus*
21. Flowers purplish or blue 22
22. Flowers blue; inflorescence subtended by a deeply divided involucre; flowers perfect .... 16. *Nigella*
22. Flowers purplish; inflorescence without an involucre; flowers unisexual .... 5. *Thalictrum*

1. *Ranunculus* L. Buttercup

1. Petals white; achenes covered by horizontal wrinkles 2
1. Petals yellow; achenes smooth or variously marked, but not with horizontal wrinkles 3
2. Leaves becoming limp after removal from the water; beak of achene less than 1 mm long, or absent .... 1. *R. trichophyllus*
2. Leaves remaining firm after removal from the water; beak of achene about 1 mm long .... 2. *R. longirostris*
3. At least some of the leaves simple and unlobed 4
3. None of the leaves simple and unlobed 12
4. All leaves simple and unlobed 5
4. At least some of the leaves lobed or divided 8
5. Leaves reniform, cordate; petals often slightly shorter than the sepals .. .. 3. *R. cymbalaria*
5. Leaves linear to lanceolate, tapering to base; petals equalling or slightly longer than the sepals 6
6. Petals 5–7 in number, 3–9 mm long; stamens 20–50 7
6. Petals 1–3 in number, 1.0–2.5 mm long; stamens 3–10 ... 6. *R. pusillus*
7. Perennial; stamens 25–50; achenes flattened, about 2 mm long .... 4. *R. ambigens*
7. Annual; stamens 12–25; achenes plump, about 1 mm long .... 5. *R. laxicaulis*
8. Petals longer than the sepals 9
8. Petals equalling or shorter than the sepals 10
9. Stamens in one or two series; fruiting head less than 6 mm thick; sepals without long white hairs .... 7. *R. harveyi*
9. Stamens in 3–5 series; fruiting head 6–10 mm thick; sepals with long white hairs .... 8. *R. rhomboideus*

10. Plants more or less fleshy; achenes with corky thickenings at base; stems often hollow .... 9. *R. sceleratus*
10. Plants not fleshy; achenes without corky thickenings at base; stems not hollow 11
11. Achenes shiny; receptacle pubescent; roots slender .. 10. *R. abortivus*
11. Achenes dull; receptacle glabrous, except sometimes near the tip; roots thickened .... 11. *R. micranthus*
12. Plants aquatic or, if creeping in mud, some of the leaves finely dissected 13
12. Plants not truly aquatics; leaves not finely divided 14
13. Achene rugose on the sides, corky-thickened at the base; beak of achene about 1.5 mm long .... 12. *R. flabellaris*
13. Achene smooth on the sides, not corky-thickened at the base; beak of achene up to 0.8 mm long .... 13. *R. gmelini*
14. Petals up to 6 mm long 15
14. Petals 6 mm long or longer 17
15. Achenes with short, hooked spines; petals 1–2 mm long .... 14. *R. parviflorus*
15. Achenes smooth; petals 2–6 mm long 16
16. Petals about equalling the sepals; achenes flat, with strongly recurved beaks; terminal lobe of leaves not stalked .... 15. *R. recurvatus*
16. Petals distinctly shorter than the sepals; achenes not flat, with nearly straight beaks; terminal lobe of leaves stalked .... 16. *R. pensylvanicus*
17. Achenes smooth on the flattened sides 18
17. Achenes papillate on the sides .... 24. *R. sardous*
18. Petals at least half as broad as long 19
18. Petals less than half as broad as long 20
19. Terminal segment of leaves not stalked .... 17. *R. acris*
19. Terminal segment of leaves stalked 20
20. Stems erect or ascending, not rooting at the nodes 21
20. Stems creeping and rooting at the nodes, at least at maturity 22
21. Beak of achene less than 1 mm long, usually curved .. 18. *R. bulbosus*
21. Beak of achene more than 1 mm long, more or less straight .... 19. *R. hispidus*
22. Achene plump, the beak up to 1.5 mm long .... 22. *R. repens*
22. Achene flattened, the beak 1.5–3.0 mm long 23
23. Achene up to 3.5 (–4.5) mm long, with a low narrow keel near the margin .... 20. *R. septentrionalis*
23. Achene 3.5–5.0 mm long, with a high broad keel near the margin .... 21. *R. carolinianus*
24. Some of the roots tuberous-thickened .... 23. *R. fascicularis*
24. All roots fibrous .... 21. *R. carolinianus*

     1. R. trichophyllus Chaix. White Water-crowfoot. May–Aug. Pond and slow streams; mostly in the n. cos. *R. aquatilis* L. var. *capillaceus* (Thuill.) DC.—G.
     2. R. longirostris Godr. White Water-crowfoot. May–Aug. Ponds and slow streams; mostly in the n. cos. *R. circinatus* Sibth.—G.
     3. R. cymbalaria Pursh. Seaside Crowfoot. May–Aug. Wet soil, rare restricted to the n. one-fourth of the state.
     4. R. ambigens Wats. Spearwort. June–Sept. Swampy woods and ditches, rare; Fulton, St. Clair, and Wabash cos.
     5. R. laxicaulis (Torr. & Gray) Darby. Spearwort. May–July. We woods and ditches; generally in s. Ill., n. to Fulton Co.

6. R. pusillus Poir. Small Spearwort. May–June. Swamps, wet woods, ditches; confined to the s. half of Ill.

7. R. harveyi (A. Gray) Britt. Harvey's Buttercup. April–May. Leaves and stems glabrous. Sandstone ravines, rare; Effingham, Jackson, and Randolph cos.

f. pilosus (Benke) Palmer & Steyerm. May. Leaves and stems pilose. Sandstone ravines, rare; Randolph Co.

8. R. rhomboideus Goldie. Prairie Buttercup. May. Prairies, rare; restricted to the extreme n. cos.

9. R. sceleratus L. Cursed Crowfoot. May–Aug. Wet meadows, ditches, and river banks; occasional throughout Ill., although rare in the s. cos.

10. R. abortivus L. Small-flowered Crowfoot. April–June. Leaves, stems, and peduncles glabrous. Fields, moist woods; common throughout Ill.

var. acrolasius Fern. April–June. Some of the leaves, stems, and peduncles pilose. Fields, moist woods; much less common than the preceding.

11. R. micranthus Nutt. Small-flowered Buttercup. March–May. Moist or dry woods; occasional in the s. half of Ill. (Including var. delitescens [Greene] Fern.)

12. R. flabellaris Raf. Yellow Water-crowfoot. April–June. Swamps and ponds; occasional throughout the state.

13. R. gmelinii DC. var. hookeri (D. Don) L. Benson. Small Yellow Water-crowfoot. July–Aug. Ponds; Cook Co. R. purshii Richards.—J.

14. R. parviflorus L. April–June. Naturalized from Europe; waste ground; Jackson Co.

15. R. recurvatus Poir. April–June. Wet woods; occasional throughout the state.

16. R. pensylvanicus L. f. Bristly Crowfoot. July–Sept. Wet ground, rare; scattered throughout the state.

17. R. acris L. Tall Buttercup. May–Aug. Naturalized from Europe; roadsides and fields; occasional in the n. half of Ill.

18. R. bulbosus L. Bulbous Buttercup. April–July. Naturalized from Europe, apparently rare; Henderson and Johnson cos.

19. R. hispidus Michx. Bristly Buttercup. April–May. Pubescence spreading; achenes 3.0–3.5 mm long (excluding the beak). Woods; occasional in the s. three-fifths of Ill.

var. marilandicus (Poir.) L. Benson. Bristly Buttercup. April–May. Pubescence generally appressed; achenes 2.0–2.5 mm long (excluding the beak). Dry woods; rare in s. Ill. Var. falsus Fern.—F, G.

20. R. septentrionalis Poir. Swamp Buttercup. April–July. Petioles and lower part of stems glabrous or appressed-pubescent. Low woods and ditches; common throughout Ill.

var. caricetorum (Greene) Fern. April–July. Petioles and lower part of stems with retrorse pubescence; much less common than the preceding.

21. R. carolinianus DC. April–May. Lowland woods, rare; Champaign, Hancock, and Union cos.

22. R. repens L. Creeping Buttercup. April–Aug. Petals 10 or fewer. Native of Europe; roadsides and fields; rare in the n. half of Ill.; also Jersey Co.

var. pleniflorus Fern. Double-flowered Creeping Buttercup. Apr.–Aug. Petals more than 10. Planted in gardens, rarely escaped into waste ground; Jackson Co.

23. R. fascicularis Muhl. Early Buttercup. April–May. Open woods
and meadows; throughout the state. (Including var. *apricus* [Greene]
Fern.)

24. R. sardous Crantz. May–July. Native of Europe; low fields and
waste places; confined to a few cos. in the southernmost part of Ill.

2. *Caltha L. Marsh Marigold*

1. C. palustris L. Marsh Marigold. April–June. Wet meadows; confined to the n. two-thirds of Ill.

3. *Delphinium L. Larkspur*

1. Pistil 1; follicle 1; cultivated annual .... 1. *D. ajacis*
1. Pistils 3–5; follicles 3–5; native perennial 2
2. Stems glabrous; follicles spreading at maturity; seeds smooth ....
.... 2. *D. tricorne*
2. Stems pubescent; follicles erect at maturity; seeds roughened 3
3. Seeds winged, with appressed scales .... 3. *D. carolinianum*
3. Seeds unwinged, with projecting scales .... 4. *D. virescens*

1. D. ajacis L. Rocket Larkspur. June–Aug. Escaped from gardens;
occasional throughout the state.

2. D. tricorne Michx. Dwarf Larkspur. April–June. Rich woods; occasional to common in the s. four-fifths of the state, absent elsewhere. Occasional white-flowered specimens (f. *albiflora* Millsp.) have
been observed.

3. D. carolinianum Walt. Wild Blue Larkspur. May–June. Dry, often
sandy soil, rare; Adams, Henderson, Macon, Mercer, Moultrie, and
Pike cos. (Our plants belong to var. *crispum* Perry.)

4. D. virescens Nutt. Prairie Larkspur. May–July. Prairies, very rare;
Hancock Co.

4. *Trautvetteria Fisch. & Mey. False Bugbane*

1. T. caroliniensis (Walt.) Vail. False Bugbane. June–July. Moist
ground along a stream, very rare; Cass Co.

5. *Thalictrum L. Meadow Rue*

1. Middle and upper leaves sessile ....
1. Middle and upper leaves petiolate .... 3. *T. dioicum*
2. Lower surface of leaflets eglandular and glabrous ....
2. Lower surface of leaflets glandular or pubescent ....
3. Lower surface of leaflets glandular .... 1. *T. revolutum*
3. Lower surface of leaflets pubescent but eglandular .. 2. *T. dasycarpum*
4. Margins of leaflets revolute, the lower surface of the leaflets conspicuously reticulate .... 1. *T. revolutum*
4. Margins of leaflets flat, the lower surface of the leaflets conspicuous
reticulate .... 2. *T. dasycarpum*

1. T. revolutum DC. Waxy Meadow Rue. Late May–June. Lower surface of leaflets glandular. Prairies and open woods; occasional
throughout the state.

f. glabrum Pennell. Waxy Meadow Rue. Late May–June. Lower surface of leaflets glabrous and eglandular. Prairies and open woods; not as common as the preceding.

2. T. dasycarpum Fisch. & Lall. Purple Meadow Rue. May–June. Leaflets firm, pubescent but eglandular beneath. Moist, wooded ravines, occasional; scattered throughout Ill.

var. hypoglaucum (Rydb.) Boivin. Meadow Rue. June–July. Leaflets thin, glabrous and eglandular beneath. Moist, wooded ravines, occasional; scattered throughout Ill. *T. hypoglaucum* Rydb.—J.

3. T. dioicum L. Early Meadow Rue. April–early May. Rich woodlands and prairies; occasional throughout Ill.

6. *Actaea L. Baneberry*

1. Pedicels much narrower than the peduncles in fruit; seeds 10 or more per berry, up to 4 mm long .... 1. *A. rubra*
1. Pedicels about as thick as the peduncles in fruit; seeds 3–9 (–10) per berry, over 4 mm long .... 2. *A. pachypoda*

1. A. rubra (Ait.) Willd. Red Baneberry. April–July (fruiting from Aug.–Oct.). Berries red. Rich woods, not common; confined to the n. cos. of Ill.

f. neglecta (Gillman) Robins. April–July (fruiting from Aug.–Oct.). Berries white. Rich woods, apparently not as common in Ill. as the preceding.

2. A. pachypoda Ell. Doll's-eyes. April–June (fruiting from July–Oct.). Rich woods; occasional throughout most of Ill. *A. alba* (L.) Mill.—G, J.

7. *Cimicifuga L. Bugbane*

1. At least the terminal leaflet cordate .... 1. *C. rubifolia*
1. All leaflets truncate or subcordate .... 2. *C. racemosa*

1. C. rubifolia Kearney. Black Cohosh. July–Sept. Rich woods, rare; Hardin, Massac, and Pope cos. *C. racemosa* (L.) Nutt.—G; *C. racemosa* var. *cordifolia* (Pursh) Gray—F; *C. cordifolia* Pursh—J.

2. C. racemosa (L.) Nutt. Black Cohosh. June–July. Woods, rare; St. Clair and Wabash cos.

8. *Hepatica Mill. Hepatica*

1. H. nobilis Schreb. var. obtusa (Pursh) Steyerm. Round-lobed Liverleaf. March–April. Lobes of the leaves, as well as the bracts, rounded at the tip. Rich woods, not common; confined to the ne. cos. of Ill. *H. americana* (DC.) Ker—F, G, J.

var. acuta (Pursh) Steyerm. Sharp-lobed Liverleaf. March–May. Lobes of the leaves, as well as the bracts, acute at the tip. Rich woods; occasional throughout the state. *H. acutiloba* DC.—F, G, J.

9. *Hydrastis Ellis Goldenseal*

1. H. canadensis L. Goldenseal. April–May. Rich woods, occasional; scattered throughout Ill.

10. Isopyrum L.

1. I. biternatum (Raf.) Torr. & Gray. False Rue Anemone. March–May. Rich woods; common throughout the state.

11. Anemonella Spach

1. A. thalictroides (L.) Spach. Rue Anemone. April–May. Dry o moist open woods; occasional throughout the state. (Including f. fa villiana Bergseng, a form with petaloid stamens.)

12. Anemone L. Anemone

1. Styles 2–4 cm long, plumose; staminodia present .... 1. A. patens
1. Styles up to 4 mm long, usually pubescent but not plumose; staminodia absent .... ?
2. Plants arising from a tuber; sepals 10–20 .... 2. A. caroliniana
2. Plants arising from rhizomes; sepals 5 (–6) .... ?
3. Leaves of the involucre sessile; beak of achene 2–5 mm long ....
.... 3. A. canadensis
3. Leaves of the involucre petiolate; beak of achene less than 2 mm long ?
4. Basal leaf solitary; plants at maturity less than 30 cm tall; achenes hir sutulous, but not woolly .... 4. A. quinquefolia
4. Basal leaves 2–several; plants at maturity more than 30 cm tall; achene woolly .... ?
5. Leaves of the involucre 5–9; fruiting heads more than twice as long a wide; styles less than 1 mm long .... 5. A. cylindric
5. Leaves of the involucre 3; fruiting heads less than twice as long as wide styles 1 mm long or longer .... 6. A. virginian

1. A. patens L. Pasque-flower. March–April. Prairies; confined t the extreme n. cos. of the state. A. ludoviciana Nutt.—J; A. patens va wolfgangiana (Bess.) Koch—G.

2. A. caroliniana Walt. Carolina Anemone. April–May. Prairie soi bluffs, not common; confined to the n. half of Ill.

3. A. canadensis L. Meadow Anemone. May–July. Open woodland moist prairies; occasional in the n. two-thirds of Ill.; also Gallatin an Jackson cos.

4. A. quinquefolia L. Wood Anemone. April–May. Rich woods, n common; confined to the n. one-fourth of Ill. (Including var. interic Fern.)

5. A. cylindrica Gray. Thimbleweed. May–Aug. Open wood prairies, occasional; limited to the n. two-thirds of Ill.; also Jackso Co.

6. A virginiana L. Tall Anemone. June–Aug. Open, usually d woods; common throughout Ill.

13. Myosurus L. Mousetail

1. M. minimus L. Mousetail. April–June. Moist ground in woods ar fields; occasional to common in the s. cos., rare northward.

14. Aquilegia L.   Columbine

1. Spurs of flower straight; flowers red or yellow .... 1. *A. canadensis*
1. Spurs of flower hooked; flowers blue, purple, pink, or white ....
.... 2. *A. vulgaris*

    1. A. canadensis L. Columbine. April–July. Rocky woods; occasional to common throughout the state. (Including var. *coccinea* [Small] Munz.)
    2. A. vulgaris L. Garden Columbine. May–July. Escaped from cultivation; rarely observed outside of gardens.

15. Helleborus L.   Hellebore

    1. H. viridis L. Green Hellebore. March–April. Escaped from cultivation; rarely observed outside of gardens.

16. Nigella L.

    1. N. damascena L. Love-in-a-mist. June–Aug. Escaped from cultivation; rarely observed outside of gardens; Jackson Co.

17. Clematis L.   Clematis

1. Inflorescence paniculate; sepals thin, white; anthers blunt 2
1. Flower solitary; sepals thick, bluish; anthers with an attenuated tip .. 3
2. Leaves primarily 3-foliolate; anthers up to 1.5 mm long; achenes with spreading hairs .... 1. *C. virginiana*
2. Leaves primarily 5-foliolate; anthers at least 2 mm long; achenes with appressed, silky hairs .... 2. *C. dioscoreifolia*
3. Tails of fruits densely plumose; only the tips of the sepals recurved ....
.... 3. *C. viorna*
3. Tails of fruits glabrous or pubescent, but not plumose; upper half of the sepals recurved 4
4. Leaves thick, conspicuously reticulate beneath; sepals over 25 mm long, the margins crisped .... 5. *C. pitcheri*
4. Leaves thin, not conspicuously reticulate beneath; sepals less than 25 mm long, the margins not crisped .... 4. *C. crispa*

    1. C. virginiana L. Virgin's Bower. July–Sept. Moist soil, particularly at the edges of woods; occasional to common throughout the state.
    2. C. dioscoreifolia Levl. & Vaniot. Virgin's Bower. July–Oct. Naturalized from Japan and Korea; roadsides and borders of woods, occasional.
    3. C. viorna L. Leatherflower. May–July. Along streams, very rare; Johnson and Richland cos.
    4. C. crispa L. Blue Jasmine. April–July. Swampy woods, wet ditches, not common; confined to s. Ill.
    5. C. pitcheri Torr. & Gray. Leatherflower. May–Sept. Woods and thickets, occasional; throughout Ill. except for the northernmost counties.

64. BERBERIDACEAE

1. Prickly shrubs; fruit a red berry .... 1. *Berberis*
1. Unarmed herbs; fruit a capsule or a blue or yellow berry 2
2. Flower solitary on each plant, white; leaves simple 3
2. Flowers in panicles, yellow-green; leaves ternately compound ....
.... 4. *Caulophyllum*
3. Leaves 7- to 9-lobed; sepals usually 6; stamens twice as many as the petals; fruit a berry .... 2. *Podophyllum*
3. Leaves 2-lobed; sepals usually 4; stamens the same number as the petals; fruit a capsule .... 3. *Jeffersonia*

1. *Berberis* L.   Barberry

1. Leaves entire; flowers solitary or in clusters of 2–4; prickles unbranched; berries dry .... 1. *B. thunbergii*
1. Leaves toothed; flowers in racemes; prickles mostly forked; berries fleshy 2
2. Leaves with up to 20 teeth on the margin; branchlets brown; petals notched; berries ovoid .... 2. *B. canadensis*
2. Leaves with 25 or more teeth on the margin; branchlets gray; petals not notced; berries ellipsoid .... 3. *B. vulgaris*

1. B. thunbergii DC. Japanese Barberry. April–May. Native of Asia, spreading from cultivation, occasional.
2. B. canadensis Mill. American Barberry. May. Dry woodlands and sandstone cliffs, very rare; Jackson and Tazewell cos.
3. B. vulgaris L. Common Barberry. May–June. Native of Europe, rarely escaped from cultivation.

2. *Podophyllum* L.   Mayapple

1. P. peltatum L. Mayapple. April–June. Mostly open woods; common throughout the state; in every co. (Including f. *polycarpum* Clute, f. *biltmoreanum* Steyerm., f. *aphyllum* Plitt, and f. *deamii* Raymond.)

3. *Jeffersonia* Bart.   Twinleaf

1. J. diphylla (L.) Pers. Twinleaf. April–May. Rich woods, not common; scattered throughout Ill. although rare in the s. one-third of the state.

4. *Caulophyllum* Michx.   Blue Cohosh

1. C. thalictroides (L.) Michx. Blue Cohosh. April–May. Rich woods, occasional; throughout Ill.

65. MENISPERMACEAE

1. Leaves deeply 3- to 7-lobed; petals none; drupe 15–25 mm long .....
.... 1. *Calycocarpum*

1. Leaves entire or with 3–7 angles or shallow lobes; petals 6–8; drupe 5–10 mm long 2
2. Drupe blue-black; stamens 12–24 .... 2. *Menispermum*
2. Drupe red; stamens 6 .... 3. *Cocculus*

1. *Calycocarpum* Nutt.   Cupseed

1. C. lyonii (Pursh) Gray. Cupseed. May–June. Swampy woodlands, rare; confined to s. Ill.

2. *Menispermum* L.   Moonseed

1. M. canadense L. Moonseed. May–July. Moist woodlands, thickets; common throughout the state.

3. *Cocculus* DC.   Snailseed

1. C. carolinus (L.) DC. Snailseed. July–Aug. Moist woods, thickets; occasional in s. Ill.

66. MAGNOLIACEAE

1. Leaves entire; petals without an orange blotch at the base within; seeds unwinged .... 1. *Magnolia*
1. Leaves 4-lobed; petals with an orange blotch at the base within; seeds winged .... 2. *Liriodendron*

1. *Magnolia* L.   Magnolia

1. M. acuminata L. Cucumber Magnolia. April–May. Rich woodlands; restricted to the s. three tiers of cos.

2. *Liriodendron* L.   Tulip Tree

1. L. tulipifera L. Tulip Tree. April–May. Rich woodlands, generally common; confined to the s. two-thirds of the state.

67. ANNONACEAE

1. *Asimina* Adans.   Pawpaw

1. A. triloba (L.) Dunal. Pawpaw. April–May. Low woods and wooded slopes; common in the s. cos., becoming rare northward.

68. LAURACEAE

Some of the leaves lobed; flowers appearing as the leaves unfold, in modified racemes; fruits blue .... 1. *Sassafras*

1. None of the leaves lobed; flowers appearing before the leaves, in axillary clusters; fruits red .... 2. *Lindera*

1. *Sassafras* Nees   Sassafras

1. S. albidum (Nutt.) Nees. Sassafras. April–May. Leaves glabrous or nearly so on the lower surface. Dry or moist woodlands, thickets, roadsides; common in the s. three-fourths of Ill., rare or absent elsewhere.

var. molle (Raf.) Fern. Red Sassafras. April–May. Leaves permanently pubescent on the lower surface. Woodlands, thickets, roadsides; occasional in the s. three-fourths of Ill.

2. *Lindera* Thunb.   Spicebush

1. L. benzoin (L.) Blume. Spicebush. March–May. Leaves glabrous. Rich woodlands; common throughout the state, except the nw. portion.

var. pubescens (Palmer & Steyerm.) Rehd. Spicebush. March–May. Leaves permanently short-hairy on both surfaces. Swampy woods; rare; Johnson and Union cos.

69. PAPAVERACEAE

1. Flowers actinomorphic; stamens numerous; sap milky or variously colored ....
1. Flowers zygomorphic; stamens 6; sap clear ....
2. Flowers basically white (pink in a rare form of *Sanguinaria*) ....
2. Flowers not white ....
3. Perianth parts 6 or more ....
3. Perianth parts 2 .... 3. *Maclea*
4. Plants without stems; petals 6 or more .... 1. *Sanguinaria*
4. Plants with stems; petals 4–6 ....
5. Leaves prickly .... 2. *Argemone*
5. Leaves not prickly .... 4. *Papaver*
6. Leaves pinnatifid ....
6. Leaves ternately divided .... 7. *Eschscholtzia*
7. Leaves prickly .... 2. *Argemone*
7. Leaves not prickly ....
8. Flowers red or orange; sap milky; capsule opening by pores along the edge .... 4. *Papaver*
8. Flowers yellow; sap yellow; capsule opening from the bottom upward
9. Capsule bristly; style distinct; petals 2–3 cm long ...... 5. *Stylophorum*
9. Capsule smooth; style inconspicuous or absent; petals about 1 cm long .... 6. *Chelidonium*
10. Corolla with two spurs, or subcordate ....
10. Corolla with a single spur ....
11. Leaves all from the base; plants erect .... 8. *Dicentra*
11. Leaves cauline and alternate; plants climbing .... 9. *Adlumia*
12. Flowers yellow or pink; capsule elongated, several-seeded 10. *Corydalis*
12. Flowers purplish, tipped with red; fruit nearly round, 1-seeded .....
.... 11. *Fumaria*

1. *Sanguinaria* L. Bloodroot

1. S. canadensis L. Bloodroot. March–April. Rich woods; common throughout the state. (Including f. *colbyorum* Benke, with pink petals, and var. *rotundifolium* [Greene] Fedde, with flowers elevated above the leaves.)

2. *Argemone* L. Prickly Poppy

. Flowers white or pink; leaves of a uniform color . . . . 1. *A. albiflora*
. Flowers yellow or orange or cream; leaves with patches of pale green . .
. . . . 2. *A. mexicana*

1. A. albiflora Hornem. Prickly Poppy. May–Sept. Naturalized from the w. U.S.; occasional; a Morgan Co. collection may represent a native population (Including *A. intermedia* Sweet.)
2. A. mexicana L. Mexican Poppy. May–Sept. Introduced from Mexico; rarely escaped; Henderson, Mason, Menard, and Stephenson cos.

3. *Macleaya* R. Br. Plume Poppy

1. M. cordata (Willd.) R. Br. Plume Poppy. June–July. Introduced from e. Asia; rarely escaped; Cook and Henry cos.

4. *Papaver* L. Poppy

. Stems glabrous; cauline leaves cordate-clasping; plants glaucous; capsule nearly spherical . . . . 1. *P. somniferum*
. Stems hirsute; cauline leaves not cordate-clasping; plants not glaucous; capsule longer than broad 2
. Capsule narrowly obovoid . . . . 2. *P. dubium*
. Capsule broadly obovoid . . . . 3. *P. rhoeas*

1. P. somniferum L. Opium Poppy. June–Aug. Introduced from Europe; rarely escaped from cultivation; Crawford Co.
2. P. dubium L. Poppy. May–Aug. Introduced from Europe; rarely escaped from cultivation; Wabash Co.
3. P. rhoeas L. Corn Poppy. May–Sept. Introduced from Europe; occasionally escaped from cultivation.

5. *Stylophorum* Nutt.

1. S. diphyllum (Michx.) Nutt. Celandine Poppy. March–May. Rich woods, not common; restricted to the s. counties; also Cook Co.

6. *Chelidonium* L. Celandine

1. C. majus L. Celandine. May–Aug. Native of Europe; occasionally escaped from cultivation.

7. *Eschscholtzia* Cham. California Poppy

1. E. californica Cham. California Poppy. June–Sept. Escaped from cultivation; rare; Kane and Wabash cos.

8. *Dicentra* Bernh.

1. Spurs of corolla spreading, subacute; flowers without an odor; pla
from granular, white tubers .... 1. *D. cuculla*
1. Spurs of corolla not spreading, rounded; flowers with a sweet od
plants from yellow corn-like tubers .... 2. *D. canaden*

    1. D. cucullaria (L.) Bernh. Dutchman's-breeches. April–May. R
woods; common throughout Ill.
    2. D. canadensis (Goldie) Walp. Squirrel-corn. April–May. R
woods, occasional; scattered throughout the state.

9. *Adlumia* Raf.

    1. A. fungosa (Ait.) Greene. Climbing Fumitory. June–Sept. Escap
from cultivation; rare; Kankakee and Ogle cos.

10. *Corydalis* Medic. *Corydalis*

1. Flowers pink, with yellow tips .... 1. *C. sempervir*
1. Flowers wholly yellow ....
2. Outer petals with a winged crest down the back ....
2. Outer petals not crested down the back ....
3. Winged crest of outer petals with 3–4 teeth; stalk of capsule 1 cm l
or longer; seeds rugulose .... 2. *C. flav*
3. Winged crest of outer petals entire; stalk of capsule up to 5 mm lo
seeds smooth ....
4. Capsules scarcely torulose, up to 15 mm long; uppermost raceme
overtopping subtending leaves .... 3. *C. micran*
4. Capsules strongly torulose, 15–25 mm long; uppermost raceme o
topping subtending leaves .... 4. *C. h*
5. Spur ⅓–⅖ the length of the corolla; sepals up to 1 mm long; se
1.3–1.7 mm long .... 5. *C. campes*
5. Spur less than one-third the length of the corolla; sepals 1.5–2.0
long; seeds 2.0–2.5 mm long ....
6. Capsules pendulous or widely spreading; margins of seeds rounded
.... 6. *C. au*
6. Capsules ascending; margins of seeds sharp .... 7. *C. mont*

    1. C. sempervirens (L.) Pers. Pink Corydalis. May–Aug. Ro
woods, rare; Cook, LaSalle, and Ogle cos.
    2. C. flavula (Raf.) DC. Pale Corydalis. April–May. Moist woo
common in the s. cos., becoming less common northward.
    3. C. micrantha (Engelm.) Gray. Slender Corydalis. May–July. Ro
woods, not common; scattered throughout the state.
    4. C. halei (Small) Fern. & Schub. April–May. Moist soil at base
cliffs, very rare; Monroe Co.
    5. C. campestris (Britt.) Buchholz & Palmer. Plains Corydalis. M
July. Prairies, rare; restricted to w. cent. Ill.; also Grundy Co.
    6. C. aurea Willd. Golden Corydalis. May–Aug. Rocky soil, very r
Cook, LaSalle, Ogle, and Winnebago cos.
    7. C. montana (Engelm.) Gray. April–May. Rocky woods, rare; c
fined to the n. one-third of Ill. *C. aurea* Willd. var. *occiden*
Engelm.—F.

11. Fumaria L.   Fumitory

1. F. officinalis L. Fumitory. May–Aug. Adventive from Europe; occasional in Ill.

70. CAPPARIDACEAE

Stamens (6–) 7–32 .... 1. *Polanisia*
Stamens 6 .... 2. *Cleome*

1. Polanisia Raf.   Clammyweed

Leaflets obovate to lance-elliptic, all or most of them over 6 mm broad; fruits 5–10 mm broad .... 1. *P. dodecandra*
Leaflets linear, up to 4 mm broad; fruits 3–4 mm broad ... 2. *P. jamesii*

1. P. dodecandra (L.) DC. Clammyweed. June–Sept. Largest petals 3.5–6.5 (–8.0) mm long; longest stamens 4–10 (–14) mm long, scarcely exceeding the petals. Gravelly soil, often along railroads, river banks; scattered throughout the state. *P. graveolens* Raf.—F, G, J.
ssp. trachysperma (Torr. & Gray) Iltis. Clammyweed. June–Sept. Largest petals (7–) 8–16 mm long; longest stamens (9–) 12–30 mm long, usually much exceeding the petals. Gravelly soil, sandy soil; scattered in all parts of the state, but not common. *P. trachysperma* Torr. & Gray—F, G, J.
2. P. jamesii (Torr. & Gray) Iltis. June–Aug. Sandy soil, rare; Carroll, JoDaviess, and Mason cos. *Cristatella jamesii* Torr. & Gray— F, G, J.

2. Cleome L.

Plants essentially glabrous; leaves 3-foliolate .... 1. *C. serrulata*
Plants pubescent, minutely spinose at the nodes; leaves 5- to 7-foliolate
.... 2. *C. hassleriana*

1. C. serrulata Pursh. July–Sept. Adventive from the w. U.S.; occasional in waste areas.
2. C. hassleriana Chod. Spider Flower. June–Sept. Rarely escaped from cultivation; Coles, Jackson, JoDaviess, Kendall, and Washington cos. *C. speciosissima* Deppe—J.

71. CRUCIFERAE

Flowers white or purple 2
Flowers yellow 38
Leaves deeply palmately divided .... 1. *Dentaria*
Leaves simple, pinnately divided, or pinnatifid 3
Flowers purplish 4
Flowers white 10

4. Plants pubescent ....
4. Plants glabrous ....
5. Some or all the leaves pinnatifid ....
5. None of the leaves pinnatifid ....
6. Plants hispid; roots thickened; petals 1.5 cm long or longer; silique 5–
mm in diameter .... 2. *Raphan*
6. Plants more or less glandular-hirtellous; roots not conspicuously thic
ened; petals less than 1.5 cm long; silique up to 2 mm in diameter ..
.... 35. *Chorispo*
7. Basal leaves cordate, on petioles longer than the blades; cauline leav
obtuse to subacute; siliques up to 4 cm long .... 10. *Cardam*
7. Basal leaves tapering to base, sessile or on petioles much shorter th
the blades; cauline leaves acuminate; siliques 5–14 cm long 3. *Hespe*
8. At least some of the upper leaves auriculate at the base . 4. *Iodanth*
8. None of the leaves auriculate at the base ....
9. Plants fleshy; leaves sinuate-dentate; flowers 4–6 mm across . 5. *Cak*
9. Plants not fleshy; some or all the leaves pinnatifid; flowers 10–15 r
across .... 6. *Eru*
10. Basal rosette present (although sometimes withered at flowering tir
....
10. Basal rosette absent ....
11. Some of the leaves pinnate or pinnatifid ....
11. None of the leaves pinnate or pinnatifid ....
12. Some or all the cauline leaves sagittate or auriculate at the base ...
12. None of the cauline leaves sagittate or auriculate at the base ......
13. Ovary and fruit triangular, broader than long .... 7. *Capse*
13. Ovary and fruit not triangular, much longer than broad ...... 8. *Ara*
14. Leaves bi- or tri-pinnate; pubescence stellate .... 9. *Descura*
14. Leaves once-pinnate or merely pinnatifid; pubescence simple (hairs
furcate in *A. lyrata*) ....
15. Petals 5–8 mm long .... 8. *Ara*
15. Petals up to 4 mm long ....
16. Basal leaves pinnate .... 10. *Cardam*
16. Basal leaves pinnatifid ....
17. Silique beaked; seeds wingless .... 10. *Cardam*
17. Silique nearly beakless; seeds winged .... 11. *Sib*
18. Ovary and fruit only 2–7 times longer than broad .... 12. *Dr*
18. Ovary and fruit much greater than seven times as long as broad ..
19. Leaves of basal rosette never exceeding 1 cm in width 13. *Arabidop*
19. Most leaves of basal rosette 1 cm broad or broader .... 8. *Ara*
20. Some or all the leaves pinnate or pinnatifid ....
20. None of the leaves pinnate or pinnatifid ....
21. Some or all the cauline leaves auriculate or sagittate .... 14. *Lepid*
21. None of the cauline leaves auriculate or sagittate ....
22. Petals to 2 mm long, or absent ....
22. Petals 4–20 mm long ....
23. Plants erect; silicle flattened .... 14. *Lepid*
23. Plants matted; silicle subglobose .... 36. *Corono*
24. Petals 4–8 mm long; plants glabrous ....
24. Petals 10–20 mm long; plants more or less hispid .... 2. *Rapha*
25. Some or all the cauline leaves simple; fruits up to three times as lon
broad .... 15. *Armor*
25. All cauline leaves pinnately compound; fruits more than three time
long as broad .... 16. *Nastur*
26. Some or all the cauline leaves clasping or sagittate ....

. None of the leaves clasping or sagittate 33

'. Petals 7–20 mm long 28

. Petals up to 6 mm long 29

. Basal leaves on petioles longer than the blades; siliques 1.5–3.5 cm long
.... 10. *Cardamine*

. Basal leaves on petioles shorter than the blades; siliques 4–10 cm long
.... 8. *Arabis*

. Cauline leaves glabrous 30

. Cauline leaves pubescent 31

. Ovary and fruit nearly orbicular, the fruit 1.0–1.8 cm in diameter ....
.... 17. *Thlaspi*

. Ovary and fruit long and slender, the fruit 5–10 cm long ..... 8. *Arabis*

. Ovary and fruit not more than three times longer than broad 32

. Ovary and fruit many times longer than broad .... 8. *Arabis*

. Fruit winged, notched at apex, 2–3 times longer than broad; petals up to
2 mm long .... 14. *Lepidium*

Fruit unwinged, not notched at apex, as broad as long; petals 3–4 mm
long .... 18. *Cardaria*

Petals 15–20 mm long .... 15. *Armoracia*

Petals less than 15 mm long 34

Leaves broadly ovate, the lowermost on petioles nearly as long as or
longer than the blades; plants with the odor of garlic ...... 19. *Alliaria*

Leaves variously shaped, but not broadly ovate, the lowermost sessile or
on short petioles; plants without the odor of garlic 35

Cauline leaves entire 36

Cauline leaves toothed 37

Petals deeply notched; pubescence of stellate hairs ..... 20. *Berteroa*

Petals entire; pubescence of bifurcate hairs .... 21. *Lobularia*

Ovary and fruit about as broad as long; petals up to 2 mm long; stems to
45 cm tall .... 14. *Lepidium*

Ovary and fruit much longer than broad; petals 4–6 mm long; stems 1–2
m tall .... 8. *Arabis*

Basal rosette present (sometimes rather withered at flowering time) 39

Basal rosette absent 44

Some or all the cauline leaves auriculate or sagittate at the base ... 40

None of the cauline leaves auriculate nor sagittate at the base 42

Basal leaves deeply pinnatifid; cauline leaves pinnatifid or coarsely dentate; petals yellow 41

Basal leaves lyrate or merely toothed; cauline leaves mostly entire;
petals creamy-yellow .... 8. *Arabis*

Petals 5 mm long or longer; ovaries and fruits several times longer than
broad .... 22. *Barbarea*

Petals up to 2 mm long; ovaries and fruits at most about three times
longer than broad .... 34. *Rorippa*

Basal leaves linear to elliptic, not pinnatifid or pinnate . 23. *Lesquerella*

Basal leaves pinnatifid or pinnate 43

Basal leaves pinnatifid; petals 6–10 mm long .... 24. *Diplotaxis*

Basal leaves bi- or tri-pinnate; petals 2.0–2.5 mm long . 9. *Descurainia*

None of the leaves pinnatifid nor pinnate 45

Some or all the leaves pinnate or pinnatifid 53

None of the leaves auriculate nor sagittate at the base 46

Some of the leaves auriculate or sagittate at the base 48

Ovary and fruit many times longer than broad; pubescence of 2- to 3-
parted hairs .... 25. *Erysimum*

Ovary and fruit about as broad as long; pubescence of stellate hairs 47

47. Petals up to 4 mm long; silicle 4-seeded; leaves up to 3 cm long . . . .
   . . . . 26. *Alysse*
47. Petals 4–7 mm long; silicle several-seeded; most of the leaves over 3 c
   long . . . . 23. *Lesquere*
48. Plants glabrous . . . .
48. Plants pubescent . . . .
49. Style absent; fruits borne on pendulous pedicels . . . . 27. *Isa*
49. Style elongate; fruits borne on ascending pedicels . . . .
50. None of the leaves on the stem entire . . . . 22. *Barbar*
50. Some or all of the leaves on the stem entire . . . .
51. Petals 7–10 mm long; cauline leaves obtuse at apex . . . .
   . . . . 28. *Conrin*
51. Petals up to 6 mm long; cauline leaves acute at apex . . . . 29. *Camel*
52. Petals 2–3 mm long; fruit 1-seeded . . . . 30. *Nes*
52. Petals 4–6 mm long; fruit several-seeded . . . . 29. *Camel*
53. Some of the leaves auriculate at the base . . . .
53. None of the leaves auriculate at the base . . . .
54. Ovary and fruit several times longer than broad . . . .
54. Ovary and fruit at most only about three times longer than broad . . .
   . . . . 34. *Rorip*
55. Fruit tapering to a slender beak; seeds globose . . . . 31. *Brass*
55. Fruit tapering to a short, thick style; seeds not globose . . 22. *Barba*
56. Plants with stellate pubescence . . . . 9. *Descura*
56. Plants glabrous or with pubescence of simple hairs . . . .
57. Racemes subtended by pinnatifid bracts . . . . 32. *Erucastr*
57. Racemes without pinnatifid bracts subtending them (except rarely
   lowest raceme) . . . .
58. Petals up to 1 cm long . . . .
58. Petals 1–2 cm long . . . .
59. Ovary and fruit conspicuously beaked . . . .
59. Ovary and fruit beakless . . . .
60. Seeds in two rows . . . . 24. *Diplot*
60. Seeds in one row . . . . 31. *Brass*
61. Ovary and fruit at least 15 times longer than broad; seeds in one row
   . . . . 33. *Sisymbr*
61. Ovary and fruit up to 7 times longer than broad; seeds in two rows
   . . . . 34. *Rori*
62. Petals with brown or purple veins . . . .
62. Petals without brown or purple veins . . . . 31. *Bras*
63. Pedicels up to 6 mm long; fruit not constricted between the seeds .
   . . . . 6. *Er*
63. Pedicels over 6 mm long; fruit more or less constricted between
   seeds . . . . 2. *Rapha*

1. Dentaria L.   Toothwort

   1. D. laciniata Muhl. Toothwort. Feb.–May. Woods, comm
throughout the state.

2. Raphanus L.   Radish

1. Petals yellowish or white; fruit up to 6 mm thick . . . . 1. *R. raphanist*
1. Petals pink or pale purple; fruit 6–10 mm thick . . . . 2. *R. sat*

1. R. raphanistrum L. Wild Radish. June–Aug. Native of Europe and Asia; occasional in waste places; Champaign, DeKalb, McHenry, Peoria, and St. Clair cos.

2. R. sativus L. Radish. May–Sept. Native of Europe and Asia; occasionally escaped from cultivation.

3. Hesperis L. Rocket

1. H. matronalis L. Rocket. May–Aug. Native of Europe and Asia; occasionally escaped from cultivation.

4. Iodanthus Torr. & Gray

1. I. pinnatifidus (Michx.) Steud. Purple Rocket. May–July. Low woods; occasional throughout the state.

5. Cakile Hill Sea Rocket

1. C. edentula (Bigel.) Hook. var. lacustris Fern. Sea Rocket. July–Sept. Shores of Lake Michigan, rare; Cook and Lake cos. C. edentula (Bigel.) Hook.—J.

6. Eruca Mill. Garden Rocket

1. E. sativa Mill. Garden Rocket. June–Sept. Native of Europe; rarely escaped from cultivation; Peoria Co.

7. Capsella Medic. Shepherd's-purse

1. C. bursa-pastoris (L.) Medic. Shepherd's-purse. Jan.–Dec. Native of Europe; naturalized in waste areas; known from every co.

8. Arabis L. Rockcress

None of the leaves sagittate nor auriculate at the base 2
Some or all of the cauline leaves sagittate or auriculate at the base . 3
Fruits strongly ascending at maturity; leaves on stem up to 5 mm broad
.... 1. A. lyrata
Fruits arching and usually pointing downward at maturity; leaves on stem 5 mm broad or broader .... 2. A. canadensis
Stems glabrous (except sometimes at the very base in A. glabra and A. drummondii) 4
Stems hairy to summit 6
At least some of the fruits over 1 mm wide, arched-recurving at maturity
.... 5. A. laevigata
Fruit never exceeding 1 mm in width, strongly erect at maturity 5
Petals creamy-yellow, about as long as the sepals; fruits terete ....
.... 3. A. glabra
Petals white, about twice as long as the sepals; fruits flat ....
.... 4. A. drummondii
Leaves and stems with simple and stellate hairs; petals 2–3 mm long; seeds wingless .... 6. A. shortii
Leaves and stems with simple hairs; petals 4–6 mm long; seeds narrowly winged .... 7. A. hirsuta

1. A. lyrata L. Sand Cress. May–July. Dunes, sandy woods, grav prairies; restricted to the n. one-third of Ill.

2. A. canadensis L. Sicklepod. May–July. Wooded slopes, moi woods; occasional throughout the state although rather rare in th se. cos.

3. A. glabra (L.) Bernh. Tower Mustard. May–July. Moist prairie limestone woods, not common; throughout the state, but very rare the s. cos.

4. A. drummondii Gray. Rock Cress. May–July. Gravelly soil, ve rare; Cook and Kane cos. A. confinis S. Wats.—J.

5. A. laevigata (Muhl.) Poir. Smooth Rock Cress. March–June. Moi woods; occasional to common throughout Ill.

6. A. shortii (Fern.) Gl. Rock Cress. April–May. Fruits stellate-pube cent. Moist woods; occasional in the n. half of the state, rare in the half. A. perstellata E. L. Br. var. shortii Fern.—F.

var. phalacrocarpa (M. Hopkins) Steyerm. Rock Cress. May. Fru glabrous. Moist woods, rare; Lake Co. A. perstellata E. L. Br. va phalacrocarpa (M. Hopkins) Fern.—F.

7. A. hirsuta (L.) Scop. var. pycnocarpa (M. Hopkins) Rollins. Ha Rock Cress. May–June. Stems hirsute to summit. Limestone cliffs a woods; not common in the n. half of the state; also Jackson Co. pycnocarpa Hopkins—J.

var. adpressipilis (M. Hopkins) Rollins. Rock Cress. May–Jur Stem with appressed hairs to summit. Limestone woods, rare; co fined to ne. Ill.; also St. Clair Co.

var. glabrata Torr. & Gray. Rock Cress. June. Stem glabrous abo the middle. Woods, very rare; Kankakee Co.

9. Descurainia Webb & Berthelot Tansy Mustard

1. Fruits up to 1 mm thick; stems and leaves grayish, not glandular-ha .... 1. D. sop

1. Fruits 1–2 mm thick; stems and leaves green, somewhat glandular-ha .... 2. D. pinnata var. brachycar

1. D. sophia (L.) Webb. Tansy Mustard. June–Aug. Native of Euro escaped into waste areas; rare and scattered.

2. D. pinnata (Walt.) Britt. var. brachycarpa (Richards.) Fern. Ta Mustard. April–June. Sandstone and limestone cliffs, along railroa occasional to common throughout the state. D. brachycarpa (R ards.) O. E. Schulz—J.

10. Cardamine L. Bitter Cress

1. None of the leaves pinnate or pinnatifid ....

1. Some or all the leaves pinnate or pinnatifid ....

2. Stems glabrous; sepals green .... 1. C. bulb

2. Stems hirsute; sepals purplish .... 2. C. dougla

3. Petioles of cauline leaves ciliate at base .... 3. C. hirs

3. Petioles of cauline leaves not ciliate at base ....

4. Terminal leaflet broader than the lateral leaflets; bases of leaflets de rent along the rachis .... 4. C. pensylvar

4. Terminal leaflet about the same width as the lateral leaflets; base leaflets not decurrent along the rachis ... 5. C. parviflora var. arenic

1. C. bulbosa (Schreb.) BSP. Spring Cress. March–June. Wet soil in woods and along streams; occasional to common throughout the state.

2. C. douglassii (Torr.) Britt. Purple Cress. April–May. Low woods; occasional in the n. two-thirds of Ill.; also Johnson Co.

3. C. hirsuta L. Spring Cress. March–April. Native of Europe and Asia; occasionally naturalized in wet areas throughout the state.

4. C. pensylvanica Muhl. Bitter Cress. March–July. Low woods, damp fields, base of moist cliffs; occasional throughout the state.

5. C. parviflora L. var. arenicola (Britt.) O. E. Schulz. Small-flowered Bitter Cress. March–July. Dry woods, wet ledges; occasional throughout the state. *C. arenicola* Britt.—J.; *C. parviflora* L.—G.

11. *Sibara Greene*

1. S. virginica (L.) Rollins. March–May. Open woods, waste ground; occasional in the s. three-fifths of Ill.; also Kankakee Co. *Arabis virginica* (L.) Poir.—G, J.

12. *Draba L. Whitlow Grass*

. No leaves present on flowering scape; petals deeply 2-cleft   1. *D. verna*
. One or more leaves present on flowering stem 2
. Leaves confined to lower part of stem 3
. Leaves all along stem 4
. Leaves dentate; pedicels pubescent .... 2. *D. cuneifolia*
. Leaves entire; pedicels glabrous .... 3. *D. reptans*
. Fruits to 5 mm long, glabrous, with up to 15 seeds   . 4. *D. brachycarpa*
. Fruits at least 6 mm long, pubescent, with 20 or more seeds ....
.... 2. *D. cuneifolia*

1. D. verna L. Vernal Whitlow Grass. Feb.–April. Seeds 40 or more; fruits more than twice as long as broad. Native of Europe and Asia; naturalized in waste areas; occasional in the s. cos., rare in the n. cos.

var. boerhaavii Van Hall. Seeds less than 40; fruits never more than twice as long as broad. Native of Europe; Jackson Co.

2. D. cuneifolia Nutt. Whitlow Grass. Feb.–May. Leaves basal and sometimes from the lower nodes. Limestone ledges, rare; Monroe and Randolph cos.

var. foliosa Mohlenbr. Leaves cauline as well as basal. Limestone ledge; Monroe Co.

3. D. reptans (Lam.) Fern. Whitlow Grass. Feb.–May. Fruits glabrous. Sand prairies, limestone ledges; occasional in the n. cos., rare in the s. cos.

var. micrantha (Nutt.) Fern. Fruits hispidulous; occasional in the n. cos.

4. D. brachycarpa Nutt. Short-fruited Whitlow Grass. March–May. Fields, lawns, prairies, woods, along railroads; common in the s. two-fifths of Ill.; also DeKalb Co., where it may be adventive.

13. *Arabidopsis Heynh. Mouse-ear Cress*

1. A. thaliana (L.) Heynh. Mouse-ear Cress. March–May. Native of Europe and Asia; naturalized in waste ground; occasional in the s. half of the state; absent from the n. half.

14. Lepidium L.   Peppergrass

1. Plants densely hairy .... 1. *L. campes*
1. Plants glabrous or puberulent ....
2. Cauline leaves auriculate at base .... 2. *L. perfoliatu*
2. Cauline leaves tapering to base ....
3. Fruits winged their entire length, 5–6 mm long; stamens 6 ....
.... 3. *L. sativu*
3. Fruits unwinged, or winged only at tip, 2–4 mm long; stamens 2 (–4)
4. Petals about as long as the sepals .... 4. *L. virginicu*
4. Petals absent ....
5. Plants with a fetid odor; basal leaves bipinnatifid .... 5. *L. rudera*
5. Plants without a fetid odor; basal leaves once-pinnatifid or coars
toothed .... 6. *L. densiflore*

1. L. campestre (L.) R. Br. Field Peppergrass. April–June. Native
Europe; naturalized in waste areas; occasional throughout the sta
2. L. perfoliatum L. Perfoliate Peppergrass. April–June. Native
Europe; naturalized in waste areas; occasional, primarily in the
cos.
3. L. sativum L. Garden Peppergrass. May–June. Native of Euro
rarely escaped from cultivation; Cook Co.
4. L. virginicum L. Common Peppergrass. Feb.–Dec. Wa
grounds, prairies, edge of woods, fields; in every co.
5. L. ruderale L. Stinking Peppergrass. May–June. Native of Euro
rarely naturalized; DeKalb Co.
6. L. densiflorum Schrad. Peppergrass. April–Oct. Native of Euro
and Asia; naturalized in waste areas; occasional throughout the sta

15. Armoracia Gaertn., Meyer, & Scherb.

1. Basal leaves divided into filiform divisions; aquatic plants   1. *A. aqua*
1. None of the leaves divided into filiform divisions; terrestrial plants . .
.... 2. *A. lapathif*

1. A. aquatica (Eat.) Wieg. Lake Cress. May–Aug. Swamps and q
streams; occasional throughout the state. *Neobeckia aquatica* (E
Greene—J.
2. A. lapathifolia Gilib. Horseradish. May–July. Native of Europe
Asia; escaped from cultivation; occasional throughout Ill. *A. rustic*
(Lam.) Gaertn.—G, J.

16. Nasturtium R. Br.   Water Cress

1. N. officinale R. Br. Water Cress. April–Oct. Cool springs
branches; occasional throughout the state. (Including var. *siifol*
[Reichenb.] Koch.)

17. Thlaspi L.   Penny Cress

1. T. arvense L. Field Penny Cress. April–June. Native of Eur
and Asia; naturalized into waste areas; occasional throughout
state.

18. Cardaria Desv.

1. C. draba (L.) Desv. Hoary Cress. April–June. Native of Europe and Asia; naturalized in waste areas; occasional in the n. half of the state, rare in the s. half.

19. Alliaria Scop.   Garlic Mustard

1. A. officinalis Andrz. Garlic Mustard. May–June. Native of Europe and Asia; naturalized into woods and waste places; occasional in the n. half of Ill.; also Coles Co.

20. Berteroa DC.

1. B. incana (L.) DC. Hoary Alyssum. May–Sept. Native of Europe; naturalized in waste areas; not common, mostly found in the n. cos.

21. Lobularia Desv.   Sweet Alyssum

1. L. maritima (L.) Desv. Sweet Alyssum. June–Aug. Native of Europe; rarely escaped from cultivation; Champaign, Cook, Hancock, and Lake cos.

22. Barbarea R. Br.   Winter Cress

All leaves pinnatifid; beak of fruit up to 1 mm long .... 1. *B. verna*
Uppermost leaves merely dentate; beak of fruit 1.5–3.0 mm long ....
.... 2. *B. vulgaris*

1. B. verna (Mill.) Aschers. Early Winter Cress. April–June. Native of Europe; rarely naturalized in Ill.; Johnson Co.
2. B. vulgaris R. Br. Yellow Rocket. April–June. Fruits strongly ascending. Native of Europe; naturalized in waste ground; rare and scattered in Ill.
var. arcuata (Opiz) Fries. Yellow Rocket. April–June. Fruits spreading. Native of Europe; naturalized in waste ground; common throughout the state. *B. vulgaris* R. Br.—J.

23. Lesquerella S. Wats.

Plants perennial; fruits densely stellate-pubescent .... 1. *L. ludoviciana*
Plants annual; fruits glabrous or nearly so .... 2. *L. gracilis*

1. L. ludoviciana (Nutt.) S. Wats. Silvery Bladder Pod. May–Aug. Sandy soil, very rare; Mason Co.
2. L. gracilis (Hook.) S. Wats. Slender Bladder Pod. May–Aug. Native of the w. U.S.; adventive along a railroad in Cook Co.

24. Diplotaxis DC.

Sepals 5–8 mm long; stems leafy throughout; fruits stipitate ....
.... 1. *D. tenuifolia*
Sepals 2.5–5.0 mm long; stems leafy only in lower half; fruits sessile ...
.... 2. *D. muralis*

1. D. tenuifolia (L.) DC. Sand Rocket. July–Sept. Native of Europ
rarely adventive in Ill.; Cook Co.
2. D. muralis (L.) DC. Wall Rocket. June–Aug. Native of Europe; n
uralized in waste areas; occasional in the n. one-fourth of Ill.

25. *Erysimum* L. *Treacle Mustard*

1. Petals well over 1 cm long, orange-yellow .... 1. *E. capitat*
1. Petals up to 1 cm long, yellow ....
2. Petals up to 5 mm long .... 2. *E. cheiranthoid*
2. Petals 5–10 mm long ....
3. Annual; fruit more than 5 cm long; plants pale green .. 3. *E. repand*
3. Perennial; fruit up to 5 cm long; plants gray-green 4. *E. inconspicu*

1. E. capitatum (Dougl.) Greene. Western Wallflower. May–J
Sandy bluffs, rare; w. cent. cos.; also Kendall and LaSalle cos.
asperum (Nutt.) DC.—G, J; *E. arkansanum* Nutt.—F.
2. E. cheiranthoides L. Wormseed Mustard. May–Sept. Native
Europe and Asia; naturalized in the n. half of Ill.
3. E. repandum L. Treacle Mustard. April–June. Native of Euro
and Asia; naturalized throughout the state.
4. E. inconspicuum (S. Wats.) MacM. May–June. Native of the
U.S.; occasionally adventive in waste areas in the n. half of Ill.

26. *Alyssum* L. *Alyssum*

1. A. alyssoides L. Pale Alyssum. May–June. Native of Europe; na
ralized in waste areas; occasional in the n. half of the state.

27. *Isatis* L. *Woad*

1. I. tinctoria L. Dyer's Woad. May–June. Native of Europe; ra
escaped from cultivation; Cook Co.

28. *Conringia* Link *Hare's-ear Mustard*

1. C. orientalis (L.) Dumort. Hare's-ear Mustard. May–Aug. Nativ
Europe; naturalized in waste places; occasional throughout the st

29. *Camelina* Crantz *False Flax*

1. Stems and leaves glabrous or with appressed pubescence; fruits
mm long .... 1. *C. sa*
1. Stems and leaves hirsutulous; fruits 4–5 (–7) mm long 2. *C. microc*

1. C. sativa (L.) Crantz. False Flax. April–Aug. Native of Europe;
uralized in waste areas; rare in the n. cos.; also Moultrie Co.
2. C. microcarpa Andrz. False Flax. April–Aug. Native of Eur
naturalized in waste areas; occasional throughout the state.

30. *Neslia* Desv. *Ball Mustard*

1. N. paniculata (L.) Desv. Ball Mustard. May–Sept. Nativ
Europe; naturalized in Ill.; DuPage Co.

31. Brassica L.   Mustard

1. *B. hirta* Moench. White Mustard. April–July. Native of Europe and Asia; rarely naturalized in Ill.; Adams, Boone, Cook, DeKalb, Henry, Kane, Kankakee, Lake, McHenry, and Will cos. *B. alba* (L.) Rabenh.—J.

2. *B. kaber* (DC.) L. C. Wheeler var. *pinnatifida* (Stokes) L. C. Wheeler. Charlock. April–July. Fruits slightly torulose, 3–4 mm thick. Native of Europe and Asia; naturalized in waste areas; occasional throughout the state. *B. kaber* (DC.) L. C. Wheeler—G, J.

var. *schkuhriana* (Reichenb.) L. C. Wheeler. Field Mustard. April–July. Fruit strongly torulose, 1.5–2.0 mm thick. Native of Europe and Asia; naturalized in waste areas; rare and scattered in Ill.

3. *B. nigra* (L.) Koch. Black Mustard. April–Oct. Native of Europe and Asia; naturalized in waste areas; occasional throughout the state.

4. *B. juncea* (L.) Coss. Indian Mustard. April–Sept. Native of Europe and Asia; naturalized in waste areas; occasional throughout the state.

5. *B. rapa* L. Field Mustard. April–Sept. Native of Europe and Asia; naturalized in waste areas; occasional throughout the state. *B. campestris* L.—J.

6. *B. oleracea* L. Cabbage. April–Sept. Native of Europe and Asia; rarely escaped from cultivation; Peoria Co.

7. *B. napus* L. Turnip. May–Sept. Native of Europe and Asia; rarely escaped from cultivation; Champaign and Hancock cos. *B. rapa* L.—J.

32. Erucastrum Presl

1. *E. gallicum* (Willd.) O. E. Schulz. May–Sept. Native of Europe; naturalized in waste areas; occasional and scattered throughout the state.

33. Sisymbrium L.

2. At least the upper leaves divided into threadlike divisions; fruits 5–10 ᵗ
long .... 2. *S. altissim*
2. Leaves divided into triangular lobes; fruits 2–4 cm long ... 3. *S. loes*

1. S. officinale (L.) Scop. Hedge Mustard. May–Oct. Fruits pub
cent. Native of Europe and Asia; naturalized in waste areas; not co
mon, but scattered throughout the state.
var. leiocarpum DC. May–Oct. Fruits glabrous or nearly so. Nat
of Europe and Asia; naturalized into waste areas; occasional throug
out the state.
2. S. altissimum L. Tumble Mustard. May–Aug. Native of Europe; n
uralized in waste areas; occasional throughout the state.
3. S. loeselii L. May–Oct. Native of Europe; naturalized in wa
areas; rare and apparently confined to the n. cos.

34. *Rorippa* Scop. Yellow Cress

1. Petals over 4 mm long, longer than the sepals ....
1. Petals absent, or up to 2 mm long, never longer than the sepals ....
2. Leaves without auricles at base; seeds up to 0.8 mm long 1. *R. sylves*
2. Leaves with auricles at base; seeds about 1 mm long .... 2. *R. sinu*
3. Petals up to 0.5 mm long; stamens 4; fruits 6–10 times longer than
pedicels; style nearly absent; seeds 175 or more per fruit ....
.... 3. *R. sessilifl*
3. Petals 1–2 mm long; stamens 6; fruits at most only 4 times longer tʰ
the pedicels; style present; seeds less than 75 per fruit ....
4. Petals 1.0–1.2 mm long; fruits 2–4 times longer than the pedicels ..
.... 4. *R. trunc*
4. Petals 1.7–2.0 mm long; fruits shorter than to up to 2 times longer tʰ
the pedicels .... 5. *R. island*

1. R. sylvestris (L.) Bess. Creeping Yellow Cress. May–Sept. Na
of Europe; naturalized in waste areas; occasional throughout
state.
2. R. sinuata (Nutt.) Hitchc. Spreading Yellow Cress. April–July. ᵗ
soil, particularly along rivers; occasional throughout the state.
3. R. sessiliflora (Nutt.) Hitchc. Sessile-flowered Yellow Cr
April–Nov. Along rivers and streams; occasional to common in tʰ
four-fifths of Ill.; also Grundy and Kane cos.
4. R. truncata (Jepson) Stuckey. May–Sept. Along rivers, rare;
Clair Co. *R. obtusa* (Nutt.) Britt.—F, G.
5. R. islandica (Oeder) Borbas. Marsh Yellow Cress. July–S
stems and leaves glabrous or nearly so; leaves pinnate or deeply
natifid, membranaceous. Mud and sand flats, rare; Jackson Co.
var. fernaldiana Butt. & Abbe. Marsh Yellow Cress. May–Nov. Stᵉ
and leaves glabrous or nearly so; leaves coarsely or shallᵒ
toothed, firm. Wet soil; common throughout the state. *R. island*
(Oeder) Borbás—J.
var. hispida (Desv.) Butt. & Abbe. Hairy Marsh Yellow Cress. ᴺ
Oct. Stems and leaves hirsutulous. Wet ground, rare; Cook
McHenry cos. *R. hispida* (Desv.) Britt.—J.

35. Chorispora DC.

1. C. tenella (Willd.) DC. May–July. Native of Asia; adventive along railroads; Cook, Greene, Kane, Kankakee, and Lake cos.

36. Coronopus Trew. Wart Cress

1. C. didymus (L.) Sm. Wart Cress. April–Oct. Native of Europe; adventive in waste ground; Cook Co.

72. RESEDACEAE

1. Reseda L. Mignonette

1. R. alba L. Mignonette. June–Oct. Rarely escaped from cultivation; Cook and DuPage cos.

73. SARRACENIACEAE

1. Sarracenia L. Pitcher-plant

1. S. purpurea L. Pitcher-plant. May–June. Bogs, rare; Cook, Lake, and McHenry cos.

74. DROSERACEAE

1. Drosera L. Sundew

Leaves spatulate; seeds obovoid, with a tight, papillate testa . . . .
. . . . 1. *D. intermedia*
Leaves suborbicular; seeds fusiform, with a loose, striate testa . . . .
. . . . 2. *D. rotundifolia*

1. D. intermedia Hayne. Narrow-leaved Sundew. July–Sept. Bogs, rare; Cook, Kane, and Kankakee cos.
2. D. rotundifolia L. Round-leaved Sundew. July–Sept. Bogs, rare; Cook, Lake, and Ogle cos.

75. CRASSULACEAE

1. Sedum L. Stonecrop

Flowers yellow 2
Flowers white, pink, or purplish 3

2. Leaves linear, to 6 mm long .... 1. *S. acr*
2. Leaves oblanceolate, 1–2 cm long .... 2. *S. sarmentosu*
3. Leaves terete, linear to linear-spatulate; petals up to 8 mm long ......
.... 3. *S. pulchellu*
3. Leaves flat, elliptic, ovate, or obovate; petals 8 mm long or longer ...
4. Flowers white; leaves entire, up to 2 cm long .... 4. *S. ternatu*
4. Flowers pink or purple; leaves dentate, at least the larger more than
cm long ....
5. Flowers pink; petals more than twice as long as the sepals ....
.... 5. *S. telephioid*
5. Flowers purple; petals at most only twice as long as the sepals ......
.... 6. *S. purpureu*

      1. S. acre L. Mossy Stonecrop. June–Aug. Native of Europe; c
casionally escaped from cultivation.
      2. S. sarmentosum Bunge. Yellow Stonecrop. June. Native
e. Asia; occasionally escaped from cultivation.
      3. S. pulchellum Michx. Widow's-cross. May–July. Dry or mo
sandstone; known only from the s. one-sixth of the state.
      4. S. ternatum Michx. Three-leaved Stonecrop. May–June. Lin
stone cliffs; moist wooded ravines; occasional and scattered throug
out Ill.
      5. S. telephioides Michx. American Orpine. Aug.–Sept. Sandsto
cliffs, rare; apparently only in the se. one-fourth of the state.
      6. S. purpureum (L.) Link. Live-forever. Aug.–Sept. Native
Europe; occasionally escaped from cultivation. *S. triphyllum* (Ha
S. F. Gray—J; *S. telephium* L. ssp. *purpureum* (Link) Schinz & Kelle
G.

76. SAXIFRAGACEAE

1. Trees or shrubs ....
1. Herbs ....
2. Leaves opposite ....
2. Leaves alternate ....
3. Leaves elliptic to oblong to narrowly ovate; flowers solitary or in cym
or racemes; stamens 10 or more ....
3. Leaves broadly ovate; flowers in flat-topped corymbs; stamens ten
less .... 3. *Hydrang*
4. Stamens 15 or more; pubescence not stellate .... 1. *Philadelp*
4. Stamens 10; pubescence stellate .... 2. *Deut*
5. Leaves palmately lobed; fruit a berry .... 4. *Ri*
5. Leaves serrulate; fruit a capsule .... 5. *
6. Leaves essentially all basal, or sometimes with 1–2 leaves on the stem
6. Leaves cauline .... 11. *Penthor*
7. Flower solitary, with gland-tipped staminodia .... 6. *Parnas*
7. Flowers several, without gland-tipped staminodia ....
8. Stem with a pair of opposite leaves; petals deeply divided .. 7. *Mit*
8. Stem with a single leaf, or leafless ....
9. Leaves longer than broad; stamens 10 .... 8. *Saxifr*
9. Leaves as broad as long; stamens 5 ....

10. Leaves pubescent; ovary 1-locular; seeds unwinged ...... 9. *Heuchera*
10. Leaves glabrous or nearly so; ovary 2-locular; seeds winged ....
.... 10. *Sullivantia*

1. *Philadelphus* L. Mock Orange

1. Flowers 1–4 in a terminal inflorescence .... 1. *P. inodorus*
1. Flowers 5–7 in both terminal and lateral racemes 2
2. Flowers faintly aromatic; calyx pubescent; lower surface of leaves
pubescent throughout .... 2. *P. pubescens*
2. Flowers strongly fragrant; calyx glabrous; lower surface of leaves pubescent only on the nerves .... 3. *P. coronarius*

    1. P. inodorus L. Scentless Mock Orange. May–June. Native of the
se. U.S.; rarely escaped from cultivation; Union Co.
    2. P. pubescens Loisel. Mock Orange. May–June. Wooded bluffs
along Ohio River, very rare; Pope Co.
    3. P. coronarius L. Sweet Mock Orange. May–June. Native of
Europe; rarely escaped from cultivation; Cook Co.

2. *Deutzia* Thunb. Deutzia

    1. D. scabra Thunb. Pride-of-Rochester. May–June. Native of China
and Japan; rarely escaped from cultivation.

3. *Hydrangea* L. Hydrangea

    1. H. arborescens L. Wild Hydrangea. June–Aug. Woods; common
in the s. four-fifths of the state, apparently absent elsewhere. (Including var. *oblonga* Torr. & Gray and var. *deamii* St. John.)

4. *Ribes* L. Gooseberry. Currant

1. Plants with spines or prickles 2
1. Plants lacking spines and prickles 4
2. Berries prickly; lobes of calyx shorter than the tube ... 1. *R. cynosbati*
2. Berries lacking prickles; lobes of calyx as long as or longer than the
tube 3
3. Flowers white to greenish-white; berries at least 10–15 mm in diameter
.... 2. *R. missouriense*
3. Flowers greenish-yellow or purplish; berries 8–10 (–12) mm in diameter
.... 3. *R. hirtellum*
4. Flowers golden-yellow, very fragrant; tube of calyx more than twice as
long as the lobes .... 4. *R. odoratum*
4. Flowers yellow-green, yellow and white, or greenish-purple; tube of
calyx up to twice as long as the lobes 5
5. Leaves and fruits with resinous dots; berries black 6
5. Leaves and fruits lacking resinous dots; berries red ..... 7. *R. sativum*
6. Flowers broadly campanulate; calyx 5–6 mm long, pubescent ....
.... 5. *R. nigrum*
6. Flowers tubular-campanulate; calyx 8–10 mm long, glabrous ....
.... 6. *R. americanum*

1. R. cynosbati L. Prickly Gooseberry. April–May. Moist, often rocky, woods; occasional in the n. half and the s. one-fourth of the state, rare or absent elsewhere. (Including var. *glabratum* Fern.)

2. R. missouriense Nutt. Missouri Gooseberry. April–May. Woods; common in the n. two-thirds of the state, occasional elsewhere.

3. R. hirtellum Michx. Northern Gooseberry. April–June. Moist woods, bogs, rare; confined to the ne. cos., except for Menard Co.

4. R. odoratum Wendl. Buffalo Currant. April–June. Native of the w. U.S.; occasionally escaped from cultivation.

5. R. nigrum L. Black Currant. May–June. Native of Europe; rarely escaped from cultivation; Lake Co.

6. R. americanum Mill. Wild Black Currant. May–June. Moist woods; common in the n. three-fifths of the state, absent elsewhere.

7. R. sativum (Reichenb.) Syme. Red Currant. May–June. Native of Europe; occasionally escaped from cultivation.

5. *Itea L.*

1. I. virginica L. Virginia Willow. May–June. Swampy woods; confined to the s. one-sixth of the state.

6. *Parnassia L.*

1. P. glauca Raf. Grass-of-Parnassus. July–Sept. Calcareous springs, dune flats; restricted to the n. one-half of the state.

7. *Mitella L.   Miterwort*

1. M. diphylla L. Bishop's-cap. May. Rich, often rocky, woods; occasional in the n. half of the state, rare in the s. half.

8. *Saxifraga L.   Saxifrage*

1. None of the leaves more than 8 cm long; calyx lobes about as long as calyx tube; petals white, at least 4 mm long .... 1. *S. virginiensis*
1. Some or all the leaves more than 8 cm long; calyx lobes at least twice as long as calyx tube; petals greenish-white or greenish-yellow, 2–3 mm long 2
2. Leaves pilose on the lower surface; petals longer than the sepals .... 2. *S. forbesii*
2. Leaves glabrous or nearly so on the lower surface; petals about as long as the sepals .... 3. *S. pensylvanica*

1. S. virginiensis Michx. Early Saxifrage. April–June. Rocky wooded ravines, rare; Hardin Co.

2. S. forbesii Vasey. Forbes' Saxifrage. May. Moist sandstone cliffs, rare; known from seven cos. in extreme s. Ill. and three cos. in extreme n. Ill.; also Monroe Co.

3. S. pensylvanica L. Swamp Saxifrage. May–June. Springs, bogs, moist meadows; occasional in the n. half of the state, absent elsewhere; also Wabash Co. (Including ssp. *interior* Burns.)

9. *Heuchera* L.   *Alumroot*

1. Calyx to 2 mm long; leaves and stems villous .... 1. *H. parviflora*
1. Calyx 2.5 mm long or longer; leaves and stems hispid or strigose ... 2
2. Leaves strigose or nearly glabrous; calyx symmetrical or nearly so .....
.... 2. *H. hirsuticaulis*
2. Leaves hispid beneath; calyx asymmetrical .... 3. *H. richardsonii*

1. H. parviflora Bartl. var. rugelii (Shuttlw.) Rosend., Butt. & Lak. Late Alumroot. July–Nov. Moist sandstone cliffs; common in the s. one-sixth of the state, absent elsewhere. *H. parviflora* Bartl.—J.

2. H. hirsuticaulis (Wheelock) Rydb. Tall Alumroot. May–June. Dry woods; occasional in the s. half of the state, absent elsewhere. (Including var. *interior* Rosend., Butt. & Lak.) *H. americana* L. var. *hirsuticaulis* (Wheelock) Rosend., Butt. & Lak.—F.

3. H. richardsonii R. Br. var. grayana Rosend., Butt. & Lak. Prairie Alumroot. May–July. Prairies; occasional in the n. three-fourths of the state, apparently absent elsewhere. (Including var. *affinis* Rosend., Butt. & Lak.)

10. *Sullivantia* Torr. & Gray

1. S. renifolia Rosend. Sullivantia. June–July. Moist shaded cliffs, rare; confined to the extreme nw. corner of Ill.

11. *Penthorum* L.

1. P. sedoides L. Ditch Stonecrop. July–Oct. Wet ground, common; in every co.

77. HAMAMELIDACEAE

1. Leaves star-shaped; flowers without petals; fruit a globose, echinate head .... 1. *Liquidambar*
1. Leaves oval to obovate to suborbicular; flowers with 4 yellow petals; fruit an obovoid capsule .... 2. *Hamamelis*

1. *Liquidambar* L.   *Sweet Gum*

1. L. styraciflua L. Sweet Gum. April–May. Low woods; occasional to common in the s. one-third of the state.

2. *Hamamelis* L.   *Witch-hazel*

1. H. virginiana L. Witch-hazel. Sept.–Nov. Woods; occasional in the n. half of the state; also Richland, Wabash, and White cos.

78. PLATANACEAE

1. *Platanus* L. Sycamore

1. P. occidentalis L. Sycamore. April–May. Moist woods and along streams; common in the s. two-thirds of the state, occasional elsewhere; in every co.

79. ROSACEAE

18. Flowers white; ovaries more than 50 per flower; receptacle bearing the achenes fleshy .... 16. *Fragaria*
19. Leaves 2- to 3-pinnate; flowers dioecious .... 17. *Aruncus*
19. Leaves once-pinnate or palmate; flowers perfect 20
20. Flowers pink, rose, or purple 21
20. Flowers white, cream, or yellow 22
21. Plants glabrous; flowers up to 8 mm across .... 18. *Filipendula*
21. Plants pubescent; flowers at least 1 cm across .... 21. *Geum*
22. Petals absent; calyx 4-parted .... 19. *Sanguisorba*
22. Petals present; calyx 5-parted 23
23. Petals white 24
23. Petals yellow 27
24. Petals strap-shaped, usually 2–3 times longer than broad; each leaf subtended by a pair of foliaceous stipules .... 20. *Gillenia*
24. Petals mostly obovate, usually less than twice as long as broad; each leaf generally not subtended by a pair of foliaceous stipules 25
25. Trailing or ascending, nearly glabrous plants; leaves pedately 3- to 5-foliolate .... 12. *Rubus*
25. Erect, pubescent plants; leaves pinnately 3- to 11-foliolate 26
26. Basal leaves 3- to 5-foliolate; petals as long as or shorter than the sepals
.... 21. *Geum*
26. Basal leaves 7- to 11-foliolate; petals longer than the sepals ....
.... 14. *Potentilla*
27. All leaves trifoliolate 28
27. At least some of the leaves 5-foliolate or more 29
28. Receptacle in fruit spongy, greatly enlarged; bractlets between the calyx lobes longer than the lobes .... 22. *Duchesnea*
28. Receptacle in fruit dry, not enlarged; bractlets between the calyx lobes about the same size as the lobes .... 14. *Potentilla*
29. Leaves pinnately compound 30
29. Leaves palmately compound 32
30. Bractlets present between calyx lobes; ovaries at least 30 per flower 31
30. Bractlets between the calyx lobes absent; ovaries 1–4 .. 23. *Agrimonia*
31. Style deciduous from the achene .... 14. *Potentilla*
31. Style persistent on the achene .... 21. *Geum*
32. Style deciduous from the achene .... 14. *Potentilla*
32. Style persistent on the achene .... 21. *Geum*

1. Physocarpus Maxim. Ninebark

1. P. opulifolius (L.) Maxim. Ninebark. May–June. Rocky slopes, rocky banks, moist swales; occasional in the n. half of the state; also Jackson, Pope, and Wabash cos. (Including var. *intermedius* [Rydb.] Robins.)

2. Spiraea L. Spiraea

1. Flowers in elongate panicles 2
1. Flowers in corymbs .... 4. *S. prunifolia*
2. Leaves woolly beneath; flowers usually rose .... 1. *S. tomentosa*
2. Leaves glabrous or nearly so beneath; flowers white 3
3. Leaves serrulate; branches yellow-brown .... 2. *S. alba*
3. Leaves coarsely toothed; branches reddish- or purplish-brown ....
.... 3. *S. latifolia*

1. S. tomentosa L. Hardhack. July–Aug. Bogs, moist thickets, rare; Cook, DuPage, Iroquois, Kankakee, Lake, and Stephenson cos. (Including var. *rosea* [Raf.] Fern.)

2. S. albà DuRoi. Meadow-sweet. July–Aug. Wet soil; occasional in the n. four-fifths of the state.

3. S. latifolia (Ait.) Borkh. Meadow-sweet. June–July. Native of ne. N.A.; rarely escaped from cultivation; Champaign and Hancock cos.

4. S. prunifolia Sieb. & Zucc. Bridal-wreath. May–June. Native of the Old World; rarely escaped from cultivation.

3. *Prunus* L.   Plum.   Cherry

1. Flowers solitary, or in corymbs or umbels 2
1. Flowers in racemes 13
2. Ovary and fruit pubescent or glaucous 3
2. Ovary and fruit neither pubescent nor glaucous 10
3. Ovary and fruit tomentose (sometimes sparsely so in *P. armeniaca*) .. 4
3. Ovary and fruit glaucous 5
4. Flowers pink; ovary densely tomentose .... 1. *P. persica*
4. Flowers white (occasionally pink); ovary usually sparsely tomentose....
   .... 2. *P. armeniaca*
5. Lobes of calyx glandular along the margins 6
5. Lobes of calyx eglandular along the margins 8
6. Flowers at least 2 cm across; lobes of calyx 3–5 mm long, reflexed from the beginning of flowering .... 3. *P. nigra*
6. Flowers 1.0–1.5 cm across; lobes of calyx less than 3 mm long, reflexed only toward the end of flowering 7
7. Flowers appearing after the expansion of the leaves ... 4. *P. hortulana*
7. Flowers appearing before the expansion of the leaves 5. *P. munsoniana*
8. Flowers up to 1 cm across; fruit 1–2 cm in diameter, the stone not compressed ....6. *P. angustifolia*
8. Flowers 1.5 cm or more across; fruit at least 2 cm in diameter, the stone compressed 9
9. Petiole with 1–2 glands near summit; leaves usually broadly rounded at the base .... 7. *P. mexicana*
9. Petiole without glands near the summit; leaves mostly narrowed at the base .... 8. *P. americana*
10. Dwarf shrubs to 2 m tall; leaves entire in the lower half; fruit 1.0–1.5 cm in diameter .... 9. *P. susquehanae*
10. Trees to 12 m tall; leaves toothed to the base; fruit either 5–7 mm in diameter or over 1.5 cm in diameter 11
11. Flowers up to 1.5 cm across; fruit 5–7 mm in diameter ....
    .... 10. *P. pensylvanica*
11. Flowers at least 2 cm across; fruit 1.5–2.5 cm in diameter 12
12. Leaves glabrous beneath; fruit sour .... 11. *P. cerasus*
12. Leaves pubescent on the nerves beneath; fruit sweet ..,. 12. *P. avium*
13. Leaves broadly rounded or cordate at base; branchlets tomentose; inflorescence 4- to 10-flowered; flowers about 1.5 cm across ....
    .... 13. *P. mahaleb*
13. Leaves more or less tapering at base; branchlets glabrous or nearly so; inflorescence more than 10-flowered; flowers up to 1 cm across ... 14
14. Leaves thin, sharply serrulate; lobes of calyx obtuse; fruit red-purple ..
    .... 14. *P. virginiana*

14. Leaves firm, crenulate-serrate; lobes of calyx acute; fruit black-purple . .
. . . . 15. *P. serotina*

1. P. persica (L.) Batsch. Peach. April. Introduced from Asia; occasionally escaped along roadsides.

2. P. armeniaca L. Apricot. May. Introduced from Europe; rarely escaped from cultivation; Champaign Co.

3. P. nigra Ait. Canada Plum. May. Mostly along rivers and streams; occasionally in the n. half of the state

4. P. hortulana Bailey. Wild Goose Plum. March–April. Thickets; occasional in the s. two-thirds of the state; rare and probably introduced in Cook, JoDaviess, LaSalle, and Stephenson cos.

5. P. munsoniana Wight & Hedrick. Wild Goose Plum. March–April. Thickets; occasional in the s. one-third of the state.

6. P. angustifolia Marsh. Chickasaw Plum. Thickets; occasional in the s. half of the state, uncommon and probably adventive in the n. half of the state.

7. P. mexicana S. Wats. Big Tree Plum. April. Base of cherty cliff; known only from Jackson Co.

8. P. americana Marsh. Wild Plum. April–May. Leaves glabrous or sparsely pubescent beneath. Thickets, woodlands; occasional throughout the state.

var. lanata Sudw. April–May. Leaves soft-pubescent beneath. Thickets, woodlands; occasional throughout the state. *P. lanata* (Sudw.) Mack. & Bush—J.

9. P. susquehanae Willd. Sand Cherry. May–June. Sandy areas, particularly along Lake Michigan; confined to the n. one-sixth of the state. *P. pumila* L.—G, J.

10. P. pensylvanica L. f. Pin Cherry. April. Sandy soil, particularly along Lake Michigan; confined to the n. one-third of the state.

11. P. cerasus L. Pie Cherry. April–May. Native of Europe; rarely escaped form cultivation; Champaign and Lawrence cos.

12. P. avium L. Sweet Cherry. April–May. Native of Europe and Asia; rarely escaped from cultivation.

13. P. mahaleb L. Mahaleb Cherry. May. Native of Europe; rarely escaped from cultivation; Kane, Sangamon, Union, and Will cos.

14. P. virginiana L. Common Chokecherry. May. Woods and thickets; occasional in the n. half of the state, rare in the s. half.

15. P. serotina Ehrh. Wild Black Cherry. May. Woods; common throughout the state; in every co.

4. *Cydonia* Mill. Quince

1. C. oblonga Mill. Common Quince. May–June. Native of Asia; established at Atwood Ridge, Union Co.

5. *Amelanchier* Medic. Shadbush

1. Stoloniferous shrub; leaves usually dentate-serrate, at least above the middle; sepals revolute at maturity; petals up to 1 cm long 1. *A. humilis*

1. Non-stoloniferous shrubs or small trees; leaves finely or sharply serrate their total length; sepals reflexed at maturity; petals 1 cm long or longer
. . . . 2

2. Leaves white-tomentose beneath at least until anthesis; pedicels of lowermost flowers in each raceme less than 2 cm long ...... 2. *A. arborea*
2. Leaves glabrous or sparsely pubescent beneath; pedicels of lowermost
flowers in each raceme usually at least 2 cm long 3
3. Leaves with up to 10 pairs of veins; ovary and usually the young fruit
pubescent .... 3. *A. interior*
3. Leaves with more than 10 pairs of veins; ovary and young fruit glabrous
.... 4. *A. laevis*

1. A. humilis Wieg. Low Shadbush. May–June. Rocky or sandy soil
restricted to the n. one-fourth of the state; also Adams Co. *A. spicata*
K. Koch—G, J.
2. A. arborea (Michx. f.) Fern. Shadbush. March–May. Wooded
slopes and bluffs; occasional throughout the state.
3. A. interior Nielsen. Shadbush. June. Bogs and wet woods, rare
Lake and Winnebago cos.
4. A. laevis Wieg. Shadbush. March–June. Wooded slopes; confined to the n. one-fourth of the state and Vermilion Co.

6. *Chaenomeles* L.   Japanese Quince

1. C. japonica L. Japanese Quince. March–April. Native of Europe
occasionally escaped from cultivation and persistent.

7. *Pyrus* L.   Pear

1. Leaves crenate, glabrous or glabrate; fruit more or less pyriform ......
.... 1. *P. communis*
1. Leaves serrate, tomentose when young; fruit subglobose   2. *P. pyrifolia*

1. P. communis L. Pear. April–May. Introduced from Europe and
Asia; escaped from cultivation throughout Ill.
2. P. pyrifolia (Burm. f.) Nakai. Chinese Pear. April. Native of Asia
escaped along a stream in Union Co.

8. *Malus* Mill.   Apple

1. Anthers yellow; most or all leaves unlobed .... 1. *M. pumila*
1. Anthers red; some of the leaves (at least those on the vegetative branches) lobed or deeply toothed ....
2. Calyx glabrous or sparsely pubescent ....
2. Calyx densely pubescent ....
3. Leaves of fruiting branches mostly obtuse at tip ..... 2. *M. angustifolia*
3. Leaves of fruiting branches mostly acute at tip .... 3. *M. coronaria*
4. Fruit up to 3.5 cm in diameter .... 4. *M. ioensis*
4. Fruit up to 5 cm in diameter or broader .... 5. *M.* X *soulardi*

1. M. pumila Mill. Apple. May. Introduced from Europe; occasionally escaped in Ill. *Pyrus malus* L.—F, G.
2. M. angustifolia (Ait.) Michx. Narrow-leaved Crab Apple. May. Low
woods, rare; Hardin and Jackson cos. *Pyrus angustifolia* Ait.—F, G.
3. M. coronaria (L.) Mill. Wild Sweet Crab Apple. April–May. Leaves
at least half as broad as long, sometimes lobulate. Woods; occasional
in the s. half of the state; less common in the n. half. *Pyrus coronaria*
L.—F, G.

var. lancifolia Rehder. Narrow-leaved Crab Apple. April–May.
Leaves less than half as broad as long, not lobulate. Woods; Gallatin,
Jackson, and Pope cos. *Pyrus coronaria* var. *lancifolia* (Rehd.)
Fern.—F.

4. M. ioensis (Wood) Britt. Iowa Crab Apple. May. Woods and thickets; occasional or common throughout the state; in every co. *Pyrus
ioensis* (Wood) Bailey—F, G.

5. M. X soulardii (Bailey) Britt. Soulard Crab. May. Reputed hybrid
between *M. pumila* and *M. ioensis;* originally cultivated at Galena, Ill.;
rarely escaped in Ill. *Pyrus* X *soulardii* Bailey—F, G.

9. *Aronia* Medic. Chokeberry

1. Axis of inflorescence and lower surface of leaves pubescent . . . .
. . . . 1. *A. prunifolia*
1. Axis of inflorescence and lower surface of leaves glabrous or nearly so
. . . . 1. *A. melanocarpa*

1. A. prunifolia (Marsh.) Rehd. Purple Chokeberry. May–June. Occasional in bogs, otherwise rare; restricted to a few cos. in the ne.
corner of the state; also Winnebago Co. *Pyrus floribunda* Lindl.—F.

2. A. melanocarpa (Michx.) Ell. Black Chokeberry. May–June. Bogs,
moist woods, sandstone ledge (in Saline Co.); occasional in the ne.
one-fourth of the state; also Saline Co. *Pyrus melanocarpa* (Michx.)
Willd.—F.

10. *Sorbus* [Tourn.] L. Mountain Ash

1. Axis of inflorescence, leaflets, and outer bud scales glabrous or nearly
so . . . . 1. *S. americana*
1. Axis of inflorescence, leaflets, and outer bud scales pubescent . . . .
. . . . 2. *S. aucuparia*

1. S. americana Marsh. Mountain Ash. May–July. Rocky woods;
known only from Ogle Co. *Pyrus americana* (Marsh.) DC.—F.

2. S. aucuparia L. European Mountain Ash. May–July. Native of
Europe and Asia; escaped to bogs and swamps in the n. one-fifth of
the state. *Pyrus aucuparia* (L.) Gaertn.—F.

11. *Crataegus* L. Hawthorn

1. Veins of larger leaves running to the sinuses as well as to the points of
the lobes. 2
1. Veins of the leaves running only to the points of the lobes 4
2. Leaves thin, early deciduous; thorns stout, some or all over 2 cm long. 3
2. Leaves thick, persisting late into the autumn; thorns slender, up to 2 cm
long . . . . 3. *C. monogyna*
3. Leaves of the flowering branchlets deltoid-cordate, usually trilobate,
glabrous at maturity; stamens 20; anthers yellow; fruit subglobose,
with dry flesh . . . . 1. *C. phaenopyrum*
3. Leaves of the flowering branchlets not deltoid-cordate, usually deeply
cut, pubescent along the veins beneath at maturity; stamens 10; anthers
red; fruit oblong, with succulent flesh . . . . 2. *C. marshallii*
4. Leaves cuneate, broadest at middle or apex 5
4. Leaves broadest at the usually rounded or truncate or cordate base 21

5. Leaves broadest toward apex 6

5. Leaves broadest at the middle 15

6. Leaves impressed-veined above 7

6. Leaves not impressed-veined above 11

7. Leaves pubescent beneath at maturity, at least on the veins, dull gray-green; fruits 12–20 mm thick .... 4. *C. punctata*

7. Leaves glabrous or nearly so at maturity, dark green or yellow-green fruits up to 12 mm thick 8

8. Some of the leaves of the flowering branches lobed 9

8. None of the leaves of the flowering branches lobed 10

9. Leaves with obscure lobes near the apex; stamens 10 or 15; anthers red leaves usually yellow-green .... 5. *C. cuneiformis*

9. Leaves with conspicuous lobes above the middle; stamens 20; anthers white or pale yellow; leaves dark green .... 6. *C. margaretta*

10. Leaves dull green, pubescent when young; corymbs pubescent; stamens 20; fruit pubescent at first, 8–10 mm thick .... 7. *C. collina*

10. Leaves yellow-green, glabrous; corymbs glabrous; stamens 10; fruit glabrous, 7–8 mm thick .... 8. *C. hannibalensis*

11. None of the leaves lobed 12

11. Some of the leaves, particularly the terminal shoot leaves, lobed (sometimes rarely lobed in *C. fecunda*) 14

12. Leaves glabrous, not reticulate-veined; petioles glabrous 9. *C. crus-galli*

12. Leaves pubescent, reticulate-veined; petioles villous 10. *C. engelmanni*

13. Leaves pubescent (at least when young); corymbs villous 11. *C. fecunda*

13. Leaves glabrous; corymbs glabrous 13

14. Terminal shoot leaves at least twice as broad as those of the flowering branches; stamens 10 or 15; anthers pale yellow ..... 12. *C. acutifolia*

14. Terminal shoot leaves usually less than twice as broad as those of the flowering branches; stamens 20; anthers pink .... 13. *C. permixta*

15. Leaves impressed-veined above 16

15. Leaves not impressed-veined above 17

16. Leaves, corymbs, and fruit glabrous .... 14. *C. succulenta*

16. Leaves, corymbs, and fruit pubescent .... 15. *C. calpodendron*

17. Petioles glandular .... 16. *C. neobushii*

17. Petioles glandless 18

18. Leaves at maturity with tufts of tomentum in the axils beneath; corymbs glabrous ....

18. Leaves at maturity glabrous beneath or at least without tufts of tomentum in the axils; corymbs pubescent 20

19. Leaves thin, scarcely lustrous above; fruit 5–8 mm thick . 17. *C. viridis*

19. Leaves thick, lustrous above; fruit 8–10 mm thick .... 18. *C. nitida*

20. Leaves glabrous; anthers pink; fruit succulent .... 19. *C. lucorum*

20. Leaves when young densely covered on the upper surface with short appressed hairs; anthers pale yellow; fruit dry .... 20. *C. faxonii*

21. Leaves completely glabrous, bluish-green; stamens usually 20 22

21. Leaves pubescent above and/or below, at least on the veins, or scabrous above, yellow-green or dark green; stamens 10 or 20. 23

22. Leaves thin; fruit succulent, not pruinose .... 21. *C. tortilis*

22. Leaves thick; fruit dry and often mealy, pruinose ...... 22. *C. pruinosa*

23. Leaves pubescent above at maturity .... 23. *C. mollis*

23. Leaves glabrous or nearly so above at maturity or scabrous above . 24

24. Leaves scabrous above, otherwise usually glabrous ....

24. Leaves usually smooth above, but with some pubescence beneath, at least on the veins ....

25. Leaves mostly broadly ovate; corymbs glabrous; fruits 10–17 mm thick
.... 24. *C. holmesiana*
25. Leaves mostly elliptic-ovate to oval; corymbs villous; fruits 7–10 mm
thick .... 25. *C. pedicellata*
26. Leaves bright yellow-green; corymbs usually pubescent (sometimes glabrous in *C. putnamiana*) 27
26. Leaves dark green or dark yellow-green; corymbs glabrous 28
27. Branchlets villous when young; stamens 5 or 10; corymbs villous; fruit
oblong, 8–12 mm thick .... 26. *C. pringlei*
27. Branchlets usually glabrous; stamens 20; corymbs usually glabrous;
fruit usually subglobose, 12–15 mm thick .... 27. *C. corusca*
28. Edge of leaf often crisped at maturity; flowers 2.4–2.6 cm wide; stamens 20; fruit subglobose .... 28. *C. coccinioides*
28. Edge of leaf not crisped; flowers 1.3–1.8 cm wide; stamens 10 or fewer;
fruit obovoid or oblongoid .... 29. *C. macrosperma*

1. C. phaenopyrum (L. f.) Medic. Washington Thorn. May–June.
Open woods; occasional in the s. one-third of the state.
2. C. marshallii Egglest. Parsley Haw. April. Wet, pin oak woods,
rare; Jackson Co.
3. C. monogyna Jacq. English Hawthorn. May. Native of Europe and
Asia; rarely escaped in Ill.; Cook Co.
4. C. punctata Jacq. Dotted Thorn. May–June. Open woods, pastures; occasional throughout the state. (Including var. *aurea* Ait., *C.
mortonis* Laughlin, and *C. sucida* Sarg.)
5. C. cuneiformis (Marsh.) Egglest. Hawthorn. May. Open woods;
occasional in the n. half of the state. (Including *C. pratensis* Sarg., *C.
peoriensis* Sarg., and *C. disperma* Ashe.) *C. disperma* Ashe—F, G.
6. C. margaretta Ashe. Hawthorn. April–May. Open woods, thickets;
occasional in the n. three-fifths of the state.
7. C. collina Chapm. Hawthorn. April–May. Low woods, rare; St.
Clair and Union cos. (Including *C. lettermanii* Sarg.)
8. C. hannibalensis Palmer. Hawthorn. May. Thickets, rare; Adams
Co. *C. cuneiformis* (Marsh.) Egglest.—J.
9. C. crus-galli L. Cock-spur Thorn. May–June. Leaves thick. Open
woods and thickets; occasional throughout the state. (Including *C. attenuata* Ashe, *C. arduennae* Sarg., *C. farwellii* Sarg., *C. acantha-
colonensis* Laughlin, and *C. pachyphylla* Sarg.)
var. barrettiana (Sarg.) Palmer. Barrett's Thorn. May–June. Leaves
thin. Thickets, rare; St. Clair Co.
10. C. engelmannii Sarg. Barberry-leaved Hawthorn. May. Open
woods, bluffs, pastures; occasional in the s. one-fifth of the state.
11. C. fecunda Sarg. Fruitful Thorn. May. Moist woods, rare; Gallatin
and St. Clair cos. (Including *C. pilifera* Sarg.) *C. engelmanni* Sarg.—J.
12. C. acutifolia Sarg. Hawthorn. May. Low woods, rare; Richland
and St. Clair cos. (Including *C. erecta* Sarg. and var. *insignis* [Sarg.]
Palmer.) *C. viridis* L.—J.
13. C. permixta Palmer. Hawthorn. May. Open woods; Peoria Co.
14. C. succulenta Link. Hawthorn. May. Dry or moist woods and
thickets; occasional in the n. half of the state, rare in the s. half.
(Including *C. macracantha* Lodd.; *C. illinoiensis* Ashe; *C. gemmosa*
Sarg.; *C. vegeta* Sarg.; *C. gaultii* Sarg.; *C. longispina* Sarg.; *C. rutila*
Sarg.; *C. laxiflora* Sarg.; *C. corporea* Sarg.; *C. leucantha* Laughlin.)
15. C. calpodendron (Ehrh.) Medic. Hawthorn. May. Woods and

thickets; occasional throughout the state. (Including *C. tomentosa* DuRoi; *C. chapmani* [Beadle] Ashe; *C. structilis* Ashe; *C. mollicula* Sarg.; *C. hispidula* Sarg.; *C. pertomentosa* Ashe; *C. chrysocarpa* Ashe; *C. whitakeri* Sarg.)

16. C. neobushii Sarg. Hawthorn. May. Rocky woods; confined to the s. one-fourth of the state.

17. C. viridis L. Green Thorn. May. Low woods; occasional in the s half of the state. (Including *C. schneckii* Ashe; *C. durifolia* Ashe; *C. ovata* Sarg.; *C. mitis* Sarg.; *C. lanceolata* Sarg.; *C. dawsoniana* Sarg. *C. pechiana* Sarg.)

18. C. nitida (Engelm.) Sarg. Hawthorn. May. Low woods; Gallatin, Henderson, and St. Clair cos. C. viridis L.—J.

19. C. lucorum Sarg. Hawthorn. May. Woods and stream banks; Cook, DuPage, Lake, and Will cos. (Including *C. apiomorpha* Sarg.) C. macrosperma Ashe—J.

20. C. faxonii Sarg. Hawthorn. May. Rocky woods and thickets; St Clair Co. C. margaretta Ashe—J.

21. C. tortilis Ashe. Hawthorn. May. Woods and thickets; Cook Co. C. macrosperma Ashe—J.

22. C. pruinosa (Wendl.) K. Koch. Hawthorn. May. Thickets and rocky woods; occasional throughout the state. (Including *C. gattingeri* Ashe; *C. virella* Ashe; *C. conjuncta* Sarg.; *C. dissona* Sarg.; *C. platycarpa* Sarg.)

23. C. mollis (Torr. & Gray) Scheele. Red Haw. May. Woods; occasional throughout the state; common northward. (Including *C. sera* Sarg.; *C. altrix* Ashe; *C. valens* Ashe; *C. venosa* Ashe; *C. nupera* Ashe; *C. verna* Ashe; *C. declivitatis* Sarg.; *C. umbrosa* Sarg.; *C. laniger* Sarg.; *C. ridgwayi* Sarg.; *C. pachyphylla* Sarg.)

24. C. holmesiana Ashe. Hawthorn. May. Woods and thickets; confined to the n. one-half of the state. (Including *C. amicta* Ashe; *C. bimoreana* Beadle; *C. intricata* Lange; *C. magniflora* Sarg.) C. coccinea L.—J.

25. C. pedicellata Sarg. Hawthorn. May. Woods and thickets; scattered in Ill. (Including *C. albicans* Ashe; *C. paucispina* Sarg.; *C. robesoniana* Sarg.) C. coccinea L.—J.

26. C. pringlei Sarg. Hawthorn. May. Thickets and woods; Lake Co. C. coccinea L.—J.

27. C. corusca Sarg. Hawthorn. May. Woods and thickets; Lake Co. (Including *C. putnamiana* Sarg.) C. coccinea L.—J; C. putnamiana Sarg.—G.

28. C. coccinioides Ashe. Hawthorn. May. Thickets and rocky woods; confined to the s. one-third of the state.

29. C. macrosperma Ashe. Hawthorn. May. Thickets and woods; occasional in the ne. one-fourth of the state. (Including *C. egani* Ashe; *C. ferrissii* Ashe; *C. demissa* Sarg.; *C. cyanophylla* Sarg.; *C. sextilis* Sarg.; *C. tarda* Sarg.; *C. depilis* Sarg.; *C. hillii* Sarg.; *C. blothra* Lauglin; *C. taetrica* Sarg.)

12. *Rubus* L. Blackberry

1. Leaves simple; flowers purple .... 1. *R. odoratus*
1. Leaves compound; flowers white ....
2. Stems without prickles or bristles, herbaceous .... 2. *R. pubescens*

2. Stems with prickles or bristles, usually at least slightly woody 3
3. Leaves white-tomentose beneath 4
3. Leaves not white-tomentose beneath 8
4. Stems glaucous; fruit purple-black at maturity .... 3. *R. occidentalis*
4. Stems not glaucous; fruit red at maturity (except *R. procerus*) 5
5. Stems, pedicels, and sepals with dense, red or purple, gland-tipped hairs
.... 4. *R. phoenicolasius*
5. Stems, pedicels, and sepals prickly, bristly, or smooth, but without
dense, red or purple, gland-tipped hairs 6
6. Plants with bristles or broad-based prickles only on the stem; fruit a
raspberry. 7
6. Plants with prickles on stem and branches of inflorescence; fruit a
blackberry .... 7. *R. procerus*
7. Sepals and pedicels without glandular hairs; prickles broad-based .....
.... 5. *R. idaeus*
7. Sepals and pedicels with glandular hairs; prickles not broad-based ....
.... 6. *R. strigosus*
8. Leaflets laciniate .... 8. *R. laciniatus*
8. Leaflets not laciniate 9
9. Stems trailing, except for the floral branches 10
9. All stems arched-ascending to erect 14
10. Stems beset with bristles, but without curved prickles .. 9. *R. hispidus*
10. Stems with curved prickles (bristles also may be intermixed in *R. trivialis*)
lis) 11
11. Stems with both bristles and prickles .... 10. *R. trivialis*
11. Stems with only prickles 12
12. Leaves velvety on the lower surface .... 11. *R. occidualis*
12. Leaves glabrous or pilose on the veins below, or sparsely pilose beneath, but never velvety beneath 13
13. Flowers two or more in a corymb; prickles (or some of them) over 2 mm
long .... 12. *R. flagellaris*
13. Flower solitary; prickles up to 2 mm long .... 13. *R. enslenii*
14. Stems bristly or setose, but without broad-based prickles ....
.... 14. *R. schneideri*
14. Stems with broad-based prickles 15
15. Pedicels and peduncles with stipitate glands 16
15. Pedicels and peduncles without stipitate glands 17
16. Inflorescence about as broad as long .... 16. *R. alumnus*
16. Inflorescence much longer than broad .... 15. *R. alleghaniensis*
17. Upper three leaflets of primary canes two or more times longer than
broad .... 17. *R. argutus*
17. Upper three leaflets of primary canes up to two times longer than broad
.... 18
18. Inflorescence a cylindrical raceme much longer than broad ....
.... 18. *R. avipes*
18. Inflorescence short-corymbiform, nearly as broad as long 19
19. Pedicels subtended by large foliaceous bracts and small stipule-like
bracts .... 19. *R. pensylvanicus*
19. Pedicels subtended only by large foliaceous bracts ... 20. *R. frondosus*

    1. R. odoratus L. Purple Flowering Raspberry. May–June. Woods;
Carroll, Cook, DeKalb, and LaSalle cos.
    2. R. pubescens Raf. Dwarf Raspberry. May–June. Bogs and
springy places; Cook, DeKalb, Lake, and Winnebago cos.

3. R. occidentalis L. Black Raspberry. May–June. Edge of woods, thickets, roadsides, bluffs; common throughout the state.

4. R. phoenicolasius Maxim. Wineberry. May–July. Native of Asia; escaped in Jersey, Lake, and Morgan cos.

5. R. idaeus L. Cultivated Raspberry. May–June. Native of Europe; occasionally escaped and persistent from cultivation.

6. R. strigosus Michx. Red Raspberry. May–June. Bogs and swampy woods; confined to the n. one-fourth of the state. *R. idaeus* L. var. *strigosus* (Michx.) Maxim.—F.

7. R. procerus P. J. Muell. Himalaya-berry. May–July. Native of Europe; escaped along a road in Randolph Co.

8. R. laciniatus Willd. Evergreen Blackberry. June–July. Native of Europe; scattered in Ill., but not common.

9. R. hispidus L. Swampy Dewberry. June–July. Bog, black oak woods, marshes, thickets; restricted to the n. one-half of the state. (Including *R. signatus* Bailey.) *R. hispidus* var. *obovalis* (Michx.) Fern.—F.

10. R. trivialis Michx. Southern Dewberry. April–May. Fields, waste ground; confined to the s. one-fourth of the state.

11. R. occidualis Bailey. Velvet-leaved Dewberry. May–June. Edge of woods; Pope Co. *R. flagellaris* L.—G.

12. R. flagellaris Willd. Dewberry. April–June. Fields, edge of woods, roadsides; common to occasional throughout most of the state. (Including *R. baileyanus* Britt.)

13. R. enslenii Tratt. Arching Dewberry. May–June. Rocky woods; Randolph Co. (Including *R. mundus* Bailey.)

14. R. schneideri Bailey. Bristly Blackberry. June. Sandy swales; Kankakee Co. (Including *R. offectus* Bailey and *R. wheeleri* Bailey.) *R. setosus* Bigel.—G, J.

15. R. allegheniensis Porter. Common Blackberry. May–June. Roadsides, thickets, woods; common throughout the state. (Including *R. impos* Bailey.)

16. R. alumnus Bailey. Blackberry. May–June. Along railroad; Jackson and Wabash cos.

17. R. argutus Link. Highbush Blackberry. May–June. Thickets, woods; common throughout the state. (Including *R. blakei* Bailey, *R. abactus* Bailey, *R. schneckii* Bailey, and *R. virilis* Bailey.) *R. ostryifolius* Rydb.—G, J.

18. R. avipes Bailey. Blackberry. May–June. Thickets, rare; confined to n. Ill.

19. R. pensylvanicus Poir. Blackberry. May–June. Fields, thickets, woods; occasional to common throughout the state. (Including *R. recurvans* Blanch. and *R. bellobatus* Bailey.)

20. R. frondosus Bigel. Blackberry. May–June. Woods and thickets; occasional throughout the state. *R. pensilvanicus* Poir.—G.

13. *Rosa* L. Rose

1. Styles united into a column ....

1. Styles free ....

2. Flowers white (rarely pinkish), up to 2.5 cm across; leaflets (5–) 7–9; stipules pectinate .... 1. *R. multiflora*

2. Flowers pink or rose, 3 cm broad or broader; leaflets 3 (–5); stipules entire or toothed .... 2. *R. setigera*

3. Achenes lining only the base of the receptacle ....

3. Achenes lining the walls and the base of the receptacle 10
4. Sepals reflexed after flowering, rarely persisting on the mature fruit . 5
4. Sepals not reflexed after flowering, usually persisting on the mature fruit
. . . . 6
5. Mature plants over 1 m tall; leaves finely toothed . . . . 3. *R. palustris*
5. Mature plants less than 1 m tall; leaves coarsely toothed . 4. *R. carolina*
6. Flowers borne singly 7
6. Flowers clustered in corymbs 8
7. Stems glabrous, except for the numerous spines . . 5. *R. pimpinellifolia*
7. Stems tomentose as well as spiny . . . . 6. *R. rugosa*
8. Leaflets glabrous, rarely more than 2 cm long . . . . 7. *R. lunellii*
8. Leaflets pubescent, usually at least 2 cm long 9
9. Prickles absent on flowering branches, although often present near base
of plant; leaflets 5–7 (–9) . . . . 8. *R. blanda*
9. Prickles present on flowering branches as well as on the lower part of
the stem; leaflets 9–11 . . . . 9. *R. suffulta*
10. Rusty resinous glands present on at least the lower surface of the leaf-
lets 11
10. Rusty resinous glands absent from the surfaces of the leaflets (glands
sometimes present on lower surface of leaflets in *R. canina*) 12
11. Leaflets with many rusty glandular dots on both surfaces; styles pubescent; calyx persistent on the fruit . . . . 10. *R. eglanteria*
11. Leaflets with few rusty glandular dots on the surfaces; styles glabrous or
nearly so; calyx falling from the fruit . . . . 11. *R. micrantha*
12. Receptacle and pedicels glabrous or nearly so . . . . 12. *R. canina*
12. Receptacle pubescent . . . . 13. *R. gallica*

1. R. multiflora Thunb. Multiflora Rose. May–June. Native of China and Japan; frequently planted and often escaped throughout the state.

2. R. setigera Michx. Prairie Rose. June–July. Leaflets glabrous beneath, or pilose on the nerves. Woods, thickets, clearings; occasional to common throughout the state.

var. tomentosa Torr. & Gray. Leaflets tomentose beneath. Woods, thickets, clearings; occasional throughout the state.

3. R. palustris Marsh. Swampy Rose. June–Aug. Swamps, bogs, moist thickets; occasional throughout the state.

4. R. carolina L. Pasture Rose. May–July. Lower surface of leaflets glabrous or sparsely pubescent. Prairies, fields, dry woods; common throughout the state. (Including var. *grandiflora* [Baker] Rehd.)

var. villosa (Best) Rehd. Lower surface of leaflets softly pubescent beneath. Prairies, fields; occasional throughout the state.

5. R. pimpinellifolia L. Burnet Rose. May–July. Native of Europe; occasionally escaped from cultivation. *R. spinosissima* L.—F, G, J.

6. R. rugosa Thunb. Rugose Rose. June–Sept. Native of Asia; rarely escaped from cultivation; Kane Co.

7. R. lunellii Greene. Lunell's Rose. June–July. Native of the w. U.S.; along a railroad; JoDaviess Co.

8. R. blanda Ait. Meadow Rose. May–June. Thickets, woods, open areas; occasional in the n. one-third of the state, very rare elsewhere.

9. R. suffulta Greene. Sunshine Rose. June–July. Thickets, woods; occasional in the n. two-thirds of the state, apparently absent elsewhere. (Including *R. relicta* Erlanson.) *R. arkansana* Porter var. *suffulta* (Greene) Cockerell—F.

10. R. eglanteria L. Sweet-brier. May–July. Native of Europe; oc casionally escaped from cultivation.

11. R. micrantha Sm. Small Sweet-brier. May–July. Native of Europe rarely escaped from cultivation; Winnebago Co.

12. R. canina L. Dog Rose. May–July. Native of Europe; occasionall escaped from cultivation.

13. R. gallica L. French Rose. June–July. Native of Europe; rarely es caped from cultivation; Piatt Co.

14. *Potentilla* L. Cinquefoil

1. Petals purple .... 1. *P. palustri*
1. Petals yellow or white ....
2. Petals white ....
2. Petals yellow ....
3. Leaflets 3, usually entire, evergreen; achenes hairy .... 2. *P. tridentat*
3. Leaflets 7–11, serrate, not evergreen; achenes glabrous ... 3. *P. argut*
4. Shrub; achenes pubescent; style lateral .... 4. *P. fruticos*
4. Herbs; achenes glabrous; style terminal ....
5. Flowers solitary ....
5. Flowers in cymes ....
6. Leaves palmately compound, with usually 5 leaflets ...... 5. *P. simple*
6. Leaves pinnately compound, with 7–25 leaflets .... 6. *P. anserir*
7. Lower leaves ternate or digitate ....
7. Lower leaves pinnate ....
8. Lower leaves with 5–7 leaflets ....
8. Lower leaves with 3 leaflets ....
9. Achenes rugose; flowers 1.5–2.5 cm across, the petals emarginat
   stems and leaves hirsute .... 7. *P. rec*
9. Achenes smooth; flowers up to 1.5 cm across, the petals entire; ste
   and leaves white-woolly or gray-tomentulose ....
10. Stem and leaves white-woolly; leaflets sometimes nearly pinnatifid ...
    .... 8. *P. argent*
10. Stem and leaves gray-tomentulose; leaflets at most coarsely denta
    .... 9. *P. intermed*
11. Plants softly villous; flowers up to 4 mm across; stamens about 10 ...
    .... 10. *P. millegra*
11. Plants rough-hirsute; flowers usually at least 6 mm across; stame
    15–20 ....
    .... 11. *P. norvegi*
12. Leaflets of the lower leaves approximate; achenes smooth ....
    .... 12. *P. riva*
12. Leaflets of the lower leaves more distant; achenes longitudinally ribb
    .... 13. *P. parado*

1. P. palustris (L.) Scop. Marsh Cinquefoil. June–July. Bogs, ra Cook, Lake, and McHenry cos.

2. P. tridentata Ait. Three-toothed Cinquefoil. June–Sept. Gra ridge, very rare; Cook Co.

3. P. arguta Pursh. Prairie Cinquefoil. June–July. Prairies; casional in the n. half of the state.

4. P. fruticosa L. Shrubby Cinquefoil. June–Aug. Interdunal pon hill prairies, boggy fens, rare; Cook, JoDaviess, Kane, Kendall, a Lake cos.

5. P. simplex Michx. Common Cinquefoil. May–June. Stems villous or hirsute; lower surface of leaflets strigose. Dry woods, prairies, fields, common; in every co.

var. calvescens Fern. Common Cinquefoil. May–June. Stems and lower surface of leaflets glabrous or strigillose. Similar habitats; occasional throughout the state.

var. argyrisma Fern. Common Cinquefoil. May–June. Stems spreading-villous; lower surface of leaflets densely silvery-silky. Dry woods; rare in the s. cos.

6. P. anserina L. Silverweed. May–Aug. Sandy beaches, interdunal ponds, meadows, gravel bars; restricted to the ne. cos. of Cook, DeKalb, Lake, McHenry, and Will.

7. P. recta L. Sulfur Cinquefoil. May–July. Native of Europe. Fields and roadsides; occasional to common throughout the state.

8. P. argentea L. Silvery Cinquefoil. May–Sept. Native of Europe; disturbed sandy soil; occasional in the n. half of the state.

9. P. intermedia L. Intermediate Cinquefoil. June–Aug. Native of Europe; adventive along railroad; Champaign Co.

10. P. millegrana Engelm. Cinquefoil. June–July. Moist soil, rare; Johnson and St. Clair cos.

11. P. norvegica L. Rough Cinquefoil. June–July. Native of Europe; disturbed soils; occasional throughout the state. *P. monspeliensis* L.—J.

12. P. rivalis Nutt. Brook Cinquefoil. June–Aug. Native of the w. U.S.; adventive once in Cook Co.

13. P. paradoxa Nutt. Cinquefoil. May–Sept. Wet soil, usually along rivers; Jackson, Randolph, and St. Clair cos.

15. *Waldsteinia Willd.*

1. W. fragarioides (Michx.) Tratt. Barren Strawberry. April–May. Sandstone ledge; Pope Co.

16. *Fragaria L. Strawberry*

1. Most or all petals less than 1 cm long; fruits up to 1.5 cm across 2
1. Most or all petals over 1 cm long; fruits over 1.5 cm across ....
.... 3. *F.* X *ananassa*

1. F. virginiana Duchesne. Wild Strawberry. April–July. Woodlands, prairies, fields; occasional to common throughout the state. (Including var. *illinoensis* [Prince] Gray.)

2. F. americana (Porter) Britt. Hillside Strawberry. May–Aug. Wooded slopes, occasional to rare; confined to the n. half of the state. *F. vesca* L. var. *americana* Porter—F, G.

3. F. X ananassa Duchesne. Cultivated Strawberry. April–May. Rarely escaped from cultivation. Reputed to be a hybrid between *F. chiloensis* (L.) Duchesne and *F. virginiana* Duchesne. *F. chiloensis* Duch. var. *ananassa* Bailey—G.

17. *Aruncus Adans. Goat's-beard*

1. A. dioicus (Walt.) Fern. Goat's-beard. May–June. Rich woods; occasional throughout the state; absent in the ne. cos., except Grundy Co. (Including var. *pubescens* [Rydb.] Fern.)

18. *Filipendula* Mill.

1. F. rubra (Hill) Robins. Queen-of-the-prairie. June–July. Spring
fens, occasional; confined to the n. half of the state.

19. *Sanguisorba* L.   Burnet

1. Leaflets up to 2.5 cm long; stamens 12 or more, drooping   . 1. *S. mino*
1. Most or all the leaflets 2.5 cm long or longer; stamens 4, erect.......
.... 2. *S. canadens*

1. S. minor Scop. Garden Burnet. May–July. Native of Europe ar
Asia; adventive in a limestone barren; Will Co.
2. S. canadensis L. American Burnet. July–Oct. Moist prairies, ve
rare; Cass, LaSalle, and Will cos.

20. *Gillenia* Moench

1. G. stipulata (Muhl.) Baill. Indian Physic. May–July. Woods; o
casional in the s. half of the state; also La Salle Co.

21. *Geum* L.   Avens

1. Flowers purple ....
1. Flowers white, cream, or yellow ....
2. Styles not jointed; flowers mostly less than 2 cm across; style 2 cm lo
or longer .... 1. *G. trifloru*
2. Styles jointed; flowers mostly at least 2 cm across; style up to 1 cm lo
.... 2. *G. riva*
3. Petals white ....
3. Petals yellow or cream ....
4. Most of the petals 5 mm long or longer; peduncles puberulent; fruiti
receptacle densely white-villous .... 3. *G. canaden*
4. Petals up to 5 mm long; peduncles hirsute; fruiting receptacle glabro
or nearly so .... 4. *G. laciniatu*
5. Receptacle stalked in the calyx; calyx not subtended by bractlets; peta
about 2 mm long, yellow .... 5. *G. vernu*
5. Receptacle sessile; calyx subtended by bractlets; petals (or some
them) more than 2 mm long, cream, orange, or deep yellow ....
6. Petals 2–4 mm long, shorter than the calyx, cream ... 6. *G. virginianu*
6. Petals 5–10 mm long, as long as the calyx, orange or deep yellow ....
.... 7. *G. strictu*

1. G. triflorum Pursh. Prairie Avens. May–June. Dry prairies; o
casional in the n. one-sixth of the state; absent elsewhere.
2. G. rivale L. Purple Avens. May–Aug. Moist soil, very rare; Ka
McHenry, and Winnebago cos.
3. G. canadense Jacq. White Avens. June–Aug. Fruiting carpel
bescent throughout. Woodlands, common; in every co. (Including v
camporum [Rydb.] Fern. & Weath.)
var. grimesii Fern. & Weath. White Avens. June–Aug. Fruiting car
hispid above, glabrous below. Woodlands, rare; confined to the
one-fifth of the state.
4. G. laciniatum Murr. Rough Avens. June–July. Achenes glabro

Meadows and thickets, not common; restricted to the n. three-fourths of the state.

var. trichocarpum Fern. Rough Avens. June–July. Achenes bristly at summit. Meadows and thickets; occasional in the n. three-fourths of the state.

5. G. vernum (Raf.) Torr. & Gray. Spring Avens. April–May. Moist woods; common in the s. half of the state, occasional or rare in the n. half.

6. G. virginianum L. Pale Avens. June–July. Dry woods, rare; restricted to the s. one-fourth of the state.

7. G. strictum Ait. Yellow Avens. June–July. Bogs, moist thickets; occasional in the n. one-half of the state; also St. Clair Co. G. aleppicum Jacq. var. strictum (Ait.) Fern.—F; G. aleppicum Jacq.—G.

22. Duchesnea Sm.

1. D. indica (Andr.) Focke. Indian Strawberry. April–June. Native of Asia; adventive in waste ground; Alexander, DuPage, and Johnson cos.

23. Agrimonia L. Agrimony

1. Leaflets (excluding smaller interposed ones) 11–17 .... 1. A. parviflora
1. Leaflets (excluding smaller interposed ones) 5–9 2
2. Leaflets and branches of the inflorescence softly pubescent 3
2. Leaflets and branches of the inflorescence sparsely hirsute or glabrous
.... 4
3. Pubescence of stems spreading; leaflets rounded at tip 2. A. microcarpa
3. Pubescence of stems incurving; leaflets pointed ...... 3. A. pubescens
4. Roots not tuberous; fruit 5–8 mm long .... 4. A. gryposepala
4. Roots tuberous; fruit 3–4 mm long .... 5. A. rostellata

1. A. parviflora Ait. Swamp Agrimony. July–Sept. Low ground; occasional throughout the state.

2. A. microcarpa Wallr. Small-fruited Agrimony. July–Aug. Dry woods, very rare; St. Clair Co.

3. A. pubescens Wallr. Soft Agrimony. July–Sept. Dry or moist woods; occasional throughout the state. (Including A. mollis [Torr. & Gray] Britt.)

4. A. gryposepala Wallr. Tall Agrimony. June–Aug. Woods and thickets; occasional in the n. two-thirds of the state, absent elsewhere.

5. A. rostellata Wallr. Woodland Agrimony. July–Sept. Woods; occasional in the s. two-thirds of the state; also Cook and Will cos.

80. LEGUMINOSAE

Plants woody (including woody vines) 2
Plants herbaceous 12
Leaves simple .... 1. Cercis
Leaves pinnately compound 3

3. Leaves twice-pinnate ....
3. Leaves once-pinnate ....
4. Some or all the leaflets over 1.5 cm broad .... 3. *Gymnocladu*
4. None of the leaflets 1.5 cm broad ....
5. Leaflets 1 cm long or longer; plants usually with large thorns; flowers i elongated spike-like racemes .... 4. *Gleditsi*
5. Leaflets (or most of them) up to 8 mm long; plants without thorns flowers in spherical heads .... 6. *Albiz*
6. Woody vines .... 7. *Wister*
6. Trees or upright shrubs ....
7. Leaflets three .... 8. *Onon*
7. Leaflets five or more ....
8. Tip of rachis developed into a spine, but soon deciduous . 2. *Caragan*
8. Tip of rachis not spinescent ....
9. Leaves even-pinnate .... 4. *Gledits*
9. Leaves odd-pinnate ....
10. Lower surface of leaflets glandular-punctate or canescent; flowers le than 1 cm long; shrubs to 7 mm tall .... 9. *Amorph*
10. Lower surface of leaflets neither glandular-punctate nor canescer flowers 1.4 cm long or longer; trees up to 15 m tall ....
11. Stipular spines usually present; stems sometimes glandular or hisp .... 10. *Robin*
11. Stipular spines absent; stems glabrous .... 11. *Cladras*
12. Leaves simple .... 12. *Crotalar*
12. Leaves compound ....
13. Leaves palmately compound, not trifoliolate ....
13. Leaves pinnately compound or trifoliolate ....
14. Leaflets 7–11 .... 13. *Lupin*
14. Leaflets 5 .... 14. *Psoral*
15. Leaves twice-pinnate .... 5. *Desmanth*
15. Leaves once-pinnate ....
16. Leaves even-pinnate ....
16. Leaves odd-pinnate ....
17. Some or all the leaf rachises ending in a tendril ....
17. None of the leaf rachises prolonged into a tendril ....
18. Calyx lobes leaf-like; lowermost stipules larger than the leaflets th subtend .... 15. *Pis*
18. Calyx lobes not leaf-like; none of the stipules larger than the leaflets
19. Wings and keel petal united; style with a terminal tuft of hairs 16. *Vi*
19. Wings free from the keel petal; style pubescent only along one side . ... 17. *Lathy*
20. Petioles with one or more conspicuous glands; corolla nearly regula .... 18. *Cas*
20. Petioles without glands; corolla pea-shaped .... 19. *Sesba*
21. Leaflets 5 or more ....
21. Leaflets 3 ....
22. Stems, petioles, and peduncles with hooked prickles ... 20. *Schran*
22. Stems, petioles, and peduncles without hooked prickles ....
23. Leaflets 5–9 ....
23. Leaflets 11-many ....
24. Climbing vines; flowers brownish-purple or greenish-white, in rat loose racemes .... 21. *Ap*
24. Upright herbs; flowers bright yellow, bright rose, or white, in umbels spikes ....

5. Flowers bright yellow, in umbels; stamens 10 .... 22. *Lotus*
5. Flowers bright rose or white, in spikes; stamens 5 .. 23. *Petalostemum*
6. Plants glandular-viscid; fruit with hooked bristles ...... 24. *Glycyrrhiza*
6. Plants not glandular-viscid; fruit without hooked bristles 27
7. Flowers in umbels .... 25. *Coronilla*
7. Flowers in spikes or elongated or shortened racemes 28
8. Flowers in dense, elongated heads, the rachis not exposed; fruits enclosed by the calyx, 1- to 2-seeded 29
8. Flowers in elongated or shortened racemes; fruits exserted, several-seeded (1-seeded in *Onobrychis*) 30
9. Flowers pink or white; stamens 10; bracts about equalling the calyx; leaflets obtuse .... 26. *Dalea*
9. Flowers rose, fading to white; stamens 5; bracts longer than the calyx; leaflets mostly acute .... 23. *Petalostemum*
0. Inflorescences terminal; flowers cream, tipped with purple............
.... 27. *Tephrosia*
0. Inflorescences axillary; flowers purple, cream, or yellow-green 31
1. Fruit strongly reticulate and occasionally toothed along the margins; corolla pale pink .... 28. *Onobrychis*
1. Fruit neither strongly reticulate nor toothed; corolla yellow-green, cream, or purple .... 29. *Astragalus*
2. Leaflets toothed 33
2. Leaflets entire 36
3. Plants spiny .... 8. *Ononis*
3. Plants not spiny 34
4. Flowers in elongated racemes .... 30. *Melilotus*
4. Flowers in heads or spikes 35
5. Stipules entire; fruit usually straight, the petals mostly persistent.......
.... 31. *Trifolium*
5. Stipules toothed (rarely entire); fruit coiled or twisted, the petals not persistent .... 32. *Medicago*
6. Foliage with a few glandular dots .... 14. *Psoralea*
6. Foliage without glandular dots 37
7. Flowers 2 cm long or longer 38
7. Flowers up to 1.6 cm long 39
8. Flower 2–3 cm long, in racemes .... 33. *Baptisia*
8. Flowers 4–6 cm long, solitary or only 1– together .... 34. *Clitoria*
9. Leaflets without stipels 40
9. Leaflets with stipels 45
0. Flowers yellow 41
0. Flowers purple, violet, rose, or white with purple markings 43
1. Flowers 1.3–1.6 cm long .... 33. *Baptisia*
1. Flowers less than 1 cm long 42
2. Stipules united to petiole, forming a sheath which encircles the stem ..
.... 35. *Stylosanthes*
2. Stipules free from petiole .... 36. *Lespedeza*
3. Uppermost leaves reduced to one leaflet; fruit a legume 1.5–2.5 cm long
.... 37. *Hosackia*
3. Uppermost leaves trifoliolate; fruit a loment less than 1 cm long or a legume opening by a lid 44
4. Fruit a legume opening by a lid; flowers in a globose head ....
.... 31. *Trifolium*
4. Fruit a loment; flowers variously arranged, but not in a globose head
.... 36. *Lespedeza*

1. *Cercis* L.   Redbud

1. C. canadensis L. Redbud. April–May. Woods; common in the s. cos., becoming less abundant northward. (Including f. *glabrifolia* L.)

2. *Caragana* Lam.   Pea-tree

1. C. arborescens Lam. Pea-tree. May. Native of Asia; escaped from cultivation and spreading in Winnebago Co.

3. *Gymnocladus* Lam.   Coffee-tree

1. G. dioica (L.) K. Koch. Kentucky Coffee-tree. May–June. Moist low woods; occasional throughout the state.

4. *Gleditsia* L.   Honey Locust

1. Legume 15 cm long or longer, with sweet pulp between the several seeds .... 1. *G. triacanthos*
1. Legume to 5 cm long, with no pulp surrounding the 1 or 2 seeds .... 2. *G. aquatica*

1. G. triacanthos L. Honey Locust. May–June. Trunk beset by stout thorns. Woods; common in the s. cos., becoming less common northward.
f. inermis (Pursh) Schneid. Thornless Honey Locust. May–June. Trunks without thorns. Occasionally found with the typical form.
2. G. aquatica Marsh. Water Locust. May–June. Swamps, rare; confined to s. Ill., extending northward to Calhoun Co. in the w. and Lawrence Co. in the e.; also Henderson Co.

5. *Desmanthus Willd.*

1. D. illinoensis (Michx.) MacM. Illinois Mimosa. June–Aug. Prairies; along levees; occasional in the s. and n. cent. cos., uncommon or absent elsewhere.

6. *Albizia Duraz.*

1. A. julibrissin Duraz. Mimosa. June–Aug. Native of Asia and Africa; occasionally escaped from cultivation in the s. cos.

7. *Wisteria Nutt.*

1. Leaflets mostly 15–19 .... 1. *W. floribunda*
1. Leaflets mostly 7–13 2
2. Pedicels 15 mm long or longer; ovary and legume velvety   2. *W. sinensis*
2. Pedicels up to 10 mm long; ovary and legume glabrous ....
.... 3. *W. macrostachya*

1. W. floribunda DC. Japanese Wisteria. April–June. Native of Japan; rarely escaped from cultivation; Jackson Co.
2. W. sinensis Sweet. Chinese Wisteria. April–May. Native of Asia; rarely escaped from cultivation; Jackson, Piatt, and Richland cos.
3. W. macrostachya Nutt. Wisteria. June–July. Swampy woods; generally confined to the s. one-fifth of the state, but extending n. to Clark and Richland cos.; adventive in Peoria and Washington cos.

8. *Ononis L. Rest Harrow*

1. O. spinosa L. Rest Harrow. June–Aug. Native of Europe; rarely persisting from cultivation; DuPage Co.

9. *Amorpha L.*

1. Some or all leaflets 1 cm broad or broader, short-petiolulate 2
1. None of the leaflets 1 cm broad, sessile or nearly so .. 3. *A. canescens*
2. Leaflets and branchlets glabrous or nearly so; legume without resinous dots .... 1. *A. nitens*
2. Leaflets and branchlets pubescent; legume with resinous dots ....
.... 2. *A. fruticosa*

1. A. nitens Boynton. Smooth False Indigo. May–June. Woods along rivers and streams, very rare; Pope Co.
2. A. fruticosa L. False Indigo. May–June. Leaflets mostly less than twice as long as broad, grayish-pubescent. Moist soil; occasional throughout the state.
var. angustifolia Pursh. False Indigo. May–June. Leaflets at least twice as long as broad, grayish-pubescent. Moist soil; occasional throughout the state. (Including var. *tennesseenis* [Shuttlew.] Palmer; var. *oblongifolia* Palmer.)
var. croceolanata (P.W. Wats.) Schneid. False Indigo. May–June. Pubescence softly tawny-villous; low woods, rare; Pope Co.
3. A. canescens Pursh. Leadplant. May–Aug. Prairies; common in the n. two-thirds of the state, less common elsewhere.

10. Robinia L. Locust

1. Branchlets neither glandular-viscid nor bristly; flowers white . . . .
. . . . 1. *R. pseudoacaci*
1. Branchlets glandular-viscid or bristly; flowers pale pink, rose, or purpl
. . . .
2. Branchlets glandular-viscid . . . . .2. *R. viscos*
2. Branchlets bristly . . . . 3. *R. hispid*

  1. R. pseudoacacia L. May–June. Black Locust. Woods and thick
ets; native in s. Ill., commonly planted elsewhere.
  2. R. viscosa Vent. Clammy Locust. May–June. Native of the se
U.S.; rarely escaped from cultivation; JoDaviess Co.
  3. R. hispida L. Bristly Locust. May–June. Native of the se. U.S.; oc
casionally escaped from cultivation, chiefly in the s. half of the state

11. Cladrastis Raf.

  1. C. lutea (Michx. f.) K. Koch. Yellow-wood. May. Wooded slope
rare; Alexander and Gallatin cos.

12. Crotalaria L. Rattlebox

1. Leaves acute at apex; flowers up to 1.5 cm long; legume less than 3 c
long . . . . 1. *C. sagittal*
1. Leaves obtuse at apex; flowers over 1.5 cm long; legume 3 cm long c
longer . . . . 2. *C. spectabil*

  1. C. sagittalis L. Rattlebox. June–Sept. Fields; edge of woods; o
casional in the s. half of the state, becoming less common northwar
  2. C. spectabilis Roth. Showy Rattlebox. Aug.–Oct. Native of th
Tropics; rarely adventive in Ill.; Alexander Co.

13. Lupinus L. Lupine

  1. L. perennis L. Wild Lupine. May–June. Sandy woods; occasion
in the n. one-fourth of the state, absent elsewhere. *L. perennis* L. va
occidentalis S. Wats.—F.

14. Psoralea L. Scurf-pea

1. Leaflets always 3; fruits variously wrinkled, often glandular-dotted . . .
1. Leaflets sometimes 5; fruits glandular-dotted, but not wrinkled . . . .
. . . . 3. *P. tenuiflo*
2. Some or all the leaflets at least 2 cm broad; peduncles shorter than t
subtending leaves . . . . 1. *P. onobrych*
2. None of the leaflets 2 cm broad; peduncles exceeding the subtendi
leaves . . . . 2. *P. psoralioides* var. *eglandulo*

  1. P. onobrychis Nutt. French Grass. June–July. Thickets, prairi
not common; in most sections of the state, except for the nw. cos.
  2. P. psoralioides (Walt.) Cory var. eglandulosa (Ell.) Freema
Sampson's Snakeroot. June–July. Dry woods, prairies; occasional
the s. one-third of the state, apparently absent elsewhere. *P. psorali*
des (Walt.) Cory—F, G.

3. P. tenuiflora Pursh. Scurf-pea. June–Sept. Prairies; occasional in the n. two-thirds of the state, absent elsewhere. (Including var. *floribunda* [Nutt.] Rydb.)

15. *Pisum* L. Garden Pea

1. P. sativum L. Garden Pea. May–June. Escaped from cultivation; Adams and Jackson cos.

16. *Vicia* L. Vetch

1. Peduncles as long as the leaflets or longer 2
1. Peduncles much shorter than the leaves or nearly absent 6
2. Pubescence on stems and axis of inflorescence spreading   1. *V. villosa*
2. Pubescence on stems and axis of the inflorescence appressed or nearly absent 3
3. Flowers white, tipped with purple; racemes rather laxly flowered ....
.... 2. *V. caroliniana*
3. Flowers purple; racemes densely flowered 4
4. Flowers less than 10 per raceme; entire inflorescence shorter than the subtending leaves .... 3. *V. americana*
4. Flowers 10 or more per raceme; entire inflorescence longer than the subtending leaves 5
5. Calyx gibbous at the base on the upper side .... 4. *V. dasycarpa*
5. Calyx merely rounded at the base on the upper side ...... 5. *V. cracca*
6. Stems more or less pubescent; leaflets all similar in shape; most or all the flowers 18 mm long or longer; legume torulose ....
.... 6. *V. sativa*
6. Stems glabrous or nearly so; upper leaflets narrower than lower ones; all of the flowers less than 18 mm long; legume flat ....
.... 7. *V. angustifolia*

1. V. villosa Roth. Winter Vetch. June–Aug. Native of Europe; in fields and along roads; common throughout the state.
2. V. caroliniana Walt. Wood Vetch. April–June. Dry, wooded slopes; occasional in the upper one-fourth of the state, absent elsewhere.
3. V. americana Muhl. American Vetch. July–Aug. Disturbed woods and disturbed prairies; occasional in the n. one-third of the state, absent elsewhere.
4. V. dasycarpa Ten. Hairy-fruited Vetch. May–Sept. Native of Europe; scattered in the state, but not common.
5. V. cracca L. Cow Vetch. May–Aug. Native of Europe; roadsides and fields; occasional throughout the state.
6. V. sativa L. Common Vetch. July–Sept. Native of Europe; roadsides and fields; scattered in Ill., but not common.
7. V. angustifolia Reich. Narrow-leaved Vetch. Native of Europe; waste ground; occasional in the extreme n. and extreme s. cos. (Including var. *segetalis* [Thuill.] W. D. J. Koch.)

17. *Lathyrus* L. Wild Pea

. Leaflets 2 2
. Leaflets 4–12 4
. Corolla pink, purple, or white 3

2. Corolla bright yellow .... 3. *L. pratensis*
3. Perennial; peduncles 4- to 10-flowered; flowers odorless 1. *L. latifolius*
3. Annual; peduncles 1- to 3-flowered; flowers fragrant .... 2. *L. odoratus*
4. Leaflets fleshy; stipules with 2 basal lobes .... 4. *L. maritimus*
4. Leaflets not fleshy; stipules with one basal lobe ....
5. Corolla yellowish-white; petiolules glabrous .... 5. *L. ochroleucus*
5. Corolla violet or purple; petiolules pubescent ....
6. Leaflets and stems glabrous ....
6. Leaflets hirtellous .... 8. *L. venosus*
7. Stem distinctly winged .... 6. *L. palustris*
7. Stem unwinged .... 7. *L. myrtifolius*

1. L. latifolius L. Everlasting Pea. June–Sept. Native of Europe; occasionally escaped from cultivation.

2. L. odoratus L. Sweet Pea. June–Sept. Native of Europe; escaped to roadsides; occasional in the s. one-fourth of the state.

3. L. pratensis L. Yellow Vetchling. June–Aug. Native of Europe; rarely escaped from cultivation.

4. L. maritimus (L.) Bigel. Beach Pea. June–Aug. Beaches along Lake Michigan; Cook and Lake cos. *L. japonicus* Willd. var. *glaber* (Ser.) Fern.—F; *L. maritimus* (L.) Bigel. var. *glaber* Ser.—G.

5. L. ochroleucus Hook. Pale Vetchling. May–July. Woods; rare in the n. one-sixth of the state; also Gallatin Co.

6. L. palustris L. Marsh Vetchling. May–Sept. Moist, open areas; occasional in the n. half of the state; also Massac, Pope, and Wabash cos.

7. L. myrtifolius Muhl. Vetchling. June–July. Moist soil; occasional in the n. four-fifths of the state.

8. L. venosus Muhl. var. intonsus Butt. & St. John. Veiny Pea. May–June. Dry prairies, woods; occasional in the n. one-fourth of the state. *L. venosus* Muhl.—J.

18. *Cassia* L.   Senna

1. Leaflets 4–18, at least 7 mm broad ....
1. Leaflets 20 or more, less than 7 mm broad ....
2. Leaflets 4–6; gland midway between lowest pair of leaflets .. 1. *C. tora*
2. Leaflets 8–18; gland near base of petiole ....
3. Leaflets obtuse; stems somewhat villous above .... 2. *C. hebecarpa*
3. Leaflets acute; stems glabrous ....
4. Ovary and fruits pubescent; legume up to 10 cm long 3. *C. marilandica*
4. Ovary and fruits glabrous; legume 10–12 cm long ... 4. *C. occidentalis*
5. Calyx 9–10 mm long; corolla 10–20 mm long; stamens 10; pedicels mm long or longer .... 5. *C. fasciculata*
5. Calyx 3–4 mm long; corolla less than 10 mm long; stamens 5; pedicels 2–4 mm long .... 6. *C. nictitans*

1. C. tora L. Sicklepod. July–Sept. Along streams and railroads, very rare; Jackson, Lake, Pulaski, and Wabash cos.

2. C. hebecarpa Fern. Wild Senna. July–Aug. Along streams, open woods; occasional throughout the state.

3. C. marilandica L. Maryland Senna. July–Aug. Roadsides, thickets; occasional throughout the state.

4. C. occidentalis L. Coffee Senna. Aug.–Sept. Native of the Tropics; rarely adventive in Ill.; Cook Co.

5. C. fasciculata Michx. Partridge Pea. July–Sept. Stems with appressed pubescence. Open areas, prairies, fields; common throughout the state.

var. robusta (Pollard) Macbr. Partridge Pea. July–Sept. Stems hirsute. Fields; rare in Ill.

6. C. nictitans L. Wild Sensitive Plant. July–Sept. Fields, roadsides, edge of woods; occasional in the s. one-fifth of the state, uncommon elsewhere.

19. *Sesbania Scop.*

1. S. exaltata (Raf.) Cory. July–Oct. Low ground, rare; Alexander and Pulaski cos.

20. *Schrankia Willd. Sensitive Brier*

1. S. uncinata Willd. Cat-claw. June–Sept. Prairies, rare; Peoria and Winnebago cos. *S. nuttallii* (DC.) Standl.—F.

21. *Apios Medic. Groundnut*

Flowers brownish-purple; standard without a spongy knob at the apex .... 1. *A. americana*
Flowers greenish-white, tinged with purple; standard with a spongy knob at the apex .... 2. *A. priceana*

1. A. americana Medic. Groundnut. July–Sept. Thickets, prairies; common throughout the state. (Including var. *turrigera* Fern.)

2. A. priceana Robins. Price's Groundnut. July–Sept. Woods, very rare; Union Co.

22. *Lotus L.*

1. L. corniculatus L. Birdsfoot-trefoil. June–Sept. Native of Europe; waste areas; occasional throughout the state.

23. *Petalostemum Michx. Prairie Clover*

Leaflets 5–7 2
Leaflets 19–27 .... 3. *P. foliosum*
Flowers white; calyx glabrous .... 1. *P. candidum*
Flowers purple or rose; calyx densely pubescent ..... 2. *P. purpureum*

1. P. candidum (Willd.) Michx. White Prairie Clover. June–July. Prairies; occasional throughout the state.

2. P. purpureum (Vent.) Rydb. Purple Prairie Clover. June–Aug. Prairies; occasional throughout the state.

3. P. foliosum Gray. Leafy Prairie Clover. July–Sept. Dry soil above streams; once known from a few counties in the ne. one-fourth of the state, now known only from Will Co.

24. Glycyrrhiza L.

1. G. lepidota Pursh. Wild Licorice. June–Aug. Native of w. U.S
waste ground; occasional in the ne. corner of the state; also St. Cla
Co.

25. Coronilla L. Crown Vetch

1. C. varia L. Crown Vetch. June–Aug. Native of Europe; ofte
planted and frequently escaped.

26. Dalea Juss.

1. D. alopecuroides Willd. Foxtail Dalea. July–Sept. Fields, roa
sides; occasional in most sections of the state.

27. Tephrosia Pers. Hoary Pea

1. T. virginiana (L.) Pers. Goat's-rue. May–Aug. Dry woods; (
casional throughout the state.

28. Onobrychis Gaertn.

1. O. viciaefolia Scop. Sainfoin. June–July. Native of Europe a
Asia; rarely escaped from cultivation.

29. Astragalus L. Milk Vetch

1. Calyx tube 2–6 mm long; fruit dry, dehiscent . . . .
1. Calyx tube 7–10 mm long; fruit fleshy, indehiscent . . . .
2. Calyx tube hirsute; corolla 1.5–2.0 cm long; fruit up to 1 cm long, lon
villous . . . . 1. A. goniat
2. Calyx tube glabrous or sparsely pubescent, not hirsute; corolla 1.0–
cm long; fruit 1–3 cm long, glabrous . . . .
3. Fruits lunate, 2–3 cm long; corolla lilac, purple, or white, about 1
long; calyx tube 2–3 mm long . . . . 2. A. distor
3. Fruits more or less straight, 1–2 cm long; corolla greenish-yello
1.2–1.5 cm long; calyx tube 4.5–6.0 mm long . . . . 3. A. canaden
4. Fruits glabrous . . . . 4. A. trichoca
4. Fruits pubescent . . . . 5. A. tennesseen

1. A. goniatus Nutt. Field Milk Vetch. May–July. Native of the
U.S.; rarely adventive; Boone Co. A. agrestis L.—G.
2. A. distortus Torr. & Gray. Bent Milk Vetch. May–June. Ro
prairies; rare in w. Ill.; adventive in Cook Co.
3. A. canadensis L. Canadian Milk Vetch. June–Aug. Prairies, thi
ets; occasional throughout the state. (Including var. longilobus F
sett.)
4. A. trichocalyx Nutt. Large Ground Plum. April–May. Ro
prairies, rare; Macoupin, Madison, and St. Clair cos.; adventive in
Co. A. mexicanus A. DC. var. trichocalyx (Nutt.) Fern.—F; A. c
sicarpus Nutt. var. trichocalyx (Nutt.) Barneby—G.
5. A. tennesseensis Gray. Ground Plum. May–June. Dry prair
rare; confined to the n. one-fourth of the state; also Mason Co.

30. Melilotus Mill. Sweet Clover

1. Flowers white .... 1. *M. alba*
1. Flowers yellow 2
2. Fruit 2.5–3.5 mm long, glabrous or nearly so, prominently reticulate ....
.... 2. *M. officinalis*
2. Fruit 4.5–6.0 mm long, pubescent, obscurely reticulate .. 3. *M. altissima*

   1. M. alba Desr. White Sweet Clover. May–Oct. Native of Europe
and Asia; naturalized in waste ground; in every co.
   2. M. officinalis (L.) Lam. Yellow Sweet Clover. June–Sept. Native of
Europe; naturalized in waste ground; in every co.
   3. M. altissima Thuill. Tall Yellow Sweet Clover. June–Sept. Native
of Europe; naturalized in waste ground; Champaign and Lake cos.

31. Trifolium L. Clover

1. Flowers yellow 2
1. Flowers white, pink, purple, or scarlet 4
2. Petiolule of middle leaflet longer than the petiolules of lateral leaflets 3
2. Petiolule of middle leaflet as long as or shorter than the petiolules of lateral leaflets .... 3. *T. aureum*
3. Flowers 15 or more per head, the head 8–15 mm thick . 1. *T. campestre*
3. Flowers up to 15 per head, the head less than 8 mm thick 2. *T. dubium*
4. Stems pubescent 5
4. Stems glabrous or nearly so 9
5. Flowers sessile or nearly so in the head 6
5. Flowers on pedicels 2 mm long or longer .... 8. *T. reflexum*
6. Flowering heads subglobose, about as broad as long 7
6. Flowering heads elongated, cylindric, longer than broad 8
7. Heads subtended by a pair of leaves; corolla 12–15 mm long; heads
usually 2–3 cm in diameter .... 4. *T. pratense*
7. Heads not subtended by a pair of leaves; corolla 5–7 mm long; heads
usually less than 2 cm in diameter .... 5. *T. fragiferum*
8. Flowers pale rose, shorter than the calyx; heads covered by silky hairs
.... 6. *T. arvense*
8. Flowers scarlet, longer than the calyx; heads not covered by silky hairs
.... 7. *T. incarnatum*
9. Flowers sessile or nearly so in the head 10
9. Flowers on pedicels 2 mm long or longer 12
9. Heads not subtended by a pair of leaves; corolla 4–6 mm long 11
9. Heads subtended by a pair of leaves; corolla 12–15 mm long ....
.... 4. *T. pratense*
. Stems repent; bracteoles as long as the calyx; corolla not resupinate ..
.... 5. *T. fragiferum*
. Stems ascending; bracteoles shorter than the calyx; corolla resupinate
.... 9. *T. resupinatum*
. Stems creeping, with basal runners; calyx teeth usually shorter than the
tube .... 10. *T. repens*
. Stems ascending, without basal runners; calyx teeth as long as the tube
or longer 13
. Calyx up to 5 mm long, the teeth less than twice as long as the tube;
stipules lanceolate .... 11. *T. hybridum*
. Calyx 6–8 mm long, the teeth at least twice as long as the tube; stipules
ovate .... 8. *T. reflexum*

1. T. campestre Schreb. Low Hop Clover. May–Sept. Native c Europe; waste ground; occasional throughout the state. *T. procum bens* L.—F, G, J.

2. T. dubium Sibth. Little Hop Clover. May–Sept. Native of Europe waste ground, not common; known only from the so. one-fourth c the state.

3. T. aureum Pollich. Yellow Hop Clover. June–Sept. Native c Europe; waste ground; Henry, Peoria, Union, and Winnebago cos. 7 *agrarium* L.—F, G, J.

4. T. pratense L. Red Clover. May–Sept. Native of Europe; natura ized in waste ground; in every co. (Including var. *sativum* [Mill Schreb.) The white-flowered f. *leucochraceum* Aschers. & Prantl ha been collected in Ill.

5. T. fragiferum L. Strawberry Clover. June–Sept. Native of Europe rarely adventive in Ill.; Cook and Jackson cos.

6. T. arvense L. Rabbit-foot Clover. May–Oct. Native of Europe; o casionally adventive in waste ground in most regions of the state.

7. T. incarnatum L. Crimson Clover. May–July. Native of Europe; o casionally adventive in the s. three-fourths of the state.

8. T. reflexum L. Buffalo Clover. May–June. Stems villous. Wood not common, but scattered in the state.

var. glabrum Lojacono. Buffalo Clover. May–June. Stems glabrou Woods; not common, but scattered in the state.

9. T. resupinatum L. Persian Clover. June–Sept. Native of Europ rarely adventive in Ill.; Cook Co.

10. T. repens L. White Clover. May–Oct. Native of Europe; adventi in lawns and waste ground; in every co.

11. T. hybridum L. Alsike Clover. June–Sept. Native of Europe; a ventive in waste ground; in every co. (Including var. *elegans* [Sav Boiss.)

32. Medicago

1. Flowers blue or purple .... 1. *M. sati*
1. Flowers yellow ....
2. Fruits spirally coiled, glabrous or spiny ....
2. Fruits not spirally coiled, pubescent .... 4. *M. lupulir*
3. Fruits glabrous; flowers about 3 mm long .... 2. *M. orbicula*
3. Fruits spiny; flowers 4–6 mm long .... 3. *M. arabi*

1. M. sativa L. Alfalfa. May–Oct. Native of Asia; commonly cultivate and often escaped; in every co. Light-flowered specimens, the res of a cross between *M. sativa* and *M. falcata,* is sometimes found.

2. M. orbicularis (L.) Bartal. Round Medic. May–July. Native Europe; rarely adventive in waste ground; Jackson Co.

3. M. arabica (L.) Huds. Spotted Medic. May–Sept. Native Europe; rarely adventive in Ill.; Jackson Co.

4. M. lupulina L. Black Medic. May–July. Native of Europe; adve tive in waste ground; in every co. (Including var. *glandulosa* Neilr.)

33. Baptisia Vent. Wild Indigo

1. Flowers yellow, 1.3–1.6 cm long .... 1. *B. tincto*
1. Flowers blue, white, or cream, 2 cm long or longer ....

2. Flowers blue; stipules longer than the petioles 3
2. Flowers white or cream; stipules shorter than the petioles 4
3. Stipe of legume no longer than the calyx; petioles 5 mm long or longer; keel petal up to 2.5 cm long .... 2. *B. australis*
3. Stipe of legume about twice as long as the calyx; petioles less than 5 mm long; keel petal 2.7–3.0 cm long .... 3. *B. minor*
4. Plants glabrous; stipules deciduous; legumes glabrous, up to 3 cm long
.... 4. *B. leucantha*
4. Plants usually pubescent; stipules persistent; legumes pubescent, usually at least 4 cm long .... 5. *B. leucophaea*

1. B. tinctoria (L.) R. Br. var. crebra Fern. Yellow Wild Indigo. June–Sept. Sandy woods, very rare and probably extinct in Ill.; Cook Co. *B. tinctoria* (L.) R. Br—G, J.

2. B. australis (L.) R. Br. Blue Wild Indigo. May–June. Woods, very rare; Kane Co.

3. B. minor Lehm. Blue Wild Indigo. June–July. Native of the sw. U.S.; rarely adventive in Ill.; Cook Co.

4. B. leucantha Torr. & Gray. White Wild Indigo. May–July. Prairies, woods; occasional throughout the state.

5. B. leucophaea Nutt. Cream Wild Indigo. May–June. Stems and leaves pubescent. Moist prairies and open woods; occasional in the n. four-fifths of the state.

var. glabrescens Larisey. May–June. Stems and leaves glabrous. Prairies; rare in w. cent. Ill.

34. *Clitoria* L.   Butterfly Pea

1. C. mariana L. Butterfly Pea. June–Aug. Dry woods, not common; confined to the s. one-sixth of the state.

35. *Stylosanthes* Sw.   Pencil-flower

1. S. biflora (L.) BSP. Pencil-flower. June–Sept. Dry woods, fields; common in the s. one-fourth of the state, rare or absent elsewhere. (Including var. *hispidissima* [Michx.] Pollard & Ball; *S. riparia* Kearney.)

36. *Lespedeza* Michx.

. Plants annuals with persistent, brown, glabrous stipules 2
. Plants perennials with deciduous, greenish, usually pubescent stipules
.... 3
. Petioles 1–3 (–5) mm long; pubescence of stems retrorse .. 1. *L. striata*
. Petioles 4 mm long or longer; pubescence of stems antrorse ....
.... 2. *L. stipulacea*
. Stems lying flat on the ground 4
. Stems erect or ascending 5
. Pubescence of stems spreading; flowers usually 6 or more in a cluster
.... 3. *L. procumbens*
. Pubescence of stems appressed or absent; flowers usually up to 6 in a cluster .... 4. *L. repens*
. Pubescence of stems spreading 6
. Pubescence of stems appressed 9

6. Peduncles longer than the subtending leaves .... 5. *L. hirt*
6. Peduncles shorter than the subtending leaves ....
7. Flowers creamy-white, with a purple spot at base .... 6. *L. capita*
7. Flowers purplish ....
8. Fruit 3–5 mm long, longer than the calyx; stipules without any appare
veins .... 7. *L. stueve*
8. Fruit 6 mm long or longer, as long as the calyx or a little longe
stipules 3-nerved .... 8. *L.* X *simula*
9. Flowers 12 mm long or longer; plants to 3 m tall ..... 9. *L. thunberg*
9. Flowers up to 10 mm long; plants at most about 1 m tall ....
10. Flowers white or yellowish, with a purple spot at base ....
10. Flowers purplish ....
11. Leaflets linear, tapering to the apex; calyx pilose .. 10. *L. leptostachy*
11. Leaflets oblanceolate, truncate or retuse at the apex; calyx sericeous
nearly glabrous .... 11. *L. cunea*
12. Peduncles longer than the subtending leaves .... 12. *L. violace*
12. Peduncles (or most of them) shorter than the subtending leaves ...
13. Leaflets oval to elliptic .... 13. *L. intermed*
13. Leaflets linear to oblong .... 14. *L. virginic*

1. L. striata (Thunb.) Hook. & Arn. Japanese Lespedeza. Aug.–O
Native of Asia; roadsides, fields; common in the s. one-third of t
state, rare or absent elsewhere.

2. L. stipulacea Maxim. Korean Lespedeza. Aug.–Oct. Native
Asia; roadsides, fields; common in the s. two-thirds of the state, ra
or absent elsewhere.

3. L. procumbens Michx. Trailing Bush Clover. July–Sept. D
woods; occasional in the s. half of the state, apparently absent els
where.

4. L. repens (L.) Bart. Creeping Bush Clover. June–Sept. D
woods; occasional in the s. one-fifth of the state, absent elsewhere

5. L. hirta (L.) Hornem. Hairy Bush Clover. Aug.–Sept. Dry wood
occasional in the extreme n. and s. cos., but virtually absent els
where.

6. L. capitata Michx. Round-headed Bush Clover. Aug.–Se
Prairies; occasional to common throughout the state. (Including v
vulgaris Torr. & Gray; var. *stenophylla* Bissell & Fern.; f. *argent*
Fern.)

7. L. stuevei Nutt. Bush Clover. Aug.–Sept. Dry woods; occasior
in the s. half of the state, rare or absent elsewhere.

8. L. X simulata Mack. & Bush. Bush Clover. July–Sept. Prai
very rare; Crawford Co. Reputed to be a hybrid between *L. capit*
and *L. virginica*.

9. L. thunbergii (DC.) Nakai. Tall Bush Clover. Aug.–Oct. Native
Asia; rarely adventive in Ill.; Coles, Jackson, and Wayne cos.

10. L. leptostachya Engelm. Prairie Bush Clover. Aug.–Sept. Prair
very rare; Cook, McHenry, and Winnebago cos.

11. L. cuneata (Dum.-Cours.) G. Don. Sericea Lespedeza. Sept.–C
Frequently planted in Ill. and often escaped in the s. one-half of
state.

12. L. violacea (L.) Pers. Violet Bush Clover. July–Sept. Dry woo
occasional in the s. half of the state, becoming less common nor
ward.

13. L. intermedia (S. Wats.) Britt. Bush Clover. Aug.–Sept.
woods; occasional in the s. half of the state, absent elsewhere.

14. L. virginica (L.) Britt. Slender Bush Clover. Aug.–Sept. Dry woods; occasional throughout the state, but more common in the s. cos.

37. Hosackia Dougl.

1. H. americana (Nutt.) Piper. Deer Vetch. June–Aug. Native of the w. U.S., rarely adventive along railroads; Cook and Greene cos. *Lotus americanus* (Nutt.) Bisch.—F.; *Lotus purshianus* Clem. & Clem.—G.

38. Phaseolus L. Kidney Bean

1. P. polystachios (L.) BSP. Wild Kidney Bean. July–Sept. Dry woods; occasional in the s. one-third of the state, absent elsewhere.

39. Dioclea

1. D. multiflora (Torr. & Gray) C. Mohr. Aug.–Sept. Moist thickets, very rare; Massac Co.

40. Desmodium Desv. Tick Trefoil

1. Flowers borne on a long, leafless peduncle .... 1. *D. nudiflorum*
1. Flowers borne on leafy stems 2
2. Leaves clustered in a whorl midway on the flowering stem.............
.... 2. *D. glutinosum*
2. Leaves scattered all along the flowering stem 3
3. Flowers white; stipe of fruit at least twice as long as the calyx ....
.... 3. *D. pauciflorum*
3. Flowers pink or purple; stipe of fruit not longer than the calyx 4
4. Petioles up to 3 mm long, or absent .... 4. *D. sessilifolium*
4. Petioles 4 mm long or longer 5
5. Stipules persistent, conspicuous 6
5. Stipules deciduous, inconspicuous 9
6. Plants lying on the ground; leaflets mostly suborbircular ....
.... 5. *D. rotundifolium*
6. Plants erect or ascending; leaflets variously shaped 7
7. Pubescence on lower surface of leaflets at least of some hooked hairs 8
7. Pubescence on lower surface of leaflets not of hooked hairs, or lower surface glabrous .... 8. *D. cuspidatum*
8. Leaflets obscurely reticulate beneath; joints of fruit angular along the lower margin; rachis of inflorescence with spreading hairs ....
.... 7. *D. canescens*
8. Leaflets conspicuously reticulate beneath; joints of fruit rounded along both margins; rachis of inflorescence with some hooked hairs
.... 6. *D. illinoense*
9. Leaflets glaucous beneath .... 9. *D. laevigatum*
9. Leaflets not glaucous beneath 10
10. Leaflets to 3 cm long, to 2 cm broad 11
10. Leaflets, or most of them, longer than 3 cm, broader than 2 cm 12
11. Stems and upper surface of leaflets glabrous or nearly so; pedicels (8–)10–20 mm long .... 10. *D. marilandicum*
11. Stems and upper surface of leaflets pilose; pedicels 4–8 (–9) mm long
.... 11. *D. ciliare*

12. Fruit with 1–2 segments; calyx 2–3 mm long .... 12. *D. rigidur*
12. Fruit with 3 or more segments; calyx 3 mm long or longer 1
13. Most of the flowers at least 9 mm long; primary bracts 4–10 mm long .
.... 13. *D. canadens*
13. Flowers up to 8 mm long; primary bracts 2–3 mm long 1
14. Leaflets velvety on the lower surface; stipules ovate-deltoid, at mos
about 2½ times longer than broad .... 14. *D. nuttal*
14. Leaflets variously pubescent to nearly glabrous on the lower surface, bu
not velvety; stipules filiform or subulate, at least 7 times longer tha
broad 1
15. Leaflets rather thick, conspicuously reticulate; terminal leaflet less tha
three times longer than broad .... 15. *D. dillen*
15. Leaflets rather thin, faintly reticulate; terminal leaflet at least three time
longer than broad .... 16. *D. paniculatur*

1. D. nudiflorum (L.) DC. Bare-stemmed Tick Trefoil. July–Aug
Woods; occasional throughout the state.

2. D. glutinosum (Muhl.) Wood. Pointed Tick Trefoil. June–Aug
Usually rich woods; occasional throughout the state.

3. D. pauciflorum (Nutt.) DC. White-flowered Tick Trefoil. July–Sep
Woods; occasional in the s. one-fourth of the stage, absent else
where.

4. D. sessilifolium (Torr.) Torr. & Gray. Sessile-leaved Tick Trefoi
July–Sept. Open woods, sand prairies; occasional throughout th
state, except for the nw. cos.

5. D. rotundifolium DC. Round-leaved Tick Trefoil. Aug.–Sept. D
woods, not common; mostly in the s. half of the state.

6. D. illinoense Gray. Illinois Tick Trefoil. July–Aug. Prairies, roac
sides; occasional throughout the state.

7. D. canescens (L.) DC. Hoary Tick Trefoil. July–Aug. Dry wood
occasional in the s. half of the state, less common elsewhere.

8. D. cuspidatum (Muhl.) Loud. Tick Trefoil. July–Sept. Stems, lea
lets, and bracts glabrous or nearly so. Dry woods; occasion
throughout the state, except for the extreme n. cos.

var. longifolium (Torr. & Gray) Schub. Tick Trefoil. Stems, leaflet
and bracts pubescent. July–Sept. Dry woods; occasional througho
the state. *D. longifolium* (Torr. & Gray) Smythe—J.

9. D. laevigatum (Nutt.) DC. Glaucous Tick Trefoil. July–Sept. D
woods; not common; apparently confined to the s. three-fourths c
the state.

10. D. marialandicum (L.) DC. Small-leaved Tick Trefoil. July–Sep
Dry woods; not common; confined to the s. two-thirds of the state.

11. D. ciliare (Muhl.) DC. Hairy Tick Trefoil. Aug.–Oct. Dry woods; o
casional in the s. half of the state.

12. D. rigidum (Ell.) DC. Stiff Tick Trefoil. July–Sept. Dry woods; n
common; restricted to the s. two-thirds of the state.

13. D. canadense (L.) DC. Showy Tick Trefoil. July–Sept. Prairies; o
casional throughout the state.

14. D. nuttallii (Schindl.) Schub. Nuttall's Tick Trefoil. Aug.–Sept. D
woods; not common; confined to the s. half of the state.

15. D. dillenii Darl. Tick Trefoil. July–Sept. Dry woods; occasion
throughout the state. *D. glabellum* (Michx.) DC.—J; *D. perplexu*
Schub.—G.

16. D. paniculatum (L.) DC. Panicled Tick Trefoil. July–Sept. D
woods; occasional throughout the state.

41. *Glycine* L. Soybean

1. G. max (L.) Merr. Soybean. June–Aug. Native of Asia; frequently cultivated and occasionally escaped throughout the state.

42. *Vigna Savi*

1. V. sinensis (L.) Endl. Cow Pea. July–Sept. Native of Asia; occasionally cultivated in Ill.; rarely escaped from cultivation; Mason Co.

43. *Pueraria* DC.

1. P. lobata (Willd.) Ohwi. Kudzu-vine. Aug.–Sept. Native of Japan; occasionally cultivated and sometimes escaping in the s. cos.

44. *Strophostyles* Ell. Wild Bean

1. Legume up to 3.5 cm long, with spreading pubescence; leaflets silky-pubescent on both surfaces; calyx tube densely pubescent, up to 1.5 mm long .... 1. *S. leiosperma*
1. Legume 4 cm long or longer, with appressed pubescence; leaflets glabrous or sparsely pubescent, at least on the upper surface; calyx tube glabrous or sparsely pubescent, 1.5–2.5 mm long 2
2. Bracteoles as long as the calyx tube; larger leaflets up to 6 cm broad; seeds 6–12 mm long .... 2. *S. helvola*
2. Bracteoles about half as long as the calyx tube; larger leaflets up to 2 cm broad; seeds 3–6 mm long .... 3. *S. umbellata*

1. S. leiosperma (Torr. & Gray) Piper. Wild Bean. July–Sept. Prairies, dry woods, fields, roadsides; occasional in the s. half of the state, becoming less common northward.
2. S. helvola (L.) Ell. Wild Bean. June–Oct. Leaflets, or some of them, lobed. Rocky woods, sand bars, roadsides, fields; occasional throughout the state.
var. missouriensis (S. Wats.) Britt. Wild Bean. June–Oct. Leaflets unlobed. Dry woods; occasional in cos. along the Mississippi River.
3. S. umbellata (Muhl.) Britt. Wild Bean. July–Oct. Dry woods, streambanks; occasional in the s. two-fifths of the state, absent elsewhere.

45. *Galactia* P. Br. Milk Pea

1. G. volubilis (L.) Britt. var. mississippiensis Vail. Milk Pea. July–Sept. Rocky woods, not common; confined to the s. one-sixth of the state.*G. volubilis* (L.) Britt.—J.

46. *Amphicarpa* Ell. Hog Peanut

1. A. bracteata (L.) Fern. Hog Peanut. Aug.–Sept. Stem with ascending white hairs; sides of legume glabrous. Woods and thickets; occasional throughout the state.
var. comosa (L.) Fern. Aug.–Sept. Stem with appressed tawny hairs; sides of legume pubescent. Woods and thickets; occasional throughout the state. *A. comosa* (L.) G. Don—J.

81. LINACEAE

1. Linum L. Flax

1. Flowers blue ....
1. Flowers yellow ....
2. Sepals acuminate; annuals .... 1. *L. usitatissimum*
2. Sepals mostly rounded; perennials .... 2. *L. perenne*
3. Petals 8–10 mm long; outer sepals glandular along the margin; leaves with a pair of stipular glands at the base .... 3. *L. sulcatum*
3. Petals up to 8 mm long; outer sepals eglandular (inner sepals glandular along the margin in *L. medium*); leaves without a pair of stipular glands at the base ....
4. Leaves mostly alternate; outer sepals lanceolate, stiff ....
4. Leaves mostly opposite; outer sepals elliptic, membranaceous .... 6. *L. striatum*
5. Inner sepals with conspicuous glandular cilia; leaves stiff .... 4. *L. medium*
5. Inner sepals eglandular or nearly so; leaves rather thin 5. *L. virginianum*

1. L.usitatissimum L. Common Flax. May–Aug. Native of Europe; adventive in waste areas; occasional throughout the state.

2. L. perenne L. Flax. May–Aug. Native of Europe; rarely escaped from cultivation; Coles Grundy, and Winnebago cos.

3. L. sulcatum Riddell. Wild Flax. May–Sept. Dry soil, hill prairies; occasional throughout the state.

4. L. medium (Planch.) Britt. var. texanum (Planch.) Fern. Wild Flax. May–Aug. Mostly dry soil; occasional in the s. half of the state, rare in the n. half where it occurs on calcareous pond shores. *L. medium* (Planch.) Britt.—J.

5. L. virginianum L. Wild Flax. June–Aug. Dry woods; occasional in the s. half of the state, rare in the n. half.

6. L. striatum Walt. Wild Flax. June–Sept. Moist soil, sometimes in woods; occasional in the s. one-fourth of the state.

82. OXALIDACEAE

1. Oxalis L. Wood Sorrel

1. Leaves all from the base of the plants; flowers purple to lavender (rarely white); plants with bulbs .... 1. *O. violacea*
1. Leaves borne along the stem; flowers yellow; plants without bulbs ...
2. Plants creeping, rooting at the nodes .... 2. *O. corniculata*
2. Stems erect or ascending, not rooting at the nodes ....
3. Flowers up to 12 mm long; seeds 1.0–1.6 mm long ....
3. Flowers 12–18 mm long; seeds 1.8–2.2 mm long .... 5. *O. grandis*
4. Pedicels of fruits deflexed; pedicels of flowers all arising from same point at tip of main axis; stipules oblong .... 3. *O. dillenii*
4. Pedicels of fruits erect or ascending; pedicels of flowers arising both from the tip of the main axis as well as along the branches of the inflorescence; stipules linear .... 4. *O. stricta*

1. O. violacea L. Purple Oxalis. April–June. Woods, bluffs, prairies; in every co.

2. O. corniculata L. Creeping Wood Sorrel. April–Oct. Native of the Tropics; in waste areas, often in greenhouses; occasional in the state.

3. O. dillenii Jacq. Yellow Wood Sorrel. May–Nov. Fields, prairies, roadsides; common in every co. *O. stricta* L.—F, G, J.

4. O. stricta L. Yellow Wood Sorrel. May–Oct. Woods, fields, roadsides; in every co. *O. europaea* Jord.—F, G; *O. cymosa* Small—J.

5. O. grandis Small. Large Wood Sorrel. June–Sept. Rich woodlands, very rare; Pope and Wabash cos.

83. GERANIACEAE

1. Leaves pinnatifid or pinnately compound .... 1. *Erodium*
1. Leaves palmately cleft or divided .... 2. *Geranium*

1. *Erodium* L'Hér. Storksbill

1. E. cicutarium (L.) L'Hér. Pin Clover. April–Oct. Native of Europe; naturalized in waste areas; not common.

2. *Geranium* L. Cranesbill

1. Petals more than 1 cm long .... 1. *G. maculatum*
1. Petals less than 1 cm long 2
2. Leaves palmately compound .... 2. *G. robertianum*
2. Leaves palmately lobed, often deeply so 3
3. Sepals awnless; stamens 5; seeds without reticulations or pits ....
.... 3. *G. pusillum*
3. Sepals awned or sharp-tipped; stamens 10; seeds reticulate or pitted 4
4. Peduncles 1-flowered .... 4. *G. sibiricum*
4. Peduncles 2-flowered 5
5. Flowers purple; most or all the leaves less than 3.5 cm broad ....
.... 5. *G. dissectum*
5. Flowers rose to pale pink to nearly white; leaves 3.5 cm broad or broader 6
6. Beak of mature style column 3–5 mm long; ultimate lobes of leaves acute .... 6. *G. bicknellii*
6. Beak of mature style column 1–2 mm long; ultimate lobes of leaves obtuse .... 7. *G. carolinianum*

1. G. maculatum L. Wild Geranium. April–June. Rich woods, common; in every co.

2. G. robertianum L. Herb Robert. July. Woods, fields, rare; Cook, DuPage, and Wabash cos.

3. G. pusillum Burm. f. Small Cranesbill. June–July. Native of Europe; adventive in waste areas, rare; Cook, Kankakee, Peoria, and Piatt cos.

4. G. sibiricum L. Siberian Cranesbill. Aug.–Sept. Native of Europe and Asia; naturalized into waste ground, rare; Champaign, Ogle, and Winnebago cos.

5. G. dissectum L. June. Native of Europe; rarely escaped into edg
of prairie; Lake Co.
6. G. bicknellii Britt. Northern Cranesbill. June–Aug. Sandy wood
fields, rare; Cook and Lake cos.
7. G. carolinianum L. Wild Cranesbill. May–Aug. Woods, field
roadsides; common in the s. half of the state, occasional in the r
half. (Including var. *confertiflorum* Fern.)

84. ZYGOPHYLLACEAE

1. Fruits with 2–4 sharp prickles .... 1. *Tribulu*
1. Fruits without prickles .... 2. *Kallstroem*

1. *Tribulus* L. Caltrop

1. T. terrestris L. Puncture-weed. June–Sept. Native of Europe; o
casional throughout the state.

2. *Kallstroemia* Scop.

1. K. intermedia Rydb. June–Sept. Native of the w. U.S.; rarely a
ventive in Ill.; Cook and St. Clair cos.

85. RUTACEAE

1. Leaflets 5–9; flowers sessile in axillary clusters; sepals absent; fruit a
ellipsoid follicle .... 1. *Xanthoxylu*
1. Leaflets 3; flowers in terminal cymes; sepals 3–5; fruit an orbicular s
mara .... 2. *Ptele*

1. *Xanthoxylum* Gmel. Prickly Ash

1. X. americanum Mill. Prickly Ash. April–May. Woodlands; o
casional in the n. half of the state, rare in the s. half. *Zanthoxylu*
americanum Mill.—G, J.

2. *Ptelea* L. Hop-tree

1. P. trifoliata L. Wafer Ash. May–July. Branchlets glabrous. Op
woods, occasional throughout the state, although less common in t
s. cos.
var. mollis Torr. & Gray. May–July. Branchlets densely pubesce
Sand dunes, rare; Cook and Lake cos. *P. trifoliata* L.—J; var. *dean*
Nieuwl.—G.

86. SIMAROUBACEAE

1. *Ailanthus Desf.* Tree-of-heaven

1. A. altissima (Mill.) Swingle. Tree-of-heaven. June–July. Native of Asia; naturalized in woods fields, and cities; occasional throughout most of Ill.

87. POLYGALACEAE

1. *Polygala L.* Milkwort

1. Flowers 1.5 cm long or longer .... 1. *P. paucifolia*
1. Flowers up to 1 cm long 2
2. At least the lowermost leaves in whorls 3
2. All leaves alternate 4
3. Leaves mostly obtuse; spikes at least 1 cm thick .... 2. *P. cruciata*
3. Leaves mostly acute; spikes up to 7 mm thick .... 3. *P. verticillata*
4. Petals conspicuously united into a cleft tube; stems glaucous ....
   .... 4. *P. incarnata*
4. Petals not united into a cleft tube; stems not glaucous 5
5. Flowers in racemes, the pedicles at least 2 mm long ... 5. *P. polygama*
5. Flowers in spikes, the pedicels absent or up to 1 mm long 6
6. Leaves serrulate; wings 3.0–3.7 mm long; plants with several stems from a thick perennial crown .... 6. *P. senega*
6. Leaves entire or nearly so; wings 4.5 mm long or longer; annuals with solitary stem .... 7. *P. sanguinea*

1. P. paucifolia Willd. Flowering Wintergreen. May–June. Moist woods, very rare; Cook Co.

2. P. cruciata L. var. aquilonia Fern. Cross Milkwort. July–Sept. Acid, sandy soils, not common; mostly in the ne. one-fourth of Ill.; also Clark and Menard cos. *P. cruciata* L.—G, J.

3. P. verticillata L. Whorled Milkwort. July–Sept. All or nearly all leaves (except near inflorescence) whorled; racemes lax; pedicels at least 0.5 mm long. Dry, often sandy soil, not common; mostly in the s. half of the state.

var. isocycla Fern. Whorled Milkwort. July–Sept. All or nearly all leaves (except near the inflorescence) whorled; racemes dense; pedicels at most 0.3 mm long. Dry, acid soil, occasional; throughout most of the state. *P. verticillata* L.—J. (Including var. *sphenostachya* Pennell.)

var. ambigua (Nutt.) Wood. Whorled Milkwort. June–Sept. Only the lowest 1–3 nodes with whorled leaves. Fields, dry woods, occasional; apparently confined to the s. half of Ill. *P. ambigua* Nutt.—G, J.

4. P. incarnata L. Pink Milkwort. July–Sept. Prairies and gravel hills, not common; in the n. half of the state; also Pope Co.

5. P. polygama Walt. var. obtusata Chod. Purple Milkwort. June–Aug. Sandy waste ground and open woods, occasional; confined to the n. half of the state. *P. polygama* Walt.—J.

6. P. senega L. Seneca Snakeroot. May–Sept. Prairies and dry

woods; occasional in the n. three-fourths of Ill., apparently absent i
the s. one-fourth. (Including var. *latifolia* Torr. & Gray.)

   7. P. sanguinea L. Field Milkwort. July–Sept. Fields, wood
prairies, occasional; throughout the state. (Including f. *viridescen*
[L.] Farw. and f. *albiflora* [Wheelock] Millspaugh.)

88. EUPHORBIACEAE

1. Leaves palmately 5- 11-lobed; plants 3 m or more tall ...... 1. *Ricin*
1. Leaves unlobed (if 3- to 5-lobed, then the plants pubescent); plants le
    than 1 m tall ....
2. Lower cauline leaves alternate, the upper opposite (*Euphorbia corollat*
    with five white petal-like structures per flower, may be sought here) ..
    .... 8. *Poinsett*
2. Cauline leaves either all opposite or all alternate, or sometimes whorl
    ....
3. Leaves alternate along the stem ....
3. Leaves opposite along the stem .... 9. *Chamaesy*
4. Uppermost leaves subtending the inflorescence whorled; latex prese
    .... 7. *Euphorb*
4. None of the leaves at the tip of the stem whorled; latex absent .....
5. Leaves sessile or nearly so; plants glabrous .... 2. *Phyllanth*
5. Leaves petiolate; plants variously pubescent ....
6. Leaves entire ....
6. Leaves variously toothed ....
7. Some of the petioles at least half as long as the blades; capsule 2- to
    celled (rarely 1-celled by abortion) .... 3. *Crot*
7. Petioles much less than half as long as the blades; capsule 1-celled
    .... 4. *Crotonop*
8. Plants twining; leaves cordate .... 5. *Tra*
8. Plants erect; leaves not cordate (except in *Acalypha ostryaefolia*, a pla
    with a 3-celled ovary) ....
9. Plants with stellate pubescence; leaves with 2 glands at base of blac
    pistillate flowers not subtended by cleft bracts .... 3. *Crot*
9. Plants pubescent, but not stellate; leaves without 2 glands at base
    blade; pistillate flowers subtended by cleft bracts .... 6. *Acalyp*

1. Ricinus L. Castor Bean

   1. R. communis L. Castor Bean. June–Sept. Native of the Tropi
occasionally escaped from cultivation.

2. Phyllanthus L.

   1. P. caroliniensis Walt. June–Oct. Mostly sandy soil; occasiona
rare in Ill.

3. Croton L. Croton

1. Leaves toothed; calyx of staminate flowers 4-parted; rudimentary pe
    in pistillate flower 5 .... 1. *C. glandulo*

1. Leaves entire; calyx of staminate flowers 5-parted; petals in pistillate flowers absent 2
2. Calyx of pistillate flowers 7- to 12-parted; some or all of the leaves cordate at the base .... 2. *C. capitatus*
2. Calyx of pistillate flowers 5-parted; leaves tapering, rounded, or barely subcordate at the base 3
3. Leaves ovate to oblong; stamens 3–8; plants monoecious; seeds pitted; capsule 7–9 mm in diameter .... 3. *C. monanthogynus*
3. Leaves linear to linear-oblong; stamens (8–) 10 (–12); plants dioecious; seeds reticulate; capsule 4–6 mm in diameter .... 4. *C. texensis*

1. C. glandulosus L. var. septentrionalis Muell.-Arg. Sand Croton. July–Oct. Sandy, often disturbed, soil; common in the s. cos., occasional in the n. cos., where it probably is adventive. *C. glandulosus* L.—J.

2. C. capitatus Michx. Capitate Croton. July–Sept. Sandy soil, limestone glades; occasional in the s. cos., less common in the cos., where it is probably adventive.

3. C. monanthogynus Michx. Croton. July–Oct. Dry fields, bluffs; occasional in the s. half of the state; much rarer in the n. half of the state.

4. C. texensis (Klotzsch) Muell.-Arg. Texas Croton. June–Sept. Dry soil, very rare; Menard co.

4. *Crotonopsis Michx. Rushfoil*

1. Fruit spiny at tip; leaves linear-lanceolate .... 1. *C. linearis*
1. Fruit without spines; leaves elliptic to lanceolate .... 2. *C. elliptica*

1. C. linearis Michx. Rushfoil. July–Sept. Sandy soil, rare; Cass, LaSalle, Mason, Menard, Mercer, Scott, and Whiteside cos.
2. C. elliptica Willd. Rushfoil. July–Sept. Dry fields, dry woods, bluffs; occasional in the s. one-third of the state.

5. *Tragia L.*

1. T. cordata Michx. July–Sept. Dry woods, bluffs, rare; Hardin, Johnson, and Pope cos.

6. *Acalypha L. Three-seeded Mercury*

1. Leaves cordate; fruit soft-echinate .... 1. *A. ostryaefolia*
1. Leaves not cordate; fruit not soft-echinate 2
2. Some or all the petioles at least ⅓ as long as the blades 3
2. Petioles at most only ¼ as long as the blades .... 5. *A. gracilens*
3. Bracts subtending pistillate flowers 5- to 9-lobed, often glandular; stems glabrous or with incurved hairs 4
3. Bracts subtending pistillate flowers 9- to 15-lobed, usually glandless; stems with spreading hairs .... 4. *A. virginica*
4. Capsule 3-seeded; seeds 1.2–2.0 mm long .... 2. *A. rhomboidea*
4. Capsule 2-seeded; seeds 2.2–3.2 mm long .... 3. *A. deamii*

1. A. ostryaefolia Riddell. Three-seeded Mercury. July–Oct. River banks, fields, bluffs, roadsides; occasional in the s. half of the state.

2. A. rhomboidea Raf. Three-seeded Mercury. July–Oct. Wood
fields, bluffs, roadsides; in every co.
3. A. deamii (Weatherby) Ahles. Large-seeded Mercury. July–Oc
Wooded river bottoms, rare; Coles and Vermilion cos. *A. rhomboide*
Raf. var. *deamii* Weatherby—F, G.
4. A. virginica L. Three-seeded Mercury. July–Oct. Fields, roadside
bluffs, woods; occasional to common in the s. three-fourths of th
state, rare elsewhere.
5. A. gracilens Gray. Slender Three-seeded Mercury. June–Sep
Leaves oblong to oblong-lanceolate, the petioles more than $^1/_{10}$
long as the blades; capsule 3-seeded. Woods, fields, roadsides; o
casional in the s. half of the state, rare and possibly adventive els
where; Grundy and Will cos.
ssp. monococca (Engelm.) Webster. Slender Three-seeded Me
cury. June–Sept. Leaves linear to linear-lanceolate, the petiol
usually not more than $^1/_{10}$ as long as the blades; capsule 1-seede
Fields and woods, not common; scattered in the s. two-thirds of th
state. *A. gracilens* Gray—J; *A. gracilens* Gray var. *monococc*
Engelm—F, G.

7. *Euphorbia* L. *Spurge*

1. Leaves entire ....
1. Leaves toothed ....
2. Glands of the involucre with petal-like appendages ....
2. Glands of the involucre without petal-like appendages ....
3. Bracts and uppermost leaves with white margins ...... 1. *E. margina*
3. Bracts and leaves green throughout .... 2. *E. corolla*
4. Cauline leaves linear to narrowly oblong-lanceolate, acute, sessi
seeds smooth; perennials with rhizomes ....
4. Cauline leaves obovate, obtuse or retuse, short-petiolate; seeds pitte
annuals or biennials ....
5. None of the leaves over 3 mm broad .... 3. *E. cypariss.*
5. Some or all the leaves over 3 mm broad .... 4. *E. es*
6. Seeds 1.8–2.0 mm long, finely pitted throughout; capsule with round
lobes .... 5. *E. commut*
6. Seeds 1.0–1.5 mm long, with 1–4 rows of large pits; capsule with
keeled lobes .... 6. *E. pep*
7. Capsule smooth; seeds ovoid .... 7. *E. helioscop*
7. Capsule warty; seeds flattened or lenticular ....
8. Seeds dark brown, 1.7–2.0 mm long, obscurely reticulate; styles long
than the ovary; cyathium with 5 oblong glands .... 8. *E. obtus*
8. Seeds reddish-brown, 1.3–1.5 mm long, distinctly reticulate; sty
shorter than the ovary; cyathium with 4 lobes and a tuft of hairs ....
.... 9. *E. spathul*

1. E. marginata Pursh. Snow-on-the-mountain. July–Oct. Native
the w. U.S.; naturalized in waste areas; occasional throughout
state.
2. E. corollata L. Flowering Spurge. May–Oct. Stem and lower s
face of the leaves glabrous or nearly so. Prairies, woods, fields, ro
sides, common; in every co. (Including var. *paniculata* [Ell.] Boiss
var. mollis Millsp. May–Oct. Stem and lower surface of leaves sc
hairy. Prairies, woods, fields; occasional in the s. tip of the state.

3. E. cyparissias L. Cypress Spurge. May–Sept. Native of Europe; naturalized along roadsides and in cemeteries; occasional throughout the state.

4. E. esula L. Leafy Spurge. June–Sept. Native of Europe; naturalized in fields and pastures; occasional in the n. half of the state.

5. E. commutata Engelm. Wood Spurge. May–June. Wooded slopes, along streams, and in gravelly soils, rare; mostly in the n. half of the state.

6. E. peplus L. Petty Spurge. June–Oct. Native of Europe; rarely adventive in Ill.; Menard Co.

7. E. helioscopia L. Wart Spurge. June–Oct. Native of Europe; adventive in waste areas; Cass and Menard cos.

8. E. obtusata Pursh. Blunt-leaved Spurge. May–July. Rich woods, wooded slopes; occasional throughout the state.

9. E. spathulata Lam. Spurge. May–July. Limestone ledge at edge of hill prairie, rare; Monroe Co. *E. dictyosperma* Fisch. & Mey.—F, G.

8. *Poinsettia* Graham   Poinsettia

. Floral bracts usually basally red; seeds not angular, sharply tuberculate; leaves glossy green .... 1. *P. cyathophora*
. Floral bracts green or basally pale, never red; seeds angular, bluntly tuberculate; leaves dull green .... 2. *P. dentata*

1. P. cyanthophora (Murr.) Kl. & Garcke. Wild Poinsettia. July–Oct. Leaves oval to lanceolate or even pandurate. Fields, roadsides; occasional throughout the state. *Euphorbia heterophylla* L.—F, G; *P. heterophylla* (L.) Kl. & Garcke—J.

var. graminifolia (Michx.) Mohlenbr. July–Oct. Leaves linear to narrowly lanceolate. Moist soil along stream; Pope and Union cos. *Euphorbia heterophylla* var. *graminifolia* (Michx.) Engelm.—F.

2. P. dentata (Michx.) Kl. & Garcke. Wild Poinsettia. June–Sept. Leaves narrowly ovate to rhombic. Fields, roadsides, prairies; occasional throughout the state. *Euphorbia dentata* Michx.—F, G.

var. cuphosperma (Engelm.) Mohlenbr. June–Sept. Leaves linear to narrowly lanceolate. Fields and roadsides; Cook and Hancock cos. *Euphorbia dentata* f. *cuphosperma* (Engelm.) Fern.—F.

9. *Chamaesyce* S. F. Gray

Leaves entire 2
Leaves toothed, at least near the apex 4
Leaves about as broad as long; style 0.2 mm long; seeds about 1 mm long .... 1. *C. serpens*
Leaves considerably longer than broad; style 0.7–1.0 mm long; seeds 1.3–2.6 mm long. 3
Seeds 2.0–2.6 mm long; capsules 3.0–3.5 mm long . 2. *C. polygonifolia*
Seeds 1.3–1.6 mm long; capsules 2.0–2.5 mm long .... 3. *C. geyeri*
Ovary and capsule strigose 5
Ovary and capsule glabrous 6
Seeds with cross-ridges; stems mostly reddish-brown; style 0.3–0.5 mm long, cleft less than halfway to the base .... 4. *C. supina*
Seeds smooth or minutely granular; stems mostly green; style 0.7 mm long, cleft halfway to the base .... 5. *C. humistrata*

6. Stems pubescent, at least near the tip; leaves toothed from tip to base
6. Stems glabrous; leaves toothed only at apex and occasionally at base
7. Stems prostrate; capsules up to 1.9 mm long .... 6. *C. vermiculat*
7. Stems ascending to erect; capsules 1.9 mm long or longer ....
.... 7. *C. maculat*
8. Leaves broadly oblong to ovate; seeds pitted and with short cross-ridge
.... 8. *C. serpyllifol*
8. Leaves linear-oblong; seeds with 3–6 long cross-ridges ....
.... 9. *C. glyptosperm*

1. C. serpens (HBK.) Small. Round-leaved Spurge. July–Sept. Mois
sandy soil; scattered but not common. *Euphorbia serpens* HBK.—
G.

2. C. polygonifolia (L.) Small. Seaside Spurge. July–Sept. Dune
and beaches, rare; Cook, Fulton, and Lake cos. *Euphorbia polygon
folia* L.—F, G.

3. C. geyeri (Engelm.) Small. Spurge. June–Sept. Sandy soil; o
casional or rare in the w. half of the state. *Euphorbia geye
Engelm.*—F, G.

4. C. supina (Raf.) Moldenke. Milk Spurge. July–Oct. Disturbed
cultivated habitats; common throughout the state. *Euphorbia supi
Raf.*—F, G.

5. C. humistrata (Engelm.) Small. Milk Spurge. July–Sept. Sanc
soil, river banks; occasional throughout the state. *Euphorbia hun
strata Engelm.*—F, G.

6. C. vermiculata (Raf.) House. Spurge. July–Sept. Rich soil, ve
rare; Lake Co. *Euphorbia vermiculata* Raf.—F, G.

7. C. maculata (L.) Small. Nodding Spurge. July–Sept. Disturb
areas; common throughout the state. *Euphorbia maculata* L.—F, G

8. C. serpyllifolia (Pers.) Small. Spurge. July–Oct. Sandy areas, ra
Cook Co. *Euphorbia serpyllifolia* Pers.—F, G.

9. C. glyptosperma (Engelm.) Small. Spurge. June–Sept. Sandy
gravelly soil; occasional in the n. half of the state; also Monroe C
Euphorbia glyptosperma Engelm.—F, G.

89. CALLITRICHACEAE

1. *Callitriche* L.   Water Starwort

1. Plants growing in water; flowers each with 2 bracts at the base ....
1. Plants growing on soil; flowers bractless .... 3. *C. terres*
2. Fruit as broad as long, rounded at the base .... 1.*C. heterophy*
2. Fruit longer than broad, narrowed to the base .... 2. *C. palus*

1. C. heterophylla Pursh. Large Water Starwort. April–Oct. Shall
water, occasional; scattered throughout the state.

2. C. palustris L. Common Water Starwort. April–Oct. Shall
water, not common; scattered throughout the state. *C. verna* L.—

3. C. terrestris Raf. Terrestrial Starwort. April–July. Moist soil;
casional in the s. two-thirds of the state, absent elsewhere. *C. defl
A. Br. var. *austini* (Engelm.) Hegelm.—F.

90. LIMNANTHACEAE

1. Floerkea Willd. False Mermaid

1. F. proserpinacoides Willd. False Mermaid. April–June. Rich woodlands; occasional in the n. half of the state, extending southward to Crawford co.

91. ANACARDIACEAE

1. Leaflets 5 or more 2
1. Leaflets 3 3
2. Plants poisonous to touch; drupes whitish; panicles axillary ....
.... 1. *Toxicodendron*
2. Plants not poisonous; drupes reddish; panicles terminal ...... 2. *Rhus*
3. Flowers appearing before or with the leaves in short spikes; drupes red, pubescent; plants not poisonous .... 2. *Rhus*
3. Flowers appearing after the leaves in axillary panicles; drupes white, glabrous or nearly so; plants poisonous to touch ... 1. *Toxicodendron*

1. Toxicodendron Mill

1. Leaflets 7 or more .... 1. *T. vernix*
1. Leaflets 3 .... 2. *T. radicans*

    1. T. vernix (L.) Kuntze. Poison Sumac. May–July. Bogs and marshes; occasional to rare in the ne. one-fourth of the state; also Coles and Woodford cos. *Rhus vernix* L.—F, G, J.
    2. T. radicans (L.) Kuntze. Poison Ivy. May–July. Fields, woods, bluffs, waste areas, common; in every co. *R. radicans* L.—F, G, J. (Including many varieties and forms.)

2. Rhus L. Sumac

. Leaflets 5 or more 2
. Leaflets 3 .... 4. *R. aromatica*
. Rachis winged; leaves entire or nearly so .... 1. *R. copallina*
. Rachis unwinged; leaves sharply serrate 3
. Branches and petioles velvety-pubescent .... 2. *R. typhina*
. Branches and petioles glabrous .... 3. *R. glabra*

    1. R. copallina L. Dwarf Sumac. July–Sept. Woodlands, fields; common in the s. one-fourth of the state, occasional or rare elsewhere. (Including var. *latifolia* Engl.)
    2. R. typhina L. Staghorn Sumac. June–July. Woods; occasional in the n. half of the state, rare and apparently adventive in the s. half.
    3. R. glabra L. Smooth Sumac. June–July. Woods, fields, roadsides, common; in every co.
    4. R. aromatica Ait. Fragrant Sumac. March–May. Flowers appearing before the leaves. Woods, bluffs, and dunes; occasional throughout the state. (Including var. *illinoensis* [Greene] Rehder.)

var. serotina (Greene) Rehder. Fragrant Sumac. April–May. Flower appearing when the leaves are partly grown; upright shrubs; leaflet pubescent or nearly glabrous beneath, not velvety. Sandy wood rare; Union Co.

var. arenaria (Greene) Fern. Fragrant Sumac. April–May. Flowe appearing when the leaves are partly grown; low, sprawling shrub leaflets velvety-pubescent beneath. Sandy areas, including dunes; a parently restricted to the n. half of the state. *R. arenaria* (Greene) N. Jones—J; *R. trilobata* Nutt. var. *arenaria* (Greene) Barkley—G.

92. AQUIFOLIACEAE

1. *Ilex* L.   Holly

1. Leaves coriaceous, evergreen .... 1. *I. opac*
1. Leaves not coriaceous, deciduous ....
2. Teeth of calyx acute, glabrous .... 2. *I. decid*
2. Teeth of calyx obtuse, ciliate .... 3. *I. verticilla*

1. *I. opaca* Ait. American Holly. May–June. Rocky, wooded slop very rare; Union Co.

2. *I. decidua* Walt. Swamp Holly. April–May. Swamps, wood slopes, and bluffs; occasional in the s. counties, becoming rare nort ward; absent in the upper two-fifths of Ill.

3. *I. verticillata* (L.) Gray. Winterberry. May–July. Swamps, edge streams, wooded slopes, and bluffs; occasional in the n. cos., rare the s. cos.

93. CELASTRACEAE

1. Leaves opposite .... 1. *Euonym*
1. Leaves alternate .... 2. *Celastr*

1. *Euonymus* L.   Spindle-tree

1. Plants trailing or climbing, rooting at the nodes ....
1. Plants erect or ascending, not rooting at the nodes ....
2. Leaves membranaceous, obovate to oblong; sepals and petals each capsule warty; plants trailing .... 1. *E. obova*
2. Leaves coriaceous, elliptic; sepals and petals each 4; capsule smoc plants climbing .... 2. *E. fortu*
3. Stems unwinged ....
3. Stems winged .... 6. *E. ala*
4. Leaves petiolate; sepals and petals each 4; capsule smooth; shrubs meters tall ....
4. Leaves sessile; sepals and petals each 5; capsule warty; shrub to meters tall .... 5. *E. america*
5. Leaves pubescent on the lower surface; flowers purple ....
.... 3. *E. atropurpur*

5. Leaves glabrous on the lower surface; flowers yellowish-green . . . .
. . . . 4. *E. europaeus*

    1. E. obovatus Nutt. Running Strawberry-bush. April–June. Rich
woods, not common; mostly in the e. half of the state.
    2. E. fortunei (Turcz.) Hand.-Maz. July–Aug. Native of Asia; adventive in low woods; Jackson Co.
    3. E. atropurpureus Jacq. Wahoo. June–July. Woods; occasional throughout the state.
    4. E. europaeus L. European Spindle-tree. May–June. Native of Europe; adventive in a floodplain; Kane Co.
    5. E. americanus L. Strawberry-bush. May–June. Rich woods, rare; Pulaski and Wabash cos.
    6. E. alatus (Thunb.) Sieb. Winged Euonymus. June–July. Native of e. Asia; rarely escaped from cultivation; Coles and DuPage cos.

2. *Celastrus L.* Bittersweet

. Leaves ovate to ovate-oblong, finely serrate; flowers in terminal panicles
or racemes . . . . 1. *C. scandens*
. Leaves suborbicular, crenate; flowers in axillary cymes 2. *C. orbiculatus*

    1. C. scandens L. Bittersweet. May–June. Woods, occasional; throughout the state.
    2. C. orbiculatus Thunb. Round-leaved Bittersweet. May–June. Native of Asia; rarely adventive in Ill.

94. STAPHYLEACEAE

1. *Staphylea L.* Bladdernut

    1. S. trifolia L. Bladdernut. April–May. Moist woods, common; in every co.

95. ACERACEAE

1. *Acer L.* Maple

Leaves pinnately compound . . . . 1. *A. negundo*
Leaves simple 2
Sinuses between adjacent lobes of the leaves rounded; flowers yellow or yellow-green, appearing at the same time as the leaves; each pair of samaras with a U-shaped sinus 3
Sinuses between adjacent lobes of the leaves angular; flowers red or bright yellow, appearing before the leaves; each pair of samaras more or less with a V-shaped sinus 6
Latex present; inflorescence pedunculate; samaras horizontally spreading . . . . 2. *A. platanoides*

3. Latex absent; inflorescence sessile; samaras upwardly curved ....
4. Ovary and young fruit pubescent; calyx at most only 3 mm long, pilos
.... 3. *A. barbatu*
4. Ovary and young fruit glabrous; calyx 4–6 mm long, glabrous or near
so ....
5. Blades of leaves flat, not drooping along the margins; bark usually gra
.... 4. *A. saccharu*
5. Blades of leaves drooping along the margins; bark usually gray-black
black .... 5. *A. nigru*
6. Leaves usually cleft more than halfway to the base; petals abser
ovaries pubescent .... 6. *A. saccharinu*
6. Leaves usually cleft less than halfway to the base; petals preser
ovaries glabrous .... 7. *A. rubru*

1. A. negundo L. Box Elder. April–May. Moist soil; comm
throughout the state. (Including var. *violaceum* [Kirsch.] Jaeg. a
var. *texanum* Pax.)

2. A. platanoides L. Norway Maple. May–July. Native of Europe; c
casionally planted, but rarely escaped from cultivation.

3. A. barbatum Michx. Southern Sugar Maple. April. Woods, c
casional; confined to the s. one-sixth of the state.

4. A. saccharum Marsh. Sugar Maple. April–May. Leaves glabro
or nearly so beneath. Rich woodlands; common throughout the sta
in every co. (Including var. *rugelii* [Pax] Rehder and f. *rugelii* [P
Palmer & Steyerm.)

var. schneckii Rehder. Schneck's Sugar Maple. April–May. Lea
densely pubescent beneath. Rich woodlands; occasional in the
one-fourth of the state.

5. A. nigrum Michx. f. Black Maple. May–June. Rich woodlands;
casional in the n. two-thirds of the state, apparently absent (althou
confused with *A. barbatum*) elsewhere.

6. A. saccharinum L. Silver Maple. Feb.–April. Bottomland woc
and along streams; common throughout the state; in every co.

7. A. rubrum L. Red Maple. March–April. Leaves sparsely pubesc
or glabrous beneath. Rocky woods, slopes, moist woods; occasio
in the s. one-third of the state, rare in the ne. corner of the state,
parently absent elsewhere.

var. rubrum L. f. tomentosum (Desf.) Dansereau. March–Ap
Leaves permanently tomentose beneath; samaras up to 2.5 cm lc
Upland slopes, rare; Union Co.

var. drummondii (H. & A.) Sarg. Drummond's Red Maple. Mar
April. Leaves permanently tomentose beneath; samaras 3 cm lonç
longer. Wooded swamps; confined to the s. one-fourth of Ill. *A. dre*
mondii H. & A.—J.

96. HIPPOCASTANACEAE

1. *Aesculus* L. Buckeye

1. Petals red, glandular along the margin; capsule smooth . 1. *A. disce*
1. Petals greenish-yellow or white marked with red; capsule prickly or c
ered with scales ....

2. Petals all similar in size and shape; stamens exserted; capsule prickly   3
2. Two of the petals very unlike the others in shape and size; stamens included; capsule covered with scales .... 4. *A. octandra*
3. Petals greenish-yellow; leaflets mostly 5 .... 2. *A. glabra*
3. Petals white marked with red; leaflets mostly 7 ... 3. *A. hippocastanum*

   1. A. discolor Pursh. Red Buckeye. April–May. Rich woodlands, rare; confined to the s. one-sixth of the state; also Richland Co. *A. pavia* L.—G.
   2. A. glabra Willd. Ohio Buckeye. April–May. Bark brown; leaflets green beneath. Rich woodlands; occasional throughout most of the state, except for the extreme n. cos. (Including var. *sargentii* Rehder.)
      var. leucodermis Sarg. Bark whitish; leaflets whitened beneath. Rich woods, very rare; Jackson Co.
   3. A. hippocastanum L. Horse Chestnut. May–June. Native of Europe and Asia; occasionally planted but rarely escaped from cultivation.
   4. A. octandra Marsh. Sweet Buckeye. May–June. Rich woodlands, very rare; Gallatin Co.

97. SAPINDACEAE

Annual climber with tendrils; petals white; leaves biternate ....
.... 1. *Cardiospermum*
Tree without tendrils; petals yellow; leaves bipinnate .. 2. *Koelreuteria*

1. *Cardiospermum* L.

   1. C. halicacabum L. Balloon-vine. July–Sept. Native of trop. Am.; rarely adventive in thickets near streams and rivers; Jackson, Monroe, Randolph, and St. Clair cos.

2. *Koelreuteria* Laxm.   China-tree

   1. K. paniculata Laxm. Golden-rain Tree. July–Sept. Native of Asia; sometimes planted and rarely escaped into waste areas.

98. BALSAMINACEAE

1. *Impatiens* L.   Jewel-weed

Flowers orange with reddish spots .... 1. *I. biflora*
Flowers pale yellow .... 2. *I. pallida*

   1. I. biflora Walt. Spotted Touch-me-not. June–Sept. Wet soil; common throughout the state. *I. capensis* Meerb.—F.
   2. I. pallida Nutt. Pale Touch-me-not. July–Sept. Wet soil, often in woodlands; occasional throughout the state.

99. RHAMNACEAE

1. Woody vines .... 1. *Berchem*
1. Trees or shrubs ....
2. Leaves 3-veined from the base; fruit a capsule .... 2. *Ceanoth*
2. Leaves pinnately veined; fruit a drupe .... 3. *Rhamn*

1. Berchemia Neck. Supple-jack

1. B. scandens (Hill) K. Koch. Supple-jack. April–June. Edge of pi
plantation; Pope Co.

2. Ceanothus L.

1. Leaves elliptic to elliptic-lanceolate, obtuse to subacute .. 1. *C. ovat*
1. Leaves ovate to ovate-oblong, acute to acuminate (obtuse in var. *p*
cheri) .... 2. *C. american*

1. C. ovatus Desf. May–June. Low dunes, sandy soil, rare; Coc
JoDaviess, Lake, Whiteside, and Winnebago cos.
2. C. americanus L. New Jersey Tea. June–Aug. Leaves acute
acuminate, glabrous or nearly so on the upper surface. Woods; c
casional throughout the state. (Including var. *intermedius* [Pursh]
Koch.)
var. pitcheri Torr. & Gray. June–Aug. Leaves obtuse at the tip, pilc
on the upper surface. Woods; not common, but throughout the sta

3. Rhamnus L. Buckthorn

1. Low shrub less than 1 m tall; petals absent .... 1. *R. alnifc*
1. Shrubs or trees, at maturity well over 1 m tall; petals present ....
2. Some of the branchlets ending in a spine; nutlets with a deep, narr
groove. .... 2. *R. cathart*
2. None of the branchlets ending in a spine; nutlets without a groove,
the groove broad and open ....
3. Winter buds with scales; flowers appearing with the leaves; nutlets w
a broad, open groove .... 3. *R. lanceol*
3. Winter buds without scales; flowers appearing after the leaves; nut!
without a furrow ....
4. Umbels peduncled; leaves more or less acute; most or all the petiole
mm long or longer; pedicels pubescent .... 4. *R. carolini*
4. Umbels sessile; leaves more or less obtuse; most or all the petioles l
than 5 mm long; pedicels glabrous or nearly so .... 5. *R. frang*

1. R. alnifolia L'Hér. Alder Buckthorn. May–July. Bogs and woo
swamps, rare; confined to the n. half of Ill.
2. R. cathartica L. Common Buckthorn. May–June. Native
Europe; naturalized into waste areas and woodlands; occasiona
common in the n. half of the state.
3. R. lanceolata Pursh. Lance-leaved Buckthorn. April–June. R
banks, bluffs, calcareous fens; rare to occasional in the n. two-th
of the state, rare or absent elsewhere. (Including var. *glabrata* Gl.*
4. R. caroliniana Walt. Carolina Buckthorn. May–June. Leaves

brate beneath at maturity. Woods, not common; in the s. one-fifth of the state. (Including var. *mollis* Fern.)

5. R. frangula L. Glossy Buckthorn. May–July. Leaves obovate. Native of Europe; naturalized in woods and bogs; occasional in the n. half of the state.

var. angustifolia Loud. June–July. Leaves narrowly lanceolate. Native of Europe; rarely naturalized in Ill.; Cook Co.

100. VITACEAE

1. Parthenocissus Planch. Virginia Creeper

1. P. tricuspidata (Sieb. & Zucc.) Planch. Boston Ivy. June–Aug. Native of Asia; rarely escaped from cultivation.

2. P. quinquefolia (L.) Planch. Virginia Creeper. June–July. Woods, along fences, common; in every co.

3. P. vitacea (Knerr) Hitchc. Virginia Creeper. May–July. Woods, thickets; occasional in the n. one-fourth of the state, absent elsewhere. *P. inserta* (Kerner) K. Fritsch—G.

2. Ampelopsis Michx.

1. A. cordata Michx. Raccoon Grape. May–July. Moist woods and thickets; common to occasional in the s. half of the state, becoming rare or absent elsewhere.

2. A. arborea (L.) Koehne. Pepper-vine. June–Aug. Moist woods and thickets, rare; Alexander, Jackson, Pulaski, Randolph, and Union cos.

3. Vitis L. Grape

1. Leaves glabrous or glabrate on the lower surface, green ....
2. Leaves felted on the lower surface, but not cobwebby or merely silver
glaucous ....1. *V. labruscan*
2. Leaves cobwebby on the lower surface or merely silvery-glaucous on t▮
lower surface ....
3. Seeds 5–8 mm long; branchlets not angular .... 2. *V. aestiva*
3. Seeds 4–5 mm long; branchlets angular .... 3. *V. cinere*
4. Leaves broader than long; tendrils lacking, or merely opposite the u▮
permost leaves .... 4. *V. rupestr*
4. Leaves longer than broad; tendrils opposite all leaves except for eve
third one ....
5. Leaves unlobed or shallowly 3-lobed .... 5. *V. vulpi*
5. Leaves, or at least those on the fertile branches, sharply 3- to 5-lobed
6. Berries black; partitions in pith at the nodes 4–5 mm thick; leav▮
scarcely or not ciliate on the margins .... 6. *V. palma*
6. Berries bluish, glaucous; partitions in pith at the nodes up to 2 m
thick; leaves ciliate on the margins .... 7. *V. ripa*

1. V. labruscana Bailey. Labruscan Vineyard Grape. May–June. C▮ tivated vineyard species, rarely escaped into waste ground; Jacks▮ Co.

2. V. aestivalis Michx. Summer Grape. May–July. Petiole tomento▮ or rusty-pubescent. Rocky woods, blufftops; occasional to comm▮ throughout the state. (Including *V. lincecumii* Buckl.)

var. argentifolia (Munson) Fern. Silver-leaved Grape. Petioles g▮ brous or nearly so. Rocky woods, occasional throughout the state.

3. V. cinerea Engelm. Winter Grape. May–July. Low woods, thic▮ ets; occasional in the s. three-fourths of the state, rare elsewhere.

4. V. rupestris Scheele. River-bank Grape. May–June. Rocky ban▮ very rare; Union Co.

5. V. vulpina L. Frost Grape. May–June. Moist soil; occasio▮ throughout the state.

6. V. palmata Vahl. Catbird Grape. May–July. Swamps, low woo▮ confined to the s. one-sixth of the state.

7. V. riparia Michx. Riverbank Grape. May–June. Petioles glabro▮ berries 8–12 mm in diameter. Low woods; occasional to comm▮ throughout the state.

var. praecox Engelm. April. Petioles glabrous; berries 6–7 mm in ameter. Low woods; occasional in the sw. cos. of the state.

var. syrticola (Fern. & Wieg.) Fern. May–June. Petioles dens▮ hairy. Low woods, rare; scattered throughout the state.

101. TILIACEAE

1. Tilia L.   Basswood

1. Leaves glabrous beneath, except in the axils of the veins ....
.... 1. *T. americ*
1. Leaves pubescent beneath with stellate hairs ....
2. Leaves green beneath; peduncles and pedicels glabrous 1. *T. americ*
2. Leaves white beneath; peduncles and pedicels pubescent ....
.... 2. *T. heteropl*

1. T. americana L. Basswood. May–July. Leaves glabrous beneath, except in the axils of the veins. Rich woods; occasional to common throughout the state.

var. neglecta (Spach) Fosberg. Basswood. May–July. Leaves pubescent beneath with stellate hairs. Rich woods, rare; in some of the n. cos. *T. neglecta* Spach—F.

2. T. heterophylla Vent. White Basswood. May–July. Low rocky woods, rare; Hardin, Massac, and Pope cos.

102. STERCULIACEAE

1. *Melochia L.*

1. M. corchorifolia L. Chocolate-weed. Aug.–Sept. Native of the s. states; adventive in Ill. along a road; Randolph Co.

103. MALVACEAE

. Some or all the leaves lobed 2
. None of the leaves lobed 13
. Calyx subtended by an involucre of bracts at the base 3
. Calyx without an involucre of bracts at the base 11
. Bracts at base of calyx 3 4
. Bracts at base of calyx 6 or more 10
. Bracts laciniate . . . . 4. *Gossypium*
. Bracts unlobed, or at least not laciniate 5
. All or most of the leaves divided nearly to the base 6
. None of the leaves divided nearly to the base 7
. Plants hispid; stems more or less procumbent . . . . 2. *Callirhoë*
. Plants with scattered pubescence, but not hispid; stems ascending . . . .
. . . . 1. *Malva*
Flowers over 2 cm across 8
Flowers less than 2 cm across . . . . 1. *Malva*
Lower leaves triangular, unlobed . . . . 2. *Callirhoë*
Lower leaves shallowly lobed or, if unlobed, not triangular 9
Bracts at base of calyx lanceolate to ovate-lanceolate; calyx lobes acute; carpels 1-seeded . . . . 1. *Malva*
Bracts at base of calyx linear; calyx lobes acuminate; carpels 2- to 4-seeded . . . . 3. *Iliamna*
Bracts triangular; fruit separating at maturity into 15–20 carpels . . . .
. . . . 7. *Althaea*
Bracts linear; fruit a 5-celled capsule . . . . 6. *Hibiscus*
Flowers unisexual; flowers white, up to 2 cm across . . . . 8. *Napaea*
Flowers bisexual; flowers blue or, if white, over 2 cm across 12
Flowers over 4 cm across . . . . 2. *Callirhoë*
Flowers up to 4 cm across . . . . 9. *Anoda*
Leaves as broad as long, some of them at least 3 cm broad 14
Leaves longer than broad, never 3 cm broad 17

14. Calyx not subtended by bracts .... 10. *Abutilo*
14. Calyx subtended by bracts 1
15. Bracts subtending the calyx 3 1
15. Bracts subtending the calyx 6 or more .... 6. *Hibiscⱔ*
16. Flowers over 2 cm broad, deep purple .... 2. *Callirhⱁ*
16. Flowers much less than 2 cm broad, pale lilac or white ...... 1. *Malⱴ*
17. Calyx subtended by 2–3 setaceous bracts .... 5. *Sphaeralce*
17. Calyx not subtended by bracts .... 11. *Siⱃ*

1. *Malva* L. Mallow

1. Flowers at least 2.5 cm across, usually broader; petals more than twiⱦ
   as long as the calyx ....
1. Flowers up to 1.5 cm across; petals at most only twice as long as tⱨ
   calyx ....
2. Petals up to four times as long as the calyx; leaves shallowly lobed, tⱨ
   tips of the lobes triangular or broadly rounded; carpels about 10, mⱶ
   nutely pubescent to nearly glabrous .... 1. *M. sylvestⱃⱶ*
2. Petals 5–8 times as long as the calyx; leaves deeply divided into narrⱶ
   divisions, the tips slender and not triangular; carpels usually 15–2ⱶ
   densely pubescent .... 2. *M. moschaⱶ*
3. Petals about twice as long as the calyx; leaves with rounded lobes
   sometimes without lobes; carpels rounded on the margins ....ⱶ
3. Petals about as long as the calyx; leaves with angular lobes; carpels wⱶ
   sharp margins .... 5. *M. rotundifoⱶ*
4. Leaves scarcely lobed, flat on the margins; stems procumbent ......
   .... 3. *M. neglecⱶ*
4. Leaves with 5–7 rounded lobes, crisped on the margins; stems erect
   .... 4. *M. verticillata* var. *crisⱶ*

1. *M. sylvestris* L. High Mallow. Aug.–Sept. Lobes of leaf triangulⱶ
stems and leaves hirsute. Native of Europe; occasionally escapⱶ
from cultivation; scattered in Ill.
   var. *mauritiana* (L.) Boiss. High Mallow. Aug.–Sept. Lobes of lⱶ
broadly rounded; stems and leaves glabrous or nearly so. Native ⱶ
Europe; escaped from cultivation; known from the ne. cos.
2. *M. moschata* L. Musk Mallow. June–Aug. Native of Europe; rarⱶ
escaped from cultivation; Cook, Lawrence, and LaSalle cos.
3. *M. neglecta* Wallr. Common Mallow. June–Sept. Native ⱶ
Europe; naturalized in waste areas, particularly farm lots; occasioⱶ
to common throughout Ill.
4. *M. verticillata* L. var. *crispa* L. Curled Mallow. July–Sept. Native
Europe; rarely escaped from cultivation in a few of the n. cos.
crispa L.—J.
5. *M. rotundifolia* L. Mallow. June–Sept. Native of Europe; natur
ized in waste areas; occasional and scattered in Ill.

2. *Callirhoë* Nutt. Poppy Mallow

1. Each flower subtended by three bracts ....
1. Flowers without subtending bracts ....
2. Some of the leaves triangular; flowers in panicles or appearing ⱶ
   bellate; carpels not rugose .... 1. *C. triangulⱶ*
2. None of the leaves triangular; flowers solitary; carpels rugose ......
   .... 2. *C. involucrⱶ*

3. Basal leaves more or less triangular; top of carpels pubescent . . . .
. . . . 3. *C. alcaeoides*
3. Basal leaves rounded; top of carpels glabrous . . . . 4. *C. digitata*

1. C. triangulata (Leavenw.) Gray. Poppy Mallow. June–Sept. Sandy soil; mostly in the n. cos.

2. C. involucrata (Torr. & Gray) Gray. Poppy Mallow. June–Aug. Native of the w. U.S.; naturalized in a few mostly cent. cos.

3. C. alcaeoides (Michx.) Gray. Poppy Mallow. May–Aug. Dry, gravelly areas, rare; Christian, Henry, and Peoria cos.

4. C. digitata Nutt. Poppy Mallow May–July. Native of the w. U.S.; rarely adventive in n. Ill.

3. *Iliamna* Greene

1. I. remota Greene. Kankakee Mallow. June–Aug. Dry banks, rare; Kankakee Co.

4. *Gossypium* L. Cotton

1. G. hirsutum L. Cotton. May–June. Native of trop. Am.; escaped from cultivation in scattered Ill. areas.

5. *Sphaeralcea* St. Hil. Globe Mallow

1. S. angusta (Gray) Fern. Globe Mallow. July–Aug. Dry soil, very rare; Grundy, LaSalle, and Rock Island cos. *Malvastrum angustum* Gray—G, J.

6. *Hibiscus* L. Rose Mallow

. Annual herbs; flowers yellow or whitish with a black or purple center 2
. Perennial herbs, shrubs, or small trees; flowers not yellow 3
. Calyx inflated, 5-winged; seeds verrucose . . . . 1. *H. trionum*
. Calyx like a spathe, splitting down one side; seeds mucilaginous . . . .
. . . . 2. *H. esculentus*
. Stems and leaves glabrous 4
. Stems and at least the lower surface of the leaves pubescent 5
. Herb 1.0–2.5 m tall; calyx glabrous or with simple hairs .. 3. *H. militaris*
. Shrub or small tree 3 m tall or taller; calyx stellate-pubescent . . . .
. . . . 4. *H. syriacus*
. Leaves glabrous or nearly so above; some of the leaves often 3-lobed; capsules glabrous . . . . 5. *H. palustris*
. Leaves soft-hairy on both surfaces; none of the leaves lobed; capsules densely hairy . . . . 6. *H. lasiocarpos*

1. H. trionum L. Flower-of-an-hour. July–Oct. Native of Europe; adventive throughout the state, although rare in the extreme s. tip.

2. H. esculentus L. Okra. July–Sept. Native of Africa; rarely escaped from cultivation; Champaign Co.

3. H. militaris Cav. Halberd-leaved Rose Mallow. July–Oct. Wet areas; occasional throughout the state, although rare in the extreme n. cos.

4. H. syriacus L. Rose-of-Sharon. July–Sept. Native of Asia; occasionally persisting from cultivation.

5. H. palustris L. Swamp Rose Mallow. July–Sept. Wet soil; Co
and LaSalle cos. *H. moscheutos* L.—J.

6. H. lasiocarpos Cav. Hairy Rose Mallow. July–Oct. Low, wet so
occasional to common in the s. half of Illinois, rare or absent in mc
of the n. cos.

7. Althaea L.

1. A. rosea (L.) Cav. Hollyhock. July–Oct. Native of Europe and As
occasionally escaped from cultivation.

8. Napaea L. Glade Mallow

1. N. dioica L. Glade Mallow. June–July. Alluvial soil along strea
and rivers; occasional in the n. half of the state, apparently abs
from the s. cos.

9. Anoda Cav.

1. A. cristata (L.) Schlecht. Aug.–Oct. Native of the sw. U.S.; rar
escaped from cultivation; Hancock and Massac cos.

10. Abutilon Mill. Indian Mallow

1. A. theophrastii Medic. Velvet-leaf. Aug.–Oct. Native of India; r
uralized in waste areas; throughout the state.

11. Sida L.

1. Leaves oblong to lance-ovate, all or most of them over 8 mm broad; c
pels 5 per flower .... 1. *S. spine*
1. Leaves linear, rarely over 7 mm broad; carpels 8 or more per flower .
.... 2. *S. ellic*

1. S. spinosa L. Prickly Sida. June–Oct. Native of trop. Am.; natu
ized into waste areas; occasional to common in the s. four-fifths
the state, rare or absent in most of the extreme n. cos.

2. S. elliottii Torr. & Gray. Aug.–Oct. Farm lots, very rare; Alexan
Co.

104. HYPERICACEAE

1. Sepals 4; petals 4 .... 1. *Ascy*
1. Sepals 5; petals 5 ....
2. Petals yellow or orange, convolute in bud; stamens 5 to many, with
glands at the base .... 2. *Hyperic*
2. Petals pinkish or flesh-colored, imbricate in bud; stamens 9, with 3 la
orange glands .... 3. *Triader*

1. *Ascyrum* L. St. Andrew's Cross

1. A. hypericoides L. St. Andrew's Cross. July–Aug. Ascending shrub with linear-oblanceolate leaves 2–4 (–5) mm broad. Dry soil, very rare; Hancock Co.

var. multicaule (Michx.) Fern. St. Andrew's Cross. July–Aug. Sprawling shrub with oblong-oblanceolate leaves 4 mm broad or broader. Dry woods, on slopes and ridges; occasional in the s. tip of Ill., absent elsewhere. *A. hypericoides* L.—G, J.

2. *Hypericum* L. St. Johns-wort

1. Either the margins or the surface of the petals streaked or dotted with black 2
1. Petals not dotted or streaked with black 3
2. Only the margins of the petals streaked with black; seeds rough; stems repeatedly branched .... 1. *H. perforatum*
2. Petals black-dotted over entire surface; seeds smooth or nearly so; stems mostly unbranched .... 2. *H. punctatum*
3. Leaves more than 2 mm in width (if less than 2 mm in width, then the leaves 3-nerved) 4
3. None of the leaves exceeding 2 mm in width 17
4. Petals 8 mm long or longer; stamens 20 or more 5
4. Petals at most only 6 mm long; stamens 5–12 (rarely 20) 13
5. Styles 5 (rarely 4) 6
5. Styles 3, often united to appear as a single beak 8
6. Flowers 4–7 cm across; capsule 2–3 cm long; some of the leaves partly clasping .... 3. *H. pyramidatum*
6. Flowers 1.5–3.0 cm across; capsule up to 1 cm long; leaves not clasping 7
7. Capsule 7–10 mm long; stem with papery whitish bark; sepals 5–15 mm long .... 4. *H. kalmianum*
7. Capsule 3.0–6.5 mm long; stem without papery whitish bark; sepals 2–5 mm long .... 5. *H. lobocarpum*
8. Stems arising from a creeping stoloniferous base 9
8. Plants without a creeping stoloniferous base, although underground rhizomes may be present 10
9. Leaves linear-lanceolate; stamens persistent .... 6. *H. adpressum*
9. Leaves elliptic-oblong; stamens deciduous .... 7. *H. ellipticum*
10. Ovary and capsule 3-locular; stems woody nearly throughout 11
10. Ovary and capsule 1-locular; stems herbaceous or slightly woody only at the base 12
11. Capsule 8–15 mm long; some or all the sepals 5 mm long or longer .... 8. *H. spathulatum*
11. Capsule 3–7 mm long; all of the sepals less than 5 mm long .... 9. *H. densiflorum*
12. Styles free; plants virgate .... 10. *H. denticulatum*
12. Styles united into a beak; plants not virgate .... 11. *H. sphaerocarpum*
13. Bracts foliaceous, resembling the foliage leaves .... 12. *H. boreale*
13. Bracts linear-setaceous, much reduced from the foliage leaves 14
14. Leaves ovate to orbicular 15
14. Leaves linear to lanceolate 16
15. Capsules 3.0–3.5 mm long; plants much branched, not virgate .... 13. *H. mutilum*

15. Capsules 4 mm long or longer; plants sparingly branched, virgate ....
.... 14. *H. gymnanthum*
16. Leaves linear, 1- to 3-nerved; some or all the sepals less than 4 mm long
.... 15. *H. canadense*
16. Leaves lanceolate, 5- to 7-nerved; some or all the sepals 4 mm long or
longer .... 16. *H. majus*
17. Leaves scale-like, at most only 3 mm long; capsule at least twice as long
as the calyx .... 17. *H. gentianoides*
17. Leaves linear, mostly 6–20 mm long; capsule about as long as the calyx
.... 18. *H. drummondii*

1. H. perforatum L. Common St. Johns-wort. June–Sept. Naturalized
from Europe; roadsides, dry pastures, and fields; scattered in all parts
of Ill.

2. H. punctatum Lam. Spotted St. Johns-wort. July–Aug. Petals 5–
mm long; uppermost leaves obtuse at the tip. Woods and roadsides
in every co.

var. pseudomaculatum (Bush) Fern. Large Spotted St. Johns-wort.
May–July. Petals 8–12 mm long; uppermost leaves acute at the tip.
Dry woods, rare; Jackson Co. *H. pseudomaculatum* Bush—G, J.

3. H. pyramidatum Ait. Giant St. Johns-wort. July–Aug. Banks of
rivers and streams; occasional in the n. half of the state; also Macoupin and St. Clair cos. *H. ascyron* L.—J.

4. H. kalmianum L. Kalm's St. Johns-wort. June–Aug. Calcareous
sand and small bogs, locally common; Cook and Lake cos.

5. H. lobocarpum Gattinger. St. Johns-wort. June–Aug. Low woods,
rare; Massac and Pope cos. *H. densiflorum* Pursh var. *lobocarpum*
(Gattinger) Svenson—F, G.

6. H. adpressum Bart. Creeping St. Johns-wort. July–Aug. Wet
ground, rare; Kankakee and Lawrence cos.

7. H. ellipticum Hook. St. Johns-wort. May–Aug. Roadsides, rare;
Fulton and St. Clair cos.

8. H. spathulatum (Spach) Steud. Shrubby St. Johns-wort. July–
Sept. Rocky stream banks and pastures; occasional in the s. half of
the state. *H. prolificum* L.—G, J.

9. H. densiflorum Pursh. St. Johns-wort. July–Sept. Swampy situations, very rare; Alexander Co.

10. H. denticulatum Walt. St. Johns-wort. June–Aug. Moist woods,
gravelly hills, rare; Jackson and Pope cos. (Including var. *recognitum*
Fern. & Schub.)

11. H. sphaerocarpum Michx. Round-fruited St. Johns-wort. June–
Sept. Leaves narrowly oblong or narrowly elliptic, flat, with lateral
nerves. Rocky woods, hill prairies, roadsides; scattered throughout
the entire state.

var. turgidum (Small) Svenson. June–Sept. Leaves linear, revolute,
without any apparent lateral nerves. Dry fields, rare; Richland Co.

12. H. boreal (Britt.) Bickn. Northern St. Johns-wort. July–Sept.
Sandy marshes, very rare; Iroquois Co.

13. H. mutilum L. Dwarf St. Johns-wort. July–Sept. Moist soil; occasional in all parts of Ill. (Including H. *mutilum* L. var. *parviflorum*
[Willd.] Fern.)

14. H. gymnanthum Engelm. & Gray. Small St. Johns-wort. June–
Sept. Moist soil; rare to occasional in the n. three-fourths of the state.

15. H. canadense L. Canadian St. Johns-wort. July–Sept. Moist sand
flats, occasional to rare; n. half of Ill.

16. H. majus (Gray) Britt. St. Johns-wort. July–Aug. Moist ground, not common; n. half of Ill.

17. H. gentianoides (L.) BSP. Pineweed. June–Oct. Sandy soil, often on dry bluffs, occasional; scattered in most parts of Ill.

18. H. drummondii (Grev. & Hook.) Torr. & Gray. Nits-and-lice. July–Sept. Bluffs, fields, and dry wooded slopes, occasional to common; s. half of Ill.

3. Triadenum Raf.

1. Leaves without punctations .... 1. *T. tubulosum*
1. Leaves punctate, at least on the lower surface 2
2. Leaves petiolate .... 2. *T. walteri*
2. Leaves sessile 3
3. Sepals obtuse, up to 5 mm long; styles up to 1.5 mm long . 3. *T. fraseri*
3. Sepals acute, 5–8 mm long; styles 2–3 mm long .... 4. *T. virginicum*

1. T. tubulosum (Walt.) Gl. Marsh St. Johns-wort. Aug.–Sept. Swampy or marshy ground in woods, rare; Alexander, Johnson, Massac, Pope, and Pulaski cos. *Hypericum tubulosum* Walt.—F.

2. T. walteri (Gmel.) Gl. Marsh St. Johns-wort. July–Sept. Wooded swamps, not common; confined to the s. one-sixth of the state. *Hypericum tubulosum* (Walt.) Gl. var. *walteri* (Gmel.) Lott—F.

3. T. fraseri (Spach) Gl. Fraser's St. Johns-wort. July–Sept. Bogs and wooded swamps, rare; confined to the n. one-half of the state. *Hypericum virginicum* L. var. *fraseri* (Spach) Fern.—F.

4. T. virginicum (L.) Raf. Marsh St. Johns-wort. July–Sept. Bogs, very rare; Lake Co.

105. ELATINACEAE

1. Sepals 2–3, obtuse; petals 2–3; plants glabrous .... 1. *Elatine*
1. Sepals 5, acuminate; petals 5; plants pubescent .... 2. *Bergia*

1. Elatine L.   Waterwort

1. E. brachysperma Gray. Waterwort. July–Oct. Shallow water, very rare; Menard and Sangamon cos. *E. triandra* Schk. var. *brachysperma* (Gray) Fassett—G.

2. Bergia L.

1. B. texana (Hook.) Seubert. June–Sept. Wet soil, very rare; St. Clair Co.

106. CISTACEAE

1. At least some of the flowers with 5 yellow petals; capsule 1-locular; style present 2

1. Flowers with 3 reddish petals; capsule partly 3-locular; style short or
none .... 3. *Leachea*
2. Some of the flowers 1.5–2.5 cm broad; style very short; leaves lanceolate
to oblanceolate .... 1. *Helianthemum*
2. None of the flowers as much as 1.5 cm broad; style slender, elongate;
leaves scale-like .... 2. *Hudsonia*

1. Helianthemum Mill. Rockrose

1. Terminal petaliferous flower solitary; seeds papillate .. 1. *H. canadense*
1. Terminal petaliferous flowers two or more; seeds not papillate ....
.... 2. *H. bicknellii*

    1. H. canadense (L.) Michx. Frostweed. June–July. Sandy woods
and prairies, occasional; in the n. half of the state; also St. Clair Co.
    2. H. bicknellii Fern. Frostweed. June–July. Sandy woods, oc-
casional; in the n. half of the state; also Madison and St. Clair cos.

2. Hudsonia L. Hudsonia. False Heather

    1. H. tomentosa Nutt. Beach Heath. May–July. Flowers sessile or on
very short bracteate pedicels. Sand dunes, hills, and blowouts, rare;
Fulton, JoDaviess, and Lee cos.
    var. intermedia Peck. Beach Heath. May–July. Flowers on leafless
pedicels 1.5–6.0 mm long. Sandy soil, rare; JoDaviess Co.

3. Lechea Kalm ex L. Pinweed

1. Outer two sepals as long as or longer than the inner three ....
1. Outer two sepals conspicuously shorter than the inner three ....
2. Stems densely pubescent with spreading hairs; three inner sepals rough
in appearance .... 1. *L. mucronata*
2. Stems sparsely to moderately pubescent with ascending hairs; three
inner sepals not appearing rough ....
3. Cauline leaves lanceolate to oblong; basal leaves elliptic-ovate; capsule
about as long as the calyx .... 2. *L. minor*
3. All leaves linear; capsule shorter than the calyx .... 3. *L. tenuifolia*
4. Leaves pubescent throughout on the lower surface .... 4. *L. stricta*
4. Leaves pubescent only on the veins of the lower surface ....
5. Seeds dark brown, smooth; pedicels mostly shorter than the calyx ...
.... 5. *L. pulchella*
5. Seeds pale brown, partly covered by a grayish membrane; pedicels as
long as or longer than the calyx .... 6. *L. intermedia*

    1. L. mucronata Raf. Hairy Pinweed. July–Sept. Sandy soil and grav-
elly slopes; occasional in the n. half of Ill., rare in the s. half. *L. villosa*
Ell.—F, G, J.
    2. L. minor L. Pinweed. July–Sept. Sandy soil and cliffs; occasional
and sparsely distributed throughout Ill.
    3. L. tenuifolia Michx. Narrow-leaved Pinweed. July–Sept. Leaves
glabrous on the upper surface. Sandy soil and cliffs; occasional to
common throughout the state.
    var. occidentalis Hodgdon. Pinweed. July–Sept. Leaves pilose on
the upper surface. Dry sandy woods, very rare; Peoria Co.

4. L. stricta Leggett. Pinweed. July–Sept. Sandy soil; sparsely scattered in the n. one-third of Ill.

5. L. pulchella Raf. Pinweed. July–Sept. Sandy soil; restricted to the n. half of Ill. *L. leggettii* Britt. & Holl.—F, G, J. (Including *L. leggettii* var. *moniliformis* [Bickn.] Hodgdon.)

6. L. intermedia Leggett. Pinweed. July–Oct. Sandy soil, rare; Cook, Lee, and Winnebago cos.

107. VIOLACEAE

1. Stamens united; sepals attached at the base, not auriculate ....
....— 1. *Hybanthus*
1. Stamens free; sepals attached above the base, auriculate ..... 2. *Viola*

1. Hybanthus Jacq.   Green Violet

1. H. concolor (T. F. Forst.) Spreng. Green Violet. April–June. Rich woods; rather common in the s. cos., becoming rare and scattered northward. *Cubelium concolor* (Forst.) Raf.—G.

2. Viola L.   Violet

1. Plants acaulescent, the leaves and peduncles seeming to arise from out of the ground 2
1. Plants with stems bearing alternate leaves and axillary flowers 25
2. Flowers basically blue 3
2. Flowers basically white 19
3. None of the leaves lobed 4
3. Some or all the leaves hastately or palmately lobed 14
4. Leaves less than half as long as broad .... 10. *V. sagittata*
4. Leaves more than half as long as broad 5
5. Style with downward turned beak .... 18. *V. odorata*
5. Style thickened at tip, not beaked nor turned downward 6
6. Spurred petals beardless within 7
6. Spurred petals bearded within 10
7. Petioles and leaves uniformly and densely villous .... 7. *V. sororia*
7. Petioles and leaves glabrous or only sporadically and sparsely pubescent 8
8. Lateral petals with hairs swollen at the tip; leaf blades somewhat pubescent on upper surface .... 2. *V. cucullata*
8. Lateral petals with hairs slender throughout; leaf blades entirely glabrous 9
9. Leaves cordate-ovate, rounded to short-pointed at the tip ....
.... 3. *V. pratincola*
9. Leaves elongate-triangular, tapering to an elongated tip ....
.... 5. *V. missouriensis*
10. Leaves and petioles glabrous (microscopically pubescent on upper surface of blade in *V. affinis*) 11
10. Leaves and petioles sparsely hirsute to pilose 12
11. Sepals obtuse; peduncles longer than petioles ..... 4. *V. nephrophylla*
11. Sepals acute; peduncles about equalling petioles .... 6. *V. affinis*

12. Leaves distinctly cordate; cleistogamous capsules on horizontal or arching peduncles 13

12. Leaves subcordate to nearly truncate; cleistogamous capsules on erec peduncles . . . . 9. *V. fimbriatula*

13. Petioles villous; leaves dark green, rather thick . . . . 7. *V. sororia*

13. Petioles hirsute, at least near the tip; leaves pale green, thin . . . .
. . . . 8. *V. septentrionalis*

14. All petals beardless . . . . 1. *V. pedata*

14. Lateral petals and sometimes the spurred petal bearded 15

15. Leaves hastately lobed 16

15. Leaves palmately lobed 1

16. Sepals ciliate . . . . 9. *V. fimbriatula*

16. Sepals glabrous . . . . 10. *V. sagittata*

17. Leaves cordate at base . . . . 11. *V. triloba*

17. Leaves cuneate or truncate at base 1

18. Leaves deeply divided into narrow segments; spurred petal villous within
. . . . 13. *V. pedatifida*

18. Leaves divided into broad segments; spurred petal glabrous within . . .
. . . . 12. *V. viarum*

19. Leaves deeply palmately cleft . . . . 1. *V. pedata*

19. Leaves unlobed 2

20. Style with downward turned beak . . . . 18. *V. odorata*

20. Style thickened at tip, not beaked nor turned downward 2

21. Lateral petals bearded 2

21. Lateral petals beardless 2

22. Peduncles and leaves glabrous; seeds brown . . . . 3. *V. pratincola*

22. Peduncles and leaves pubescent; seeds buff . . . . 15. *V. incognita*

23. Leaves cordate at base . . . . 14. *V. macloskeyi* ssp. *pallens*

23. Leaves tapering to base 2

24. Leaves linear, lanceolate, or oblanceolate . . . . 16. *V. lanceolata*

24. Leaves oblong to ovate . . . . 17. *V. primulifolia*

25. Stipules not foliaceous; tip of style slender or only slightly swollen . 2

25. Stipules foliaceous; tip of style enlarged into a hollow, globose tip . 3

26. Flowers blue . . . . 22. *V. conspersa*

26. Flowers white or yellow 2

27. Flowers white 2

27. Flowers yellow 2

28. Stipules toothed . . . . 21. *V. striata*

28. Stipules entire . . . . 20. *V. canadensis* var. *rugulosa*

29. Leaves and at least the upper half of the stem densely pubescent . . . .
. . . . 19. *V. pubescens*

29. Leaves and stems glabrous or sparsely pubescent . . . .
. . . . 19. *V. pubescens* var. *eriocarpa*

30. Base of stipules pinnately divided; upper leaves distinctly toothed . . 3

30. Base of stipules palmately divided; upper leaves entire or nearly so . .
. . . . 25. *V. rafinesquii*

31. Petals and sepals subequal, the petals pale yellow . . . . . . 24. *V. arvensis*

31. Petals 2–3 times as long as the sepals, variously colored . . . .
. . . . 23. *V.* X *wittrockiana*

1. *V. pedata* L. Birdfoot Violet. April–June. Petals lilac or violet
Cherty slopes, borders of woods; occasional throughout Ill. (Including var. *lineariloba* DC.)

f. alba Thurb. April–June. Petals basically white. Dry woods; Hardin
Co. *V. pedata* L. var. *lineariloba* DC. f *alba* (Thurb.) Britt.—F.

2. V. cucullata Ait. Marsh Blue Violet. April–June. Moist soil in woods and marshes; occasional in the n. half of Ill., apparently rare in the s. half.

3. V. pratincola Greene. Common Blue Violet. March–May. Flowers basically blue or purple. Woods, common; known from every Ill. co. *V. papilionacea* Pursh—F, G, J.

f. albiflora (Glover) Mohl. Confederate Violet. March–May. Flowers basically white, with purplish veins. Woods, occasional; scattered in Ill. *V. priceana* Pollard—J; *V. papilionacea* var. *priceana* (Pollard) Alexander—G; *V. papilionacea* f. *albiflora* Glover—F.

4. V. nephrophylla Greene. Northern Blue Violet. May–June. Usually moist soil; seemingly confined to ne. Ill.

5. V. missouriensis Greene. Missouri Violet. March–May. Low woods, occasional; scattered throughout the state.

6. V. affinis LeConte. Woodland Blue Violet. April–June. Low woods, occasional; scattered throughout Ill.

7. V. sororia Willd. Woolly Blue Violet. March–May. Woods, common; in every co.

8. V. septentrionalis Greene. Northern Blue Violet. May–June. Woods, rare; confined to the northernmost tier of cos.

9. V. fimbriatula Smith. Sand Violet. April–May. Dry woods, sandy soil, occasional; apparently restricted to the n. half of the state; also Richland Co.

10. V. sagittata Ait. Arrow-leaved Violet. March–June. Moist or dry woods, occasional; throughout the state.

11. V. triloba Schwein. Cleft Violet. April–June. Leaves with 3-5 usually shallow lobes. Woods and wooded slopes; occasional in the s. half of Ill., becoming rarer northward. *V. falcata* Greene—J.

var. dilatata (Ell.) Brainerd. Cleft Violet. April–June. Leaves with 5–7 very deep lobes. Woods; occasional in the s. two-thirds of Ill. *V. falcata* Greene—J.

12. V. viarum Pollard. May. Dry, gravelly hilltops, very rare; Kankakee and Peoria cos.

13. V. pedatifida G. Don. Prairie Violet. May–June. Prairies and dry woods; occasional in the n. half of Ill., becoming rare southward.

14. V. macloskeyi Lloyd ssp. pallens (Banks) M. S. Baker. Smooth White Violet. April–May. Moist soil, rare; confined to the n. one-eighth of Ill. (Including *V. blanda* Willd.) *V. pallens* (Banks) Brainerd—F, G, J.

15. V. incognita Brainerd. White Violet. May–June. Wooded slopes, rare; Cook, JoDaviess, and McHenry cos. (Including var. *forbesii* Brainerd.)

16. V. lanceolata L. Lance-leaved Violet. May–June. Leaf blades lanceolate, 3½ to 5 times as long as broad. Wet soil, occasional; confined to the n. half of the state.

var. vittata (Greene) Russell. Narrow-leaved Violet. May–June. Leaf blades linear, 6–15 times as long as broad. Moist sandy soil, rare; Kankakee Co. *V. vittata* Greene—G.

17. V. primulifolia L. Primrose Violet. May–June. Moist soil, rare; Kankakee Co.

18. V. odorata L. Sweet Violet. April–May. Introduced from Europe and occasionally escaped from cultivation; scattered throughout Ill.

19. V. pubescens Ait. Downy Yellow Violet. April–May. Leaves and at least the upper half of the stem densely pubescent; basal leaves 1–2. Rich woods, rare; confined to the n. half of the state. (Including var. *peckii* House, with glabrous capsules.)

var. eriocarpa (Schwein.) Russell. Smooth Yellow Violet. April–May
Leaves and stems glabrous or puberulent; basal leaves 5 or more
Woods, common; in every Ill. co. *V. eriocarpa* Schw.—G, J; *V. penn*
sylvanica Michx.—F. (Including *V. pennsylvanica* var. *leiocarpa* [Ferr
& Wieg.] Fern., with glabrous capsules.)

20. V. canadensis L. var. rugulosa (Greene) C. L. Hitchcock. Whit
Violet. April–June. Rich woods, rare; Cook, JoDaviess, Lake, and Wir
nebago cos. *V. rugulosa* Greene—F, G, J.

21. V. striata Ait. Cream Violet. April–June. Moist soil in wood
meadows, and fields, common in the s. half of Ill., becoming rare i
the n. half.

22. V. conspersa Reichenb. Dog Violet. April–June. Moist wood
rare; Cook, Lake, and Richland cos.

23. V. X wittrockiana Gams. Pansy. May–Aug. Native of Europe ar
occasionally escaped from cultivation in a few areas. *V. tricolor* L.—
G, J.

24. V. arvensis Murr. Wild Pansy. April–Aug. Naturalized fro
Europe; rarely found as an adventive in Ill.

25. V. rafinesquii Greene. Johnny-jump-up. March–May. Naturalize
from Europe and Asia; commonly escaped into waste areas in the
three-fifths of Ill.; nearly absent in the n. cos. *V. kitaibeliana* Roem.
Schultes var. *rafinesquii* (Greene) Fern.—F.

108. PASSIFLORACEAE

1. *Passiflora* L. Passion-flower

1. Leaves shallowly 3-lobed, the lobes obtuse, entire; flowers greenis
   yellow, up to 2.5 cm across; fruit purple .... 1. *P. lute*
1. Leaves deeply 3- to 5-lobed, the lobes acute to acuminate, serrulat
   flowers usually purple and white, 4 cm or more across; fruit yellow ..
   .... 2. *P. incarna*

1. P. lutea L. var. glabriflora Fern. Small Passion-flower. May–Sep
Woods and thickets; occasional to common in the s. half of the stat
rare or absent elsewhere. *P. lutea* L.—J.

2. P. incarnata L. Large Passion-flower; Maypops. May–Sept. D
soil, often in fields; occasional in the s. one-third of the state.

109. LOASACEAE

1. *Mentzelia* L.

1. Petals 5, 1.0–1.5 cm long; stamens about 20; capsules up to 1 cm lor
   containing about 9 seeds .... 1. *M. oligosperr*
1. Petals 10, 2–5 cm long; stamens numerous; capsule 2.5–4.0 cm lor
   containing several seeds ....
2. Petals 2–3 cm long; outer filaments dilated; capsule about 3 cm long
   .... 2. *M. nu*

2. Petals 4–5 cm long; all filaments filiform; capsule about 4 cm long . . . . .
. . . . 3. *M. decapetala*

    1. M. oligosperma Nutt. Stickleaf. June–July. Limestone ledges, rare; confined to cos. bordering the Mississippi River.
    2. M. nuda Torr. & Gray. July–Aug. Native of the w. U.S.; adventive along a railroad; Cook Co.
    3. M. decapetala (Pursh) Urban & Gilg. July–Sept. Native of the w. U.S.; adventive along a railroad; Grundy Co.

110. CACTACEAE

1. *Opuntia* Mill. Prickly-pear Cactus

1. All roots fibrous . . . . 1. *O. compressa*
1. Some of the roots tuberous thickened . . . . 2. *O. macrorhiza*

    1. O. compressa (Salisb.) Macbr. Prickly-pear. May–July. Sandy soil, bluff-tops; occasional throughout the state, except for the e. cent. cos. where it is rare or absent. *O. humifusa* Raf.—F; *O. rafinesquii* Engelm.—J.
    2 O. macrorhiza Engelm. Prickly-pear. June–July. Sand prairie, very rare; Whiteside Co.

111. THYMELAEACEAE

1. *Dirca* L. Leatherwood

    1. D. palustris L. Leatherwood. April–May. Rich, shaded woods; uncommon or rare throughout the state.

112. ELAEAGNACEAE

. Leaves opposite; stamens 8; flowers unisexual . . . . 1. *Shepherdia*
. Leaves alternate; stamens 4; flowers bisexual . . . . 2. *Elaeagnus*

1. *Shepherdia* Nutt.

    1. S. canadensis (L.) Nutt. Canadian Buffalo-berry. May–July. Shores of Lake Michigan, rare; Cook and Lake cos.

2. *Elaeagnus* L. Oleaster

. Fruit yellow; branchlets and leaves with silvery scales 1. *E. angustifolia*
. Fruit pink or red; branchlets and leaves with brown and silvery scales . .
. . . . 2. *E. umbellata*

1. E. angustifolia L. Russian Olive. May–June. Native of Europe an
Asia; rarely escaped from cultivation.

2. E. umbellata Thunb. May–June. Native of e. Asia; rarely escape
from cultivation; Williamson Co.

113. LYTHRACEAE

1. Plants more or less shrubby; leaves, or some of them, whorled ......
.... 1. *Decode*
1. Plants herbaceous; leaves opposite, or the uppermost alternate (rare
but sometimes whorled in *Lythrum salicaria*) ....
2. Plants viscid-pubescent; petals 6; calyx tube spurred at base   2. *Cuphe*
2. Plants glabrous or variously pubescent, but not viscid ....
3. Calyx with 5, 6, or 7 lobes or teeth, the tube cylindric ...... 3. *Lythru*
3. Calyx with 4 lobes or teeth, the tube hemispheric ....
4. Plants usually in water; petals absent; leaves up to 3 mm broad 4. *Pep*
4. Plants usually not in water; petals 4; leaves over 3 mm broad ......
5. Flower usually one per leaf axil; leaves tapering to a short petiole or su
sessile base; capsule dehiscing septicidally .... 5. *Rota*
5. Flowers often 2 or more per leaf axil; leaves auriculate at sessile bas
capsule dehiscing irregularly .... 6. *Ammanr*

1. *Decodon* J. F. Gmel.

1. D. verticillatus (L.) Ell. Swamp Loosestrife. July–Aug. Swamp
marshes, bogs; uncommon but scattered throughout the state. (
cluding var. *laevigatus* Torr. & Gray.)

2. *Cuphea* P. Br.

1. C. petiolata (L.) Koehne. Clammy Cuphea. July–Oct. Usually c
soil; occasional in the s. three-fourths of the state, rare or abse
elsewhere.

3. *Lythrum* L.   Loosestrife

1. Flower solitary, axillary, up to 1.2 cm broad; stamens 4 ... 1. *L. alatu*
1. Flowers in terminal spikes, usually at least 1.2 cm broad; stamens 8 (–
.... 2. *L. salica*

1. L. alatum Pursh. Winged Loosestrife. June–Aug. Moist grour
occasional to common throughout the state.
2. L. salicaria L. Purple Loosestrife. June–Aug. Native of Euro
occasionally escaped into moist areas. (Including var. *tomentos*
[Mill.] DC.)

4. *Peplis* L.

1. P. diandra Nutt. Water Purslane. June–Aug. Shallow water;
common but scattered in the state.

5. *Rotala L.*

1. R. ramosior (L.) Koehne. Tooth-cup. July–Sept. Wet ground; occasional throughout the state.

6. *Ammannia L.*

1. Flowers sessile .... 1. *A. coccinea*
1. Flowers pedicellate .... 2. *A. auriculata*

1. A. coccinea Rottb. Long-leaved Ammannia. July–Aug. Wet soil; occasional throughout the state, except for the extreme n. cos.
2. A. auriculata Willd. Scarlet Loosestrife. May–June. Wet soil, rare; Cass and Jackson cos.

114. NYSSACEAE

1. *Nyssa L. Tupelo*

1. Most or all petioles 3 cm long or longer; pistillate flowers borne singly; fruit 2–3 cm long .... 1. *N. aquatica*
1. None of the petioles over 2.5 cm long; pistillate flowers 2 or more in a cluster; fruit up to 1.5 cm long .... 2. *N. sylvatica*

1. N. aquatica L. Tupelo Gum. April–May. Swamps, not common; in the s. one-sixth of the state; also Crawford, Richland, and Wabash cos.
2. N. sylvatica Marsh. Sour Gum. April–June. Leaves obtuse or acute; lower surface of leaves not papillose. Bogs and swamps, not common; in the extreme s. tip of the state; also Cook and Kankakee cos.

var. caroliniana (Poir.) Fern. Sour Gum. April–June. Leaves acuminate; lower surface of leaves papillose. Mostly upland woods, occasional; confined to the s. one-third of the state.

115. MELASTOMACEAE

1. *Rhexia L. Meadow Beauty*

1. Stems 4-angled; plants with tubers; leaves rounded at the sessile base; neck of hypanthium shorter than the body at maturity ... 1. *R. virginica*
1. Stems cylindrical; plants without tubers; leaves narrowed to a short-petiolate base; neck of hypanthium longer than the body at maturity .....
.... 2. *R. mariana*

1. R. virginica L. Meadow Beauty. July–Sept. Sandy soils; not common but scattered throughout the state.
2. R. mariana L. Meadow Beauty. June–Sept. Sandy fields, rare; Mason, Massac, and Pope cos.

116. ONAGRACEAE

1. Calyx 2-lobed; petals 2; stamens 2 .... 1. *Circae*
1. Calyx 4-lobed; petals 4 (or absent); stamens 4–12 ....
2. Leaves opposite ....
2. Leaves alternate or basal ....
3. Leaves entire .... 2. *Ludwigi*
3. Leaves serrate .... 3. *Epilobiur*
4. Petals yellow or greenish ....
4. Petals white or pink ....
5. Calyx divided to the tip of the ovary, persistent on the fruit ....
5. Calyx not divided to the tip of the ovary, deciduous from the fruit .....
.... 5. *Oenother*
6. Stamens 8–12 .... 4. *Jussiae*
6. Stamens 4 .... 2. *Ludwigi*
7. Petals at least 1 cm long ....
7. Petals less than 1 cm long ....
8. Calyx tube prolonged beyond the ovary; seeds without a coma ......
.... 5. *Oenothe*
8. Calyx tube not prolonged beyond the ovary; seeds with a coma ......
.... 3. *Epilobiu*
9. Fruit indehiscent, with 1–4 seeds; seeds without a coma ..... 6. *Gau*
9. Fruit dehiscent, with several seeds; seeds usually with a coma ......
.... 3. *Epilobiu*

1. *Circaea* L.   Enchanter's Nightshade

1. Stems firm, often 40 cm tall or taller; calyx lobes (1.8–) 2.0–2.5 mm lon
bristles of fruit stiff; fruit 2-locular .... 1. *C. quadrisulca*
1. Stems weak, up to 30 cm tall; calyx lobes up to 1.2 mm long; bristles
fruit weak; fruit 1-locular .... 2. *C. alpir*

    1. C. quadrisulcata (Maxim.) Franch. & Sav. var. canadensis (L
Hara. Enchanter's Nightshade. June–Aug. Moist woods; comme
throughout the state. *C. latifolia* Hill—J.
    2. C. alpina L. Small Enchanter's Nightshade. June–July. Cool r
vines, very rare; JoDaviess, Kane, and Lake cos.

2. *Ludwigia* L.   False Loosestrife

1. Leaves opposite; stems more or less trailing or creeping   1. *L. palustr*
1. Leaves alternate; stems erect ....
2. Flowers sessile, greenish, to 3 mm broad ....
2. Flowers short-pedicellate, yellow, over 1 cm across ... 4. *L. alternifo*
3. Capsule up to 5 mm high, about twice the length of the calyx lobe
bracts at base of calyx 3–4 mm long .... 2. *L. polycar*
3. Capsule 6–10 mm high, about four times the length of the calyx lobe
bracts at base of calyx absent or up to 1 mm long .... 3. *L. glandulo*

    1. L. palustris (L.) Ell. var. americana (DC.) Fern. & Grisc. Mar
Purslane. July–Aug. Wet soil, particularly around ponds, alo
streams, and in ditches; common throughout the state. *L. palust*
(L.) Ell.–J.

2. L. polycarpa Short & Peter. False Loosestrife. June–Sept. Wet soil, particularly in ditches; occasional throughout the state.

3. L. glandulosa Walt. False Loosestrife. June–Sept. Swamps, low woods, not common; confined to the s. one-sixth of the state.

4. L. alternifolia L. Seedbox. June–Aug. Stems, leaves, and calyx glabrous. Wet ground; occasional throughout the state.

var. pubescens Palmer & Steyerm. Seedbox. June–Aug. Stems, leaves, and calyx pubescent. Wet ground; occasional in the southernmost cos.

3. Epilobium L. Willow Herb

1. Calyx tube not prolonged beyond ovary; petals not notched .... 1. *E. angustifolium*
1. Calyx tube prolonged beyond ovary; petals notched 2
2. Petals at least 1 cm long; stems softly hirsute; stigma deeply 4-lobed; seeds smooth .... 2. *E. hirsutum*
2. Petals up to 1 cm long; stems variously pubescent but not softly hirsute; stigma entire or slightly notched; seeds papillose 3
3. Leaves entire 4
3. Leaves serrulate or denticulate 5
4. Stems and leaves velvety-pubescent; most or all the leaves more than 4 mm wide .... 3. *E. strictum*
4. Stems and leaves merely canescent; most or all the leaves less than 4 mm wide .... 4. *E. leptophyllum*
5. Inflorescence and capsules glandular-pubescent; seeds short-beaked; coma white .... 5. *E. adenocaulon*
5. Inflorescence and capsules not glandular; seeds beakless; coma reddish-brown .... 6. *E. coloratum*

1. E. angustifolium L. Fireweed. June–Aug. In dunes and bogs, after a fire; confined to the n. one-fifth of the state.

2. E. hirsutum L. Hairy Willow Herb. Aug. Native of Europe; adventive in springy habitats and borders of ditches; Cook, DuPage, and Lake cos.

3. E. strictum Muhl. Downy Willow Herb. July–Sept. Boggy and springy habitats, not common; confined to the n. one-third of the state.

4. E. leptophyllum Raf. Bog Willow Herb. Aug.–Sept. Boggy and springy habitats, not common; confined to the n. one-half of the state.

5. E. adenocaulon Haussk. Northern Willow Herb. July–Sept. Bogs and calcareous springy areas; occasional in the n. one-fourth of the state. *E. glandulosum* Lehm. var. *adenocaulon* (Haussk.) Fern.—F.

6. E. coloratum Muhl. Cinnamon Willow Herb. Aug.–Sept. Bogs, marshes, wet ground; occasional throughout the state.

4. Jussiaea L. Primrose Willow

. Stems and leaves glabrous 2
. Stems and leaves pubescent .... 3. *J. leptocarpa*
. Plants creeping; petals 5; capsule cylindrical .... 1. *J. repens*
. Plants more or less erect; petals 4; capsule 4-sided .... 2. *J. decurrens*

1. J. repens L. var. glabrescens Ktze. Creeping Primrose Willo
May–Oct. In shallow water or on muddy banks and shores; occasior
to common in the s. half of the state. *J. repens* L.—G, J. (Including
diffusa Forsk.)

2. J. decurrens (Walt.) DC. Erect Primrose Willow. July–Sept. W
ground, rare; confined to the s. one-sixth of the state.

3. J. leptocarpa Nutt. Hairy Primrose Willow. July–Sept. W
ground, particularly along the Mississippi River; confined to the
one-fifth of the state.

5. *Oenothera* L. Evening Primrose

1. Petals white or rosy ....
1. Petals yellow (sometimes fading reddish) ....
2. Stems with white peeling epidermis; calyx tube 2–4 cm long; capsule
   ribbed, 2.5–3.5 cm long .... 1. *O. nutta*
2. Stems without white peeling epidermis; calyx tube 1–2 cm long; caps
   4-winged and 8-ribbed, to 2 cm long .... 2. *O. specic*
3. Leaves linear or broader, never filiform, more than 1 mm broad .... 13. *O. linifc*
3. Leaves filiform, up to 1 mm broad .... 13. *O. linifc*
4. Some or all the petals 4 cm long or longer; capsule at least 5 cm lc
   .... 3. *O. missourien*
4. None of the petals 4 cm long; capsule up to 4 cm long ....
5. Some or all the leaves sinuate-pinnatifid .... 4. *O. lacini*
5. None of the leaves sinuate-pinnatifid ....
6. Ovary and capsule sharply 4-angled ....
6. Ovary and capsule more or less terete ....
7. Petals up to 1 cm long; flower buds nodding; anthers to 2.5 mm long
   .... 5. *O. peren*
7. Petals at least 1 cm long; flower buds erect; anthers 4 mm long
   longer ....
8. Capsule glandular-pubescent .... 6. *O. tetrag*
8. Capsule eglandular ....
9. Stems with appressed hairs; calyx tube 1.5–2.5 cm long ....
   .... 7. *O. pilos*
9. Stems with appressed hairs; calyx tube 0.5–1.5 cm long 8. *O. frutic*
10. Stigma deeply 4-lobed; main stem leaves mostly more than 5 mm br
    ....
10. Stigma shallowly lobed or entire; main stem leaves up to 5 mm broa
    .... 12. *O. serru*
11. Capsules at least 4 mm thick at base; most of the main stem leaves
    mm broad or broader ....
11. Capsules up to 3 mm thick at base; most of the main stem leaves u
    10 mm broad .... 11. *O. rhombipe*
12. Petals 3.5 mm wide or wider .... 9. *O. bier*
12. Petals up to 3 mm wide .... 10. *O. cruc*

1. O. nuttallii Sweet. White Evening Primrose. June–July. Nativ
w. U. S.; rarely adventive in Illinois. *O. albicaulis* Nutt.—J.

2. O. speciosa Nutt. Showy Evening Primrose. June–July. Nativ
w. U.S.; occasionally adventive along roadsides.

3. O. missouriensis Sims. Missouri Primrose. May–Aug. Rc
glades, very rare; St. Clair Co.

4. O. laciniata Hill. Ragged Evening Primrose. May–July. Fields, prairies, roadsides; occasional throughout the state.

5. O. perennis L. Small Sundrops. June–Aug. Moist, open ground; Cook, Lake, Will, and Winnebago cos.

6. O. tetragona Roth. Four-angled Sundrops. June–Aug. Dry woods, rare; confined to the s. one-sixth of the state.

7. O. pilosella Raf. Prairie Sundrops. May–July. Prairies, fields; occasional throughout the state.

8. O. fruticosa L. var. linearis (Michx.) S. Wats. Shrubby Sundrops. May–Aug. Rocky woods, rare; confined to the s. one-sixth of the state.

9. O. biennis L. Evening Primrose. June–Oct. Plants variously pubescent, but not white-hairy. Fields, prairies, waste ground, common; in every co. (Including var. *pycnocarpa* [Atkinson & Bartlett] Wieg.)

var. canescens Torr. & Gray. Gray Evening Primrose. June–Oct. Plants white-hairy. Fields, prairies, waste ground; occasional throughout the state. *O. strigosa* (Rydb.) Mack. & Bush—F. (Including var. *hirsutissima* Gray.)

10. O. cruciata Nutt. Evening Primrose. July–Aug. Native in the N.E. states; adventive in Ill. in dry soil; Boone, Cook, and Kane cos. *O. parviflora* L.—G.

11. O. rhombipetala Nutt. Sand Primrose. June–Sept. Disturbed sandy soils; occasional throughout the state.

12. O. serrulata Nutt. Toothed Evening Primrose. May–Sept. Native of w. U.S.; rarely adventive in waste ground; Peoria and Randolph cos.

13. O. linifolia Nutt. Thread-leaved Sundrops. May–July. Sandstone cliffs, prairies; occasional in the s. one-fifth of the state.

6. Gaura L.

- Ovary and capsule sessile 2
- Ovary and capsule stipitate . . . . 4. *G. filipes*
- Petals 5.5–15.0 mm long; calyx lobes 6–13 mm long. 3
- Petals 1.5–3.0 mm long; calyx lobes 2.0–3.5 mm long . . 3. *G. parviflora*
- Stems densely villous; about 50 percent of the pollen grains fertile . . . .
. . . . 1. *G. biennis*
- Stems strigulose to short-villous; about 90 percent of the pollen grains fertile . . . . 2. *G. longiflora*

1. G. biennis L. Butterfly-weed. July–Sept. Clearings in woodland and waste places; occasional in the n. two-thirds of the state, apparently absent elsewhere.

2. G. longiflora Spach. July–Oct. Waste places, often along roadsides and railway embankments; occasional throughout the state. (Including *G. biennis* L. var. *pitcheri* Torr. & Gray.)

3. G. parviflora Dougl. Small-flowered Gaura. June–July. Native of w. U.S.; adventive in fields, roadsides, edge of woods; occasional in the n. half of the state, less common elsewhere.

4. G. filipes Spach. Slender Gaura. July–Sept. Dry woods, rare; Hardin Co. (Including var. *major* Torr. & Gray.)

117. HALORAGIDACEAE

1. Leaves whorled or scattered, crowded; flowers 4-merous; fruit 4-angle
.... 1. *Myriophyllu*
1. Leaves alternate, rather remote; flowers 3-merous; fruit 3-angled .....
.... 2. *Proserpinac*

1. *Myriophyllum* L.   Water Milfoil

1. Leaves verticillate and scattered; fruit with 4 dorsal tuberculate or und
late ridges ....
1. Leaves all verticillate; fruit without dorsal ridges or with smooth dors
ridges ....
2. Bracts entire to pectinate; each section of the 4-angled fruit flattene
slightly papillose with a smooth dorsal ridge with undulate margins ..
.... 1. *M. hippuroid*
2. Bracts deeply toothed; each section of the 4-angled fruit rounde
smooth, with a tuberculate dorsal ridge .... 2. *M. pinnatu*
3. Bracts deeply pinnatifid; stamens 8 .... 3. *M. verticillatu*
3. Bracts entire to serrate to dentate; stamens 4 ....
4. Bracts much exceeding the flower; fruit with smooth dorsal ridges ...
.... 4. *M. heterophyllu*
4. Bracts not exceeding the flower; fruit slightly tuberculate, without dors
ridges .... 5. *M. exalbesce*

    1. M. hippuroides Nutt. Mare's-tail Milfoil. June–Oct. Shallow wat
very rare; Cook Co. *M. hippuroides* Nutt. var. *cheneyi* Fassett—G.
    2. M. pinnatum (Walt.) BSP. Rough Water Milfoil. June–Oct. Root
in muddy shores or in shallow waters of ponds and lakes; occasior
throughout the state.
    3. M. verticillatum L. var. pectinatum Wallr. Whorled Water Milf
June–Sept. Shallow water, not common; scattered throughout t
state. *M. verticillatum* L.—J.
    4. M. heterophyllum Michx. Water Milfoil. June–Sept. Quiet wate
occasional throughout the state.
    5. M. exalbescens Fern. Spiked Water Milfoil. July–Sept. Quiet w
ters; occasional in the ne. cos., rare elsewhere.

2. *Proserpinaca* L.   Mermaid-weed

    1. P. palustris L. Mermaid-weed. July–Oct. Shallow water a
shores; occasional throughout the state. (Including var. *crebra* Fe
& Grisc.)

118. HIPPURIDACEAE

1. *Hippuris* L.   Mare's-tail

    1. H. vulgaris L. Mare's-tail. June–Sept. Shores or shallow wat
very rare; Kane, Lake, and McHenry cos.

119. ARALIACEAE

1. Leaves simple, palmately lobed, evergreen .... 1. *Hedera*
1. Leaves compound 2
2. Leaves alternate or basal, the ultimate divisions pinnately arranged ....
.... 2. *Aralia*
2. Leaves whorled, the leaflets palmately arranged .... 3. *Panax*

1. Hedera L.   Ivy

1. H. helix L. English Ivy. June–Sept. Native of Europe; commonly planted but rarely escaping from cultivation; Jackson Co.

2. Aralia L.   Spikenard

1. Shrub or small tree; parts of stem and leaves coarsely prickly ....
.... 1. *A. spinosa*
1. Herbs; leaves and stems without prickles, but sometimes bristly 2
2. Stems bristly .... 2. *A. hispida*
2. Stems without bristles 3
3. Leaflets 3–7 per leaf; peduncle appearing to arise directly from the ground .... 3. *A. nudicaulis*
3. Leaflets 9–21 per leaf; peduncle arising from the stem . 4. *A. racemosa*

1. A. spinosa L. Hercules' Club; Devil's Walking-stick. July–Sept. Wooded slopes, occasional; confined to the s. one-third of the state.
2. A. hispida Vent. Bristly Sarsaparilla. June–Aug. Sandy soil, rare; Cook and Lake cos.
3. A. nudicaulis L. Wild Sarsaparilla. May–June. Rich wooded slopes, dune woods, rocky woods, bogs; occasional in the n. one-half of the state; also Macoupin Co.
4. A. racemosa L. American Spikenard. June–Aug. Rich woods, rocky woods; occasional throughout the state.

3. Panax L.   Ginseng

1. P. quinquefolius L. Ginseng. June–July. Rich woods, rocky woods; occasional throughout the state.

120. UMBELLIFERAE

. All leaves simple 2
. Some or all the leaves compound 5
. Leaves spiny along the margins, mostly linear to lanceolate 1. *Eryngium*
. Leaves not spiny along the margins, mostly ovate or oblong 3
. Leaves sessile, perfoliate .... 2. *Bupleurum*
. Leaves petiolate, not perfoliate 4
. Plants prostrate; leaves entire or lobulate; flowers bluish, borne in heads
.... 1. *Eryngium*
. Plants erect; leaves crenate; flowers yellow or purplish, borne in umbels
.... 3. *Thaspium*

  5. Ovaries and fruits prickly . . . .
  5. Ovaries and fruits glabrous, pubescent, or warty, not prickly . . . .
  6. Stems glabrous; leaves palmately divided . . . . 4. *Sanicul*
  6. Stems pubescent; leaves pinnately divided . . . .
  7. Bracts of the primary involucre linear, entire . . . . 5. *Toril*
  7. Bracts of the primary involucre pinnately divided . . . . 6. *Daucu*
  8. All leaves once-divided (basal ones sometimes undivided) . . . .
  8. Leaves, or most of them, doubly compound 2
  9. Leaflets entire 1
  9. Leaflets toothed 1
 10. Leaves palmately compound . . . . 7. *Cynosciadiu*
 10. Leaves pinnately compound 1
 11. Leaflets capillary, up to 1 mm broad . . . . 8. *Ptilimniu*
 11. Leaflets lanceolate to linear, more than 1 mm broad 1
 12. Leaflets cross-septate; fruits up to 4 mm long . . . . . . 9. *Limnosciadiu*
 12. Leaflets not cross-septate; fruits 4.5–6.0 mm long . . . . 10. *Oxypo*
 13. Stems pubescent . . . . 11. *Heracleu*
 13. Stems glabrous . . . .
 14. All cauline leaves ternate, with 3 leaflets . . . .
 14. Some of the cauline leaves with more than 3 leaflets . . . .
 15. Flowers white . . . .
 15. Flowers purple or yellow . . . .
 16. Leaflets doubly toothed; fruits cylindrical . . . . 12. *Cryptotaen*
 16. Leaflets singly toothed; fruits flat . . . . 13. *Falcar*
 17. Fruit winged; central flower of the ultimate umbel pedicellate . . . .
 . . . . 3. *Thaspiu*
 17. Fruit unwinged; central flower of the ultimate umbel sessile . . 14. *Ziz*
 18. Leaflets palmately arranged . . . . 13. *Falcar*
 18. Leaflets pinnately arranged . . . .
 19. Leaflets sparingly toothed, with no more than 8 teeth . . . . 10. *Oxypo*
 19. Leaflets many-toothed . . . .
 20. Flowers yellow; plants of fields, roadsides, and waste ground . . . .
 . . . . 15. *Pastina*
 20. Flowers white; plants of swamps and wet woods . . . .
 21. Fruit 2.5–3.0 mm long, with prominent ribs; leaflets serrate . . . 16. *Siu*
 21. Fruit about 2 mm long, with inconspicuous ribs; leaflets sometimes
     ciniately toothed . . . . 17. *Ber*
 22. Bulblets borne in the axils of the upper leaves . . . . 29. *Cicu*
 22. Bulblets absent . . . .
 23. Stems pubescent only at the nodes . . . .
 23. Stems glabrous or pubescent, but the hairs not confined to the nod
 . . . .
 24. Flowers yellow . . . . 3. *Thaspiu*
 24. Flowers white . . . . 18. *Osmorh*
 25. Ultimate leaf segments capillary, linear, or narrowly elliptic, usually le
     than 3.5 mm broad . . . .
 25. Ultimate leaf segments broadly elliptic or broader, usually more than
     mm broad . . . .
 26. Flowers yellow . . . .
 26. Flowers white . . . .
 27. Sheaths of the petioles of the main cauline leaves at least 3 cm long
 . . . . 19. *Foenicul*
 27. Sheaths of the petioles of the main cauline leaves up to 3 cm lo
 . . . . 20. *Aneth*

8. Leaves all basal or nearly so .... 21. *Erigenia*
8. Leaves cauline 29
9. Bracts of the primary involucre numerous, often pinnately divided .. 30
9. Bracts of the primary involucre absent or few, not pinnately divided   31
0. Two finely divided stipules inserted beneath the sheathing petioles ....
.... 22. *Carum*
0. Stipules absent .... 8. *Ptilimnium*
1. Flowers, or some of them, zygomorphic .... 23. *Coriandrum*
1. Flowers actinomorphic 32
2. Ovaries and fruits warty .... 24. *Spermolepis*
2. Ovaries and fruits not warty 33
3. Umbel with 1–6 rays; involucral bracts elliptic to oblong ....
.... 25. *Chaerophyllum*
3. Umbel with at least 7 rays; involucral bracts linear or absent 34
4. Ultimate leaflets up to 1.5 mm broad; all leaves pinnately divided .....
.... 22. *Carum*
4. Ultimate leaflets 1.5–3.0 mm broad; at least the upper leaves ternately
divided .... 26. *Perideridia*
5. Flowers yellow or purple 36
5. Flowers white 39
6. Leaflets entire .... 27. *Taenidia*
6. Leaflets toothed or lobed 37
7. Leaflets of cauline leaves cuneate at base, rarely more than 10 mm
broad .... 28. *Polytaenia*
7. Leaflets of cauline leaves more or less rounded at base, usually more
than 10 mm broad 38
8. Fruit winged; central flower of the ultimate umbel pedicellate ....
.... 3. *Thaspium*
8. Fruit unwinged; central flower of the ultimate umbel sessile .. 14. *Zizia*
9. Leaves all or nearly all basal .... 21. *Erigenia*
9. Leaves borne along the stem 40
0. Fruits twice as long as broad or longer, at least 1 cm long ....
.... 18. *Osmorhiza*
0. Fruits never twice as long as broad, up to 1 cm long 41
1. Major leaflets toothed, but not divided into narrow segments 42
1. Major leaflets dissected into additional segments 44
2. Uppermost leaf sheaths longer than the blades .... 30. *Angelica*
2. Uppermost leaf sheaths shorter than the blades 43
3. Basal leaves with 9 leaflets; petals incurved at tip .... 31. *Aegopodium*
3. Basal leaves usually with more than 9 leaflets; petals not incurved at tip
.... 29. *Cicuta*
4. Fruits winged; central umbel barely or not overtopped by lateral branches .... 32. *Conioselinum*
4. Fruits unwinged, but conspicuously ribbed; central umbel much over-
topped by lateral branches 45
5. Primary involucre absent; stems not spotted with purple .. 33. *Aethusa*
5. Primary involucre composed of linear bracts; stems usually spotted with
purple .... 34. *Conium*

1. *Eryngium* L.

. Stems erect; leaves linear, with spinulose margins; corolla white ....
.... 1. *E. yuccifolium*

1. Stems prostrate; leaves lanceolate to ovate, not spinulose on th
margins; corolla blue .... 2. *E. prostratur*

1. E. yuccifolium Michx. Rattlesnake Master. July–Aug. Prairies
openings in woods; occasional throughout the state.
2. E. prostratum Nutt. Eryngo. May–Nov. Edge of pond, very rare
Pope Co.

2. *Bupleurum* L.

1. Bupleurum rotundifolium L. Thoroughwax. May–July. Native c
Europe; rarely escaped from cultivation; St. Clair Co.

3. *Thaspium* Nutt.   Meadow Parsnip

1. Stems pubescent at the nodes; leaflets ciliate along the margins; basa
leaves 2- 3-ternate .... 1. *T. barbinoo*
1. Stems glabrous; leaf margins not ciliate; basal leaves toothed or on
once-ternate .... 2. *T. trifoliatu*

1. T. barbinode (Michx.) Nutt. Hairy Meadow Parsnip. April–Jun
Moist woods, usually near streams; occasional in the n. half of th
state, rare or absent elsewhere.
2. T. trifoliatum (L.) Gray. Meadow Parsnip. April–June. Flowe
purple. Prairies, rocky woods, thickets; not common, but scattere
throughout the state.
var. flavum Blake. Yellow Meadow Parsnip. April–June. Flowers ye
low; occasional throughout the state.

4. *Sanicula* L.   Snakeroot

1. Styles much longer than the bristles of the fruit; plants with rhizomes
1. Styles shorter than or merely equalling the bristles of the fruit; plan
without rhizomes ....
2. Fruits sessile, at least 5 mm long; sepals of staminate flowers firm, 1
mm long .... 1. *S. marilandi*
2. Fruits stipitate, up to 4 mm long (excluding the stipe); sepals of stami
ate flowers soft, up to 1 mm long .... 2. *S. gregar*
3. Pedicels of staminate flowers less than twice as long as the flower
fruits stipitate, 3–5 mm long .... 3. *S. canadens*
3. Pedicels more than twice as long as the flowers; fruits sessile, at lea
5.5 mm long .... 4. *S. trifolia*

1. S. marilandica L. Black Snakeroot. May–July. Wooded slope
occasional in the n. two-thirds of the state, rare or absent elsewhe
2. S. gregaria Bickn. Common Snakeroot. May–June. Woods; c
casional to common throughout the state.
3. *S. canadensis* L. *Canadian Black Snakeroot. May–Aug. Woo*
occasional throughout the state. (Including var. *grandis* Fern.)
4. S. trifoliata Bickn. Large-fruited Black Snakeroot. June–Ju
Woods, not common; confined to the n. half of the state.

5. Torilis Adans. Hedge Parsley

1. T. japonica (Houtt.) DC. Hedge Parsley. June–Aug. Native of Europe and Asia; naturalized in waste ground; occasional throughout the state. (Including *T. arvensis* Link.)

6. Daucus L. Carrot

1. Primary involucral bracts bipinnatifid .... 1. *D. pusillus*
1. Primary involucral bracts once-pinnate .... 2. *D. carota*

1. D. pusillus Michx. Small Wild Carrot. April–June. Woods, rare; Jackson and Perry cos.
2. D. carota L. Wild Carrot; Queen-Anne's-Lace. May–Oct. All flowers white except for the central purple one or ones. Native of Europe; common throughout the state.
f. epurpuratus Farw. May–Oct. All flowers white or pinkish. Native of Europe; waste ground; occasional in the s. two-thirds of the state.
f. roseus Millsp. May–Oct. All flowers rose or purple. Native of Europe; waste ground, rare; Jackson Co.

7. Cynosciadium DC.

1. C. digitatum DC. May–June. Swampy woods, very rare; Jackson Co.

8. Ptilimnium Raf. Mock Bishop's-weed

. Main leaves alternate or opposite; fruits about 1.5 mm long ....
.... 1. *P. costatum*
. Main leaves verticillate; fruits 2–4 mm long .... 2. *P. nuttallii*

1. P. costatum (Ell.) Raf. Mock Bishop's-weed. July–Sept. Swampy ground, rare; confined to the s. one-fourth of the state.
2. P. nuttallii (DC.) Britt. Mock Bishop's-weed. June–Aug. Swampy ground, rare; Jackson, Pulaski, Randolph, St. Clair, and Union cos.

9. Limnosciadium Math. & Const.

1. L. pinnatum (DC.) Math. & Const. May–June. Native to the w. U.S.; rarely adventive in our w. cos. *Cynosciadium pinnatum* DC.—F.

10. Oxypolis Raf. Cowbane

1. O. rigidior (L.) Coulter & Rose. Cowbane. July–Sept. Leaflets lanceolate to oblong, at least 4 mm broad. Wet prairies; occasional throughout the state.
var. ambigua (Nutt.) Robins. Cowbane. July–Sept. Leaflets linear, up to 4 mm broad. Wet ground, rare; confined to the s. half of the state.

11. Heracleum L. Cow Parsnip

1. H. maximum Bartr. Cow Parsnip. May–Aug. Low woods, ric woods; occasional in the n. two-thirds of the state, rare elsewhere. *I lanatum* Michx.—G, J.

12. Cryptotaenia DC. Honewort

1. C. canadensis (L.) DC. Honewort. May–Aug. Rocky woods, lc ground; occasional throughout the state.

13. Falcaria Bernh. Sickleweed

1. F. sioides (Wibel) Aschers. Sickleweed. July–Sept. Native Europe; rarely adventive in Ill.; Schuyler Co.

14. Zizia Koch Golden Alexanders

1. All of the basal leaves simple .... 1. *Z. apte*
1. All or nearly all of the basal leaves ternately compound .... 2. *Z. aur*

1. Z. aptera (Gray) Fern. Heart-leaved Meadow Parsnip. April–Jur Prairies, rocky woods, not common; confined to the n. one-fifth of state.
2. Z. aurea (L.) Koch. Golden Alexanders. April–June. Moist woo prairies; occasional in the n. three-fourths of the state, less comm in the s. one-fourth.

15. Pastinaca L. Parsnip

1. P. sativa L. Parsnip. May–Oct. Native of Europe; naturalized waste ground and along roads; common in the n. three-fourths of state, occasional in the s. one-fourth, but in every co.

16. Sium L. Water Parsnip

1. S. suave Walt. Water Parsnip. July–Sept. Marshes, ponds, woods, swamps; occasional throughout the state.

17. Berula Hoffm.

1. B. pusilla (Nutt.) Fern. Water Parsnip. July–Sept. Marshes, ra Kane, Kendall, McHenry, Mason, Peoria, Tazewell, and Woodford c B. erecta Cowles—G; *B. incisa* (Torr.) G. N. Jones—J.

18. Osmorhiza Raf. Sweet Cicely

1. Styles 2 mm long or longer, exceeding the petals; roots strongly an scented .... 1. *O. longist*
1. Styles up to 1.5 mm long, shorter than the petals; roots weakly an scented .... 2. *O. clayt*

1. O. longistylis (Torr.) DC. Anise-root. April–June. Stems glabr or nearly so. Rich woods; occasional throughout the state.

var. villicaulis Fern. Sweet Cicely. April–June. Stems densely villous. Rich woods; less common than the preceding, but scattered throughout the state.

2. O. claytonii (Michx.) Clarke. Sweet Cicely. April–June. Rich woods; occasional throughout the state.

19. Foeniculum Mill. Fennel

1. F. vulgare Mill. Fennel. May–Sept. Native of Europe; occasionally cultivated but rarely escaped; Madison Co.

20. Anethum L. Dill

1. A. graveolens L. Dill. June–Aug. Native of Asia; occasionally cultivated and escaped; scattered throughout the state.

21. Erigenia Nutt.

1. E. bulbosa (Michx.) Nutt. Harbinger-of-Spring; Pepper-and-Salt. Feb.–April. Rich woods; occasional throughout the state.

22. Carum L. Caraway

1. C. carvi L. Caraway. May–July. Native of Europe; occasionally cultivated and escaped.

23. Coriandrum L. Coriander

1. C. sativum L. Coriander. May–June. Native of Europe; occasionally cultivated but rarely escaped; Jackson and Will cos.

24. Spermolepis Raf.

1. S. inermis (Nutt.) Math. & Constance. May–June. Sandy soil; occasional in the nw. one-fourth of the state, rare or absent elsewhere.

25. Chaerophyllum L. Chervil

- Pedicel of fruit the same diameter throughout; fruits broadest near the middle .... 1. *C. procumbens*
- Pedicel of fruit wider at the top than at the bottom; fruits broadest below the middle .... 2. *C. tainturieri*

1. C. procumbens (L.) Crantz. Wild Chervil. April–June. Rich woods, along streams, along railroads and highways; common throughout the state. (Including var. *shortii* Torr. & Gray.)

2. C. tainturieri Hook. Wild Chervil. March–May. Fields, roadsides, waste ground, not common; confined to the s. one-third of the state.

26. Perideridia Reichenb.

1. P. americana (Nutt.) Reichenb. April–July. Floodplains, thickets, rocky woods; occasional in the n. half of the state, becoming rare in the s. half.

27. Taenidia Drude   Yellow Pimpernel

1. T. integerrima (L.) Drude. Yellow Pimpernel. May–July. Prairies, dry, often rocky, woods; occasional throughout the state.

28. Polytaenia DC.

1. P. nuttallii DC. Prairie Parsley. April–June. Prairies, rocky woods, occasional throughout the state.

29. Cicuta L.   Water Hemlock

1. Upper leaves bearing bulblets in their axils; leaflets linear 1. C. bulbifera
1. None of the leaves bearing bulblets in their axils; leaflets narrowly lanceolate or broader .... 2. C. maculata

   1. C. bulbifera L. Bulblet Water Hemlock. July–Sept. Marshes, swamps, not common; restricted to the n. half of the state; also Union Co.
   2. C. maculata L. Water Hemlock. May–Sept. Marshes, wet prairies, moist woods; occasional to common throughout the state.

30. Angelica L.   Angelica

1. Uppermost leaf sheaths inflated to at least 2 cm broad ....
.... 1. A. atropurpurea
1. Uppermost leaf sheaths slander, not more than 1 cm broad ....
.... 2. A. venenosa

   1. A. atropurpurea L. Angelica. May–Aug. Woodlands, thickets, marshes; occasional in the n. half of the state, rare elsewhere.
   2. A. venenosa (Greenway) Fern. Wood Angelica. May–July. Prairies, rich woods, rocky woods, not common; confined to the n. one-fourth of the state.

31. Aegopodium L.   Goutweed

1. A. podagraria L. Goutweed. May–Aug. Native of Europe; frequently cultivated and rarely escaped; DuPage and Lake cos.

32. Conioselinum Hoffm.   Hemlock Parsley

1. C. chinense (L.) BSP. Hemlock Parsley. July–Sept. Shady springy ground, very rare; Cook and Kane cos.

33. Aethusa L.

1. A. cynapium L. Fool's Parsley. June–Aug. Native of Europe; rarely escaped from cultivation; Cook and Kendall cos.

34. Conium L.   Poison Hemlock

1. C. maculatum L. Poison Hemlock. May–Aug. Native of Europe; naturalized in waste ground, fields, thickets, and along roads; occasional throughout the state.

121. CORNACEAE

1. *Cornus* L. Dogwood

. Flowers surrounded by 4 conspicuous, petal-like bracts; fruits red ... 2
. Flowers not subtended by 4 conspicuous, petal-like bracts; fruit blue or
  white 3
. Herbs to 30 cm tall; flowers pedicellate; drupes globose, soft ....
  .... 1. *C. canadensis*
. Trees to 12 m tall; flowers sessile; drupes ellipsoid, hard .. 2. *C. florida*
. Leaves alternate .... 3. *C. alternifolia*
. Leaves opposite 4
. Leaves with spreading or curled pubescence on the lower surface .. 5
. Leaves with appressed-pubescence or glabrous on the lower surface 7
. Leaves broadly ovate to suborbicular, with 6–9 pairs of veins; drupes
  light blue .... 4. *C. rugosa*
. Leaves elliptic to lance-ovate, with 3–4 pairs of veins; drupes white .. 6
. Leaves finely pubescent above; branchlets red; pith white ....
  .... 5. *C. stolonifera*
. Leaves scabrous above; branchlets gray; pith brown . 6. *C. drummondii*
. Leaves conspicuously whitish or glaucous beneath; drupes white (bluish
  in *C. obliqua*) 8
. Leaves green or rufescent beneath; drupes blue 10
Young twigs glabrous or nearly so; fruit white 9
. Young twigs densely hairy; fruit blue .... 8. *C. obliqua*
Twigs red; cymes flat-topped, broader than high ..... 5. *C. stolonifera*
Twigs gray; cymes elongated, as high as broad .... 7. *C. racemosa*
Pith white; leaves glabrous beneath .... 9. *C. foemina*
Pith tawny; leaves with reddish pubescence beneath . 10. *C. amomum*

1. *C. canadensis* L. Bunchberry. May–June. Boggy woods, rare; Cook, Lake, LaSalle, and Ogle cos.
2. *C. florida* L. Flowering Dogwood. April–May. Rocky woods, wooded slopes, low woods; occasional to common in the s. half of the state. (Including f. *rubra* [Weston] Palmer & Steyerm.)
3. *C. alternifolia* L. f. Alternate-leaved Dogwood. May–June. Moist woods and slopes, not common; mostly in the n. half of the state.
4. *C. rugosa* Lam. Round-leaved Dogwood. May. Shaded, rocky slopes, not common; confined to the n. one-fourth of the state.
5. *C. stolonifera* Michx. Red Osier. June–July. Leaves glabrous or appressed-pubescent beneath. Open marshes, calcareous fens; occasional in the n. half of the state, rare elsewhere.
var. *baileyi* (Coult. & Evans) Drescher. Bailey's Dogwood. June–July. Leaves densely soft-pilose beneath. Dunes, rare; Cook and Lake cos. *C. baileyi* Coult. & Evans—J.
6. *C. drummondii* C. A. Mey. Rough-leaved Dogwood. May–June. Rocky woods, prairies, low ground; occasional to common in the s. three-fourths of the state, rare elsewhere.
7. *C. racemosa* Lam. Gray Dogwood. May–July. Moist woods, upland woods, prairies, roadsides; occasional to common throughout the state.
8. *C. obliqua* Raf. Pale Dogwood. May–July. Swamps, low woods, wet prairies; occasional to common in most of Ill.
9. *C. foemina* Mill. Stiff Dogwood. May–June. Swamps, low woods, occasional; restricted to the s. one-half of the state.

10. C. amomum Mill. Dogwood. June–July. Moist thickets, not com
mon; confined to the extreme s. tip of the state.

122. ERICACEAE

1. Plants herbaceous ....
1. Plants woody (*Epigaea* only slightly so) ....
2. Plants devoid of chlorophyll; pollen borne singly .... 1. *Monotro*
2. Plants with chlorophyll; pollen borne in tetrads ....
3. Leaves on stem; filaments hairy; flowers in corymbs .... 2. *Chimaph*
3. Leaves basal or nearly so; filaments glabrous; flowers in racemes ...
.... 3. *Pyr*
4. Ovary superior ....
4. Ovary inferior ....
5. Leaves entire ....
5. Leaves toothed ....
6. Plants upright ....
6. Plants trailing ....
7. Leaves deciduous; anthers awnless; flowers over 1.5 cm long, bright p
or rose; capsule elongated .... 4. *Rhododend*
7. Leaves evergreen; anthers awned; flowers up to 1 cm long, pale pink
white; capsule depressed-globose .... 5. *Androme*
8. Fruit a capsule; flowers salverform; anthers awnless; leaves corda
anthers splitting lengthwise .... 6. *Epig*
8. Fruit a drupe; flowers urn-shaped; anthers awned; leaves cunea
anthers opening by a terminal pore .... 7. *Arctostaphy*
9. Erect tree to 15 m tall; flowers in terminal racemes; leaves deciduou
.... 8. *Oxydendr*
9. Trailing or ascending shrub to 35 cm tall; flowers in axillary leafy
cemes or solitary; leaves evergreen ....
10. Fruit dry; leaves scurfy beneath .... 9. *Chamaedap*
10. Fruit fleshy; leaves not scurfy beneath .... 10. *Gaulth*
11. Leaves with glandular dots; winter twigs predominantly black ......
.... 11. *Gaylussa*
11. Leaves without glandular dots; winter twigs green, gray, or brown ..
.... 12. *Vaccin*

1. *Monotropa* L.   Indian Pipe

1. Flowers several per stem; plants somewhat pubescent   1.*M. hypopit*
1. Flower solitary; plants glabrous .... 2. *M. unif*

1. M. hypopithys L. Pinesap. June–Oct. Rich woods, not comm
scattered throughout the state. *M. lanuginosa* Michx.—J.
2. M. uniflora L. Indian Pipe. May–Sept. Rich woods; occasi
throughout the state.

2. *Chimaphila* Pursh   Pipsissewa

1. C. umbellata (L.) Bart. var. cisatlantica Blake. Pipsissewa. Ju
Aug. Dry woods, very rare; Lake, McHenry, and Winnebago cos
corymbosa Pursh—J.

3. Pyrola L. Wintergreen

. Flowers borne along one side of raceme; style straight . 1. *P. secunda*
. Flowers borne spirally in a raceme; style curved downward 2
. Leaves thin, dull, the blades longer than the petiole; lobes of calyx up to 2 mm long .... 2. *P. elliptica*
. Leaves coriaceous, shiny, the blades about the same length as the petiole; lobes of calyx over 2 mm long .... 3. *P. americana*

    1. P. secunda L. One-sided Pyrola. June–Aug. Moist woods, very rare; Cook and Winnebago cos.
    2. P. elliptica Nutt. Shinleaf. June–Aug. Moist woods, rare; confined to extreme n. Ill.
    3. P. americana Sweet. Wild Lily-of-the-Valley. June–Aug. Shaded, mossy, wooded slope, very rare; Ogle Co. *P. rotundifolia* L. var. *americana* (Sweet) Fern.—F, G.

4. Rhododendron L. Azalea

. Pedicels, sepals, corolla, and capsules glandular; flowers fragrant .....
.... 1. *R. prinophyllum*
. Pedicels, sepals, corolla, and capsules not glandular; flowers essentially without an odor .... 2. *R. periclymenoides*

    1. R. prinophyllum (Small) Millais. Pink Azalea. May. Cherty slopes, rare; Alexander and Union cos. *R. roseum* (Loisel.) Rehd.—F, G, J.
    2. R. periclymenoides (Michx.) Shinners. Pink Azalea. May. Cherty slopes, rare; Union Co. *R. nudiflorum* (L.) Torr.—F, G.

5. Andromeda L. Bog Rosemary

    1. A. glaucophylla Link. Bog Rosemary. May–June. Bogs, very rare; Lake and McHenry cos.

6. Epigaea L. Trailing Arbutus

    1. E. repens L. Trailing Arbutus. April–May. Woods, very rare; "northern Illinois," according to the collector. (Including var. *glabrifolia* Fern.)

7. Arctostaphylos Adans. Bearberry

    1. A. uva-ursi (L.) Spreng. var. coactilis Fern. & Macbr. Kinnikinnick. June–Aug. Sand dunes and black oak woods, rare; n. one-fourth of Ill.; also Peoria Co. *A. uva-ursi* (L.) Spreng.—J.

8. Oxydendrum DC. Sourwood

    1. O. arboreum (L.) DC. Sourwood. June–July. Naturalized from the se.; Cook Co.

9. *Chamaedaphne Moench   Leatherleaf*

1. C. calyculata (L.) Moench var. angustifolia (Ait.) Rehder. Leathe
leaf. May. Bogs, rare; Cook, Kane, Lake, and McHenry cos. *C. calyc*
lata (L.) Moench—G, J.

10. *Gaultheria L. Wintergreen*

1. G. procumbens L. Checkerberry. June–Aug. Sandy soil of woo
and bogs, rare; Cook, Lake, LaSalle, and Ogle cos.

11. *Gaylussacia HBK.   Huckleberry*

1. G. baccata (Wang.) K. Koch. Black Huckleberry. May–Jun
Rocky woods and cliffs; occasional in the n. and s. cos., apparen
absent from the cent. cos.

12. *Vaccinium L.   Blueberry*

1. Corolla urceolate to campanulate, 5-lobed; stems upright; leaves event
ally deciduous ....
1. Corolla deeply 4-lobed; stems prostrate; leaves evergreen ....
.... 7. *V. macrocarp*
2. Stamens long-exserted from the corolla 1. *V. stamineum* var. *neglectu*
2. Stamens included in the corolla ....
3. Flowers in leafy bracted racemes; anthers awned; pedicels jointed; be
dry .... 2. *V. arboreu*
3. Flowers solitary, in racemes, or in glomerules, but not conspicuou
bracteate; anthers awnless; pedicels not jointed; berry juicy ....
4. Leaves with marginal spinulose teeth 3. *V. angustifolium* var. *laevifolit*
4. Leaves entire or serrulate, but without spinulose teeth ....
5. Corolla green, greenish-purple, or purple; leaves up to 5 cm long; l
shrubs up to 1 m tall ....
5. Corolla white, sometimes tinged with pink; some or all the leaves ove
cm long; tall shrubs over 1 m in height .... 6. *V. corymbos*
6. Branchlets densely pubescent; berries strongly glaucous, sour......
.... 4. *V. myrtilloi*
6. Branchlets glabrous or somewhat pubescent; berries faintly glauco
(strongly so in a few specimens), sweet .... 5. *V. vacilla*

1. V. stamineum L. var. neglectum (Small) Deam. Deerberry. M
June. Edge of sandstone cliff, very rare; Pope Co. *V. neglect*
(Small) Fern.—G.
2. V. arboreum Marsh. Farkleberry. May–June. Bracteal lea
much smaller than foliage leaves. Sandstone cliffs, locally abunda
restricted to the s. three tiers of cos., plus Jersey and Randolph c
var. glaucescens (Greene) Sarg. Farkleberry. May–June. Bract
leaves similar to foliage leaves. Sandstone cliffs, occasional; in th
one-eighth of Ill.
3. V. angustifolium Ait. var. laevifolium House. Low-bush Blueber
May–June. Sandy, open woods, slopes of dunes, bogs; confined
the n. one-fourth of the state. *V. angustifolium* Ait.—G, J.
4. V. myrtilloides Michx. Canada Blueberry. May–June. Tamar

bogs, sandy or rocky slopes; Lake, LaSalle, Ogle, and Winnebago cos.

5. V. vacillans Torr. Low-bush Blueberry. April–May. Sandstone cliffs, open sandy woods; occasional in the s. one-sixth of the state, becoming rare or absent northward. (Including var. *crinitum* Fern.)

6. V. corymbosum L. High-bush Blueberry. May–June. Swamps and bogs; Cook, Lake, LaSalle, McHenry, Stephenson, and Winnebago cos.

7. V. macrocarpon Ait. American Cranberry. June–Aug. Bogs; Cook, Lake, McHenry, and Will cos. *Oxycoccus macrocarpus* (Ait.) Pursh—J.

123. PRIMULACEAE

. Leaves entire or toothed 2
. Leaves deeply dissected into threadlike divisions . . . . 9. *Hottonia*
. Leaves all basal 3
. Leaves cauline 5
. Lobes of corolla reflexed; anthers exserted as a cone; calyx deeply divided; leaves entire, more than 2.5 cm long . . . . 1. *Dodecatheon*
. Lobes of corolla spreading; anthers included; calyx tubular; leaves dentate or, if entire, less than 2.5 cm long 4
. Leaves usually more than 2 cm long, dentate; corolla tube at least as long as the calyx, open at the throat; perennial . . . . 2. *Primula*
. Leaves up to 2 cm long, entire; corolla tube shorter than the calyx, closed at the throat; annual . . . . 3. *Androsace*
. Leaves, or most of them, alternate 6
. Leaves, or most of them, opposite and whorled (occasional alternate leaves may be found in *Trientalis* [which also always has some whorled leaves], *Lysimachia quadrifolia,* and *L. terrestris*) 7
Ovary superior; flowers generally 4-merous; pedicels absent or less than 5 mm long; flower solitary . . . . 4. *Centunculus*
Ovary inferior; flowers 5-merous; pedicels 1–2 cm long; flowers in racemes . . . . 5. *Samolus*
Flowers scarlet (rarely blue; if white, the petals rounded at tip); capsule circumscissile . . . . 6. *Anagallis*
Flowers yellow or, if white, the petals usually pointed at the tip; capsule splitting lengthwise 8
Flowers yellow (white in *L. clethroides*); petals usually 5 or 6 . . . .
. . . . 7. *Lysimachia*
Flowers white; petals usually 7 . . . . 8. *Trientalis*

1. Dodecatheon L. Shooting-star

Leaves tapering to the petiole, oblanceolate 2
Leaves abruptly contracted to a distinct petiole, the blades broadly ovate . . . . 3. *D. frenchii*
Capsule pale brown to yellow, with thin papery walls . . . .
. . . . 1. *D. amethystinum*
Capsule dark reddish-brown, with rather thick, woody walls . . . .
. . . . 2. *D. meadia*

1. D. amethystinum Fassett. Shooting-star. May–June. Bluffs an
wooded slopes, rare; Carroll and JoDaviess cos. *D. radicatum*—G.

2. D. meadia L. Shooting-star. April–June. Woods, bluffs, meadow
and prairies; occasional to common throughout Ill. (Including f. *albu*
Macbr. and var. *brachycarpum* [Small] Fassett.)

3. D. frenchii (Vasey) Rydb. French's Shooting-star. April–Ma
Under overhanging sandstone cliffs, locally abundant; confined to th
s. one-eighth of Ill. The type is from Jackson Co. *D. meadia* L. va
frenchii Vasey—F, G.

2. *Primula* L. Primrose

1. P. mistassinica Michx. Bird's-eye Primrose. May–June. Lim
stone cliffs, rare; JoDaviess Co.

3. *Androsace* L.

1. A. occidentalis Pursh. March–April. Cliffs, sandy soil, roc
woods; scattered in most parts of Ill.

4. *Centunculus* L. Chaffweed

1. C. minimum L. Chaffweed. May–Aug. Moist soil, rare; mostly
the cent. cost.

5. *Samolus* L. Brookweed

1. S. parviflorus Raf. Brookweed. May–Sept. Moist soil; occasio
throughout Ill., except for the nw. cos.

6. *Anagallis* L. Pimpernel

1. A. arvensis L. Scarlet Pimpernel. June–Aug. Naturalized fr
Europe; waste ground, not common; scattered in Ill. (Including
caerulea [Schreb.] Baumg.)

7. *Lysimachia* [Tourn.] L. Loosestrife

1. Corolla yellow . . . .
1. Corolla white . . . . 12. *L. clethroi*
2. Leaves epunctate; corolla lobes erose and apiculate; staminodia pres
. . . .
2. Leaves punctate; corolla lobes entire; staminodia absent . . . .
3. Lateral nerves of blades obscure; blades firm . . . . 1. *L. quadrifl*
3. Lateral nerves of blades conspicuous; blades thin . . . .
4. Plants decumbent, rooting at the nodes . . . . 2. *L. radic*
4. Plants erect, not rooting at the nodes . . . .
5. Blades ovate to ovate-lanceolate, rounded to subcordate at base . . . .
. . . . 3. *L. cili*
5. Blades narrow-lanceolate, elliptic, or linear, tapering to base (ra
somewhat rounded) . . . . 4. *L. lanceo*
6. Flowers in dense axillary racemes; corolla lobes linear . 5. *L. thyrsif*

6. Flowers in terminal panicles or racemes or, if axillary, solitary or in panicles; corolla lobes lanceolate to orbicular (rarely linear in *L.* X *commixta*) 7
7. Stems creeping; plants often evergreen .... 6. *L. nummularia*
7. Stems erect; plants not evergreen 8
8. Corolla without dark markings, crateriform 9
8. Corolla streaked with dark markings, rotate or saucer-shaped 10
9. Calyx dark glandular along the margin; corolla lobes not glandular-ciliate; flowers in terminal or axillary panicles .... 7. *L. vulgaris*
9. Calyx not dark glandular along the margin; corolla lobes glandular-ciliate; flowers in the axils of the upper leaves .... 8. *L. punctata*
10. Leaves in whorls of 3–7 .... 9. *L. quadrifolia*
10. Leaves opposite or alternate 11
11. Most or all the leaves (except the lowermost scale-like ones) opposite; style 3–4 mm long; capsule 2.8–3.5 mm in diameter .... 10. *L. terrestris*
11. Most or all the leaves (except the lowermost scale-like ones) alternate; style 5–6 mm long; capsule about 2 mm in diameter   11. *L.* X *commixta*

1. L. quadriflora Sims. Loosestrife. June–Aug. Marshes, moist prairies, bogs; occasional in the n. cos., becoming rare in the s. cos.

2. L. radicans Hook. Creeping Loosestrife. June–Aug. Swampy woods, rare; Pulaski and St. Clair cos.

3. L. ciliata L. Fringed Loosestrife. June–Aug. Moist woods, bottomlands, thickets; common throughout the state.

4. L. lanceolata Walt. Loosestrife. June–Aug. Basal rosettes from slender rhizomes; leaves pale beneath. Moist woods, stream banks, thickets; common throughout the state.

var. hybrida (Michx.) Gray. July–Aug. Basal rosettes not from slender rhizomes; leaves green beneath. Swampy woods, thickets, meadows; occasional throughout the state. *L. hybrida* Michx.—F.

5. L. thyrsiflora L. Tufted Loosestrife. May–July. Bottomland woods, swamps, marshes, bogs; occasional in the n. half of Ill.; also Wabash Co.

6. L. nummularia L. Moneywort. June–Aug. Native of Europe; commonly naturalized in moist, shady areas; common throughout the state.

7. L. vulgaris L. June–Sept. Native of Europe; occasionally escaped into moist fields.

8. L. punctata L. Dotted Loosestrife. May–June. Native of Europe; rarely escaped along wood's edge; Champaign and Cook cos.

9. L. quadrifolia L. Whorled Loosestrife. May–Aug. Woods, roadsides, fields; confined to the n. one-fourth of the state.

10. L. terrestris (L.) BSP. Swamp Candle. June–Aug. Swamps, bogs; confined to the n. one-third of Ill.; Union Co.

11. L. X commixta Fern. June–Aug. Wet areas, rare; scattered in Ill. Reported to be a hybrid between *L. terrestris* (L.) BSP. and *L. thyrsiflora* L.

12. L. clethroides Duby. July–Aug. Native of Japan; rarely escaped from cultivation; Cook, Lake, and McHenry cos.

8. *Trientalis* L.   Star-flower

1. T. borealis Raf. Star-flower. June–July. Moist woods, rare; Cook, Lake, LaSalle, Ogle, and Winnebago cos.

9. *Hottonia L.*

1. H. inflata Ell. Featherfoil. June–Aug. Swamps, rare; Jackso
Johnson, and Union cos.

124. SAPOTACEAE

1. *Bumelia Sw. Buckthorn*

1. Branchlets and lower surface of leaves rusty-woolly; pedicels pube
cent; fruit oblongoid to globose ..... 1. *B. lanuginosa* var. *oblongifo*
1. Branchlets and lower surface of leaves silky at first, becoming smooth
maturity; pedicels glabrous; fruit ovoid .... 2. *B. lycioid*

1. B. lanuginosa (Michx.) Pers. var. oblongifolia (Nutt.) R. B. Cla
Woolly Buckthorn. June–July. Dry woods, rare; Hardin, Pulaski, a
Monroe cos. *B. lanuginosa* (Michx.) Pers.—J.
2. B. lycioides (L.) Gaertn. f. Southern Buckthorn. June–Aug. Mo
woods or rocky cliffs, rare; apparently confined to the one-fourth
the state.

125. EBENACEAE

1. *Diospyros L. Persimmon*

1. D. virginiana L. Common Persimmon. May–June. Dry woo
fields, roadsides, clearings; occasional to common in the s. tw
thirds of the state, apparently absent in the n. one-third. (Includi
var. *pubescens* [Pursh] Dippel, with pubescent leaves and branch
var. *platycarpa* Sarg., with large, depressed-globose, early-ripeni
fruits, and f. *atra* Sarg., with dark purple fruits.)

126. STYRACACEAE

1. Calyx and corolla 4-parted; ovary inferior; fruit an elongated, winged n
.... 1. *Hale*
1. Calyx and corolla 5-parted; ovary superior; fruit a globose drupe ....
.... 2. *Sty*

1. *Halesia Ellis Silverbell*

1. H. carolina L. Silverbell Tree. April–May. Moist woods, rare; M
sac Co.

2. *Styrax L. Storax*

1. Leaves glabrous or puberulent beneath; flowers 3–4 in a cluster ....
.... 1. *S. america*

1. Leaves densely white-hairy beneath; flowers more than 4 in a raceme .. .. 2. *S. grandifolia*

1. S. americana Lam. Storax. April–May. Swampy woods, rare; confined to extreme s. Ill. and Lawrence Co.

2. S. grandifolia Ait. Big Leaf Snowbell Bush. May. Along stream in woods, very rare; Alexander Co.

127. OLEACEAE

1. Leaves pinnately compound; fruit a samara; trees .... 1. *Fraxinus*
1. Leaves simple; fruit a capsule, drupe, or berry 2
2. Leaves somewhat serrulate; petals absent; fruit an ellipsoid drupe .... 2. *Forestiera*
2. Leaves entire; petals 4; fruit a capsule or subglobose berry 3
3. Leaves truncate or subcordate; petioles 1–2 cm long; flowers lilac; fruit a capsule .... 3. *Syringa*
3. Leaves cuneate; petioles less than 1 cm long; flowers white; fruit a berry .... 4. *Ligustrum*

1. *Fraxinus* [Tourn.] L. Ash

. Leaflets sessile (except for the terminal one) .... 1. *F. nigra*
. Leaflets petiolulate 2
. Twigs 4-sided; body of samara flat .... 2. *F. quadrangulata*
. Twigs more or less terete or angular, not 4-sided; body of samara terete 3
. Leaflets with petiolules winged nearly to base; leaves green on both sides .... 3. *F. pennsylvanica*
. Leaflets with petiolules unwinged; leaves paler or brownish beneath   4
. Leaves whitened beneath; body of fruit winged less than ⅓ its length .. .. 4. *F. americana*
. Leaves brownish beneath; body of fruit winged about ½ its length .... 5. *F. tomentosa*

1. F. nigra Marsh. Black Ash. May–June. Moist woods, springy slopes; occasional in the n. half of the state; also Jersey, Lawrence, and Wabash cos.

2. F. quadrangulata Michx. Blue Ash. March–April. Moist woods, limestone cliffs; occasional in the n. three-fifths of the state, rare elsewhere.

3. F. pennsylvanica Marsh. Red Ash. April–May. Branchlets and lower surface of leaves velvety-pubescent; leaflets entire; samaras 4–7 cm long. Moist woods; occasional to common throughout the state.

var. austinii Fern. Green Ash. April–May. Branchlets and lower leaf surface velvety-pubescent; leaflets serrate; samaras less than 4 cm long. Moist woods, rare; confined to the s. half of the state.

var. subintegerrima (Vahl) Fern. Green Ash. April–May. Branchlets and lower leaf surface essentially glabrous; leaflets serrate. Moist woods, common; throughout the state. *F. lanceolata* Borkh.—J.

4. F. americana L. White Ash. April–May. Branchlets and petioles glabrous. Woods; common throughout the state.

var. biltmoreana (Beadle) J. Wright. Hairy White Ash. April–Ma
Branchlets and petioles velvety-pubescent. Rich woods, rare; co
fined to the s. one-third of the state.

5. F. tomentosa Michx. f. Pumpkin Ash. April–May. Swamps ar
low woods; occasional or rare in the s. two-fifths of the state.

2. *Forestiera Poir.* Swamp Privet

1. F. acuminata (Michx.) Poir. Swamp Privet. March–April. Swamp
low woods; occasional or rare in the s. three-fifths of the state.

3. *Syringa L.* Lilac

1. S. vulgaris L. Common Lilac. May–June. Native of Europe; rare
escaped in Ill.

4. *Ligustrum L.* Privet

1. L. vulgare L. Common Privet. June–July. Native of Europe; rare
escaped in Ill.

128. LOGANIACEAE

1. Leaves ovate-lanceolate to ovate; flowers large and showy, red and y
low; stamens 5 .... 1. *Spige*
1. Leaves linear or subulate; flowers small, white; stamens 4 ....
.... 2. *Polypremu*

1. *Spigelia L.*

1. S. marilandica L. Indian Pink. May–June. Woods; occasional
the s. one-fourth of the state.

2. *Polypremum L.*

1. P. procumbens L. June–Oct. Moist, sandy soil, very rare; Al
ander Co.

129. GENTIANACEAE

1. Leaves whorled; flowering stem 1 m or more tall; each lobe of the
rolla with a large nectariferous gland .... 1. *Swe*
1. Leaves opposite, alternate, or reduced to scales ....
2. Leaves all scale-like .... 2. *Barto*
2. Some or all the leaves not scale-like ....
3. Lowermost leaves scale-like; calyx lobes 2 .... 3. *Obol*
3. None of the leaves scale-like; calyx lobes 4–5 ....
4. Lobes of corolla longer than the tube; stem 4-angled .... 4. *Sab*

4. Lobes of corolla as long as or shorter than the tube; stem terete 5
5. Corolla red or pink, the tube up to 2 mm in diameter 5 . 5. *Centaurium*
5. Corolla blue, purple, or white, the tube at least 5 mm in diameter ......
.... 6. *Gentiana*

1. Swertia L. Columbo

1. S. caroliniensis (Walt.) Kuntze. American Columbo. May–June. Rich woods; occasional in the s. half of the state; also Cook Co. *Frasera caroliniensis* Walt.—J.

2. Bartonia Muhl.

1. Most or all the scale leaves opposite .... 1. *B. virginica*
1. Most or all the scale leaves alternate .... 2. *B. paniculata*

1. B. virginica (L.) BSP. Yellow Bartonia. Aug.–Oct. Bogs, shaded sandstone cliffs, rare; in the n. one-fourth of the state; also Clark Co.
2. B. paniculata (Michx.) Muhl. Screw-stem. Aug.–Sept. Shaded sandstone cliffs, very rare; Johnson Co.

3. Obolaria L. Pennywort

1. O. virginica L. Pennywort. April–May. Rich woods, rare; confined to the s. one-fourth of the state; also Crawford Co.

4. Sabatia Adans. Rose Gentian

1. Tube of calyx 2–3 mm long; leaves at middle of stem clasping ....
.... 1. *S. angularis*
1. Tube of calyx 4–8 mm long; leaves at middle of stem not clasping .....
.... 2. *S. campestris*

1. S. angularis (L.) Pursh. Marsh Pink. June–Sept. Moist soil; occasional in the s. two-thirds of the state, rare or absent elsewhere. (Including f. *albiflora* [Raf.] House.)
2. S. campestris Nutt. Prairie Rose Gentian. July–Sept. Prairies, very rare; DuPage, Peoria, and Washington cos.

5. Centaurium Hill Centaury

1. C. pulchellum (Sw.) Druce. Centaury. June–Sept. Native of Europe; rarely adventive in Ill.; Cook Co.

6. Gentiana L. Gentian

1. Corolla with petaloid outgrowths between the lobes 2
1. Corolla lacking petaloid outgrowths between the lobes 5
2. Corolla white or yellowish-white; margins of leaves glabrous 1. *G. alba*
2. Corolla blue or bluish-purple; margins of leaves ciliolate 3
3. Stems puberulent; anthers free; tube of corolla open at the top ....
.... 2. *G. puberulenta*
3. Stems glabrous; anthers coherent in a ring; tube of corolla nearly closed at the top 4

4. Lobes of corolla smaller than the appendages between them ....
.... 3. *G. andrewsii*
4. Lobes of corolla longer than the apendages between them ....
.... 4. *G. saponaria*
5. Corolla 2–6 cm long, the lobes fringed or dentate at the tip; seeds papillose ....
5. Corolla 1.0–2.0 (–2.5) cm long, the lobes neither fringed nor toothed seeds smooth .... 7. *G. quinquefolia*
6. Uppermost leaves narrowly ovate or ovate, rounded at the base; fringe of corolla lobes uniformly at least 2 mm long .... 5. *G. crinita*
6. Uppermost leaves linear to linear-lanceolate; fringe of corolla lobes 2 mm long at the edges but reduced to short teeth at the summit ......
.... 6. *G. procera*

1. G. alba Gray. Pale Gentian. Aug.–Oct. Prairies, rich wooded slopes not common; scattered throughout the state. *G. flavida* Gray—F, G, J.

2. G. puberulenta Pringle. Downy Gentian. Sept.–Oct. Prairies; occasional in the n. two-thirds of the state, rare or absent elsewhere. *G. puberula* Michx.—F, G, J.

3. G. andrewsii Griseb. Closed Gentian. Aug.–Oct. Moist wood prairies; occasional in the n. half of the state, rare in the s. half.

4. G. saponaria L. Soapwort Gentian. Aug.–Oct. Moist prairies sandy woods; occasional in the ne. cos.; also Gallatin and Pope cos.

5. G. crinita Froel. Fringed Gentian. Aug.–Nov. Marshes, not common; confined to the n. one-fourth of the state.

6. G. procera Holm. Small Fringed Gentian. Aug.–Oct. Marshes, n common; confined to the n. one-fourth of the state.

7. G. quinquefolia L. var. occidentalis (Gray) Hitchc. Stiff Gentian Aug.–Oct. Meadows, prairies, calcareous woods; occasional in the two-thirds of the state, absent elsewhere. *G. quinquefolia* L.—J.

130. MENYANTHACEAE

1. Leaves trifoliolate .... 1. *Menyanthes*
1. Leaves simple .... 2. *Nymphoides*

1. *Menyanthes* L. Buckbean

1. M. trifoliata L. var. minor Raf. Buckbean. May–June. Bog marshes, not common; confined to the ne. cos.; also Peoria Co. *trifoliata* L.—G, J.

2. *Nymphoides* Hill Floating Heart

1. N. peltata (Gmel.) Kuntze. Yellow Floating Heart. June–Sept. N tive of Europe; rarely adventive in Ill. Montgomery Co.

131. APOCYNACEAE

1. Leaves alternate .... 1. *Amsonia*
1. Leaves opposite 2
2. Leaves evergreen; flowers solitary in the axils; corolla salverform ......
.... 2. *Vinca*
2. Leaves deciduous; flowers in cymes; corolla funnelform or campanulate
.... 3
3. Plants erect; corolla campanulate, white or pink .... 3. *Apocynum*
3. Plants twining; corolla funnelform, greenish-yellow .4. *Trachelospermum*

1. *Amsonia* Walt.

1. A. tabernaemontana Walt. Blue Star. April–June. Leaves elliptic to ovate, most of them at least 3 cm broad. Rocky woods, thickets; occasional in the s. three-fourths of the state.

var. salicifolia (Pursh) Woodson. Blue Star. May–June. Leaves lanceolate, to 3 cm broad, glaucous on the lower surface. Rocky woods; occasional in the s. half of the state.

var. gattingeri Woodson. Blue Star. May–July. Leaves lanceolate, to 3 cm broad, green on the lower surface. Rich woods, not common; confined to the s. and w. cos. of the state.

2. *Vinca* L.   Periwinkle

1. Leaves to 3 cm long; calyx lobes glabrous; flower to 3 cm broad ....
.... 1. *V. minor*
1. Leaves over 3 cm long; calyx lobes ciliate; flower over 3 cm broad .....
.... 2. *V. major*

1. V. minor L. Common Periwinkle; Myrtle. April–May. Native of Europe; occasionally escaped from cultivation.
2. V. major L. Large Periwinkle. April–May. Native of Europe; rarely escaped from cultivation; Pope Co.

3. *Apocynum* L.   Dogbane, Indian Hemp

Corolla 4–10 mm long, pink or pink-tinged, rarely white; seeds 2.5–4.0 mm long 2
Corolla 2–4 mm long, white or greenish-white; seeds 4–6 mm long .. 3
Flowers pendulous; corolla lobes recurved; seeds 2.5–3.0 mm long ....
.... 1. *A. androsaemifolium*
Flowers ascending; corolla lobes not recurved; seeds 3–4 mm long ....
.... 2. *A. X medium*
Leaves petiolate; corolla greenish-white; follicles over 10 cm long; coma of seed at least 2.5 cm long .... 3. *A. cannabinum*
Leaves sessile or nearly so; corolla white; follicles up to 10 cm long; coma of seed up to 2 cm long .... 4. *A. sibiricum*

1. A. androsaemifolium L. Spreading Dogbane. May–July. Woods, prairies; occasional to common throughout the state.
2. A. X medium Greene. Intermediate Dogbane. May–Aug. Woods,

rare; Peoria Co. Reputedly the hybrid between *A. androsaemifolium*
L. and *A. cannabinum* L.

3. A. cannabinum L. Indian Hemp; Dogbane. May–Aug. Stems,
leaves, and branches of the inflorescence glabrous or nearly so.
Prairies, fields, rocky woods, common; in every co.

var. pubescens (Mitchell) A. DC. Hairy Dogbane. May–Aug. Stems,
leaves, and branches of the inflorescence pubescent. Prairies, fields,
rocky woods, common; in every co. *A. pubescens* R. Br.—J.

4. A. sibiricum Jacq. Indian Hemp. June–Aug. Leaves rounded at
the sessile base. Prairies, fields, rocky woods, common; scattered
throughout the state.

var. cordigerum (Greene) Fern. Indian Hemp. June–Aug. Leaves
cordate at the clasping base. Moist fields, not common; scattered
Ill.

4. *Trachelospermum* Lemaire   Climbing Dogbane

1. T. difforme (Walt.) Gray. Climbing Dogbane. May–July. Swamps,
wet woods; confined to the s. one-fourth of the state.

132. ASCLEPIADACEAE

1. Plants erect . . . . 1. *Asclepi*
1. Plants climbing or twining . . . .
2. Leaves cordate . . . .
2. Leaves rounded at the base . . . . 2. *Cynanchu*
3. Flowers whitish; plants glabrous . . . . 2. *Cynanchu*
3. Flowers greenish-yellow or reddish-purple; plants pubescent 3. *Matel*

1. *Asclepias* L.   Milkweed

1. Most or all leaves alternate . . . .
1. Most or all leaves opposite or whorled . . . .
2. Leaves glabrous . . . .
2. Leaves pubescent . . . .
3. Corolla lobes spreading; flowers green and purple; leaves at least 1 cm
broad . . . . 1. *A. viri*
3. Corolla lobes reflexed; flowers green and white; leaves up to 5 mm
broad . . . . 2. *A. stenophy*
4. Latex present; flowers green and white; hood without a horn . . . .
. . . . 3. *A. hirte*
4. Latex absent; flowers orange or yellow; hood with an incurved horn
. . . . 4. *A. tuberc*
5. Leaves filiform to narrowly linear, up to 3 mm broad . . . 5. *A. verticill*
5. Leaves broader, more than 3 mm broad . . . .
6. Hood without a horn; corolla green, the hood green or purple . . . . .
6. Hood with a horn; corolla not green or, if green, the hood white . . . .
7. Umbel solitary; stems and leaves hirsute . . . . 6. *A. lanugin*
7. Umbels several; stems and leaves glabrous or pilose . . 7. *A. viridifl*
8. Leaves sessile . . . .
8. Leaves petiolate . . . .

9. Leaves obtuse, oblong or elliptic, broadest at or near the middle, clasping .... 8. *A. amplexicaulis*
9. Leaves subacute to acute, lanceolate to ovate, broadest at or near the base, not clasping .... 9. *A. meadii*
10. Leaves pubescent, at least on the lower surface 11
10. Leaves glabrous 13
11. Veins of leaf ascending; hoods yellow .... 10. *A. ovalifolia*
11. Veins of leaf horizontally spreading; hoods not yellow 12
12. Flowers dark purplish-red; follicles without soft spines ....
.... 11. *A. purpurascens*
12. Flowers variously pale purple and green, or mixed with white; follicles usually with soft spines .... 12. *A. syriaca*
13. Never more than six pairs or whorls of leaves on the stem 14
13. At least seven pairs of leaves on the stem 15
14. Some of the leaves usually whorled; corolla pale pink, the lobes 4–6 mm long .... 13. *A. quadrifolia*
14. None of the leaves usually whorled; corolla white with a purple center, the lobes 8–12 mm long .... 14. *A. variegata*
15. Leaves broadly rounded or subcordate at base; lobes of corolla at least 1 cm long .... 15. *A. sullivantii*
15. Leaves tapering to base; lobes of corolla up to 1 cm long 16
16. Lobes of corolla greenish, 6–8 mm long, the hoods white or pink ......
.... 16. *A. exaltata*
16. Lobes of corolla rose-purple to white, 3–5 mm long, the hoods rose-purple or white 17
. Flowers white; follicles glabrous; seeds not comose ... 17. *A. perennis*
. Flowers rose-purple (rarely white); follicles puberulent; seeds comose ..
.... 18. *A. incarnata*

1. A. viridis Walt. Green-flowered Milkweed. May–July. Prairies; occasional in the s. one-third of the state. *Asclepiodora viridis* (Walt.) Gray—F, G, J.

2. A. stenophylla Gray. Narrow-leaved Green Milkweed. June–July. Dry upland woods, very rare; Adams Co. *Acerates angustifolia* (Nutt.) Decne.—G; *Asclepias angustifolia* Nutt.—J.

3. A. hirtella (Pennell) Woodson. Tall Green Milkweed. May–Aug. Prairies, fields; occasional throughout the state. *Acerates hirtella* Pennell—G.

4. A. tuberosa L. var. interior (Woodson) Shinners. Butterfly-weed. May–Sept. Woods, prairies, common; in every co. (Including f. *lutea* [Clute] Steyerm.) *A. tuberosa* L.—F, J.

5. A. verticillata L. Horsetail Milkweed. May–Sept. Dry rocky woods, prairies, fields, occasional to common; in every co.

6. A. lanuginosa Nutt. Woolly Milkweed. May–June. Rocky or gravelly prairies, rare; confined to the n. one-sixth of the state. *Acerates lanuginosa* (Nutt.) Dec.—G.

7. A. viridiflora Raf. Green Milkweed. May–Aug. Prairies, sands, gravels, and fields; occasional throughout the state. (Including var. *lanceolata* [Ives] Torr.) *Acerates viridiflora* (Raf.) Eat.—G.

8. A. amplexicaulis Sm. Sand Milkweed. May–July. Prairies, sandy areas; occasional in the n. three-fourths of the state, rare elsewhere.

9. A. meadii Torr. Mead's Milkweed. May–June. Mesic, virgin prairies, rare; Cook, Fulton, Hancock, Henderson, Peoria, Saline, and Tazewell cos.

10. A. ovalifolia Decne. Oval Milkweed. May–June. Prairies and dr
woods, rare; Cook, Kankakee, Lake, and McHenry cos.

11. A. purpurascens L. Purple Milkweed. May–July. Woodland bo
ders, prairies; occasional throughout the state.

12. A. syriaca L. Common Milkweed. May–Aug. Follicle with spin
processes up to 3 mm long. Fields, edge of prairies, waste grounc
common in Ill.

var. kansana (Vail) Palmer & Steyerm. Common Milkweed. May
Aug. Follicle with spiny processes 3–10 mm long. Fields, wast
ground; occasional and scattered in Ill.

13. A. quadrifolia Jacq. Whorled Milkweed. May–July. Rocky ope
woods; occasional in the s. two-thirds of the state, absent elsewhere

14. A. variegata L. Variegated Milkweed. May–July. Rocky ope
woods, occasional; restricted to the s. one-fourth of the state.

15. A. sullivantii Engelm. Prairie Milkweed. June–July. Moist prairie
occasional in the n. two-thirds of the state, rare elsewhere.

16. A. exaltata L. Poke Milkweed. June–July. Woodland borders, u
common; scattered in most parts of the state.

17. A. perennis Walt. White Milkweed. May–Sept. Wet wood
swamps; occasional in the s. one-fourth of the state; also along th
Wabash River in se. Ill. and Richland Co.

18. A. incarnata L. Swamp Milkweed. June–Aug. Marshes, bogs, an
ditches, common; in every co.

2. Matelea Aubl. Climbing Milkweed

1. Flowers greenish-yellow; calyx and pedicels glabrous or nearly so; fol
cles smooth, angular; blooming usually in July and August . . . .
. . . . 1. M. gonocarp
1. Flowers rose, maroon, or rarely cream; calyx and pedicels pubescer
follicles rounded, muricate; blooming usually in May and June . . . . . .
2. Petals 1.5–2.5 mm wide, rose or rarely cream . . . . 2. M. obliqu
2. Petals 3–6 mm wide, maroon . . . . 3. M. decipie

1. M. gonocarpa (Walt.) Shinners. Climbing Milkweed. July–Au
Rocky woods, rare; confined to the s. one-sixth of the state. Gon
lobus gonocarpos (Walt.) Perry —F, G, J.

2. M. obliqua (Jacq.) Woodson. Climbing Milkweed. May–Jur
Rocky woods, very rare; Pope Co. Gonolobus obliquus (Jac
Schultes—F, G, J.

3. M. decipiens (Alex.) Woodson. May–June. Low floodplain woo
natural levee of Big Muddy River, very rare; Williamson Co. Gon
lobus decipiens (Alex.) Perry—F, G, J.

3. Cynanchum L.

1. Leaves cordate; flowers white . . . . 1. C. lae
1. Leaves rounded at base; flowers dark purple . . . . 2. C. nigr

1. C. laeve (Michx.) Pers. Blue Vine. July–Sept. Moist woods, fiel
thickets, roadsides; occasional to common in the s. two-thirds of
state, rare or absent elsewhere. Ampelamus albidus (Nutt.) Britt.-
J.

2. C. nigrum (L.) Pers. Black Swallow-wort. June–Sept. Native

Europe; rarely adventive in Ill.; Christian, Cook, DuPage, Kane, and Peoria cos.

133. CONVOLVULACEAE

1. Plants with chlorophyll 2
1. Plants without chlorophyll .... 6. *Cuscuta*
2. Leaves linear, up to 3 (–6) mm broad; style deeply 2-cleft ... 1. *Stylisma*
2. Leaves oblong to ovate, never linear, over 6 mm broad; style undivided or merely notched at the apex 3
3. Stigmas 2, non-capitate, without lobes 4
3. Stigma 1, capitate or with 2 or 3 lobes .... 5. *Ipomoea*
4. Stigmas elliptic, oblong, or flattened .... 2. *Jacquemontia*
4. Stigmas filiform or subulate 5
5. Calyx not concealed by 2 large bracts; fruit 2-locular .. 3. *Convolvulus*
5. Calyx concealed by 2 large bracts; fruit 1-locular .... 4. *Calystegia*

1. *Stylisma* Raf.

1. S. pickeringii (Torr.) Gray var. pattersonii (Fern. & Schub.) Myint. Aug. Sandy prairies, very rare; Cass, Henderson, and Mason cos. The original collection by *H. N. Patterson* in 1873 from Henderson Co., and a subsequent collection from the same area by V. H. Chase in 1934, are from the type locality. *Breweria pickeringii* (Torr.) Gray var. *pattersoni* Fern. & Schub.—F; *S. patersonii* (Fern. & Schub.) G. N. Jones—J.

2. *Jacquemontia* Choisy

1. J. tamnifolia (L.) Griseb. Aug.–Oct. Native of trop. Am.; adventive along a railroad; Grundy Co.

3. *Convolvulus* L.   Bindweed

1. C. arvensis L. Field Bindweed. May–Sept. Native of Europe; naturalized in waste areas; occasional to common throughout the state.

4. *Calystegia* R. Br.   Bindweed

. Flowers double .... 1. *C. pubescens*
. Flowers single 2
. Plants upright, either erect or ascending .... 2. *C. spithamaea*
. Plants trailing or climbing .... 3. *C. sepium*

1. C. pubescens Lindl. California Rose. May–Sept. Native of Asia; rarely adventive in Ill.; Champaign, DuPage, and Wabash cos. *Convolvulus japonicus* Thunb.—G, J; *Convolvulus pellitus* Ledeb. f. *anestius* Fern.—F.

2. C. spithamaea (L.) Pursh. Dwarf Bindweed. May–July. Dry soil; occasional in the n. four-fifths of the state. *Convolvulus spithamaeus* L.—F, G, J.

3. C. sepium (L.) R. Br. var. americana (Sims) Mohlenbr. American Bindweed. June–Aug. Moist soil, fields, waste areas; common throughout the state. *Convolvulus sepium* L.—F, G; *Convolvulus americanus* (Sims) Greene—J; *Convolvulus sepium* L. var. *americanus* Sims—G.

var. repens (L.) Mohlenbr. Trailing Bindweed. July. Apparently adventive in Ill. in a waste area; Lake Co. *Convolvulus repens* L.—J Convolvulus sepium L. var. *repens* (L.) Gray—F, G.

var. fraterniflora (Mack. & Bush) Mohlenbr. Bindweed. June–Aug. Roadsides, fields; occasional in the s. three-fourths of the state; also DeKalb Co. *Convolvulus fraterniflorus* Mack. & Bush—J; *Convolvulus sepium* L. var. *fraterniflorus* Mack. & Bush—F, G.

5. *Ipomoea* L. *Morning-glory*

1. Flowers scarlet; stamens and style exserted .... 1. *I. coccinea*
1. Flowers white, pink, purple, or blue; stamens and style included 2
2. Perennials with glabrous or puberulent stems; sepals glabrous, obtuse to subacute; seeds pubescent .... 2. *I. pandurata*
2. Annuals with pubescent stems; sepals pubescent, acute to acuminate; seeds glabrous or nearly so ....
3. Calyx lobes linear-lanceolate, with long-tapering tips; corolla essentially 3.0–4.5 cm long, sky blue (when fresh) .... 3. *I. hederacea*
3. Calyx lobes oblong to lanceolate, acute to short-acuminate; corolla either less than 3 cm long or more than 4.5 cm long, not sky blue (when fresh) ....
4. Corolla less than 3 cm long, essentially white; ovary and capsule 2-locular .... 4. *I. lacunosa*
4. Corolla more than 4.5 cm long, usually not white; ovary and capsule 3-locular .... 5. *I. purpurea*

1. I. coccinea L. Red Morning-glory. July–Oct. Native of trop. Am.; occasionally naturalized in moist areas, particularly in the s. half of the state. *Quamoclit coccinea* (L.) Moench—G, J.

2. I. pandurata (L.) G. F. W. Mey. Wild Sweet Potato Vine. June–Oct. Generally in disturbed areas and thickets; occasional to common in the s. three-fourths of the state, less common elsewhere.

3. I. hederacea (L.) Jacq. Ivy-leaved Morning-glory. June–Oct. Native of trop. Am.; naturalized in waste areas; occasional to common throughout the state. (Including var. *integriuscula* Gray, with unlobed leaves.)

4. I. lacunosa L. Small White Morning-glory. July–Oct. Fields and near streams; occasional in the s. three-fifths of Ill., apparently absent elsewhere. (Including f. *purpurata* Fern.)

5. I. purpurea (L.) Roth. Common Morning-glory. July–Oct. Native of trop. Am.; naturalized in disturbed areas; occasional throughout the state.

6. *Cuscuta* L. *Dodder*

1. Sepals free to base ....
1. Sepals united below into a tube ....
2. Flowers pedicellate, borne in rather loose cymes or panicles; seeds 1.4–1.5 mm long .... 1. *C. cuspidata*

2. Flowers sessile, borne in dense glomerules; seeds 1.7 mm long or longer 3
3. Bracts at base of sepals appressed; lobes of corolla obtuse; seeds 2.5–2.6 mm long .... 2. *C. compacta*
3. Bracts at base of sepals with recurved tips; lobes of corolla acute; seeds 1.7–1.8 mm long .... 3. *C. glomerata*
4. Most of the flowers with 4-lobed corollas 5
4. Most of the flowers with 5-lobed corollas 7
5. Corolla lobes erect; flowers sessile or on pedicels up to 0.5 mm long 6
5. Corolla lobes inflexed; flowers on pedicels usually at least 1 mm long .. .. 6. *C. coryli*
6. Scales absent, or reduced to minute teeth along the filaments; lobes of corolla acute, about as long as the corolla tube; seeds 1.3–1.4 mm long .... 4. *C. polygonorum*
6. Scales toothed from base to apex; lobes of corolla obtuse to subacute, shorter than the corolla tube; seeds 1.6–1.7 mm long   5. *C. cephalanthi*
7. Lobes of corolla obtuse, erect or spreading .... 7. *C. gronovii*
7. Lobes of corolla acute, the tips inflexed 8
8. Lobes of calyx obtuse; pedicels shorter than the flowers 9
8. Lobes of calyx acute; pedicels as long as or longer than the flowers ... . 10. *C. indecora*
9. Lobes of corolla acuminate; scales about half as long as corolla tube; seeds 1.0–1.2 mm long .... 8. *C. pentagona*
9. Lobes of corolla acute; scales about as long as corolla tube; seeds 1.5–1.6 mm long .... 9. *C. campestris*

1. C. cuspidata Engelm. Dodder. July–Oct. Moist or dry areas; scattered throughout the state, but apparently not common.

2. C. compacta Juss. Dodder. July–Oct. Moist areas; scattered throughout the state, but not common.

3. C. glomerata Choisy. Dodder. July–Oct. Low, wet areas; occasional and scattered throughout the state.

4. C. polygonorum Engelm. Dodder. July–Sept. Moist areas; occasional and scattered throughout the state.

5. C. cephalanthi Engelm. Dodder. Aug.–Oct. Low, wet areas; occasional and scattered throughout the state.

6. C. coryli Engelm. Dodder. July–Oct. Moist or dry areas; scattered throughout the state, but apparently not common.

7. C. gronovii Willd. Dodder. July–Oct. Low, wet areas; common throughout the state.

8. C. pentagona Engelm. Dodder. June–Oct. Dry, mostly disturbed, areas; occasional to common throughout the state.

9. C. campestris Yuncker. Dodder. June–Oct. Mostly dry areas; scattered but not common.

10. C. indecora Choisy. Dodder. July–Sept. Calyx lobes about half as long as corolla tube. Low areas; apparently only in the s. two-thirds of the state, but not common.

var. neuropetala (Engelm.) Hitchc. Dodder. July–Sept. Calyx lobes about as long as corolla tube. Low areas; rare in the s. half of the state.

134. POLEMONIACEAE

1. Leaves pinnately compound or deeply pinnatifid ....
1. Leaves simple, entire ....
2. Leaves pinnately compound with lanceolate to oval leaflets; flower:
blue, campanulate, to 1.5 cm long .... 1. *Polemoniun*
2. Leaves deeply pinnatifid, the segments filiform; flowers red or pink, nar
rowly funnelform, at least 2.5 cm long .... 2. *Gili*
3. All leaves alternate .... 3. *Collomi*
3. At least the lower cauline leaves opposite ....
4. Usually all leaves opposite; calyx actinomorphic; corolla 1.5 cm long c
longer .... 4. *Phlo*
4. Uppermost leaves alternate; calyx slightly zygomorphic; corolla 8–1
mm long .... 5. *Microster*

1. *Polemonium* L. Greek Valerian

1. P. reptans L. Jacob's-ladder. April–June. Moist or sometimes d
woods; common throughout the state.

2. *Gilia* R. & P.

1. G. rubra (L.) Heller. Standing Cypress. June–Aug. Native of the
U.S.; rarely escaped from cultivation; Clark and Woodford cos.

3. *Collomia* Nutt.

1. C. linearis Nutt. June–Aug. Native of the w. U.S.; adventive alon
railroads; occasional in the n. half of the state.

4. *Phlox* L. Phlox

1. Petals deeply notched or emarginate at the tip ....
1. Petals not notched or emarginate at the tip ....
2. Leaves narrowly linear; plants diffuse and much branched; petals deep
notched .... 1. *P. bific*
2. Leaves lanceolate to narrowly ovate; plants erect or ascending, not d
fusely branched; petals emarginate .... 2. *P. divaricat*
3. Calyx lobes longer than calyx tube; stamens and style about half as lor
as corolla tube ....
3. Calyx lobes equalling or shorter than calyx tube; stamens and sty
about as long as corolla tube ....
4. Leaves lanceolate to elliptic to narrowly ovate; plants often with steri
leafy shoots .... 2. *P. divarica*
4. Leaves linear to narrowly lanceolate; plants without sterile leafy shoo
.... 3. *P. pilos*
5. Leaves with conspicuous lateral veins and reticulations; calyx teeth su
ulate .... 4. *P. panicula*
5. Leaves without conspicuous lateral veins and reticulations; calyx tee
lanceolate ....
6. Flowers in panicles usually longer than broad; stems usually purpl
spotted .... 5. *P. macula*

6. Flowers in corymbs usually nearly as broad as long; stems green ......
.... 6. *P. glaberrima*

1. P. bifida Beck. Cleft Phlox. March–June. Some of the pubescence glandular. Sandy soil, limestone cliffs, occasional; throughout
the state, except for some of the cent. cos.

var. stellaria (Gray) Wherry. Cleft Phlox. March–May. None of the
pubescence glandular. Sandy areas, rare; Jackson Co.

2. P. divaricata L. Blue Phlox. May–June. Corolla lobes emarginate.
Woods; rare in the n. cos.

ssp. laphamii (Wood) Wherry. Blue Phlox. April–June. Corolla lobes
entire at tip. Woods, common; throughout the state. *P. divaricata*
L.—J.

3. P. pilosa L. Downy Phlox. May–Aug. Leaves pubescent, with
some of the pubescence glandular. Dry woods and prairies; occasional throughout the state.

ssp. fulgida (Wherry) Wherry. May–Aug. Leaves pubescent, but
none of the pubescence glandular. Dry woods and prairies; occasional throughout the state. *P. pilosa* L.—J; *P. pilosa* var. *fulgida*
Wherry—F.

ssp. sangamonensis Levin & Smith. Sangamon Phlox. May–Aug.
Leaves glabrous or nearly so. Roadsides, fields, woodlands; Champaign and Piatt cos.

4. P. paniculata L. Garden Phlox. July–Sept. Rich woods; occasional in the s. four-fifths of the state.

5. P. maculata L. Wild Sweet William. June–Aug. Moist woods; occasional in the n. half of the state.

6. P. glaberrima L. ssp. interior (Wherry) Wherry. Smooth Phlox.
May–Aug. Woods, prairies; occasional throughout the state. *P. glaberrima* L.—J; *P. glaberrima* var. *interior* Wherry—F.

5. Microsteris Greene

1. M. gracilis (Dougl.) Greene. July–Aug. Native of the West Coast;
rarely adventive in Ill.; Macon Co.

135. HYDROPHYLLACEAE

1. Leaves entire, with axillary spines; styles 2, free .... 1. *Hydrolea*
1. Leaves toothed, pinnately lobed, or pinnately compound; styles 2-cleft
.... 2
2. Some or all the cauline leaves opposite .... 2. *Ellisia*
2. All the cauline leaves alternate 3
3. Leaves palmately lobed .... 3. *Hydrophyllum*
3. Leaves pinnately lobed or compound 4
4. Lobes of corolla fimbriate .... 4. *Phacelia*
4. Lobes of corolla entire 5
5. Flowers up to 5 mm broad .... 4. *Phacelia*
5. Flowers 1 cm broad or broader 6
6. Branches of inflorescence glandular-pubescent .... 4. *Phacelia*
6. Branches of inflorescence lacking glandular-pubescence ....
.... 3. *Hydrophyllum*

1. *Hydrolea* L.

1. H. uniflora Raf. June–Sept. Swampy woods, ditches, rare; confined to the s. one-sixth of the state. *H. affinis* Gray—J.

2. *Ellisia* L.

1. E. nyctelea L. April–June. Wet ground; occasional in the n. three-fourths of the state, rare elsewhere.

3. *Hydrophyllum* L. Waterleaf

1. Leaves palmately lobed .... ?
1. Leaves pinnately lobed or compound .... ?
2. Calyx with a reflexed appendage between two adjacent lobes; branches of inflorescence bristly hairy, with the hairs over 1 mm long ....
.... 1. *H. appendiculatum*
2. Calyx not appendaged between the lobes, or with minute teeth branches of inflorescence pubescent with hairs up to 1 mm long .....
.... 2. *H. canadense*
3. Leaves 9- to 13-parted; stems hirsute .... 3. *H. macrophyllum*
3. Leaves 3- to 7-parted; stems glabrous or sparsely pubescent ....
.... 4. *H. virginianum*

1. H. appendiculatum Michx. Great Waterleaf. April–July. Rich woods; occasional throughout the state.
2. H. canadense L. Broad-leaf Waterleaf. May–July. Rich woods; occasional to rare throughout the state.
3. H. macrophyllum Nutt. Large-leaf Waterleaf. May–June. Rich woods; confined to the s. one-fourth of the state.
4. H. virginianum L. Virginia Waterleaf. April–July. Rich woods; occasional to common throughout the state.

4. *Phacelia* Juss.

1. Lobes of corolla fimbriate .... 1. *P. purshii*
1. Lobes of corolla not fimbriate ....
2. Flowers up to 5 mm broad .... 2. *P. ranunculacea*
2. Flowers at least 1 cm broad .... 3. *P. bipinnatifida*

1. P. purshii Buckley. Miami Mist. April–June. Woods, thickets; occasional in the s. one-half of the state.
2. P. ranunculacea (Nutt.) Constance. April–May. Rich woods; rare in the s. one-third of the state; also Adams Co.
3. P. bipinnatifida Michx. April–June. Rich, often rocky woods; occasional in the s. half of the state, rare elsewhere.

136. BORAGINACEAE

1. Stems and leaves glabrous ....
1. Stems and leaves pubescent ....

2. Leaves conspicuously veined, oblong to oval; flowers blue or rarely pinkish, 2–3 cm long .... 1. *Mertensia*
2. Leaves inconspicuously veined, linear to linear-oblong; flowers white, with a yellow eye, rarely blue, up to 6 mm long .... 2. *Heliotropium*
3. Ovary unlobed .... 2. *Heliotropium*
3. Ovary 4-lobed 4
4. Stamens exserted .... 3. *Echium*
4. Stamens included 5
5. Fruits echinate or prickly 6
5. Fruits smooth, wrinkled, rugose, reticulate, but not echinate or prickly 8
6. Corolla 8–12 mm broad; fruit 8–12 mm broad .... 4. *Cynoglossum*
6. Corolla 2–3 mm broad; fruit 3–5 mm broad 7
7. Racemes bracted throughout; pedicels not deflexed .... 5. *Lappula*
7. Racemes bracted only at the base; pedicels deflexed ...... 6. *Hackelia*
8. Corolla rotate .... 7. *Borago*
8. Corolla funnelform, salverform, or tubular 9
9. Leaves decurrent by wings along the stem .... 8. *Symphytum*
9. Leaves not decurrent 10
10. Corolla blue 11
10. Corolla greenish-white, white, orange, or yellow 13
11. All flowers bracted at the base; calyx tube hispid, the hairs neither glandular nor hooked .... 9. *Anchusa*
11. None of the flowers or only the lowermost flowers bracted at the base; calyx tube with appressed hairs or hooked or glandular hairs 12
12. All leaves petiolate .... 2. *Heliotropium*
12. None or only the basal leaves petiolate .... 10. *Myosotis*
13. All flowers subtended by bracts 14
13. None or only a few of the flowers subtended by bracts 17
14. Nutlets pubescent, neither white and shiny nor brown and pitted ....
.... 2. *Heliotropium*
14. Nutlets glabrous, white and shiny or brown and pitted 15
15. Leaves without lateral veins .... 11. *Lithospermum*
15. Leaves with distinct lateral veins 16
16. Stems spreading-hirsute; corolla 1–2 cm long .... 12. *Onosmodium*
16. Stems pubescent, but not spreading hirsute; corolla less than 1 cm long
.... 11. *Lithospermum*
17. Corolla yellow; leaves lanceolate to ovate-lanceolate .... 13. *Amsinckia*
17. Corolla white; leaves linear to oblong 18
18. Corolla 1–3 mm broad, with yellow appendages in the throat; nutlets obovoid; calyx tube with some glandular or hooked hairs   10. *Myosotis*
18. Corolla 5–10 mm broad, without yellow appendages in the throat; nutlets ovoid; calyx tube without glandular or hooked hairs . 14. *Plagiobothrys*

1. *Mertensia* Roth

1. M. virginica (L.) Pers. Bluebells. March–June. Rich woods; occasional throughout the state. (Including f. *rosea* Steyerm.)

2. *Heliotropium* L. Heliotrope

1. Plants glabrous throughout .... 1. *H. curassavicum*
1. Plants pubescent 2
2. Leaves linear, sessile or nearly so; flowers subtended by bracts ....
.... 2. *H. tenellum*

2. Leaves ovate or oval, petiolate; flowers bractless 3
3. Flowers blue; plants hirsute or hispid .... 3. *H. indicum*
3. Flowers white; plants hoary-pubescent .... 4. *H. europaeum*

      1. H. curassavicum L. Seaside Heliotrope. May–Sept. Native of the s. U.S. and trop. Am.; probably adventive in Ill.; bottomlands, very rare; St. Clair Co.

      2. H. tenellum (Nutt.) Torr. Slender Heliotrope. June–Aug. Limestone ledges, very rare; Monroe Co.

      3. H. indicum L. Indian Heliotrope. July–Nov. Native of Asia; moist disturbed areas; occasional in the s. half of the state, apparently absent elsewhere.

      4. H. europaeum L. European Heliotrope. June–Sept. Native of Europe; rarely escaped from cultivation; Cook Co.

3. *Echium* L. Viper's Bugloss

      1. E. vulgare L. Blueweed. May–Sept. Native of Europe; adventive in disturbed areas, particularly along railroads; occasional in the n. half of the state, absent elsewhere.

4. *Cynoglossum* L. Hound's-tongue

1. Corolla pale blue to white; stem rough-hairy; leaves only on the lower half of the stem .... 1. *C. virginianum*
1. Corolla reddish-purple; stem softly hairy; leaves borne all along the stem
.... 2. *C. officinale*

      1. C. virginianum L. Wild Comfrey. April–June. Rich woods, wooded slopes; occasional in the s. one-third of the state, rare or absent elsewhere.

      2. C. officinale L. Common Hound's-tongue. May–July. Native of Europe; adventive in pastures, fields, roadsides; occasional to common in the n. two-thirds of the state, rare elsewhere.

5. *Lappula* Moench Stickseed

1. Each margin of the nutlet with two rows of prickles; fruits mostly 3–4 mm long .... 1. *L. echinata*
1. Each margin of the nutlet with one row of prickles; fruits mostly 2–3 mm long .... 2. *L. redowskii*

      1. L. echinata Gilib. Beggar's Lice. May–Sept. Native of Europe; adventive in waste areas; occasional in the n. half of the state; also Coles and St. Clair cos. *L. myosotis* Moench—J.

      2. L. redowskii (Hornem.) Greene var. occidentalis (Wats.) Rydb. Stickseed. May–Sept. Native of the w. U.S.; rarely adventive in waste ground; McHenry and Rock Island cos. *L. occidentalis* (S. Wats.) Greene—J.

6. *Hackelia* Opiz Stickseed

1. Fruits globose; basal leaves cordate at base; cauline leaves ovate-oblong .... 1. *H. virginiana*

1. Fruits pyramidal; basal leaves cuneate at base; cauline leaves linear to oblong .... 2. *H. americana*

    1. H. virginiana (L.) I. M. Johnston. Stickseed. June–Sept. Moist or dry woods; occasional throughout the state.
    2. H. americana (Gray) Fern. Stickseed. June–July. Rocky slopes, rare; JoDaviess Co.

7. *Borago* L. Borage

    1. B. officinalis L. Borage. July–Sept. Native of Europe; occasionally cultivated but rarely escaped; Champaign, Cook, and Hancock cos.

8. *Symphytum* L. Comfrey

    1. S. officinale L. Common Comfrey. June–Aug. Native of Europe; occasionally escaped from cultivation into waste ground.

9. *Anchusa* L. Alkanet

    1. A. officinalis L. Common Alkanet. May–Oct. Native of Europe; rarely adventive in waste ground; Kane Co.

10. *Myosotis* L. Forget-me-not

1. Corolla white, 1–2 mm across; lobes of calyx 2-lipped .. 1. *M. virginica*
1. Corolla blue, sometimes with a yellow center, 2–9 mm across; lobes of calyx equal or nearly so 2
  2. Corolla 5–9 mm broad, blue with a yellow center 3
  2. Corolla 2–4 mm broad, blue but usually without a yellow center 4
    3. Calyx with appressed hairs, the lobes much shorter than the tube ......
.... 2. *M. scorpioides*
    3. Calyx with hooked hairs, the lobes about as long as the tube ....
.... 3. *M. sylvatica*
    4. All flowers bractless; pedicels longer than the calyx .... 4. *M. arvensis*
    4. At least the lower flowers bracted; pedicels shorter than the calyx .....
.... 5. *M. stricta*

    1. M. virginica (L.) BSP. Scorpion Grass. April–May. Fruiting pedicels more or less erect, the lowest to 20 mm apart. Dry woods, moist woods, fields; occasional to common throughout the state. *M. verna* Nutt.—F, G, J.
    var. macrosperma (Engelm.) Fern. Scorpion Grass. April–May. Fruiting pedicels spreading, the lowest 20 mm or more apart. Dry woods, moist woods, fields; occasional in the s. one-third of the state. *M. macrosperma* Engelm.—F, G, J.
    2. M. scorpioides L. Forget-me-not. April–Oct. Native of Europe; frequently cultivated, rarely escaped into waste ground.
    3. M. sylvatica Hoffm. Garden Forget-me-not. June–July. Native of Europe; commonly cultivated but rarely escaped; Jackson Co.
    4. M. arvensis (L.) Hill. Field Scorpion Grass. June–Aug. Native of Europe; rarely adventive in Ill.; Cook Co.

364 BORAGINACEAE

5. M. stricta Link. Small-flowered Forget-me-not. May–Aug. Native of Europe; rarely adventive in Ill.; St. Clair Co.

11. *Lithospermum* L. Gromwell

1. Corolla white, greenish-white, or pale yellow, up to 8 mm long 2
1. Corolla bright yellow or orange, 15 mm long or longer 4
2. Leaves linear, the lateral veins not evident; flowers white; nutlets brown or gray, pitted .... 1. *L. arvense*
2. Leaves lanceolate to ovate, the lateral veins prominent; flowers greenish-white to pale yellow; nutlets white, shiny 3
3. Most or all the cauline leaves 2 cm broad or broader, ovate...........
.... 2. *L. latifolium*
3. Most or all the cauline leaves up to 2 cm broad, lanceolate ....
.... 3. *L. officinale*
4. Leaves linear, acute, most of them not exceeding 5 mm in width; corolla tube at least twice as long as the calyx, the lobes toothed 4. *L. incisum*
4. Leaves linear-lanceolate to oblong, more or less obtuse, most of them more than 5 mm in width; corolla tube less than twice as long as the calyx, the lobes entire 5
5. Lobes of calyx during flowering up to 6 mm long, lengthening to 8 mm during fruiting; stems softly pubescent; corolla tube glabrous within at the base .... 5. *L. canescens*
5. Lobes of calyx during flowering up to 11 mm long, lengthening to 1 mm long during fruiting; stems somewhat rough-pubescent; corolla tube pubescent within at the base .... 6. *L. caroliniense*

1. L. arvense L. Corn Gromwell. April–June. Native of Europe; adventive in fields and waste ground; in every co.
2. L. latifolium Michx. American Gromwell. April–June. Rich woods occasional throughout the state.
3. L. officinale L. European Gromwell. June–July. Native of Europe rarely adventive in Ill.; Lake Co.
4. L. incisum Lehm. Yellow Puccoon. April–June. Prairies; occasional in the n. half of the state, becoming rare in the s. half.
5. L. canescens (Michx.) Lehm. Hoary Puccoon. April–June. Prairies; occasional to common throughout the state.
6. L. caroliniense (Walt.) MacM. Hairy Puccoon. April–June. Sand prairies, sandy woods; occasional to common in the n. half of the state, less common in the s. half. (Including *L. croceum* Fern.)

12. *Onosmodium* Michx. Marbleseed

1. Nutlets constricted at the base .... 1. *O. hispidissimum*
1. Nutlets not constricted at the base 2
2. Nutlets not strongly pitted, usually 3.5–4.5 mm long . 2. *O. occidentale*
2. Nutlets strongly pitted, usually not more than 3 mm long .. 3. *O. mo*

1. O. hispidissimum Mack. Marbleseed. May–July. Rocky prairie rocky woods, often in calcareous areas, not common; scattered throughout the state. (Including var. *macrospermum* Mack. & Bush
2. O. occidentale Mack. Marbleseed. May–July. Rocky prairie rocky woods, not common; confined to the w. half of the state.

3. O. molle Michx. Marbleseed. May–July. Limestone ridges, rare; confined to the s. one-sixth of the state.

13. Amsinckia Lehm.   Fiddle-neck

1. Calyx lobes with long, white hairs .... 1. *A. barbata*
1. Calyx lobes without long, white hairs .... 2. *A. spectabilis*

1. A. barbata Greene. Tarweed. June–July. Native of w. N.Am.; rarely adventive in waste ground.

2. A. spectabilis Fisch. & Mey. Fiddle-neck. June–July. Native of the w. U.S.; rarely adventive in Ill. along railroads; Champaign and De-Kalb cos.

14. Plagiobothrys Fisch. & Mey.

1. P. hirtus (Greene) I. M. Johnston var. figuratus (Piper) I. M. Johnston. June–Aug. Native of the w. U.S.; rarely adventive in Ill.; St. Clair Co. *Allocarya figurata* Piper—J.

137. VERBENACEAE

1. Corolla 2-lipped; calyx teeth 4; flowers in a dense, rounded head ......
.... 1. *Lippia*
1. Corolla nearly regular; calyx lobes 5; flowers various, but not in a dense, rounded head .... 2. *Verbena*

1. Lippia L.

1. Leaves acute to acuminate at the apex .... 1. *L. lanceolata*
1. Leaves broadly rounded at the apex .... 2. *L. cuneifolia*

1. L. lanceolata Michx. Fog-fruit. May–Sept. Wet soil; common throughout the state; in every co. (Including var. *recognita* Fern. & Grisc.) *Phyla lanceolata* (Michx.) Greene—G, J.

2. L. cuneifolia (Torr.) Steud. Hoary Fog-Fruit. May–Sept. Native to the w. U.S.; rarely adventive in Ill.; Menard Co. *Phyla cuneifolia* Torr.) Greene—G.

2. Verbena L.   Vervain

1. Corolla at least 1 cm long; calyx at least 8 mm long 2
1. Corolla up to 8 mm long; calyx up to 5 mm long 3
2. Corolla tube about twice as long as the calyx; calyx lobes deltoid, short
.... 1. *V. peruviana*
2. Corolla tube slightly longer than the calyx; calyx lobes setaceous, elongated .... 2. *V. canadensis*
3. Stems often spreading or sprawling; bracts slightly to much longer than the flowers 4
3. Stems erect; bracts not longer than the flowers 6
4. Stems spreading or sprawling; bracts 5 mm long or longer 5

4. Stems more or less erect; bracts 3–4 mm long .... 5. *V.* X *perriana*
5. Bracts 6 mm long or longer; inflorescence 1 cm broad or broader ....
.... 3. *V. bracteata*
5. Bracts about 5 mm long; inflorescence up to 1 cm broad   4. *V.* X *deami*
6. Leaves up to 2 cm broad ....
6. Leaves, or most of them, more than 2 cm broad ....
7. Flowers up to 6 mm broad; stem glabrous or sparsely strigose ....
.... 6. *V. simple*
7. Flowers up to 10 mm broad; stem densely hairy ....
8. Leaves lanceolate .... 7. *V.* X *blanchard*
8. Leaves elliptic to elliptic-ovate ....8. *V.* X *moechin*
9. Leaves sessile or on petioles less than 5 mm long; spikes 1–3 per stem
7–10 mm thick .... 9. *V. strict*
9. Leaves petiolate, the petioles at least 5 mm long; spikes usually mor
than 3 per stem, mostly less than 7 mm thick ....
10. Some or all the flowers overlapping, the calyx of one reaching beyon
the base of the calyx of the flower above ....
10. None or few of the flowers overlapping, the calyx of one at most reach
ing only to the base of the calyx of the flower above ....
11. Calyx 3–4 mm long; leaves sparsely pubescent .... 10. *V. hasta*
11. Calyx 4–5 mm long; leaves softly pubescent ....11. *V.* X *rydberg*
12. Flowers white ....
12. Flowers blue .... 14. *V. engelmanr*
13. Stems hirtellous or nearly glabrous; each flower not reaching th
base of the one immediately above ....12. *V. urticifol*
13. Stems long-hairy; each flower reaching just to the base of the caly
of the flower immediately above .... 13. *V.* X *illici*

1. **V. peruviana** Britt. Peruvian Vervain. June–Aug. Native of S.Arr
frequently cultivated but rarely escaped; Kane and Kankakee cos.

2. **V. canadensis** Britt. Rose Verbena. March–Oct. Rocky wood
edge of fields and prairies; occasional in the s. two-thirds of the stat
rare or absent elsewhere.

3. **V. bracteata** Lag. & Rodr. Creeping Vervain. April–Oct. Was
ground; occasional to common throughout the state.

4. **V.** X **deamii** Moldenke. July–Aug. Known only from Hendersc
Peoria, and Wash cos. Reputed to be a hybrid between *V. bractea*
and *V. stricta*. The type is from Peoria Co.

5. **V.** X **perriana** Moldenke. June–Aug. Dry soil; Adams, Cas
Monroe, Wabash, and Woodford cos. Reputed to be a hybrid betwee
V. bracteata and *V. urticifolia*.

6. **V. simplex** Lehm. Narrow-leaved Vervain. May–Sept. Prairie
fields, waste ground; occasional and scattered throughout the stat

7. **V.** X **blanchardii** Moldenke. July–Aug. Dry fields; Cook Co. F
puted to be a hybrid between *V. hastata* and *V. simplex*. The type
from Cook Co.

8. **V.** X **moechina** Moldenke. June–July. Dry ground, particularly
pastures; Adams, Hardin, Peoria, and Winnebago cos. Reputed to
a hybrid between *V. simplex* and *V. stricta*.

9. **V. stricta** Vent. Hoary Vervain. May–Sept. Prairies, pastur
fields, common; in every co. (Including f. *albiflora* Wadmond anc
roseiflora Benke.)

10. **V. hastata** L. Blue Vervain. June–Oct. Wet woods, wet prairi
wet waste ground, common; in every co. (Including f. *rosea* Chene

11. **V.** X **rydbergii** Moldenke. June–Sept. Wet ground; scatte

throughout the state. Reputed to be a hybrid between *V. hastata* and *V. stricta*. The type is from St. Clair Co.

12. V. urticifolia L. White Vervain. June–Oct. Leaves hirsute on the lower surface; nutlets about 2 mm long, corrugated on back. Fields, thickets, disturbed woods, common in every co.

var. leiocarpa Perry & Fern. June–Oct. Leaves velutinous on the lower surface; nutlets about l.5 mm long, not corrugated on the back. Low ground; Cook and Kane cos.

13. V. X illicita Moldenke. June–Sept. Low ground; scattered throughout the state. Reputed to be a hybrid between *V. stricta* and *V. urticifolia*.

14. V. X engelmannii Moldenke. July–Sept. Woods, fields; scattered throughout the state. Reputed to be a hybrid between *V. hastata* and *V. urticifolia*.

138. PHRYMACEAE

1. *Phryma* L. Lopseed

1. P. leptostachya L. Lopseed. June–Sept. Rich woods, occasional; scattered throughout the state.

139. LABIATAE

Lobes of corolla almost equal, not bilabiate 2
Lobes of corolla bilabiate 7
All flowers borne in the axils of the leaves 3
Some or all the flowers in terminal inflorescences 4
Flowers white; fertile stamens 2 .... 4. *Lycopus*
Flowers pink or purple; fertile stamens 4 .... 3. *Mentha*
Leaves entire or nearly so 5
Leaves toothed 6
Calyx more or less equally 5-cleft; stamens included; ovary 5-lobed ....
.... 1. *Isanthus*
Calyx more or less bilabiate; stamens excluded; ovary 4-lobed ....
.... 2. *Trichostema*
Corolla 4-lobed .... 3. *Mentha*
Corolla 5-lobed .... 28. *Perilla*
Ovary 4-lobed, the style not basal 8
Ovary deeply 4-parted, the style basal 9
Upper lip of corolla seemingly absent; leaves petiolate; stems at least 2.5 dm tall; creeping stolons absent .... 5. *Teucrium*
Upper lip of corolla very short; leaves, or the upper ones, sessile; stems less than 2.5 dm tall; creeping stolons present .... 6. *Ajuga*
Calyx crested on the upper side 10
Calyx not crested on the upper side 11
Calyx bilobed, the tips rounded .... 7. *Scutellaria*
Calyx 5-lobed, the tips pointed .... 16. *Ocimum*
Stamens included within the corolla tube .... 8. *Marrubium*
Stamens exserted or ascending under the lip, not confined to the tube of the corolla 12

12. Anther-bearing stamens 2 .... 1:

12. Anther-bearing stamens 4 .... 1:

13. Stamens exserted beyond the corolla 1

13. Stamens included 1

14. Flowers in axillary verticils or clusters or cymes; flowers not yellow (
      yellow, then with purple spots) 1

14. Flowers in terminal panicles; corolla light yellow ...... 12. *Collinsoni*

15. Calyx very unequally lobed .... 11. *Blephil*

15. Calyx equally 5-lobed 1

16. Corolla up to 1 cm long; flowers in loose cymes .... 9. *Cunil*

16. Corolla more than 1 cm long; flowers in dense capitate clusters .....
      .... 10. *Monard*

17. Bracts longer than the clusters of flowers they subtend .. 13. *Hedeom*

17. Bracts shorter than the clusters of flowers they subtend 1

18. Flowers mostly up to 10 in a rather loose whorl .... 14. *Salvi*

18. Flowers mostly more than 10 in dense whorls .... 11. *Blephil*

19. Leaves entire 2

19. Leaves serrate, crenate, dentate, or lobed 2

20. Plants forming dense mats; leaves obtuse at the apex. .... 15. *Thymu*

20. Plants erect; leaves acute to acuminate at the apex 2

21. Leaves ovate; upper lip of corolla 4-lobed .... 16. *Ocimu*

21. Leaves linear to lanceolate; upper lip of corolla 2-lobed, emarginate,
      entire 2

22. Calyx bilabiate (nearly equally 5-lobed in *S. hortensis*); stamens strong
      curved .... 17. *Sature*

22. Calyx more or less equally 5-lobed; stamens straight or nearly so ....
      .... 18. *Pycnanthemu*

23. Some or all the cauline leaves cordate ....

23. None of the cauline leaves cordate ....

24. Flowers yellow, greenish-yellow, or white dotted with purple ....

24. Flowers purple, rose-purple, or blue ....

25. Stem glabrous or nearly so .... 19. *Agastac*

25. Stem pubescent ....

26. Stem densely canescent; calyx 5-toothed, the tube puberulent; coro
      white, dotted with purple .... 20. *Nepe*

26. Stem more or less hirsute; calyx 4-parted, the tube hirsute; coro
      yellow-white .... 21. *Synand*

27. Stems creeping or trailing .... 22. *Glechon*

27. Stems erect ....

28. All flowers borne in axillary clusters .... 23. *Lamiu*

28. Some or all the flowers in terminal inflorescences ....

29. Lower leaf surface densely short-white-hairy .... 19. *Agastac*

29. Lower leaf surface glabrous or pubescent, but not densely short-whi
      hairy ....

30. Calyx 15-nerved; inflorescence crowded, the flower clusters touch
      .... 19. *Agastac*

30. Calyx 5- to 10-nerved; inflorescence interrupted, most or all the flo
      clusters separated .... 24. *Stach*

31. All flowers borne in axillary clusters ....

31. Some or all the flowers borne in terminal inflorescences ....

32. Flowers and fruits on pedicels 1–2 mm long .... 25. *Meli*

32. Flowers and fruits sessile ....

33. Cauline leaves often lobed (except *L. marrubiastrum*); upper lip of
      rolla woolly .... 26. *Leonu*

33. Cauline leaves merely toothed or entire; upper lip of corolla not woolly
.... 27. *Galeopsis*
34. Calyx teeth spinescent .... 26. *Leonurus*
34. Calyx teeth not spinescent 35
35. Flowers and fruits on pedicels 36
35. Flowers and fruits sessile 37
36. Calyx teeth more or less equal; stamens 4 .... 29. *Physostegia*
36. Calyx 2-lipped; stamens 2 .... 28. *Perilla*
37. Calyx with lobes more or less equal, not bilabiate 38
37. Calyx bilabiate 40
38. Flowers in dense or loose inflorescences longer than broad 39
38. Flowers in dense heads as broad as or broader than long ....
.... 18. *Pycnanthemum*
39. Flowers much longer than the subtending bracts; stems glabrous;
leaves sessile .... 29. *Physostegia*
39. Flowers not longer than or only a little longer than the subtending
bracts; stems pubescent or, if glabrous, the leaves petiolate 24. *Stachys*
40. Bracts spinescent .... 30. *Dracocephalum*
40. Bracts not spinescent 41
41. Flowers in heads as broad as or broader than long; corolla less than 1
cm long .... 18. *Pycnanthemum*
41. Flowers in heads longer than broad; corolla 1 cm long or longer.......
.... 31. *Prunella*

1. *Isanthus* Michx.   False Pennyroyal

1. I brachiatus (L.) BSP. False Pennyroyal. July–Oct. Rocky woods, prairies; occasional throughout the state.

2. *Trichostema* L.   Blue Curls

1. T. dichotomum L. Blue Curls. Aug.–Oct. Woods, edge of fields, not common; scattered in the state.

3. *Mentha* L.   Mint

. All flowers in axils of the leaves 2
. At least some of the flowers in terminal inflorescences 4
. Uppermost leaves not more than three times longer than the cluster of
flowers .... 1. *M.* X *cardiaca*
. Uppermost leaves at least five times longer than the cluster of flowers 3
. Stems glabrous, or at least not retrorsely hairy .... 2. *M.* X. *gentilis*
. Stems retrorsely hairy .... 3. *M. arvensis*
. Leaves sessile or nearly so; spikes not interrupted 5
. Leaves petiolate; spikes more or less interrupted 7
. Plants glabrous or nearly so .... 4. *M. spicata*
. Plants downy-pubescent 6
. Spikes dense, mostly 1.2 cm thick or thicker ... 5. *M.* X *alopecuroides*
. Spikes very slender, mostly less than 1 cm thick ..... 6. *M. rotundifolia*
. Calyx lobes pubescent 8
. Calyx lobes glabrous .... 8. *M. citrata*
. Leaves sharply serrate, not crisped .... 7. *M.* X *piperita*
. Leaves lacerate, crisped .... 9. *M. crispa*

1. M. X cardiaca Gerarde. Little-leaved Mint. July–Oct. Native c Europe; occasionally adventive in Ill.; Champaign, Christian, Coles Lake, Whiteside, and Woodford cos. Reputed to be a hybrid betwee *M. spicata* and *M. arvensis.*

2. M. X gentilis L. Red Mint. July–Sept. Native of Europe; rarely ac ventive in Ill.; Henry Co. Reputed to be a hybrid between *M. spicat* and *M. arvensis.*

3. M. arvensis L. Field Mint. July–Sept. Leaves cuneate at base, th petioles longer than the cluster of flowers. Waste ground, very rare.

var. villosa (Benth.) S. R. Stewart. Field Mint. July–Sept. Marshe: low ground; occasional to common in most of the state. *M. canader sis* L.—J.

4. M. spicata L. Spearmint. June–Oct. Native of Europe; frequent planted and occasionally escaped; throughout the state.

5. M. X alopecuroides Hull. Foxtail Mint. July–Oct. Native (Europe; rarely adventive in Ill.; Cook, Effingham, Fayette, Kane, Pia and Schuyler cos. Reputed to be a hybrid between *M. longifolia* (L Huds. and *M. rotundifolia.*

6. M. rotundifolia (L.) Huds. Apple Mint. June–Sept. Native Europe; rarely adventive in Ill.; Hancock Co.

7. M. X piperita L. Peppermint. June–Oct. Native of Europe; o casionally adventive in moist waste ground. Reputed to be a hybr between *M. spicata* and *M. aquatica* L.

8. M. citrata Ehrh. Bergamot Mint. July–Oct. Native of Europe; n commonly adventive in moist ground.

9. M. crispa L. Curly Mint. July–Sept. Native of Europe; adventive Cook and Lake cos.

4. *Lycopus* L. Water Horehound

1. Some or all the leaves pinnatifid to pinnate . . . . 1. *L. american*
1. Leaves merely toothed, never pinnatifid . . . .
2. Calyx teeth broadly triangular, obtuse to subacute, up to 1 mm long
2. Calyx teeth narrowly triangular to subulate, acute, 1–2 mm long . . . .
3. Stem puberulent; base of plant not tuberous; stamens more or less cluded . . . . 2. *L. virginic*
3. Stem glabrous or nearly so; base of plant tuberous; stamens exsert . . . . 3. *L. uniflor*
4. Middle and lower leaves petiolate; base of plant not tuberous . . . .
. . . . 4. *L. rubel*
4. Middle and lower leaves sessile; base of plant tuberous . . . . 5. *L. asp*

1. L. americanus Muhl. Common Water Horehound. June–Oct. V ground, common; in every co.

2. L. virginicus L. Bugle Weed. July–Oct. Wet ground, occasior scattered throughout the state.

3. L. uniflorus Michx. Northern Bugle Weed. Aug.–Sept. Marsh calcareous fens, around lakes; occasional in the n. half of the sta also Pope Co.

4. L. rubellus Moench. Stalked Water Horehound. July–Oct. St and lower leaf surface glabrous or nearly so. Low woods, meadows, not common; scattered throughout the state.

var. arkansanus (Fresn.) Benner. July–Oct. Stem and lower leaf s face pubescent. Swampy woods, rare; Jackson and Union cos.

5. L. asper Greene. Rough Water Horehound. July–Aug. Usually in stagnant water, rare; confined to the n. one-fourth of the state.

5. *Teucrium* L.   Germander

1. T. canadense L. var. virginicum (L.) Eat. American Germander. June–Sept. Calyx and bracts without glandular hairs; lower leaf surface appressed-pubescent. Low woods, wet prairies, wet ditches and fields, common; in every co.

var. occidentale (Gray) McClintock & Epling. Gray Germander. July–Sept. Calyx and bracts glandular-pubescent; lower leaf surface spreading-pubescent. Low woods, wet prairies, wet fields; occasional in the n. three-fourths of the state. (Including *T. occidentale* Gray var. *boreale* [Bickn.] Fern.) *T. occidentale* Gray—F, J.

6. *Ajuga* L.   Bugleweed

Leaves glabrous or slightly pubescent; plants stoloniferous 1. *A. reptans*
Leaves densely soft-pubescent; plants tufted, without stolons . . . .
. . . . 2. *A. genevensis*

1. A. reptans L. Carpet Bugle. May–July. Native of Europe; frequently planted, but rarely escaped; Jackson, Kendall, and Lake cos.
2. A. genevensis L. Geneva Bugleweed. May–July. Native of Europe; occasionally cultivated, seldom escaped; Cook, DuPage, Jackson, Lake, and McHenry cos.

7. *Scutellaria* L.   Skullcap

Flowers solitary in the axils of the leaves or of the reduced leaf-like bracts 2
Flowers 2-several in racemes 4
Corolla at least 1.5 cm long . . . . 1. *S. epilobiifolia*
Corolla up to 1.2 cm long 3
Most of the leaves 2 cm long or longer; stolons without moniliform tubers; stems (at maturity) glabrous . . . . 2. *S. nervosa*
Most of the leaves up to 2 cm long; stolons with moniliform tubers; stems (at maturity) pubescent . . . . 3. *S. parvula*
Corolla up to 1 cm long; racemes produced from the axils of the leaves, lateral . . . . 4. *S. lateriflora*
Corolla 1 cm long or longer; most of the racemes terminal 5
Leaves cordate at the base . . . . 5. *S. ovata*
Leaves rounded to cuneate at the base 6
Calyx with glandular hairs . . . . 6. *S. elliptica*
Calyx with eglandular hairs . . . . 7. *S. incana*

1. S. epilobiifolia Muhl. Marsh Skullcap. June–Sept. Marshes; confined to the n. half of the state. *S. galericulata* L.—G, J.
2. S. nervosa Pursh. Veiny Skullcap. April–July. Low woods; occasional in the s. three-fourths of the state; also Carroll and JoDaviess cos. (Including var. *calvifolia* Fern. and f. *alba* Steyerm.)
3. S. parvula Michx. Small Skullcap. May–July. Pubescence glandular; lower leaf surface with sessile glands and long hairs. Rocky

woods, prairies, fields, limestone barrens; occasional throughout tʰ
state.

var. australis Fassett. Small Skullcap. May–July. Pubescence glaⁿ
dular; lower leaf surface with long hairs only. Prairies, rocky wooᵈ
occasional in the s. cos. *S. australis* (Fassett) Epling—G.

var. leonardii (Epling) Fern. Small Skullcap. May–July. Pubescenᵗ
eglandular. Prairies, rocky woods; scattered throughout the state.
leonardii Epling—G, J.

4. S. lateriflora L. Mad-dog Skullcap. June–Oct. Marshes, swamᵖ
woods, borders of rivers and streams; occasional to commᵐ
throughout the state.

5. S. ovata Hill. Heart-leaved Skullcap. May–Oct. Some or all tʰ
leaves over 4 cm long; plants generally taller than 25 cm; uppermᵒ
bracts longer than the calyx. Rocky woods; occasional in the s. oⁿ
third of the state.

var. versicolor (Nutt.) Fern. Heart-leaved Skullcap. May–Oct. Soᵐ
or all the leaves over 4 cm long; plants generally taller than 25 c
uppermost bracts shorter than the calyx. Rocky woods, rich wooᵉ
occasional throughout the state. *S. ovata* Hill—J.

var. rugosa (Wood) Fern. Heart-leaved Skullcap. May–Oct. Noneᵗ
the leaves 4 cm long; plants generally less than 25 cm tall. Limestᵒ
woods, rare; Monroe and Randolph cos.

6. S. elliptica Muhl. Hairy Skullcap. May–July. Dry, rocky woods;
stricted to the s. one-sixth of the state; also Sangamon Co. (Includᵢ
var. *hirsuta* [Short] Fern.)

7. S. incana Biehler. Downy Skullcap. June–Sept. Dry, rocky wooᵈ
occasional in the s. three-fourths of the state.

8. *Marrubium* L. Horehound

1. M. vulgare L. Common Horehound. May–Sept. Native of Eurᵒ
and Asia; naturalized in fields, pastures, and along roads; occasioⁿ
to common throughout the state.

9. *Cunila* L. Dittany

1. C. origanoides (L.) Britt. Dittany. July–Nov. Dry woods, sandstᵉ
cliffs; occasional to common in the s. one-third of the state.

10. *Monarda* L. Wild Bergamot

1. Only one head of flowers on each stem or stem-branch . . . .
1. Heads of flowers two or more on each stem or stem-branch . . . . . .
2. Leaves sessile or nearly so . . . . 1. *M. bradburi*
2. Leaves on petioles 5 mm long or longer . . . .
3. Bracts bright red or whitish; tip of upper lip of corolla beardless . .
3. Bracts green or pink-tinged; tip of upper lip of corolla bearded . . . .
   . . . . 4. *M. fistul*
4. Bracts bright red; corolla red; throat of calyx without a beard . . . . . .
   . . . . 2. *M. didy*
4. Bracts whitish; corolla white or pink; throat of calyx bearded . . . . . .
   . . . . 3. *M. clinopᵉ*
5. Corolla yellow, with purple spots; calyx teeth triangular . 5. *M. punc*
5. Corolla white or pink; calyx teeth subulate . . . . 6. *M. citrioᵈ*

1. M. bradburiana Beck. Monarda. April–June. Dry woods, bluffs, roadsides; occasional in the s. half of the state, rare in the n. half. *M. russeliana* Nutt.—F.

2. M. didyma L. Oswego Tea. June–Sept. Native of the e.U.S.; escaped from cultivation into woodlands; Cook, Hancock, Lake, and Wabash cos.

3. M. clinopodia L. Bee Balm. June–July. Woods; scattered in the s. three-fourths of the state.

4. M. fistulosa L. Wild Bergamot. May–Aug. Dry woods, fields, prairies, along roads; common throughout the state. (Including var. *mollis* [L.] Benth.)

5. M. punctata L. var. villicaulis Pennell. Horsemint. June–Oct. Stems villous or pilose. Sandy fields and woods, dunes, prairies; occasional in the n. half of the state, rare southward.

var. lasiodonta Gray. Horsemint. June–Oct. Stems canescent-pubescent. Sandy soil; Madison and St. Clair cos. *M. punctata* L. ssp. *occidentalis* Epling—G.

6. M. citriodora Cerv. Lemon Mint. May–Aug. Native of the w. U.S.; known as an escape from cultivation in Cook Co.

11. *Blephilia* Raf. *Pagoda Plant*

. At least the upper leaves cuneate to the nearly sessile base 1. *B. ciliata*
. Upper leaves petiolate, rounded at the base .... 2. *B. hirsuta*

1. B. ciliata (L.) Benth. Pagoda Plant. May–Aug. Open woods, fields, prairies; occasional throughout the state.

2. B. hirsuta (Pursh) Benth. Pagoda Plant. May–Sept. Rich woods; occasional throughout the state.

12. *Collinsonia* L.

1. C. canadensis L. Richweed. July–Sept. Rocky woods; occasional in the s. tip of the state; also Champaign and Clark cos.

13. *Hedeoma* Pers.

Leaves linear, entire, sessile; plants faintly aromatic ..... 1. *H. hispida*
Leaves oblong-ovate to elliptic, more or less toothed, petiolate; plants strongly aromatic .... 2. *H. pulegioides*

1. H. hispida Pursh. Rough Pennyroyal. May–July. Rocky woods, prairies; occasional in the n. half of the state, rare in the s. half of the state.

2. H. pulegioides (L.) Pers. American Pennyroyal. July–Sept. Rocky woods, fields, roadsides; occasional to common throughout the state.

14. *Salvia* L. *Sage*

Most leaves in a basal rosette 2
Most leaves cauline 3
Whorls of flowers more or less separated; leaves lyrate-pinnatifid; calyx not viscid ....1. *S. lyrata*

2. Whorls of flowers very remote; leaves crenate; calyx viscid . . . .
. . . . 2. *S. pratens*
3. Leaves not more than 2 cm broad; upper lip of calyx without teeth . .
3. Leaves more than 2 cm broad; upper lip of calyx 3-toothed . . . .
4. Leaves denticulate or serrate; corolla 15 mm long or longer; flowers 6
more per whorl . . . . 3. *S. azurea* var. *grandiflo.*
4. Leaves entire or sparsely serrate; corolla 8–12 mm long; flowers 1–3 p
whorl . . . . 4. *S. reflex*
5. Upper surface of leaves glabrous; calyx much longer than pedicel
fruit . . . . 5. *S. sylvestr*
5. Upper surface of leaves pubescent; calyx about as long as the pedic
in fruit . . . . 6. *S. verticilla*

1. S. lyrata L. Cancer-weed. April–June. Rich, open woods; c
casional in the s. one-sixth of the state.
2. S. pratensis L. Meadow Sage. June–July. Native of Europe; e
caped to an open woods in Lake Co.
3. S. azurea Lam. var. grandiflora Benth. Blue Sage. Native of t
w. U.S. Prairies, edge of woods; Champaign, DuPage, Johnson, Pia
Pope, and Tazewell cos. *S. pitcheri* Torr.—G, J.
4. S. reflexa Hornem. Rocky Mountain Sage. May–Oct. Native of
w. U.S. Dry woods, pastures, fields; occasional throughout the sta
except for the southernmost cos.
5. S. sylvestris L. Wild Sage. June–Sept. Native of Europe; adve
tive in pastures and along roads; Cook and Winnebago cos.
6. S. verticillata L. June–Sept. Native of Europe; adventive alon
railroad; Scott Co.

15. Thymus L. Thyme

1. T. serpyllum L. Creeping Thyme. July–Aug. Native of Europe;
casionally cultivated but rarely escaped in Ill.; Hancock and Hend
son cos.

16. Ocimum L. Basil

1. O. basilicum L. Basil. Aug.–Sept. Native of Africa and Asia;
casionally cultivated but rarely escaped in Ill.; Hancock Co.

17. Satureja L. Savory

1. Annual from fibrous roots; all leaves lance-linear; style branches equ
. . . . 1. *S. horten*
1. Perennial from stolons; leaves from stolons elliptic to ovate; s
branches very unequal . . . . 2. *S. arkans.*

1. S. hortensis L. Summer Savory. Aug.–Oct. Native of Europe
Asia; rarely escaped from cultivation; DuPage and Peoria cos.
2. S. arkansana (Nutt.) Briq. Low Calamint. May–Oct. Calcarec
fens, rocky soil, sand flats; occasional in the ne. one-fourth of
state. *Clinopodium arkansanum* (Nutt.) G. N. Jones—J; *Satureja*
bella (Michx.) Briq. var. *angustifolia* (Torr.) Svenson—G.

18. Pycnanthemum Michx.   Mountain Mint

1. Bracts whitened, at least on the upper surface 2
1. Bracts green 5
2. Calyx regular, the teeth up to 1 mm long .... 1. *P. muticum*
2. Calyx bilabiate, the lower teeth 1–2 mm long 3
3. Lower and middle cauline leaves green on the upper surface; calyx teeth acute to acuminate 4
3. Lower and middle cauline leaves whitened on the upper surface; calyx teeth obtuse .... 4. *P. albescens*
4. Corolla rose to purple; at least some part of the lower leaf surface hirsute .... 2. *P. pycnanthemoides*
4. Corolla white with purple spots; lower leaf surface not hirsute ....
.... 3. *P. incanum*
5. Stems glabrous; largest leaves up to 5 (–5.5) mm broad 5. *P. tenuifolium*
5. Stems pubescent, at least on the angles; largest leaves (5–) 6–40 mm broad 6
6. Outermost bracts glabrous or nearly so on the upper surface 7
6. Outermost bracts pubescent on the upper surface .... 8. *P. pilosum*
7. Calyx teeth 1–2 mm long, pilose throughout; leaves short-petiolate .....
.... 6. *P. torrei*
7. Calyx teeth up to 1 mm long, pubescent only at the tip; leaves sessile ..
.... 7. *P. virginianum*

1. P. muticum (Michx.) Pers. Mountain Mint. July–Sept. Low woods, very rare; Henderson and Wabash cos.

2. P. pycnanthemoides (Leavenw.) Fern. Mountain Mint. July–Aug. Open woods; confined to the s. one-fifth of the state.

3. P. incanum (L.) Michx. Gray Mountain Mint. July–Sept. Dry woods; confined to the s. one-fourth of the state. (Including var. *loomisii* [Nutt.] Fern.)

4. P. albescens Torr. & Gray. White Mountain Mnt. July–Sept. Cherty slopes, very rare; Union Co.

5. P. tenuifolium Schrad. Slender Mountain Mint. June–Sept. Woods, fields, prairies; occasional to common throughout the state. *P. flexuosum* [Walt.] BSP.—G, J.

6. P. torrei Benth. Mountain Mint. June–Oct. Dry woods, very rare; Alexander Co.

7. P. virginianum (L.) Dur. & Jacks. Common Mountain Mint. July–Sept. Marshes, calcareous fens, prairies; occasional to common in the n. half of the state, becoming less common southward.

8. P. pilosum Nutt. Hairy Mountain Mint. July–Sept. Dry woods, prairies; occasional in the cent. cos., less common to rare in the n. and s. cos.

19. Agastache Clayt.   Giant Hyssop

Corolla yellow; calyx teeth obtuse to subacute .... 1. *A. nepetoides*
Corolla purple or blue; calyx teeth acute to acuminate 2
Stems pubescent; corolla purple .... 2. *A. scrophulariaefolia*
Stems glabrous; corolla blue .... 3. *A. foeniculum*

1. A. nepetoides (L.) Ktze. Yellow Giant Hyssop. July–Sept. Open woods; occasional throughout the state.

2. A. scrophulariaefolia (Willd.) Ktze. Purple Giant Hyssop. Jul
Sept. Open woods; occasional in the n. half of the state; also Waba
Co. (Including var. *mollis* [Fern.] Heller.)

3. A. foeniculum (Pursh) Ktze. Blue Giant Hyssop. June–Sept. D
soil, very rare; Menard Co.

20. Nepeta L. Catnip

1. N. cataria L. Catnip. June–Sept. Native of Europe; naturalized
fields, open woods, along roads and railroads; occasional to comm
throughout the state.

21. Synandra Nutt.

1. S. hispidula (Michx.) Baill. Synandra. May–June. Rich woo
rare; Jackson and Williamson cos.

22. Glechoma L. Ground Ivy

1. G. hederacea L. Ground Ivy. April–July. Corolla more than 1.5 c
long. Native of Europe; naturalized in moist ground; not comm
DuPage, Lawrence, and Peoria cos.

var. micrantha Moricand. Ground Ivy. April–July. Corolla up to
cm long. Native of Europe; naturalized in moist ground and law
occasional throughout the state, becoming abundant northward.
heterophylla Waldst. & Kit.—J.

23. Lamium L. Dead Nettle

1. At least the upper leaves sessile .... 1. *L. amplexica*
1. All leaves petiolate. ....
2. Flowers up to 15 mm long .... 2. *L. purpure*
2. Flowers more than 15 mm long .... 3. *L. maculat*

1. L. amplexicaule L. Henbit. Feb.–Nov. Native of Europe, Asia, a
Africa; naturalized in waste ground; occasional to common throu
out the state. (Including f. *albiflorum* D. M. Moore.)

2. L. purpurem L. Purple Dead Nettle. April–Oct. Native of Eur
and Asia; naturalized in waste ground; occasional in the s. half of
state, uncommon in the n. half.

3. L. maculatum L. Spotted Dead Nettle. May–Aug. Native
Europe; adventive in waste ground; Cook, DuPage, and Kane cos.

24. Stachys L. Hedge Nettle

1. Stems pubescent on the sides ....
1. Stems glabrous on the sides, although sometimes pubescent on
angles ....
2. Calyx teeth triangular, shorter than the tube; leaves deeply cordat
the base .... 1. S. ridd
2. Calyx teeth lance-subulate, as long as the tube; leaves rounded or sl
lowly cordate at the base .... 2. S. palus
3. Leaves conspicuously nerved; margins of leaves sharply toothed ..

3. Leaves obscurely nerved; margins of leaves low-toothed . . . .
. . . . 5. *S. hyssopifolia* var. *ambigua*
4. Calyx uniformly pubescent throughout . . . . 3. *S. clingmanii*
4. Calyx glabrous or merely with setae on the nerves . . . . . 4. *S. tenuifolia*

1. S. riddellii House. Heart-leaved Hedge Nettle. June–July. Rich woods, very rare; Hardin Co.

2. S. palustris L. var. pilosa (Nutt.) Fern. Woundwort. June–Sept. Leaves obtuse to subacute, oblong to oval, sessile or on petioles less than 1 cm long. Moist soil, rare; Hancock Co.

var. homotricha Fern. Woundwort. June–Sept. Leaves acuminate, lanceolate, sessile or on petioles less than 1 cm long. Wet prairies, swampy or marshy soil; occasional to common in the n. half of the state, rare elsewhere. Including var. *nipigonesis* Jennings.) *S. arenicola* Britt.—J.

var. phaneropoda Weath. Woundwort. July–Sept. Leaves acuminate, on petioles 1 cm long or longer. Moist soil, rare; confined to extreme w. Ill.

3. S. clingmanii Small. Hedge Nettle. July–Sept. Rich, rocky woods, rare; Alexander, Hardin, and Massac cos.

4. S. tenuifolia Willd. Smooth Hedge Nettle. June–Sept. Leaves glabrous on both surfaces; calyx glabrous. Moist soil; occasional throughout the state.

var. hispida (Pursh) Fern. Hairy Hedge Nettle. June–Sept. Leaves hispid, at least on the veins beneath; calyx setose. Low woods, swamps, marshes; occasional in the n. half of the state, rare elsewhere. (Including var. *platyphylla* Fern.) *S. hispida* Pursh—G, J.

5. S. hyssopifolia Michx. var. ambigua Gray. Hyssop Hedge Nettle. June–Aug. Moist soil; occasional in the cent. cos., rare in the n. and s. cos. *S. aspera* Michx.—J.

25. *Melissa* L. Balm

1. M. officinalis L. Balm. June–Aug. Native of Europe; rarely adventive in Ill.; Jackson and Wabash cos.

26. *Leonurus* L. Motherwort

Some of the leaves palmately lobed; corolla bearded within . . . .
. . . . 1. *L. cardiaca*
All leaves unlobed; corolla beardless . . . . 2. *L. marrubiastrum*

1. L. cardiaca L. Motherwort. May–Aug. Native of Europe and Asia; naturalized in disturbed, shaded areas; occasional in the n. half of the state, less common in the s. half.

2. L. marrubiastrum L. Lion's-tail. June–Aug. Native of Europe and Asia; naturalized in disturbed soil; occasional in the n. two-thirds of the state.

27. *Galeopsis* L. Hemp Nettle

Stems bristly-pubescent at the nodes and swollen beneath the nodes . .
. . . . 1. *G. tetrahit*
Stems appressed-pubescent at the nodes and not swollen beneath the nodes . . . . 2. *G. ladanum*

1. G. tetrahit L. Common Hemp Nettle. June–Sept. Native of Europ and Asia; naturalized in waste places; Boone, Cook, and Hendersc cos.

2. G. ladanum L. Red Hemp Nettle. June–Sept. Native of Europ rarely adventive in waste ground; Cook Co.

28. Perilla L.

1. P. frutescens L. Beefsteak Plant. Aug.–Oct. Native of Asia; di turbed soil, particularly in damp areas; common in the s. one-four of the state, less common northward. (Including var. *crispa* [Benth Deane.)

29. Physostegia Benth. False Dragonhead

1. Leaves more or less undulate ....1. *P. intermed*
1. Leaves definitely serrate ....
2. Leaves (at least the upper) broadly rounded at base, the teeth rare more than 1 mm long; corolla rarely longer than 1.5 cm 2. *P. parviflo*
2. Leaves cuneate or subcuneate at base, the teeth regularly more than mm long; corolla 1.5–3.0 cm long ....
3. Upper leaves abruptly reduced in size; spike appearing pedunculate
3. Upper leaves gradually reduced in size; spike appearing sessile ......
....5. *P. specio*
4. Broadest leaves never exceeding a width of 1 cm; flowers remote ....
....3. *P. angustifo*
4. At least some of the leaves over 1 cm broad; some of the flowers ov lapping ....4. *P. virginia*

    1. P. intermedia (Nutt.) Engelm. & Gray. False Dragonhead. Ma July. Edge of slough, rare; Adams Co. *Dragocephalum intermedi* Nutt.—G.

    2. P. parviflora Nutt. Small-flowered False Dragonhead. July–C Moist prairies, rare; Kane, Lake, and Lee cos. *Draocephalum nutta Britt*—G.

    3. P. angustifolia Fern. Narrow-leaved False Dragonhead. Jur Sept. Low prairies; occasional in the n. half of the state.

    4. P. virginiana (L.) Benth. False Dragonhead. May–Sept. Moist s particularly prairies; occasional throughout the state. (Including *candida* Benke.)

    5. *P. speciosa* (Sweet) Sweet. False Dragonhead. May–Sept. L prairies; occasional throughout the state. *P. virginiana* var. *specio* (Sweet) Gray—F; *Dracocephalum formosius* (Lunell) Rydb.—G.

30. Dracocephalum L. Dragonhead

    1. D. parviflorum Nutt. American Dragonhead. May–Aug. Native the w. U.S. Dry soil, along railroads; not common, confined to the half of the state. *Moldavica parviflora* (Nutt.) Adans.—G.

31. Prunella L. Self-heal

    1. P. vulgaris L. Self-heal. May–Sept. At least the upper lea rounded at the base. Native of Europe; naturalized in lawns, fie and waste ground; occasional throughout the state.

var. lanceolata (Bart.) Fern. Self-heal. May–Sept. At least the upper leaves cuneate at the base. Disturbed woods, pastures, meadows, common; in every co.

140. SOLANACEAE

1. Plants woody, at least at base, often climbing or trailing 2
1. Plants herbaceous throughout, not climbing or trailing 3
2. At least some of the leaves with lobes near the base; flowers in terminal or axillary cymes .... 1. *Solanum*
2. None of the leaves lobed near the base; flowers solitary in the axils .... 2. *Lycium*
3. Leaves compound (the deeply pinnatifid, yellow-flowered, densely prickly *Solanum rostratum* might be sought here) 4
3. Leaves simple, although often deeply lobed 5
4. Flowers yellow; stems pilose .... 3. *Lycopersicum*
4. Flowers white, purplish, or blue; stems glabrous or appressed pubescent .... 1. *Solanum*
5. Stems and sometimes the leaves prickly .... 1. *Solanum*
5. Stems and leaves without prickles 6
6. None of the leaves petiolate, the uppermost clasping .. 4. *Hyoscyamus*
6. Some or all the leaves petiolate, the uppermost not clasping (although broadly rounded and sessile in some species of *Petunia*) 7
7. Leaves pinnatifid at least halfway to midvein .... 1. *Solanum*
7. Leaves entire or toothed, but never divided halfway to midvein 8
8. Corolla pale blue, campanulate .... 5. *Nicandra*
8. Corolla white or yellow (sometimes purplish in *Solanum americanum*), rotate or funnelform 9
9. Stamens exserted .... 1. *Solanum*
9. Stamens included 10
10. Corolla 7 cm long or longer; fruit prickly .... 6. *Datura*
10. Corolla up to 3 cm long; fruit not prickly 11
11. Flowers campanulate or rotate; fruit a berry enclosed in an inflated calyx .... 7. *Physalis*
11. Flowers funnelform; fruit a capsule 12
12. Flowers greenish-yellow; stamens all alike .... 8. *Nicotiana*
12. Flowers variously colored, but usually not greenish-yellow; one of the stamens considerably smaller than the others .... 9. *Petunia*

1. *Solanum* L. Nightshade

Plants prickly 2
Plants not prickly 5
Corolla yellow; calyx prickly .... 1. *S. rostratum*
Corolla purple or white; calyx not prickly 3
Stems hirsute, the stellate hairs sessile .... 2. *S. carolinense*
Stems with short tomentum, the stellate hairs stipitate 4
Leaves linear-lanceolate to narrowly oblong, silvery-gray; calyx lobes linear .... 3. *S. elaeagnifolium*
Leaves ovate, green; calyx lobes ovate .... 4. *S. torreyi*
Plants woody, climbing; berry red .... 5. *S. dulcamara*
Plants herbaceous, not climbing; berry green or black 6

6. Leaves compound .... 6. *S. tuberosur*
6. Leaves simple, although sometimes deeply pinnatifid ....
7. Leaves deeply pinnatifid; berry green .... 7. *S. triflorur*
7. Leaves entire; berry black .... 8. *S. americanur*

1. S. rostratum Dunal. Buffalo-bur. July–Oct. Native of the w. U.S
naturalized in waste ground; occasional throughout the state.
2. S. carolinense L. Horse-nettle. June–Oct. Open woods and wast
places; in every co. (Including f. *albiflorum* [Kuntze] Benke.)
3. S. elaeagnifolium Cav. Silvery Horse-nettle. July–Oct. Native o
the w. U.S.: adventive in waste areas; Adams, Cook, and Crawfo
cos.
4. S. torreyi Gray. July–Sept. Native of the w. U.S.; adventive
waste areas; Henry Co.
5. S. dulcamara L. Bittersweet Nightshade. June–Oct. Native
Europe; naturalized in waste areas, thickets, and boggy areas; cor
mon in the n. half of the state, occasional in the s. half. (Including va
villosissimum Desv. and f. *albiflorum* House).
6. S. tuberosum L. Potato. June–Aug. Native of S. Am.; rarely e
caped from cultivation.
7. S triflorum Nutt. June–Aug. Cut-leaved Nightshade. Native of tl
w. U.S.; adventive in waste ground; occasional in the n. cos.
8. S. americanum Mill. Black Nightshade. May–Nov. Woods, was
ground; in every co. *S. nigrum* L.—G, J.

2. *Lyceum L.   Matrimony Vine*

1. Leaves gray-green; lobes of calyx obtuse .... 1. *L. halimifoliu*
1. Leaves dark green; lobes of calyx acute .... 2. *L. chinen*

1. L. halimifolium Mill. Common Matrimony Vine. June–Sept. Nati
of Europe; occasionally naturalized in waste ground; scatter
throughout the state.
2. L. chinense Mill. Chinese Matrimony Vine. June–Sept. Native
Asia; rarely adventive; Champaign Co.

3. *Lycopersicum Mill.*

1. L. esculentum Mill. Tomato. May–Sept. Native of trop. Am.; rar
escaped from cultivation.

4. *Hyoscyamus L.   Henbane*

1. H. niger L. Black Henbane. June–Aug. Native of Europe; rar
escaped in Ill.; McHenry and Peoria cos.

5. *Nicandra Adans.   Apple-of-Peru*

1. N. physalodes (L.) Pers. Apple-of-Peru. July–Sept. Native of Pe
naturalized in waste ground; scattered throughout the state.

6. *Datura L.   Jimsonweed*

1. Plants glabrous; leaves coarsely toothed; corolla to 10 cm long .....
.... 1. *D. stramoni*

1. Plants pubescent; leaves entire or undulate; corolla 15–20 cm long ....
.... 2. *D. innoxia*

    1. D. stramonium L. Jimsonweed. July–Oct. Stem green; corolla white. Native of Asia; naturalized in waste areas; in every co.
    var. tatula (L.) Torr. Purple Jimsonweed. July–Oct. Stem purple; corolla purplish. Native of Asia; naturalized in waste areas; occasional throughout the state.
    2. D. innoxia Mill. Aug.–Oct. Native of trop. Am.; naturalized in waste areas; occasional throughout the state. *D. meteloides* DC.—G.

7. *Physalis L. Ground Cherry*

1. Flowers yellow; calyx during fruiting green or brownish 2
1. Flowers white; calyx during fruiting red or scarlet .... 14. *P. alkekengi*
2. Stems glabrous or nearly so, never uniformly villous, hispid, hirsute, or glandular-hairy 3
2. Stems uniformly villous, hispid, hirsute, or glandular-hairy 8
3. Annuals without rhizomes; leaves mostly conspicuously dentate 4
3. Perennials with rhizomes; leaves entire or merely repand-dentate ... 6
4. Corolla with black spots near base; peduncles up to 6 mm long; calyx often puberulent, with deltoid teeth; berry purple .... 1. *P. ixocarpa*
4. Corolla without a darkened center; peduncles 10 mm long or longer; calyx usually glabrous, with acute to attenuate teeth; berry yellow ... 5
5. Calyx in fruit strongly 10-angled; peduncles 1–2 cm long 2. *P. angulata*
5. Calyx in fruit obscurely 10-angled; peduncles nearly all over 2 cm long
.... 3. *P. pendula*
6. Leaves rather thick, lanceolate to linear, on petioles up to 3 cm long; berry yellow .... 4. *P. longifolia*
6. Leaves membranaceous, ovate to ovate-oblong, at least some of them on petioles up to 6 cm long; berry red or purple 7
7. Leaves opaque; fruiting calyx ovoid, to 3 cm long, sunken at the base ..
.... 5. *P. subglabrata*
7. Leaves translucent; fruiting calyx pyramidal, to 6 cm long, deeply sunken at the base .... 6. *P. macrophysa*
8. Pubescence of stem glandular-viscid 9
8. Pubescence of stem not glandular-viscid 12
9. Annuals without rhizomes; fruiting calyx abruptly acuminate 10
9. Perennials with rhizomes; fruiting calyx merely acute 11
9. Fruiting calyx long-tapering, 3–4 cm long; reticulations between the lateral nerves obscure .... 7. *P. barbadensis*
9. Fruiting calyx short-tapering, 2–3 cm long; reticulations between the lateral nerves prominent .... 8. *P. pruinosa*
. Leaves more or less tapering to the base, most of them less than 6 cm long; berry red .... 9. *P. virginiana*
. Leaves more or less rounded or cordate at the base, many of them over 6 cm long; berry yellow .... 10. *P. heterophylla*
. Peduncles during flowering up to 5 mm long; annuals without rhizomes
.... 11. *P. pubescens*
. Peduncles during flowering 1 cm long or longer; perennials with rhizomes 13
. Stems villous; calyx in fruit shallowly or deeply sunken at the base; leaves mostly oblong to ovate-lanceolate to ovate. 14
. Stems hispid or hirsute; calyx in fruit not sunken at the base; leaves lanceolate to oblanceolate .... 13. *P. lanceolata*

14. Leaves more or less tapering to the base .... 1!

14. Leaves more or less rounded or cordate at the base 10. *P. heterophylla*

15. Some or all the hairs branched; calyx in fruit mostly 4 cm long or longer shallowly sunken at the base .... 12. *P. pumila*

15. All the hairs unbranched; calyx in fruit mostly less than 4 cm long deeply sunken at the base .... 9. *P. virginiana*

1. P. ixocarpa Brotero. Tomatillo. July–Aug. Native of the sw. U.S. Mex., and S. Am.; rarely adventive in Ill.; DuPage, Lake, and Wabash cos.

2. P. angulata L. Ground Cherry. July–Sept. Native of the s. U.S. infrequently escaped to waste areas in the cent. cos.

3. P. pendula Rydb. Ground Cherry. July–Sept. Native of the sw U.S.; rarely adventive in Ill.; Alexander, Cook, DuPage, Jackson, an Union cos.

4. P. longifolia Nutt. Ground Cherry. June–Aug. Native of the w U.S.: rarely adventive in Ill.; Jackson Co.

5. P. subglabrata Mack. & Bush. Smooth Ground Cherry. June Sept. Disturbed areas; common throughout the state. *P. longifolia* Nutt.—G.

6. P. macrophysa Rydb. Large-fruited Ground Cherry. June–July Native of the sw. U.S.; adventive in waste ground; Champaign an Peoria cos.

7. P. barbadensis Jacq. Ground Cherry. June–Sept. Native of trop Am.; rarely escaped in Ill. Jackson Co.

8. P. pruinosa L. Ground Cherry. June–Oct. Disturbed areas, o casional; scattered throughout the state.

9. P. virginiana Mill. Ground Cherry. May–Aug. Woods, disturbed areas; occasional throughout the state.

10. P. heterophylla Nees. Ground Cherry. May–Sept. At least some the pubescence on the stem glandular. Woods, disturbed areas; o casional throughout the state.

var. ambigua (Gray) Rydb. Ground Cherry. May–Sept. Stems with out glandular hairs; leaves thick, dentate. Waste areas; occasional common throughout the state.

var. nyctaginea (Dunal) Rydb. Ground Cherry. May–Sept. Stems without glandular hairs; leaves thin, usually entire. Waste areas, n common; apparently confined to the s. half of the state.

11. P. pubescens L. Annual Ground Cherry. June–Oct. Disturbed areas, occasional; throughout the state.

12. P. pumila Nutt. Dwarf Ground Cherry. June–Aug. Dry hillside very rare; Peoria Co.

13. P. lanceolata Michx. Narrow-leaved Ground Cherry. May–Sept Native of the w. U.S.; rarely adventive in Ill. Champaign, Mason, an Union cos.

14. P. alkekengi L. Chinese Lantern. July–Sept. Native of Asia; rare escaped from cultivation.

8. *Nicotiana* L. Tobacco

1. N. rustica L. Wild Tobacco. Sept.–Oct. Native of Peru; adventive in waste areas; Cook, Menard, and Peoria cos.

9. *Petunia* Juss. *Petunia*

1. Corolla white, salverform .... 1. *P. axillaris*
1. Corolla red, violet, or purple, funnelform 2
2. Corolla 3–4 cm long .... 2. *P. violacea*
2. Corolla 5–9 cm long .... 3. *P. X hybrida*

    1. P. axillaris (Lam.) BSP. White Petunia. July–Sept. Native of S. Am.; rarely escaped from cultivation; Jackson and Lake cos.
    2. P. violacea Lindl. Violet Petunia. June–Aug. Native of S. Am.; occasionally escaped from cultivation.
    3. P. X hybrida Vilm. Garden Petunia. June–Aug. Native of S. Am.; rarely escaped from cultivation.

141. SCROPHULARIACEAE

1. Tree with ovate, cordate leaves up to 30 cm long and nearly as broad or broader; corolla to 6 cm long, violet and yellow .... 1. *Paulownia*
1. Herbs with various leaves but not as above; corolla various, but not as above 2
2. Leaves whorled .... 2. *Veronicastrum*
2. Leaves opposite or alternate 3
3. All leaves opposite (excluding bracteal leaves) 4
3. At least some of the leaves alternate (excluding bracteal leaves) 22
4. Anther-bearing stamens 2 5
4. Anther-bearing stamens 4 7
5. Calyx and corolla lobes each 4; fruit heart-shaped .... 3. *Veronica*
5. Calyx and corolla lobes each 5; fruit subacute or rounded at tip, not heart-shaped 6
6. Sterile stamens 2; bractlets absent .... 5. *Lindernia*
6. Sterile stamens absent, or minute; each flower subtended by 2 bractlets .... 4. *Gratiola*
7. One elongated sterile stamen present 8
7. Sterile stamen absent, or represented only by a gland as broad as long 9
8. Flowers sessile .... 6. *Chelone*
8. Flowers pedicellate .... 7. *Penstemon*
9. Some or all of the leaves pinnatifid or pinnately lobed (excluding *Gerardia auriculata* which has 2 small basal lobes) 10
9. Leaves toothed or entire, not pinnatifid 13
10. Corolla lavender, to 5 mm long; plants never exceeding 20 cm in height .... 8. *Conobea*
10. Corolla yellow, at least 15 mm long; plants more than 20 cm tall ... 11
11. Stamens included 12
11. Stamens exserted .... 11. *Seymeria*
12. Corolla bilabiate; calyx asymmetrical; capsule flattened . 9. *Pedicularis*
12. Corolla not bilabiate; calyx usually symmetrical; capsule globose to ovoid .... 10. *Gerardia*
13. Flowers 1–2 in the axils of the cauline leaves 14
13. Flowers 3 or more in spikes, racemes, panicles, or corymbs 18
14. Leaves entire 15
14. Leaves toothed 17
15. Leaves orbicular .... 12. *Bacopa*
15. Leaves filiform to linear to elliptic 16

16. Calyx lobes 5; corolla slightly 2-lipped; seeds usually more than 4 pe capsule .... 10. *Gerardi*
16. Calyx lobes 4; corolla strongly 2-lipped; seeds 2–4 per capsule .... 13. *Melampyrur*
17. Calyx 5-parted nearly to base; flowers up to 1 cm long, white with purpl lines .... 12. *Bacop*
17. Calyx tubular, with 5 lobes; flowers over 1 cm long, violet, or, if less tha 1 cm long, yellow .... 14. *Mimulu*
18. Flowers sessile 1
18. Flowers pedicellate 2
19. Corolla campanulate; leaves usually with a pair of lobes at the base otherwise entire .... 10. *Gerardi*
19. Corolla salverform; leaves toothed .... 15. *Buchner*
20. Stems creeping; sterile stamens absent .... 16. *Mazu*
20. Stems erect; sterile stamen reduced to a gland or scale 2
21. All leaves petiolate; corolla purple or greenish-purple; coarse perennial .... 17. *Scrophular*
21. Middle and upper leaves sessile; corolla blue and white or violet-purp .... 18. *Collins*
22. Corolla strongly spurred at base 2
22. Corolla not spurred at base (occasionally saccate at base in *A tirrhinum*) 2
23. Stems glabrous 2
23. Stems pubescent 3
24. Plants trailing or twining; leaves suborbicular, lobed or crenulate .... 19. *Cymbalar*
24. Plants erect; leaves linear to lanceolate, entire .... 20. *Linar*
25. Stems spreading or prostrate; fruit glabrous; leaves ovate to oblong . ... 21. *Kickx*
25. Stems erect; fruit puberulent; leaves linear to linear-lanceolate .... 22. *Chaenorrhinu*
26. Stamens 5; corolla more or less actinomorphic .... 23. *Verbascu*
26. Stamens 2 or 4; corolla zygomorphic .... *
27. Leaves and lower part of the stem glabrous; leaves entire .... 24. *Antirrhinu*
27. Leaves and stems pubescent; leaves toothed or lobed ....
28. Leaves merely toothed; calyx about as long as the greenish-yellow c rolla .... 25. *Wulfen*
28. Some or all the leaves pinnatifid or deeply lobed; calyx shorter than t scarlet, red, or yellow corolla ....
29. Margins of the lobed leaves toothed; leaves divided into 12 or mo lobes .... 9. *Pedicula*
29. Margins of the lobed leaves entire; leaves divided into 3–7 lobes .... 26. *Castille*

1. *Paulownia* Sieb. & Zucc.

1. *P. tomentosa* (Thunb.) Steud. Princess Tree. April–May. Native China; adventive in the s. one-third of the state.

2. *Veronicastrum* Fab.

1. *V. virginicum* (L.) Farw. Culver's-root. June–Sept. Woo prairies; occasional throughout the state. (Including f. *villosum* [Ra Pennell.)

3. Veronica L. Speedwell

1. Flowers solitary in the axils of the cauline leaves 2
1. Flowers in racemes or spikelike inflorescences (rarely solitary in *V. serpyllifolia*) 6
2. Corolla white; leaves glabrous .... 1. *V. peregrina*
2. Corolla blue or lilac; leaves pubescent 3
3. Corolla 8–12 mm across; capsule 6–10 mm long .... 2. *V. persica*
3. Corolla up to 5 mm across; capsule less than 5 mm long 4
4. Leaves reniform-suborbicular, 3- to 7-lobed, broader than long; capsule not compressed .... 3. *V. hederaefolia*
4. Leaves ovate to oval, or the upper ones linear to lanceolate, longer than broad; capsule compressed 5
5. Pedicels 1–2 mm long; seeds smooth .... 4. *V. arvensis*
5. Pedicels 4–12 mm long; seeds rugulose .... 5. *V. polita*
6. Stems glabrous (or with upcurved hairs less than 0.5 mm long in *V. serpyllifolia*). 7
6. Stems pubescent 10
7. Racemes terminal; leaves up to 1.5 cm long .... 6. *V. serpyllifolia*
7. Racemes axillary; leaves, or most of them, over 1.5 cm long 8
8. Leaves linear to linear-lanceolate; axis of raceme zigzag 7. *V. scutellata*
8. Leaves lanceolate to oblong to ovate; axis of raceme straight 9
9. Leaves sessile, sometimes clasping; pedicels glandular-pubescent .... 8. *V. comosa*
9. Leaves petiolate; pedicels eglandular .... 9. *V. americana*
10. Racemes terminal; leaves acuminate .... 10. *V. longifolia*
10. Racemes axillary; leaves obtuse to subacute 11
11. Leaves petiolate; raceme densely flowered; pedicels shorter than the calyx .... 11. *V. officinalis*
11. Leaves sessile or very short-petiolate; raceme loosely flowered; pedicels about as long as the calyx or longer .... 12. *V. chamaedrys*

1. V. peregrina L. White Speedwell. April–Aug. Stem and capsules glabrous. Fields, disturbed soil, common; in every co.
var. xalapensis (HBK.) St. John. April–Aug. Stem and capsules glandular-pubescent. Fields, disturbed soil, occasional but not as common as the preceding.
2. V. persica Poir. Bird's-eye Speedwell. April–Aug. Native of Europe and Asia; occasionally naturalized in Ill.; Coles, DuPage, Jackson, Lake, Peoria, and Wabash cos.
3. V. hederaefolia L. Ivy-leaved Speedwell. March–April. Native of Europe; adventive in a grassy meadow; Jackson and Union cos.
4. V. arvensis L. Corn Speedwell. March–Aug. Native of Europe; fields, woods, waste ground, common; in every co.
5. V. polita Fries. Speedwell. March–May. Native of Europe; adventive in lawns, occasional; Jackson, Lawrence, Piatt, Randolph, and Saline cos. *V. didyma* Tenore—J; *V. agrestis* L.—G.
6. V. serpyllifolia L. Thyme-leaved Speedwell. April–July. Native of Europe; adventive in lawns and meadows; scattered throughout the state.
7. V. scutellata L. Marsh Speedwell. April–July. Moist soil, often around ponds, rare; confined to the n. half of the state.
8. V. comosa Richter. Water Speedwell. June–Sept. Wet ditches, shores; occasional in the n. half of the state, rare or absent elsewhere. *V. catenata* Pennell—J; *V. salina* Schur.—G.

386 SCROPHULARIACEA

9. V. americana (Raf.) Schwein. American Brooklime. June–Aug
Swampy ground, rare; confined to the n. half of the state.

10. V. longifolia L. Garden Speedwell. May–Sept. Native of Europe
rarely escaped from cultivation; Cook Co.

11. V. officinalis L. Common Speedwell. May–Sept. Native of Europe
adventive in grassy areas and open woods; Cook, DuPage, and Rich
land cos.

12. V. chamaedrys L. Germander Speedwell. May–July. Native o
Europe; rarely adventive in Ill.; DuPage Co.

4. Gratiola L. Hedge Hyssop

1. Leaves toothed; sterile stamens absent or minute; corolla pale yellow o
white; annuals ....
1. Leaves entire or remotely toothed; sterile stamens 2; corolla golden
yellow .... 3. *G. aure*
2. Pedicels to 10 mm long (in fruit); stems glabrous, more or less fleshy
.... 1. *G. virginiar*
2. Pedicels over 10 mm long (in fruit); stems more or less pubescent, n
fleshy .... 2. *G. neglec*

1. G. virginiana L. Round-fruited Hedge Hyssop. April–Oct. Moi
soil; occasional to uncommon throughout the state.

2. G. neglecta Torr. Clammy Hedge Hyssop. May–Oct. Moist so
occasional throughout the state.

3. G. aurea Muhl. Goldenpert. June–Sept. Wet ground, very rar
Cook Co.

5. Lindernia All. False Pimpernel

1. Pedicels of all flowers longer than the subtending leaves ....
.... 1. *L. anagallide*
1. Pedicels of the lowermost flowers shorter than the subtending leaves
.... 2. *L. dub*

1. L. anagallidea (Michx). Pennell. Slender False Pimpernel. Jun
Oct. Moist soil; occasional throughout the state, except for the n
cos. where it is rare or absent.

2. L. dubia (L.) Pennell. False Pimpernel. June–Oct. All pedice
shorter than the subtending leaves. Moist ground; common throug
out the state.

var. riparia (Raf.) Fern. False Pimpernel. June–Oct. Uppermost p
dicels longer than the subtending leaves. Moist ground; comm
throughout the state. *L. dubia* (L.) Pennell—J.

6. Chelone L. Turtlehead

1. Corolla greenish-yellow to creamy-white; leaves sessile or on petioles
to 5 mm long .... 1. *C. glab*
1. Corolla rose-purple; leaves on petioles 5 mm long or longer ....
.... 2. *C. obliq*

1. C. glabra L. White Turtlehead. July–Oct. Wet soil; occasional
the n. half of the state, rare in the s. half. (Including var. *linifo*
Coleman, and f. *tomentosa* [Raf.] Pennell.)

2. C. obliqua L. var. speciosa Pennell & Wherry. Pink Turtlehead. Aug.–Oct. Low woods, swampy meadows, not common; known from the s. one-third of the state; also Adams, Fulton, Hancock, Henderson, and Tazewell cos.; adventive in DuPage Co. *C. obliqua* L.—G, J.

7. Penstemon Mitchell Beard-tongue

1. Corolla 3.5–5.0 cm long 2
1. Corolla up to 3 cm long 3
2. Leaves glaucous, broadly ovate, entire .... 1. *P. grandiflorus*
2. Leaves green, narrowly ovate, sharply toothed .... 2. *P. cobaea*
3. Corolla glandular-pubescent within .... 3. *P. tubaeflorus*
3. Corolla eglandular within 4
4. The two lips of the corolla nearly equal in size and shape; sterile filament sparsely to moderately pubescent 5
4. The two lips of the corolla unequal in size and shape; sterile filament densely bearded 8
5. Anthers pubescent with short stiff hairs 6
5. Anthers glabrous .... 7. *P. calycosus*
6. Sepals during anthesis 5–8 mm long; capsules 9–12 mm long .... 4. *P. digitalis*
6. Sepals during anthesis 2–5 mm long; capsules 6–9 mm long 7
7. Leaves acuminate, sharply serrate; sepals at fruiting time 5 mm long or longer .... 5. *P. alluviorum*
7. Leaves obtuse to subacute, entire or sparsely toothed; sepals at fruiting time 3–4 mm long .... 6. *P. deamii*
8. Corolla up to 2.3 cm long; stems without glandular pubescence 9
8. Corolla more than 2.3 cm long; stems glandular-pubescent .... 10. *P. hirsutus*
9. Leaves pubescent; throat of corolla longer than the tube 8. *P. pallidus*
9. Leaves glabrous or nearly so; throat of corolla about as long as the tube .... 9. *P. arkansanus*

1. P. grandiflorus Nutt. Large-flowered Bear-tongue. May–June. Sandy soil, very rare; Henderson Co.; also adventive in McHenry Co.
2. P. cobaea Nutt. June. Native s. and w. of Ill.; in a flat, dry meadow in Ill.; Kane Co.
3. P. tubaeflorus Nutt. Beard-tongue. May–June. Prairies, dry woods; not common in the s. half of the state, rare in the n. half.
4. P. digitalis Nutt. Foxglove Beard-tongue. May–July. Woods, thickets; occasional throughout the state.
5. P. alluviorum Pennell. Lowland Beard-tongue. May–June. Swampy woods, rare; Union Co.
6. P. deamii Pennell. Deam's Beard-tongue. May–June. Dry woods, rare; Jackson, Monroe, Pope, Union, and Wabash cos.
7. P. calycosus Small. Smooth Beard-tongue. May–July. Wooded slopes, edge of woods; occasional throughout the state.
8. P. pallidus Small. Pale Beard-tongue. April–July. Dry woods, prairies; common in the s. half of the state, occasional in the n. half.
9. P. arkansanus Pennell. Beard-tongue. April–June. Wooded slopes, very rare; Jackson and Randolph cos. *P. pallidus* Small—G.
10. P. hirsutus (L.) Willd. Hairy Beard-tongue. June–July. Gravelly prairies, wooded slopes, not common; Cook, Grundy, Hardin, Union, and Will cos.

8. Conobea Aubl.

1. C. multifida (Michx.) Benth. May–Oct. Moist soil, fields, prairies; occasional in the s. four-fifths of the state, rare or absent elsewhere. *Leucospora multifida* (Michx.) Nutt.—G, J.

9. Pedicularis L. Lousewort

1. Most of the cauline leaves opposite, sessile; corolla creamy or yellowish white; flowers appearing from August to October ...... 1. *P. lanceolata*
1. Most of the cauline leaves alternate, petiolate; corolla yellow; flowers appearing from April to June .... 2. *P. canadensis*

    1. P. lanceolata Michx. Swamp Wood Betony. Aug.–Oct. Calcareous fens, wet meadows; occasional in the n. half of the state, uncommon in the s. half.
    2. P. canadensis L. Lousewort. April–June. Woods, prairies; occasional throughout the state. (Including var. *dobbsii* Fern.)

10. Gerardia L. Gerardia

1. Corolla yellow; some or all the leaves pinnately divided ....
1. Corolla pink or purple; all leaves entire, or merely 3-lobed ....
2. Stems glabrous and glaucous .... 1. *G. flava*
2. Stems pubescent and usually not glaucous ....
3. Corolla 4.5 cm long or longer; pubescence not glandular; capsule rusty pubescent 1.2–2.5 cm long .... 2. *G. grandiflora*
3. Corolla 3–4 cm long; pubescence glandular; capsule glandular-pubescent, up to 1.2 cm long .... 3. *G. pedicularia*
4. Leaves with two basal lobes; stems retrorsely hispid ... 4. *G. auriculata*
4. Leaves without lobes; stems glabrous or scabrous ....
5. Plants yellow-green, usually not turning black upon drying; calyx strongly reticulate-nerved; seeds yellow ....
5. Plants deep green or purplish, usually turning black upon drying; calyx weakly reticulate-nerved, or nerveless; seeds dark brown to black ...
6. Branches with solitary flowers; stems terete .... 5. *G. gattingeri*
6. Branches with 2 or more flowers; stems more or less angled ....
.... 6. *G. skinneriana*
7. Pedicels more than twice as long as the calyx .... 7. *G. tenuifolia*
7. Pedicels less than twice as long as the calyx ....
8. Capsule ellipsoid, longer than broad; pedicels 5–8 mm long 8. *G. aspera*
8. Capsule globose or subglobose, about as broad as long; pedicels to mm long ....
9. Axillary fascicles of leaves as long as the main leaves; stem scabrous .... 9. *G. fasciculata*
9. Axillary fascicles of leaves much shorter than the main leaves or lacking; stem smooth or nearly so ....
10. Calyx up to half as long as the capsule; style to 1 cm long; flowers usually up to 2 cm long .... 10. *G. paupercula*
10. Calyx half as long as the capsule or longer; style 1.5–2.0 cm long; flowers usually over 2 cm long .... 11. *G. purpurea*

    1. G. flava L. Smooth False Foxglove. June–Sept. Rocky woods; occasional in the s. half of the state, less common in the n. half. (Including var. *macrantha* [Pennell] Fern.) *Aureolaria flava* (L.) Farw.—G, J.

2. G. grandiflora Benth. var. pulchra (Pennell) Fern. Yellow False Foxglove. July–Sept. Woods; occasional in the n. two-thirds of the state, absent elsewhere. *Aureolaria grandiflora* (Benth.) Pennell—G, J.

3. G. pedicularia L. var. ambigens Fern. Clammy False Foxglove. Aug.–Sept. Sandy woods; occasional in the ne. cos., absent elsewhere. *Aureolaria pedicularia* (L.) Raf.—G, J.

4. G. auriculata Michx. Auriculate False Foxglove. Aug.–Sept. Dry prairies; occasional in the n. three-fourths of the state, absent elsewhere. *Tomanthera auriculata* (Michx.) Raf.—G.

5. G. gattingeri Small. Round-stemmed False Foxglove. Aug.–Oct. Dry, often rocky, woods; uncommon to occasional throughout the state.

6. G. skinneriana Wood. Pale False Foxglove. July–Sept. Dry prairies, rocky woods; not common but scattered throughout the state.

7. G. tenuifolia Vahl. Slender False Foxglove. Aug.–Oct. Calyx up to 1 mm long; axillary fascicles of leaves rarely present. Moist soil; occasional throughout the state.

var. macrophylla Benth. Aug.–Oct. Calyx 1–2 mm long; auxillary fascicles of leaves commonly present. Moist soil; occasional throughout the state. (Including var. *parviflora* Nutt.)

8. G. aspera Dougl. Rough False Foxglove. Aug.–Sept. Prairies, not common but scattered throughout the state.

9. G. fasciculata Ell. False Foxglove. Aug.–Sept. Edge of woods, sandy open ground, not common; confined to the s. one-sixth of the state.

10. G. paupercula (Gray) Britt. False Foxglove. Aug.–Sept. Moist soil, rare; scattered throughout the state. *G. purpurea* L. var. *paupercula* Gray—G. (Including var. *borealis* [Pennell] Deam.)

11. G. purpurea L. False Foxglove. July–Sept. Moist soil; occasional in the n. three-fourths of the state, less common in the s. one-fourth.

11. Seymeria Pursh

1. S. macrophylla Nutt. Mullein Foxglove. June–Sept. Rich woods, rocky slopes, dry woods, thickets; occasional throughout the state. *Dasistoma macrophylla* (Nutt.) Raf.—G, J.

12. Bacopa Aubl.   Water Hyssop

1. Leaves entire, more or less orbicular; pedicels bractless . . . . 1. *B. rotundifolia*
1. Leaves serrate, oblanceolate; pedicels with 2 bracts . . 2. *B. acuminata*

1. B. rotundifolia (Michx.) Wettst. Water Hyssop. May–Sept. Wet ground or in shallow water; occasional in the s. half of the state, rare or absent elsewhere.

2. B. acuminata (Walt.) Robins. July–Aug. Wet roadside ditch, very rare; Wabash Co. *Mecardonia acuminata* (Walt.) Small—G, J.

13. Melampyrum L.   Cow-wheat

1. M. lineare Desr. var. latifolium Bart. Cow-wheat. June–Aug. Bogs, marshes, dunes, very rare; Cook Co. *M. lineare* Desr.—J.

14. *Mimulus* L. Monkey-flower

1. Corolla yellow, up to 1.2 cm long; leaves oval to suborbicular, entire o
denticulate .... 1. *M. glabratus*
1. Corolla purple, at least 2 cm long; leaves lanceolate to elliptic to ovate
serrate 2
2. Leaves petiolate; stem winged .... 2. *M. alatus*
2. Leaves sessile or even clasping; stem wingless .... 3. *M. ringens*

1. M. glabratus HBK. var. fremontii (Benth.) Grant. Yellow Monkey
flower. June–Oct. Around springs, rare; confined to the n. half of the
state. *M. geyeri* Torr.—J.
2. M. alatus Ait. Winged Monkey-flower. June–Sept. Wet ground
common in the s. half of the state, becoming rare in the n. half. A
hybrid between this species and *M. ringens* L. has been collected i
Cass Co.
3. M. ringens L. Monkey-flower. June–Sept. Most or all the caulin
leaves rounded or clasping at the base. Wet ground; occasional in th
n. three-fourths of the state, rare elsewhere.
var. minthodes (Greene) Grant. Monkey-flower. June–Sept. Most o
all the cauline leaves tapering to the base. Wet ground; occasional i
the westernmost cos. of the state.

15. *Buchnera* L.

1. B. americana L. Blue Hearts. July–Sept. Prairies, fields, not com
mon; scattered in Ill.

16. *Mazus* Lour.

1. M. japonicus (Thunb.) Kuntze. July–Oct. Native of Asia; adventiv
in lawns; Cook Co.

17. *Scrophularia* L. Figwort

1. Petioles unwinged; sterile stamen purple or brown; flowering from July
October ....1. *S. marilandic*
1. Petioles winged; sterile stamen greenish-yellow; flowering from May
June .... 2. *S. lanceolat*

1. S. marilandica L. Late Figwort. July–Oct. Usually rich woods; o
casional throughout the state. (Including f. *neglecta* [Rydb.] Pennel
2. S. lanceolate Pursh. Early Figwort. May–June. Woods, fenc
rows; occasional in the n. half of the state, otherwise very rare.

18. *Collinsia* Nutt.

1. Lower lip of corolla blue; at least some of the upper leaves claspin
capsule with 2–4 seeds ....1. *C. vern*
1. Lower lip of corolla violet or purple; none of the leaves clasping; ca
sule with 6 or more seeds .... 2. *C. violace*

1. C. verna Nutt. Blue-eyed Mary. April–June. Rich woods; o
casional throughout the state, but often abundant where found.

2. C. violacea Nutt. Violet Collinsia. May–June. Sandy hillside, rare; Shelby Co.

19. *Cymbalaria* Hill

1. C. muralis Gaertn. Kenilworth Ivy. June–Oct. Native of Europe and Asia; occasionally escaped from cultivation; Champaign and Kane cos.

20. *Linaria* Mill. *Toadflax*

1. Corolla yellow or yellow and orange, 2–4 cm long 2
1. Corolla blue or violet, up to 1.2 cm long .... 3. *L. canadensis*
2. Leaves linear to linear-lanceolate; corolla 2–3 cm long, yellow and orange .... 1. *L. vulgaris*
2. Leaves ovate to oblong; corolla 3.5–4.0 cm long, yellow 2. *L. dalmatica*

1. L. vulgaris Hill. Butter-and-Eggs. May–Nov. Native of Europe; commonly escaped from cultivation throughout the state.
2. L. dalmatica (L.) Mill. June–Sept. Native of Europe; rarely escaped from cultivation; Winnebago Co.
3. L. canadensis (L.) Dumort. Blue Toadflax. April–Sept. Flowers 5–10 mm long; seeds smooth. Sandy soil; occasional in the n. half of the state, uncommon in the s. half.
var. texana (Scheele) Pennell. South Blue Toadflax. April–Sept. Flowers 10–17 mm long; seeds rugulose. Sandy soil, not common; confined to the s. tip of the state. *L. texana* Scheele—J.

21. *Kickxia* Dumort. *Cancerwort*

1. K. elatine (L.) Dumort. Canker-root. May–Oct. Native of Europe; adventive in waste ground; not common but scattered in Ill.

22. *Chaenorrhinum* Reichenb.

1. C. minus (L.) Lange. Dwarf Snapdragon. June–Sept. Native of Europe; adventive in Ill., particularly along railroads.

23. *Verbascum* L. *Mullein*

. Plants pubescent, not glandular; capsule densely tomentose; flowers in spikes 2
. Plants glabrous or glandular-pubescent; capsule puberulent or glabrous; flowers in racemes 3
. Flowers up to 2 (–2.5) cm across; spike crowded; leaves usually not clasping .... 1. *V. thapsus*
. Flowers over 2.5 cm across; spike interrupted, at least below; some of the leaves usually clasping .... 2. *V. phlomoides*
. Pedicels 1 cm long or longer, longer than the capsules .. 3. *V. blattaria*
. Pedicels up to 5 mm long, shorter than the capsules ... 4. *V. virgatum*

1. V. thapsus L. Woolly Mullein. May–Sept. Native of Europe; commonly adventive in Ill.; in every co.

2. V. phlomoides L. Clasping Mullein. July–Aug. Native of Europe
rarely adventive in Ill.; Cook and DuPage cos.

3. V. blattaria L. Moth Mullein. May–Sept. Native of Europe; often
adventive in Ill.; scattered throughout the state. (Including f. *erubes-
cens* Brugger, and f. *albiflora* [Don] House.)

4. V. virgatum Stokes. Mullein. Native of Europe; rarely adventive i
Ill.; Coles and Pulaski cos.

24. Antirrhinum L. Snapdragon

1. Perennial; leaves oblong to ovate; corolla at least 2 cm long   1. *A. maju*
1. Annual; leaves lanceolate; corolla up to 1.2 cm long .... 2. *A. orontiur*

1. A. majus L. Common Snapdragon. June–Oct. Native of Europe
rarely escaped from cultivation; Jackson Co.

2. A. orontium L. Lesser Snapdragon. Sept. Native of Europe; e
caped into a cultivated oat field; DuPage Co.

25. Wulfenia Jacq.

1. W. bullii (Eat.) Barnh. Kitten Tails. May–June. Sandy soil, prairie
not common; primarily along the Illinois and Mississippi rivers in th
nw. one-fourth of the state. *Synthyris bullii* (Eat.) Heller—J; *Bessey*
bullii (Eat.) Rydb.—G.

26. Castilleja Mutis Paintbrush

1. Bracts scarlet, orange, or yellow; corolla up to 3.5 cm long ....
.... 1. *C. coccine*
1. Bracts green; corolla 4 cm long or longer .... 2. *C. sessiliflo*

1. C. coccinea L. Indian Paintbrush. April–July. Bracts scarlet
orange. Prairies, sandy woods; occasional in the n. three-fourths
the state, absent elsewhere.

f. lutescens Farw. Bracts yellow. Prairies, sandy woods; scatter
in the n. three-fourths of the state.

2. C. sessiliflora Pursh. Downy Yellow Painted Cup. May–Ju
Prairies, rare; DuPage, Lake, McHenry, and Winnebago cos.

142. BIGNONIACEAE

1. Vines; leaves compound ....
1. Trees; leaves simple .... 3. *Catal*
2. Leaflets 7–11; tendrils absent .... 1. *Camp.*
2. Leaflets 2; tendrils present .... 2. *Bignor*

1. Campsis Lour. Trumpet Creeper

1. C. radicans (L.) Seem. Trumpet Creeper. June–Aug. Roadsid
fields, edge of woods; native and common in the s. half of the sta
adventive and rarer in the n. half.

2. *Bignonia* L. Cross-vine

1. B. capreolata L. Cross-vine. April–June. Low woods, wooded slopes; confined to the s. one-fourth of the state.

3. *Catalpa* Scop.

1. Lowest corolla lobe not notched; fruit 5–10 mm in diameter; crushed leaves with an unpleasant odor .... 1. *C. bignonioides*
1. Lowest corolla lobe notched; fruit 12 mm in diameter or broader; crushed leaves without an unpleasant odor .... 2. *C. speciosa*

   1. C. bignonioides Walt. Catalpa. May–June. Native to the se. U.S.; escaped along roads, railroads, and streams; occasional throughout the state.
   2. C. speciosa Warder. Catalpa. May–June. Low woods; native in the s. one-fourth of the state, escaped from cultivation in most parts of Ill.

143. MARTYNIACEAE

1. *Proboscidea* Schmidel Unicorn-plant

   1. P. louisianica (Mill.) Thell. Proboscis-flower. June–Sept. Low ground, particularly along rivers; occasional throughout the state, where it probably is naturalized.

144. OROBANCHACEAE

1. Calyx deeply split on the lower side, 3- to 4-toothed on the upper side; stamens exserted .... 1. *Conopholis*
1. Calyx more or less equally 4- or 5-toothed; stamens included 2
2. Corolla 5-lobed; flower solitary .... 3. *Orobanche*
2. Corolla 4-lobed; flowers spicate 3
3. Calyx 4-lobed .... 3. *Orobanche*
3. Calyx 5-lobed 4
4. Corolla up to 1 cm long .... 2. *Epifagus*
4. Corolla 1.5–2.0 cm long .... 3. *Orobanche*

1. *Conopholis* Wallr. Squaw-root

   1. C. americana (L. f.) Wallr. Cancer-root. May–July. Wooded ravines; parasitic on oak roots; rare; scattered in the n. two-thirds of the state.

2. *Epifagus* Nutt. Beech-drops

   1. E. virginiana (L.) Bart. Beech-drops. Aug.–Oct. Parasitic on beech roots; restricted to the s. one-fourth of the state; also Clark, Crawford, and Lawrence cos.

3. *Orobanche* L. Broomrape

1. Flowers in spikes; each flower subtended by 1–3 bracts; corolla 4-lobed
.... 2
1. Flower solitary, not subtended by bracts; corolla 5-lobed 3
2. Calyx 4-lobed; each flower subtended by 3 bracts .... 1. *O. ramosa*
2. Calyx 5-lobed; each flower subtended by 1–2 bracts . 2. *O. ludoviciana*
3. Corolla creamy-white to lilac; calyx lobes lance-subulate, longer than
the tube .... 3. *O. uniflora*
3. Corolla purple; calyx lobes deltoid, equalling or shorter than the tube .
.... 4. *O. fasciculata*

1. O. ramosa L. Broomrape. Aug. Native of Europe; parasitic on hemp roots, very rare; Champaign Co.

2. O. ludoviciana Nutt. Broomrape. Aug.–Sept. Parasitic on roots of various members of the Compositae, rare; Lee, Menard, White, and Wabash cos.

3. O. uniflora L. One-flowered Broomrape. May–June. Parasitic on roots of various plants; scattered throughout the state.

4. O. fasciculata Nutt. Clustered Broomrape. April–Aug. Parasitic on roots of various members of the Compositae, not common; confined to the n. half of the state.

145. LENTIBULARIACEAE

1. *Utricularia* L. Bladderwort

1. Leaves entire, rarely seen .... 1. *U. cornuta*
1. Leaves pinnately divided, conspicuous ....
2. Fruiting pedicels ascending or erect ....
2. Fruiting pedicels arched-recurving ....
3. Corolla up to 1.2 cm long; leaf segments terete .... 2. *U. gibba*
3. Corolla 1.2 cm long or longer; leaf segments flat ..... 3. *U. intermedia*
4. Corolla 1.5–2.5 cm long; leaf segments terete .... 4. *U. vulgaris*
4. Corolla 5–8 mm long; leaf segments flat .... 5. *U. minor*

1. U. cornuta Michx. Horned Bladderwort. July–Sept. Bogs and wet sands, rare; Cook and Lake cos.

2. U. gibba L. Humped Bladderwort. Aug.–Oct. Swamps, ponds, lakes, and ditches, not common; in s. cos. along the Mississippi River; also Cook, Lake, and McHenry cos.

3. U. intermedia Hayne. Flat-leaved Bladderwort. July–Aug. Shallow water, rare; Lake, McHenry, Ogle, and Tazewell cos.

4. U. vulgaris L. Common Bladderwort. July–Aug. Shallow water, not common; scattered throughout the state.

5. U. minor L. Small Bladderwort. June–Aug. Shallow water, very rare; Lake and McHenry cos.

146. ACANTHACEAE

1. Corolla 2-lipped, less than 2 cm long; stamens 2 2
1. Corolla nearly actinomorphic, not 2-lipped, at least 2.5 cm long; stamens
4 .... 3. *Ruellia*
2. Petioles at least 1 cm long; flowers bracteate .... 1. *Dicliptera*
2. Petioles mostly less than 1 cm long; flowers ebracteate ..... 2. *Justicia*

1. *Dicliptera* Juss.

1. D. brachiata (Pursh) Spreng. Aug.–Oct. Bottomland woods, rare;
Massac Co.

2. *Justicia* L. Water Willow

1. Leaves linear to lanceolate .... 1. *J. americana*
1. Leaves elliptic to oblong .... 2. *J. ovata*

1. J. americana (L.) Vahl. Water Willow. May–Oct. In and along
streams, occasional; throughout the state. *Dianthera americana* L.—J.
2. J. ovata (Walt.) Lindau. Water Willow. May–June. Swamps, bottomland woods, rare; Alexander and Pulaski cos. *J. humilis* Michx.
var. *lanceolata* (Chapm.) Gl.—G.

3. *Ruellia* L. Wild Petunia

1. Calyx lobes not more than 1.5 mm broad; stems and leaves pubescent
throughout 2
1. Calyx lobes at least 2 mm broad; stems and leaves glabrous or sparsely
pubescent .... 4. *R. strepens*
2. Leaves sessile, or with petioles at most 3 mm long ...... 1. *R. humilis*
2. Leaves with petioles longer than 3 mm 3
3. Flowers pedunculate, solitary or in cymes; capsules puberulent ....
.... 2. *R. pedunculata*
3. Flowers sessile or subsessile in glomerules; capsules glabrous ....
.... 3. *R. caroliniensis*

1. R. humilis Nutt. Wild Petunia. May–Oct. Corolla up to 4.5 cm
long, the tube up to 2.5 cm long. Prairies, dry woods, bluffs; occasional throughout the state. (Including f. *alba* [Steyerm.] Fern., f.
grisea Fern., and var. *frondosa* Fern.)
var. longiflora (Gray) Fern. Wild Petunia. May–Oct. Corolla 5 cm
long or longer, the tube 3 cm long or longer. Prairies, dry woods,
bluffs; less common than the preceding but scattered throughout the
state. (Including var. *expansa* Fern.)
2. R. pedunculata Torr. Wild Petunia. May–Sept. Rocky woods; confined to the s. one-fourth of the state.
3. R. caroliniensis (Walt.) Steud. var. dentata (Nees) Fern. Wild Petunia. May–Oct. Dry woods, not common; confined to the s. one-
fourth of the state. (Including var. *salicina* Fern.)
4. R. strepens L. Smooth Ruellia. May–Oct. Corolla 3 cm long or
longer, the tube open. Rich or low woods; occasional in the s. four-
fifths of the state. (Including f. *alba* Steyerm.)

ACANTHACEAE

396

f. cleistantha (Gray) McCoy. Smooth Ruellia. May–Oct. Corolla less
than 3 cm long, the tube more or less closed. Rich or low woods,
scattered in the s. one-third of the state.

147. PLANTAGINACEAE

1. Plantago L. Plantain

1. Leaves all basal ....
1. Leaves opposite .... 12. *P. indica*
2. Leaves linear, never broader than 1 cm ....
2. Leaves lanceolate, elliptic, oval, or ovate, some or all usually over 1 cm
   broad ....
3. Bracts much exceeding the flowers, very conspicuous ... 1. *P. aristata*
3. Bracts shorter than to barely exceeding the flowers, not conspicuous
4. Bracts, sepals, and inflorescence tomentose .... 2. *P. purshii*
4. Bracts, sepals, and inflorescence glabrous or nearly so ....
5. Capsule with 4 seeds; sepals obovate .... 3. *P. pusilla*
5. Capsule with 10 or more seeds; sepals ovate .... 4. *P. heterophylla*
6. Leaves cordate at the base, with lateral veins arising from the mid-vein;
   flowering stalks hollow .... 5. *P. cordata*
6. Leaves rounded or cuneate at the base, not cordate; flowering stalk
   solid ....
7. Leaves lanceolate, oblanceolate, or elliptic; seeds 1–3 (–4) per capsule
7. Leaves ovate or oval; seeds usually 4 or more per capsule 10
8. Spikes ellipsoid; leaves lanceolate .... 6. *P. lanceolata*
8. Spikes cylindrical; leaves obovate to elliptic ....
9. Flowers fragrant; corolla spreading, leaving the capsule exposed ....
   .... 7. *P. media*
9. Flowers not fragrant; corolla ascending, closed over the concealed capsule ....
10. Leaves mostly entire; sepals obtuse; seeds pale brown, up to 2 mm long
    .... 8. *P. virginica*
10. Leaves with a few coarse teeth; sepals attenuate at tip; seeds red,
    2.5–3.0 mm long. .... 9. *P. rhodosperma*
11. Seeds flat; leaves densely canescent; seeds never more than 4 per cap-
    sule .... 7. *P. media*
11. Seeds turgid; leaves glabrous or nearly so on at least one surface, at
    least not densely canescent; seeds usually more than 4 per capsule
12. Base of petioles purplish; sepals acute or subacute, 2.5–3.0 mm long
    .... 10. *P. rugelii*
12. Base of petioles green; sepals obtuse, up to 2 mm long ... 11. *P. major*

1. P. aristata Michx. Bracted Plantain. May–Nov. Pastures, fields,
upland woods, cliffs, waste ground; common throughout the state.
2. P. purshii Roem. & Schultes. Salt-and-pepper Plant. May–Aug.
Sandy soil, not common; known from the n. one-half of the state; also
Christian Co.
3. P. pusilla Nutt. Small Plantain. April–June. Leaves more or less
entire; spikes up to 6 cm long; seeds 1.2 mm long. Fields, pastures,
sandstone cliffs; occasional in the s. half of the state.

var. major Engelm. Small Plantain. April–June. Leaves usually toothed; spikes more than 6 cm long; seeds 1.7–1.8 mm long. Sandstone cliffs, rare; confined to the s. one-fourth of the state.

4. P. heterophylla Nutt. Small Plantain. April–May. Sandy soil, very rare; Pulaski and Union cos.

5. P. cordata Lam. Heartleaf Plantain. April–July. Along streams in woods; at one time scattered throughout most of the state, but apparently rare now.

6. P. lanceolata L. Buckhorn. April–Oct. Native of Europe; naturalized on waste ground; in every co.

7. P. media L. Hoary Plantain. June–Sept. Native of Europe; rarely adventive in Ill.; roadside; Cook Co.

8. P. virginica L. Dwarf Plantain. April–June. Fields, roadsides, sandstone cliffs; occasional to common throughout the state.

9. P. rhodosperma Dcne. Red-seeded Plantain. May–June. Native to the sw. U.S.; rarely adventive in Ill.; field; Williamson Co.

10. P. rugelii Dcne. Rugel's Plantain. May–Oct. Fields, woods, waste ground, common; in every co.

11. P. major L. Common Plantain. May–Oct. Native of Europe; naturalized in waste ground; common in the n. one-third of the state, less common elsewhere.

12. P. indica L. Whorled Plantain. July–Oct. Native of Europe; adventive in waste ground; Champaign, Cook, Kane, Kankakee, Lake, McHenry, and Winnebago cos.

148. RUBIACEAE

1. Shrubs; flowers in dense, globose, pedunculate heads 1. *Cephalanthus*
1. Herbs; flowers solitary, or 1–3 together, or in cymes or, if in heads, the head not pedunculate 2
2. Leaves in whorls of 4–8 3
2. Leaves opposite 4
3. Flowers pink or blue, subtended by an involucre .... 2. *Sherardia*
3. Flowers white, yellow, greenish, or maroon, not subtended by an involucre .... 3. *Galium*
4. Flowers sessile 5
4. Flowers pedicellate 6
5. Flowers 1–3 per axil .... 4. *Diodia*
5. Flowers several in an axillary glomerule .... 5. *Spermacoce*
6. Leaves evergreen, about as broad as long; fruit a red berry; flowers in pairs, with their ovaries united .... 6. *Mitchella*
6. Leaves deciduous, longer than broad; fruit a capsule; flowers solitary or in cymes, their ovaries not united .... 7. *Houstonia*

1. *Cephalanthus* L.   Buttonbush

1. C. occidentalis L. Buttonbush. June–Aug. Branchlets and leaves glabrous. Wet ground; occasional throughout the state.

var. pubescens Raf. Buttonbush. June–Aug. Branchlets and leaves soft-pubescent. Wet ground; less common than the preceding.

2. *Sherardia* L.

1. S. arvensis L. Field Madder. May–Sept. Native of Europe; rarel
adventive in waste ground; Jackson Co.

3. *Galium* L.   Bedstraw

1. Fruits and ovaries bristly or pubescent ....
1. Fruits and ovaries smooth ....
2. Leaves in whorls of 4 ....
2. Leaves in whorls of 6–8 ....
3. Leaves up to 1 cm long; flowers solitary in the axils .... 1. *G. virgatun*
3. Leaves, or some of them, over 1 cm long; flowers in cymes or panicles
4. Stems and leaves glabrous, or the leaf margins slightly scabrous ....
4. Stems and leaves pilose ....
5. Leaves linear-lanceolate; flowers bright white, in panicles 2. *G. boreal*
5. Leaves ovate-lanceolate to lanceolate; flowers greenish-white, yellowish
   or becoming purplish, in cymes ....
6. Leaves ovate-lanceolate; flowers greenish-white, the lobes of the coroll
   usually more or less hairy .... 3. *G. circaezan*
6. Leaves lanceolate; flowers yellow, becoming purple, the lobes of the cc
   rolla glabrous .... 4. *G. lanceolatu*
7. All flowers sessile; leaves 3-nerved .... 3. *G. circaezan*
7. Some or all flowers pedicellate; leaves 1-nerved .... 5. *G. pilosur*
8. Stems smooth or retrorsely scabrous; margins of leaves with ascendin
   cilia .... 6. *G. triflorur*
8. Stems retrorsely bristly; margins of leaves retrorsely scabrous ....
   .... 7. *G. aparin*
9. Plants erect 1
9. Plants matted or loosely ascending 1
10. Leaves in whorls of 4 .... 2. *G. borea*
10. Leaves in whorls of 6–8 1
11. Corolla yellow; leaves roughish .... 8. *G. veru*
11. Corolla white; leaves smooth .... 9. *G. mollug*
12. Corolla 3-lobed; upper part of stems scabrous, but not retrorsely brist
    1
12. Corolla 4-lobed; stems glabrous or retrorsely bristly ....
13. Pedicels smooth; leaves in whorls of 4–6 on the same plant ....
    .... 10. *G. tinctoriu*
13. Pedicels scabrous; leaves mostly in whorls of 4 .... 11. *G. trifidu*
14. Stems retrorsely bristly ....
14. Stems smooth ....
15. Leaves mostly in whorls of 6; flowers white .... 12. *G. asprellu*
15. Leaves mostly in whorls of 4; flowers yellow ... 13. *G. pedemontanu*
16. Leaves in whorls of 6, mostly linear .... 14. *G. concinnu*
16. Leaves in whorls of 4, mostly elliptic-oblong to oblanceolate ....
17. Leaves conspicuously reflexed; inflorescence over-topped by later
    branches; fruits up to 1.5 mm in diameter .... 15. *G. labradoricu*
17. Leaves ascending; inflorescence not over-topped by lateral branche
    fruits more than 2 mm in diameter .... 16. *G. obtusu*

1. G. virgatum Nutt. Dwarf Bedstraw. April–June. Limestone ledg
rare; Monroe Co.
2. G. boreale L. Northern Bedstraw. May–July. Moist meadov
prairies, and fields; occasional in the n. one-fourth of the state, a

sent elsewhere. (Including var. *intermedium* DC. and var. *hyssopifolium* [Hoffm.] DC.)

3. G. circaezans Michx. Wild Licorice. May–July. Rich, often rocky, woods; occasional to common throughout the state. (Including var. *hypomalacum* Fern.)

4. G. lanceolatum Torr. Wild Licorice. June–July. Rich woods, rare; Cook and Kane cos.

5. G. pilosum Ait. Hairy Bedstraw. June–Aug. Sandy woods; occasional to common throughout the state.

6. G. triflorum Michx. Sweet-scented Bedstraw. May–Sept. Moist woods; occasional to common throughout the state.

7. G. aparine L. Goosegrass. May–July. Woods; common throughout the state. (Including var. *vaillantii* [DC.] W. D. J. Koch.)

8. G. verum L. Yellow Bedstraw. June–Aug. Native of Europe; rarely adventive in waste ground; DuPage and LaSalle cos.

9. G. mollugo L. White Bedstraw. June–Sept. Native of Europe; rarely adventive in waste ground; Champaign, Cook, DuPage, Henry, and Kane cos.

10. G. tinctorium L. Stiff Bedstraw. May–Sept. Swamps, marshes; occasional in the ne. corner of the state, rare in the extreme s. cos., absent elsewhere.

11. G. trifidum L. Small Bedstraw. July–Aug. Moist soil, not common; apparently confined to the ne. cos. and a few cos. in cent. Ill. along the Illinois River.

12. G. asprellum Michx. Rough Bedstraw. July–Aug. Moist thickets; occasional in the n. one-fourth of the state.

13. G. pedemontanum All. Bedstraw. May–June. Native of Europe; adventive in waste ground; Champaign Co.

14. G. concinnum Torr. & Gray. Shining Bedstraw. June–July. Dry woods; common throughout the state.

15. G. labradoricum Wieg. Bog Bedstraw. Bogs, marshes, rare; Boone, Lake, and McHenry cos.

16. G. obtusum Bigel. Wild Madder. May–July. Moist ground; common throughout the state.

4. *Diodia* L. Buttonweed

. Corolla 7–10 mm long; fruits 7–10 mm long .... 1. *D. virginiana*
. Corolla 4–6 mm long; fruits 2.5–4.0 mm long .... 2. *D. teres*

1. D. virginiana L. Large Buttonweed. June–Aug. Swamps, wet woods, not common; confined to the s. one-fourth of the state.

2. D. teres Walt. Rough Buttonweed. July–Aug. Fields, roadsides, woods; occasional to common throughout the state. (Including var. *setifera* Fern. & Grisc.)

5. *Spermacoce* L. Buttonweed

1. S. glabra Michx. Smooth Buttonweed. June–Oct. Wet ground; occasional in the s. half of the state, rare or absent elsewhere.

6. *Mitchella* L. Partridge-berry

1. M. repens L. Partridge-berry. Rocky woods, swampy woods; occasional in the s. and n. cos., rare in the cent. cos.

7. *Houstonia* L.   Bluets

1. Flower one per peduncle; stamens included; corolla salverform .....
1. Flowers in corymbs, cymes, or panicles; stamens exserted; corolla funnelform ....
2. Corolla lobes 2.5–5.0 mm broad, very pale blue .... 1. *H. caerulea*
2. Corolla lobes to 2.5 mm broad, usually dark purple ....
3. Calyx lobes 2–4 mm long, about as long as the corolla tube 2. *H. minima*
3. Calyx lobes 1–2 mm long, up to ⅔ as long as the corolla tube ....
   .... 3. *H. pusilla*
4. Calyx lobes shorter than calyx tube; middle and upper stipules setiform
   .... 4. *H. nigricans*
4. Calyx lobes longer than calyx tube; middle and upper stipules deltoid or
   rounded ....
5. Lowermost leaves 3- to 5-nerved .... 5. *H. purpurea*
5. Lowermost leaves 1-nerved .... 6. *H. longifolia*

1. H. caerulea L. Bluets. April–June. Sandy soil of fields, prairies
and open woods; not common but scattered in all parts of the state
except the nw. cos; very rare in the s. cos.

2. H. minima Beck. Tiny Bluets. April–May. Rocky woods, prairies,
fields; occasional in the s. and w. cos., apparently absent elsewhere.

3. H. pusilla Schoepf. Star Violet, Small Bluets. March–April. Rocky
woods, fields; occasional in the s. one-fourth of the state. *H. patens*
Ell.—F. (Including f. *albiflora* Standl.)

4. H. nigricans (Lam.) Fern. Narrow-leaved Bluets. May–Oct. Rocky
ledges, hill prairies; confined to the s. cos. along the Mississippi
River.

5. H. purpurea L. Broad-leaved Bluets. May–June. Leaves ovate-lanceolate to broadly ovate, mostly 10 mm broad or broader. Rocky
woods, prairies; occasional in the s. half of the state.

var. calycosa Gray. Bluets. May–June. Leaves broadly lanceolate,
mostly less than 12 mm broad. Rocky woods, occasional; confined to
the s. one-fourth of the state. *H. lanceolata* (Poir.) Britt.—F, J.

6. H. longifolia Gaertn. Long-leaved Bluets. April–July. Cauline
leaves oblong to oblanceolate, mostly 2–5 mm broad; basal leaves
eciliate; capsules as high as or higher than broad. Rocky woods,
prairies, fields; occasional in the southernmost and northernmost
cos., absent elsewhere.

var. ciliolata (Torr.) Torr. Bluets. April–July. Cauline leaves oblong
to oblanceolate, mostly 2–5 mm broad; basal leaves ciliate; capsules
as high as or higher than broad. Rocky woods, prairies, fields; occasional in the southernmost and northernmost cos., absent elsewhere. *H. canadensis* Willd.—F, G.

var. tenuifolia (Nutt.) Wood. Bluets. April–Oct. Cauline leaves linear
1.0–2.5 mm broad; capsules broader than high. Rocky woods; occasional in the s. one-fourth of the state. *H. tenuifolia* Nutt.—F, G, J.

149. CAPRIFOLIACEAE

1. Small trees or erect or climbing shrubs ....
1. Creeping or ascending herbs ....

2. Leaves pinnately compound .... 1. *Sambucus*
2. Leaves simple 3
3. Mature leaves entire (leaves of vigorous shoots rarely pinnatifid) 4
3. Leaves toothed or lobed 5
4. Corolla tubular or funnelform, mostly zygomorphic; berries several-seeded .... 2. *Lonicera*
4. Corolla campanulate, actinomorphic; drupes 2-seeded ....
.... 3. *Symphoricarpos*
5. Corolla tubular; inflorescence 3-flowered; fruit a several-seeded capsule
.... 4. *Diervilla*
5. Corolla spreading; inflorescence almost always more than 3-flowered; fruit a 1-seeded drupe .... 5. *Viburnum*
6. Stems creeping; leaves evergreen; flowers pedicellate; fruit a 1-seeded capsule .... 6. *Linnaea*
6. Stems ascending; leaves deciduous; flowers sessile; fruit a 2- to 5-seeded drupe .... 7. *Triosteum*

1. *Sambucus* L. Elderberry

1. Cymes umbel-like, flat; twigs with white pith; fruit dark purple ....
.... 1. *S. canadensis*
1. Cymes panicle-like, ovoid; twigs with brown pith; fruit usually bright red
.... 2. *S. pubens*

1. *S. canadensis* L. Elderberry. June–July. Woodlands, along roads, common; in every co. (Including var. *submollis* Rehder.)
2. *S. pubens* Michx. Red-berried Elder. April–May. Moist woods, not common; Boone, Cook, Lake, LaSalle, McHenry, and Winnebago cos.

2. *Lonicera* L. Honeysuckle

1. Some of the leaves connate 2
1. None of the leaves connate 5
2. Corolla at least 3.5 cm long, red, the lobes more or less equal ....
.... 1. *L. sempervirens*
2. Corolla up to 3 cm long, white, yellow, orange, or purple, the lobes 2-lipped 3
3. Uppermost connate leaves broader than long, glaucous on the upper surface .... 2. *L. prolifera*
3. Uppermost connate leaves longer than broad, not glaucous 4
4. Corolla orange-yellow or yellow; leaves gray on the lower surface ......
.... 3. *L. flava*
4. Corolla yellow tinged with purple or brick-red; leaves whitened on the lower surface .... 4. *L. dioica*
5. Plants twining or climbing; corolla mostly 3 cm long or longer ....
.... 5. *L. japonica*
5. Plants erect or ascending; corolla up to 2 cm long 6
6. Peduncles much longer than the petioles 7
6. Peduncles shorter than the petioles .... 10. *L. maackii*
7. Corolla pubescent on the outside 8
7. Corolla glabrous on the outside 9
8. Bractlets about half as long as the ovaries; filaments pubescent ....
.... 6. *L. xylosteum*
8. Bractlets about as long as the ovaries; filaments glabrous 7. *L. morrowi*

9. Peduncles longer than the flowers .... 8. *L. tatarica*
9. Peduncles mostly shorter than the flowers .... 9. *L. X bella*

1. L. sempervirens L. Trumpet Honeysuckle. April–July. Escaped from cultivation, mostly along roads; occasional and scattered throughout the state.
2. L. prolifera (Kirchn.) Rehd. Grape Honeysuckle. May–June. Wooded slopes, rocky banks; occasional to common in the n. three fifths of the state.
3. L. flava Sims. Yellow Honeysuckle. April–May. Rocky woods sandstone cliffs, rare; Jackson and Pope cos.
4. L. dioica L. Red Honeysuckle. May–June. Woods, not common Cook, Kane, and Lake cos.
5. L. japonica Thunb. Japanese Honeysuckle. May–June. Native of Asia; woods, thickets; very common in the s. one-third of the state less common elsewhere.
6. L. xylosteum L. European Fly Honeysuckle. May. Native of Europe and Asia; rarely escaped from cultivation; Coles and Cook cos.
7. L. morrowi Gray. May–June. Native of Japan; rarely escaped from cultivation; Carroll, Cook, and Lake cos.
8. L. tatarica L. Tartarian Honeysuckle. May–June. Native of Europe and Asia; occasionally escaped from cultivation in the n. half of the state, much rarer in the s. half.
9. L. X bella Zabel. May–June. Native of Asia; occasionally escaped from cultivation in the n. one-fourth of the state; also Macon and Peoria cos. Reputed to be a hybrid between *L. morrowi* and *L. tatarica*.
10. L. maackii Maxim. Amur Honeysuckle. May–June. Native of Asia escaped into woodlands; DuPage and Kankakee cos.

3. *Symphoricarpos* Duham. Snowberry

1. Corolla 3–4 mm long, greenish and purplish; drupes coral-pink to purple
.... 1. *S. orbiculatus*
1. Corolla 5–9 mm long, pink; drupes white or greenish-white ....
2. Flowers pedicellate; style included .... 2. *S. albus*
2. Flowers sessile; style exserted .... 3. *S. occidentalis*

1. S. orbiculatus Moench. Coralberry. July–Aug. Disturbed woods pastures; occasional throughout the state.
2. S. albus (L.) Blake var. laevigatus (Fern.) Blake. Garden Snowberry. Native of the w. U.S.; rarely escaped from cultivation. S. rivularis Suksd.—J.
3. S. occidentalis Hook. Wolfberry. June–Aug. Dry open ground; occasional in the n. one-fifth of the state.

4. *Diervilla* Mill. Bush Honeysuckle

1. D. lonicera Mill. Bush Honeysuckle. May–June. Sandy woods; occasional in the n. one-fourth of the state; also Coles and Piatt cos.

5. *Viburnum* L. Viburnum

1. Leaves more or less 3-lobed, palmately nerved 2
1. Leaves serrulate to dentate, pinnately nerved 4
2. Leaves dotted beneath; young branchlets more or less pilose; drupes purple-black at maturity .... 1. *V. acerifolium*
2. Leaves not dotted beneath; young branchlets glabrous or nearly so; drupes red or orange 3
3. Petiole with convex glands; stipules broadened or thickened at the tip .... 2. *V. trilobum*
3. Petiole with concave glands; stipules slender-tipped ...... 3. *V. opulus*
4. Pubescence of stellate hairs; winter buds without scales . 4. *V. lantana*
4. Pubescence of simple hairs or absent; winter buds with 1–2 pairs of outer scales 5
5. Leaves serrulate; petioles flat; stone of drupe flat; buds with 1 pair of outer scales 6
5. Leaves dentate; petioles terete; stone of drupe grooved or ridged; buds with 2 pairs of outer scales 8
6. Petiole and midvein of leaf on the lower surface rufous-pubescent .... 5. *V. rufidulum*
6. Petioles and midvein of leaf on the lower surface glabrous or without rufous pubescence 7
7. Leaves obtuse to acute at the apex; margins of the petioles neither undulate nor revolute .... 6. *V. prunifolium*
7. Leaves acuminate at the apex; margins of the petioles undulate and revolute .... 7. *V. lentago*
8. Leaves deeply cordate; leaves with 40 or more teeth; bark peeling off .. .. 8. *V. molle*
8. Leaves subcordate, rounded, or subcuneate at the base; leaves with up to 40 (–44) teeth; bark not peeling off 9
9. Petioles on leaves subtending the inflorescence up to 7 mm long .... 9. *V. rafinesquianum*
9. Petioles on leaves subtending the inflorescence 1.5 cm long or longer 10
10. Stipules present, subulate; branches of inflorescence and lower surface of leaves somewhat pubescent .... 10. *V. dentatum*
10. Stipules absent; branches of inflorescence and lower surface of leaves glabrous except on the nerves .... 11. *V. recognitum*

1. V. acerifolium L. Maple-leaved Arrowwood. May–June. Moist woods, occasional; confined to the ne. one-fourth of the state.
2. V. trilobum Marsh. High-bush Cranberry. June–July. Moist woods, rare; restricted to the n. one-third of the state.
3. V. opulus L. European High-bush Cranberry. June–July. Native of Europe; frequently cultivated and sometimes escaped, particularly in the ne. cos.
4. V. lantana L. Wayfaring Tree. May–June. Native of Europe; rarely escaped from cultivation; DuPage, Kane, and Kendall cos.
5. V. rufidulum Raf. Southern Black Haw; Rusty Nannyberry. April–May. Rocky woods; occasional in the s. one-fourth of the state, extending northward in cos. along the Mississippi River to Pike Co.
6. V. prunifolium L. Black Haw; Nannyberry. April–May. Low woods, wooded slopes; occasional throughout the state.

7. V. lentago L. Nannyberry. May–June. Moist woods; occasional i
the n. half of the state; apparently adventive in Jackson Co.

8. V. molle Michx. Arrowwood. May–June. Rocky banks of stream:
dry hillsides, rare; Adams and Peoria cos.

9. V. rafinesquianum Schultes. Downy Arrowwood. May–Jun
Wooded slopes, rocky woods; occasional in the n. half of the stat
extending southward to Clark and Coles cos.

10. V. dentatum L. var. deamii (Rehd.) Fern. Southern Arrowwoo
May–June. Low woods, rare; restricted to the s. one-fourth of th
state.

11. V. recognitum Fern. Smooth Arrowwood. May–June. Woo
lands, along streams; occasional in the s. half of the state; also Coo
DuPage, and Lake cos. *V. dentatum* L. var. *indianense* (Reh
Gl.—G.

6. *Linnaea* Gron. ex. L.   Twinflower

1. L. americana Forbes. Twinflower. June–Aug. Bogs, very rar
Cook Co.

7. *Triosteum* L.   Horse Gentian

1. Stems setose-hispid; calyx lobes glabrous except for the bristly marg
.... 1. *T. angustifoliu*
1. Stems glabrous or villous or glandular-pubescent; calyx lobes pube
cent throughout ....
2. At least 3 or more pairs of leaves connate at the base; style exserted; a
rolla yellow, green, or purplish .... 2. *T. perfoliatu*
2. None or only up to 2 (–3) pairs of leaves connate at the base; style
cluded; corolla orange or red ....
3. Stem with some glandular pubescence .... 3. *T. aurantiac*
3. Stem without glandular pubescence .... 4. *T. illinoer*

1. T. angustifolium L. Yellow-flowered Horse Gentian. April–M
Low ground, rocky woods, not common; confined to the s. one-fou
of the state.

2. T. perfoliatum L. Late Horse Gentian. May–July. Dry woo
thickets; occasional in the n. two-thirds of the state, less commor
the s. one-third.

3. T. aurantiacum Bickn. Early Horse Gentian. May–June.
woods, rich woods; occasional in the n. two-thirds of the state
perfoliatum L. var. *aurantiacum* (Bickn.) Wieg.—G.

4. T. illinoense (Wieg.) Rydb. Illinois Horse Gentian. May–June.
woods, rich woods; occasional throughout the state. *T. perfoliatur*
var. *illinoense* Wieg.—G; *T. aurantiacum* Bickn. var. *illinoense* (Wi
Palmer & Steyerm.—F. The type was collected in Henderson Co.

150. ADOXACEAE

1. *Adoxa* L.

1. A. moschatellina L. Moschatel. May–July. Moist cliffs, rare; Jc
viess Co.

151. VALERIANACEAE

1. Cauline leaves, or some of them, pinnately compound; calyx divided into 5–15 setiform lobes .... 1. *Valeriana*
1. Cauline leaves entire or toothed, not pinnately compound; calyx lobes reduced to short teeth, or absent .... 2. *Valerianella*

1. *Valeriana* L. Valerian

1. Flowers unisexual, white; leaves parallel-veined; root tuberous ....
.... 1. *V. ciliata*
1. Flowers perfect, pink or rose; leaves pinnately veined; roots fibrous . 2
2. Corolla pale pink, 1 cm long or longer; most of the basal leaves simple
.... 2. *V. pauciflora*
2. Corolla rose, 4–5 mm long; all of the basal leaves pinnately compound
.... 3. *V. officinalis*

1. V. ciliata Torr. & Gray. Valerian. May–June. Wet prairies, calcareous fens; occasional in the n. one-fifth of the state. *V. edulis* Nutt. ssp. *ciliata* (Torr. & Gray) F. G. Mey.—G.

2. V. pauciflora Michx. Pink Valerian. May–June. Low woods; occasional in the s. half of the state; also Vermilion Co.

3. V. officinalis L. Garden Heliotrope. June–July. Native of Europe; occasionally cultivated but infrequently escaped; Cook, Kane, and Lake cos.

2. *Valerianella* Mill.

1. Corolla 1.5–2.0 mm long; bracts and bractlets spinulose-ciliate 2
1. Corolla 3–5 mm long; bracts and bractlets eciliate, except sometimes at the tip 3
2. Corolla lobes bluish; stamens included or shortly exserted 1. *V. olitoria*
2. Corolla lobes white; stamens long-exserted .... 2. *V. radiata*
3. Fruits ovoid-ellipsoid, about 1 mm broad .... 3. *V. intermedia*
3. Fruits orbicular or globose, 2–3 mm broad 4
4. Fruits flattened, 3.0–3.5 mm long, about 3 mm broad ... 4. *V. patellaria*
4. Fruits not flattened, 2.0–2.5 mm long, about 2 mm broad 5. *V. umbilicata*

1. V. olitoria (L.) Poll. Corn Salad. May. Native of Europe; rarely adventive in waste ground; LaSalle Co.

2. V. radiata (L.) Dufr. Corn Salad. April–May. Seed-bearing cavity of fruit broader than the combined width of the two empty cells; fruits 1.4–1.5 mm long, 0.7–0.8 mm broad. Wet fields, prairies, low woods; occasional to common in the s. half of the state.

var. missouriensis Dyal. Corn Salad. April–May. Seed-bearing cavity of fruit not as broad as the combined width of the two empty cells; fruits 1.7–1.8 mm long, about 1.2 mm broad. Wet ground, rare; Jackson Co.

3. V. intermedia Dyal. Corn Salad. May–June. Wet ground, rare; Kankakee, LaSalle, and Will cos. *V. radiata* (L.) Dufr. var. *intermedia* (Dyal) Gl.—G.

4. V. patellaria (Sulliv.) Wood. Corn Salad. May–June. Wet ground, rare; LaSalle Co.

5. V. umbilicata (Sulliv.) Wood. Corn Salad. May–June. Wet ground rare; Kankakee and LaSalle cos.

152. DIPSACACEAE

1. Dipsacus L.   Teasel

1. Leaves pinnatifid or bipinnatifid, ciliate on the margins   1. D. laciniatu
1. Leaves toothed, prickly on the margins .... 2. D. sylvestr

1. D. laciniatus L. Cut-leaved Teasel. July–Sept. Native of Europ adventive in waste ground; Cook, DuPage, Jackson, Johnson, Kan Lake, and Will cos.
2. D. sylvestris Huds. Common Teasel. June–Oct. Native of Europ occasionally adventive in waste ground throughout the state.

153. CUCURBITACEAE

1. Leaves pinnately cleft, often more than halfway to the midvein ......
.... 1. Citrull
1. Leaves unlobed or shallowly lobed ....
2. Plants covered with soft, sticky hairs .... 2. Lagena
2. Plants glabrous or pubescent, but without soft, sticky hairs ....
3. Ovaries and fruits not spinulose ....
3. Ovaries and fruits spinulose ....
4. Corolla more than 5 cm long and more than 5 cm broad . 3. Cucurb
4. Corolla up to 3.5 cm long and up to 3.5 cm broad ....
5. Corolla yellow, 2.5–3.5 cm long and broad .... 4. Cucun
5. Corolla white or greenish, at most only 1 cm long and broad ....
.... 5. Meloth
6. Plants pubescent; corolla 5-lobed; ovary 1-locular; fruit 1-seeded ...
.... 6. Sicy
6. Plants glabrous or nearly so; corolla 6-lobed; ovary 2-locular; fruit seeded .... 7. Echinocys

1. Citrullus Neck   Melon

1. C. vulgaris Schrad. Watermelon. May–Sept. Native of Africa; casionally escaped from cultivation, but seldom persisting.

2. Lagenaria Seringe   Gourd

1. L. vulgaris Seringe. Gourd. May–Sept. Native of the E. He sphere; rarely escaped from cultivation; Hancock Co.

3. Cucurbita L.   Gourd

1. Leaves stiff, thick, often nearly twice as long as broad   1. C. foetidiss
1. Leaves flexible, thin, never twice as long as broad ....2. C. p

1. C. foetidissima HBK. Missouri Gourd. May–Sept. Native of the w. U.S.; occasional as an adventive along railroads.

2. C. pepo L. var. ovifera (L.) Alef. Pear Gourd. May–Sept. Native of trop. Am.; moist soil, very rare; Union Co.

4. Cucumis L. Cucumber

1. C. sativus L. Cucumber. May–Sept. Native of Asia; rarely escaped from cultivation.

5. Melothria L.

1. M. pendula L. Creeping Cucumber. June–Sept. Rocky soil, rare; Alexander and Pope cos.

6. Sicyos L. Bur Cucumber

1. S. angulatus L. Bur Cucumber. July–Sept. Moist soil of fields and woods, occasional; scattered throughout the state.

7. Echinocystis Torr. & Gray

1. E. lobata (Michx.) Torr. & Gray. Wild Balsam-apple. July–Sept. Moist soil; occasional in the n. three-fourths of Ill., apparently absent in the s. one-fourth.

154. CAMPANULACEAE

- Flowers actinomorphic; corolla tube not split down one side 2
- Flowers zygomorphic; corolla tube split down one side ..... 3. *Lobelia*
- Leaves sessile, often clasping; flowers solitary and sessile in the axils ..
.... 1. *Specularia*
- Leaves petiolate, or at least not clasping at the base; flowers in spikes or racemes or, if solitary, then pedicellate .... 2. *Campanula*

1. Specularia Fab. Venus' Looking-glass

- Bracts suborbicular to ovate; seeds up to 0.7 mm long 2
- Bracts linear to lanceolate; seeds 0.7–1.0 mm long ... 3. *S. leptocarpa*
- Capsule with pores near apex .... 1. *S. biflora*
- Capsule with pores midway between base and apex ... 2. *S. perfoliata*

1. S. biflora (R. & P.) Fisch. & Mey. April–June. Venus' Looking-glass. Dry, often disturbed, soil; confined to the s. one-sixth of the state.

2. S. perfoliata (L.) A. DC. April–Aug. Venus' Looking-glass. Dry, often disturbed, soil, common; probably in every co.

3. S. leptocarpa (Nutt.) Gray. May–July. Fields, rare; Logan and Will cos.

2. Campanula L. Bellflower

1. Cauline leaves linear to narrowly lanceolate (ovate, cordate basal leave
may be present at base of plant in *C. rotundifolia*) ....
1. Cauline leaves oblong-lanceolate to ovate ....
2. Stems glabrous or rarely closely puberulent; corolla 1.5 cm long o
longer .... 1. *C. rotundifoli*
2. Stems scabrous with retrorse hairs on the angles; corolla up to 1.2 c
long ....
3. Corolla white, 5–8 mm long; capsule up to 2 mm long 2. *C. aparinoide*
3. Corolla bluish, 10–12 mm long; capsule 3–5 mm long ... 3. *C. uliginos*
4. Corolla campanulate; capsule with pores nearly basal ....
.... 4. *C. rapunculoide*
4. Corolla rotate; capsule with pores nearly apical ...... 5. *C. american*

1. *C. rotundifolia* L. Bellflower. June–Sept. Stems and caulir
leaves smooth. Woods, hill prairies, sandstone cliffs; occasional
the n. half of the state; also Jackson Co. *C. intercedens* Witasek—J
var. *velutina* A. DC. Bellflower. July–Sept. Stems and cauline leave
closely puberulent; crevices of cliffs; JoDaviess Co.

2. *C. aparinoides* Pursh. Marsh Bellflower. June–Aug. Marshes ar
bogs; occasional in the n. half of the state.

3. *C. uliginosa* Rydb. Marsh Bellflower. June–Aug. Marshes ar
bogs; occasional in the n. half of the state. *C. aparinoides* Pursh v&
grandiflora Holz—G.

4. *C. rapunculoides* L. European Bellflower. July–Sept. Native
Europe; adventive in waste areas; apparently confined to the n. h&
of the state.

5. *C. americana* L. American Bellflower. June–Oct. Woods, com
mon; in every co. (Including var. *illinoensis* [Fresn.] Farw.)

3. Lobelia L. Lobelia

1. Flowers 1.5 cm. long or longer ....
1. Flowers up to 1.5 cm long ....
2. Flowers red or deep rose ....
2. Flowers blue (rarely white) ....
3. Flowers red; calyx glabrous or puberulent ....1. *L. cardina*
3. Flowers deep rose; calyx hirsute .... 2. *L. cardinalis X siphiliti*
4. Auricles at base of calyx 2–5 mm long; stems glabrous or sparsely h
sute; flowers 2.0–3.3 cm long .... 3. *L. siphilit*
4. Auricles at base of calyx less than 2 mm long; stems densely puberule
throughout; flowers 1.5–2.5 cm long .... 4. *L. pubert*
5. Cauline leaves linear to narrowly lanceolate; lower lip of corolla g
brous .... 5. *L. kalr*
5. Cauline leaves oblong, lanceolate, or obovate ....
6. Lower part of stems villous or hirsute; capsules inflated, complet
enclosed by the calyx .... 6. *L. infl&*
6. Stems glabrous or short-pubescent; capsules not inflated, partly &
serted from the calyx .... 7. *L. spic&*

1. *L. cardinalis* L. Cardinal-flower. July–Sept. Wet ground; o
casional to common throughout the state.

2. *L. cardinalis X siphilitica* Schneck. Sept. Wabash Co.

COMPOSITAE

409

3. L. siphilitica L. Blue Cardinal-flower. Aug.–Oct. Wet ground; occasional to common throughout the state. (Including var. *ludoviciana* A. DC.)

4. L. puberula Michx. Downy Lobelia. Aug.–Oct. Wet ground; occasional in the s. one-third of the state. (Including var. *simulans* Fern.)

5. L. kalmii L. Kalm's Lobelia. July–Sept. Springy areas and dunes; occasional in the n. one-third of the state.

6. L. inflata L. Indian Tobacco. June–Oct. Woods, fields, disturbed areas; in every co.

7. L. spicata Lam. Spiked Lobelia. July–Aug. Appendages between calyx lobes up to 1 mm long or absent; leaves mostly spreading. Dry woods, prairies; occasional throughout the state. (Including var. *hirtella* Gray and var. *campanulata* McVaugh.)

var. leptostachys (A. DC.) Mack. & Bush. Spiked Lobelia. June–Aug. Appendages between calyx lobes 1–5 mm long; leaves mostly ascending. Dry soil; occasional throughout the state. *L. leptostachys* A. DC.—J.

155. COMPOSITAE

1. Heads with both ray- and disc-flowers or with disc-flowers only; sap not milky 2
1. Heads with ray-flowers only; sap milky 75
2. Heads with both ray- and disc-flowers 3
2. Heads with disc-flowers only (some plants with small and inconspicuous rays are also keyed out here) 40
3. Pappus of capillary bristles 4
3. Pappus of rigid awns, small chaffy scales, reduced to a mere crown, or absent 11
4. Rays yellow (white in one species of *Solidago*) 5
4. Rays not yellow 10
5. Involucral bracts in one series 6
5. Involucral bracts in more than one series 7
6. Leaves opposite; involucral bracts with 3–7 conspicuous glands; rays small, inconspicuous .... 23. *Dyssodia*
6. Leaves alternate; involucral bracts glandless; rays conspicuous ....
.... 45. *Senecio*
7. Pappus double .... 26. *Heterotheca*
7. Pappus single 8
8. Leaves white woolly beneath .... 42. *Inula*
8. Leaves not white woolly beneath 9
9. Heads numerous; involucre 2.5–9.0 mm high .... 27. *Solidago*
9. Heads few; involucre 1–2 cm high .... 28. *Haplopappus*
10. Involucral bracts in 3–5 series, either foliaceous or with chartaceous base and herbaceous tip .... 31. *Aster*
10. Involucral bracts in 1–2 series, neither foliaceous nor with chartaceous base and herbaceous tip .... 32. *Erigeron*
11. Rays yellow or orange 12
11. Rays not yellow or orange 29
12. Pappus of 2-several awns or bristles (sometimes deciduous) 13
12. Pappus of a few short teeth, scales, or absent 18

13. Involucral bracts with recurved tips, often resinous ...... 24. *Grindel*

13. Involucral bracts without recurved tips, never resinous 1

14. Pappus of 2 (–3) smooth awns .... 14. *Verbesin*

14. Pappus of 2-several awns (if two, then barbed) 1

15. Awns of pappus barbed or chaffy 1

15. Awns of pappus not barbed nor chaffy .... 22. *Gaillard*

16. Leaves alternate .... 25. *Xanthocephalu*

16. Leaves opposite or whorled 1

17. Achenes beaked; ray-flowers orange .... 17. *Cosmo*

17. Achenes not beaked; ray-flowers yellow .... 16. *Bider*

18. Leaves, at least the lower, opposite or whorled, not all basal 1

18. Leaves all alternate or basal 2

19. Achenes flattened 2

19. Achenes not at all or scarcely flattened 2

20. Rays usually 8; involucral bracts in two series; disc-flowers fertile ....
.... 15. *Coreops*

20. Rays numerous; involucral bracts in more than two series; disc-flowe
sterile .... 2. *Silphiu*

21. Leaves deeply angulate lobed or lyrate-pinnatifid .... 1. *Polymr*

21. Leaves entire or serrate ....

22. Involucral bracts obtuse; ray-flowers fertile, persistent .... 7. *Heliops*

22. Involucral bracts acute to acuminate; ray-flowers sterile, deciduous ..
.... 12. *Helianth*

23. Stems scapose ....

23. Stems leafy ....

24. Leaves large, ovate, toothed; scapes 2–3 m tall; heads several ......
.... 2. *Silphiu*

24. Leaves small, spatulate, entire; scapes 5–15 cm tall; heads solitary ...
.... 20. *Hymenox*

25. Receptacle chaffy ....

25. Receptacle naked .... 21. *Heleniu*

26. Stem usually winged .... 14. *Verbesin*

26. Stem never winged ....

27. Receptacle flat or convex .... 12. *Helianth*

27. Receptacle conic or columnar ....

28. Achenes flattened, sharp margined or winged .... 11. *Ratibi*

28. Achenes 4-sided, not sharp margined nor winged .... 9. *Rudbeck*

29. Leaves opposite ....

29. Leaves alternate and/or basal ....

30. Leaves deeply angulate lobed or lyrate-pinnatifid .... 1. *Polymr*

30. Leaves serrate ....

31. Leaves ovate, petiolate .... 18. *Galinso*

31. Leaves lanceolate, sessile .... 8. *Eclip*

32. Leaves all basal .... 29. *Be*

32. Leaves cauline and basal ....

33. Leaves dissected, incised, or coarsely and irregularly toothed ......

33. Leaves entire, regularly serrate, or dentate ....

34. Rays 4–6; heads numerous .... 34. *Achil*

34. Rays 10–30; heads few ....

35. Leaves cut into filiform divisions .... 33. *Anther*

35. Leaves incised or coarsely and irregularly toothed .36. *Chrysanthem*

36. Rays purple or purple with yellow ....

36. Rays white or pink ....

37. Rays reflexed, entirely purple .... 10. *Echinac*

7. Rays not reflexed, purple, often with yellow toward tip .. 22. *Gaillardia*
8. Rays numerous .... 30. *Boltonia*
8. Rays 1–5 39
9. Pappus of two slender awns; leaves entire or serrate ... 14. *Verbesina*
9. Pappus of 2–3 inconspicuous scales; leaves dentate .... 3. *Parthenium*
0. Pappus of capillary bristles 41
0. Pappus of rigid awns, small chaffy scales, reduced to a mere crown, or
   absent 64
1. Flowers pink, purple, blue, or yellow (rarely white) 42
1. Flowers white or cream 55
2. Twining herbaceous vine .... 47. *Mikania*
2. Erect herb 43
3. Leaves opposite or whorled, cauline 44
3. Leaves alternate or basal 45
4. Involucral bracts with 3–7 conspicuous glands; flowers yellow ....
   .... 23. *Dyssodia*
4. Involucral bracts glandless; flowers not yellow .... 46. *Eupatorium*
5. Leaves prickly 46
5. Leaves not prickly 49
6. Heads 1-flowered, in capitate clusters .... 52. *Echinops*
6. Heads many-flowered, distinct 47
7. Pappus bristles plumose .... 55. *Cirsium*
7. Pappus bristles not plumose 48
8. Receptacle with numerous bristles .... 54. *Carduus*
8. Receptacle without bristles .... 56. *Onopordum*
. Involucral bracts pectinate or tipped with a rigid spine .. 57. *Centaurea*
. Involucral bracts not pectinate nor tipped with a rigid spine 50
. Involucral bracts with hooked tips .... 53. *Arctium*
. Involucral bracts without hooked tips 51
. Flowers yellow; involucral bracts in one series .... 45. *Senecio*
. Flowers not yellow; involucral bracts in two series 52
. Pappus bristles plumose or barbellate .... 49. *Liatris*
. Pappus bristles not plumose nor barbellate 53
. Pappus double, the outer series shorter .... 50. *Vernonia*
. Pappus single, subequal 54
. Heads 2- to 5-flowered, aggregated in dense clusters subtended by sev-
  eral foliaceous bracts .... 51. *Elephantopus*
. Heads many-flowered, not aggregated in dense clusters subtended by
  foliaceous bracts .... 39. *Pluchea*
. Leaves prickly 56
. Leaves not prickly 57
. Heads 1-flowered, in capitate clusters .... 52. *Echinops*
. Heads many-flowered, distinct .... 55. *Cirsium*
Involucral bracts scarious 58
Involucral bracts not scarious 59
Leaves mostly basal, cauline leaves small .... 40. *Antennaria*
Leaves all or mostly cauline .... 41. *Gnaphalium*
Involucral bracts in one series (often with a few small bracts at base of
head) 60
Involucral bracts in more than one series 61
Pappus smooth .... 43. *Erechtites*
Pappus scabrous .... 44. *Cacalia*
Leaves, at least lower, opposite or whorled .... 46. *Eupatorium*
Leaves all alternate 62

62. Involucral bracts striate; achenes nearly round; leaves resinous-dotte
.... 48. *Brickell.*
62. Involucral bracts not striate; achenes flattened; leaves not resinous-do
ted ....
63. Involucral bracts in 3–5 series, either foliaceous or with chartaceo
base and herbaceous tip; rays broad, few .... 31. *Aste*
63. Involucral bracts in 1–2 series, neither foliaceous nor with chartaceo
base and herbaceous tip; rays narrow, numerous .... 32. *Erigerc*
64. Flowers green or greenish ....
64. Flowers yellow or white ....
65. Staminate and pistillate flowers in separate heads ....
65. Staminate and pistillate flowers in same head, or all flowers perfect
66. Involucre of pistillate heads tuberculate or with straight spines ......
.... 5. *Ambros*
66. Involucre of pistillate heads with hooked spines .... 6. *Xanthiu*
67. Heads numerous, in spikes, racemes, or panicles ....
67. Heads few to solitary .... 35. *Matricar*
68. Heads in long terminal bracted spikes; leaves opposite, at least t
lower entire or serrate .... 4. *I*
68. Heads in panicles or racemes; leaves alternate, lobed or incised, or c
casionally entire or toothed .... 38. *Artemis*
69. Crushed plant smelling like pineapple .... 35. *Matricar*
69. Crushed plant not smelling like pineapple ....
70. Flowers white ....
70. Flowers yellow or yellowish ....
71. Leaves alternate, pinnatifid .... 19. *Hymenopapp*
71. Leaves opposite, entire or lobed .... 13. *Melanthe*
72. Leaves opposite ....
72. Leaves alternate or basal ....
73. Pappus of 2–4 awns; achenes flattened; disc-flowers fertile   16. *Bide*
73. Pappus absent; achenes scarcely flattened; disc-flowers sterile .....
.... 1. *Polymr*
74. Leaves pinnately dissected .... 37. *Tanacet*
74. Leaves shallowly lobed, spiny margined .... 58. *Cnic*
75. Pappus of capillary bristles only (an outer series of small scales oft
overlooked) ....
75. Pappus not entirely of capillary bristles ....
76. Pappus plumose ....
76. Pappus not plumose, of simple capillary bristles ....
77. Leaves grass-like; plume-branches of pappus interwebbed..........
.... 66. *Tragopog*
77. Leaves not grass-like; plume branches of pappus not interwebbed ..
78. Stems leafy .... 65. *Pic*
78. Stems scapose ....
79. Receptacle chaffy .... 63. *Hypochae*
79. Receptacle naked .... 64. *Leontoc*
80. Achenes round or angled, scarcely flattened ....
80. Achenes strongly flattened ....
81. Heads usually several on each stem; stems leafy ....
81. Heads solitary; stems scapose ....
82. Flowers whitish or purplish; heads pendent .... 72. *Prenanth*
82. Flowers yellow to orange-red; heads not pendent ....
83. Achenes filiform-beaked .... 70. *Pyrrhopapp*
83. Achenes more or less narrowed toward apex, but scarcely beaked .

34. Pappus white .... 71. *Crepis*
34. Pappus sordid to tawny .... 73. *Hieracium*
35. Achenes smooth or nearly so, 10-nerved; leaves entire .. 61. *Microseris*
35. Achenes muricate toward apex, 4- to 5-nerved; leaves lobed to toothed
     (rarely subentire) .... 67. *Taraxacum*
36. Achenes filiform-beaked or narrowed toward apex, only rarely beakless;
     flowers blue or purple .... 69. *Lactuca*
36. Achenes beakless; flowers yellow .... 68. *Sonchus*
37. Flowers blue (rarely white or pink); pappus of minute narrow scales only
     .... 59. *Cichorium*
37. Flowers yellow or orange; pappus various, but not composed of minute
     narrow scales only, or absent 88
38. Involucre with short bracts at base imitating an outer series 60. *Lapsana*
38. Involucre without short bracts at base .... 62. *Krigia*

1. *Polymnia* L.   Leafcup

1. Leaves pinnately lobed; achenes 3-angled, not striate . 1. *P. canadensis*
1. Leaves subpalmately lobed; achenes slightly flattened laterally, striate ..
    .... 2. *P. uvedalia*

   1. P. canadensis L. Leafcup. June–Nov. Moist or dry woods; common throughout Ill. (Including f. *radiata* [Gray] Fassett.)
   2. P. uvedalia (L.) L. Bear's-foot; Yellow Leafcup. July–Sept. Rich
woods; occasional in s. Ill., extending n. to St. Clair Co; also Vermilion
Co. (Including var. *densipilis* Blake.)

2. *Silphium* L.   Rosinweed

1. Leaves connate-perfoliate; stem conspicuously 4-angled ....
    .... 1. *S. perfoliatum*
1. Leaves not connate-perfoliate; stem nearly round or only obtusely an-
    gled 2
2. Leaves deeply pinnatifid or bipinnatifid .... 2. *S. laciniatum*
2. Leaves entire or toothed 3
3. Leaves chiefly basal, large, cordate, long-petiolate; stems 1–3 m tall ....
    .... 3. *S. terebinthinaceum*
3. Leaves cauline, relatively small, lanceolate to ovate, sessile; stems
    0.5–1.5 m tall .... 4. *S. integrifolium*

   1. S. perfoliatum L. Cup-plant. July–Aug. Open woods and low
ground; common throughout Ill.
   2. S. laciniatum L. Compass-plant; Rosinweed. July–Aug. Prairies;
common throughout Ill.
   3. S. terebinthinaceum Jacq. Prairie-dock. July–Sept. Prairies; common throughout Ill. (Including var. *lucy-brauniae* Steyerm.)
   4. S. integrifolium Michx. Rosinweed. July–Aug. Prairies; common
throughout Ill. (Including var. *deamii* Perry.)

3. *Parthenium* L.

Leaves pinnately dissected .... 1. *P. hysterophorus*
Leaves toothed .... 2. *P. integrifolium*

1. *P. hysterophorus* L. Santa Maria. Aug.–Oct. Adventive from trop
Am. Waste ground, rare; Cook Co.

2. *P. integrifolium* L. American Feverfew; Wild Quinine. July–Sep
Prairies and dry woods; common throughout Ill.

4. *Iva* L.   Marsh-elder

1. Heads in racemose spikes, bracteate .... 1. *I. annu*
1. Heads in paniculate spikes, ebracteate.   .... 2. *I. xanthifol*

1. *I. annua* L. Marsh-elder; Sumpweed. Aug.–Oct. Moist ground; oc
casional throughout Ill., chiefly in the s. half. *I. ciliata* Willd—F, G. J

2. *I. xanthifolia* Nutt. Burweed Marsh-elder. July–Sept. Moist wast
ground; occasional in the n. half of Ill.

5. *Ambrosia* L.   Ragweed

1. Pistillate involucre with only 1 row of tubercles or erect spines near ape
....

1. Pistillate involucre with several rows of spines on surface ....
.... 5. *A. tomentos*
2. Staminate heads sessile in solitary terminal spikes ..... 1. *A. bidenta*
2. Staminate heads pedicellate in 1-several slender racemes ....
3. Leaves palmately 3- to 5-lobed or entire .... 2. *A. trific*
3. Leaves pinnatifid or bipinnatifid ....
4. Leaves petiolate, bipinnatifid, glabrous or nearly so above; involuc
with 5–7 spines .... 4. *A. artemisiifol*
4. Leaves sessile or subsessile, pinnatifid, harsh above; involucre unarm
or with blunt tubercles .... 3. *A. psilostach*

1. *A. bidentata* Michx. Ragweed. July–Oct. Fields and was
ground; common in s. half of Ill.

2. *A. trifida* L. Giant Ragweed. July–Oct. Petioles of at least t
upper leaves more or less wing margined; ribs or fruits ending
short spines. Fields and waste ground; in every co. (Including f. *i
tegrifolia* [Muhl.] Fern.)

var. *texana* Scheele. Giant Ragweed. July–Oct. Petioles wingle
ribs of fruits ending in blunt to almost obsolete tubercles. Advent
from sw. Waste ground; occasional in s. Ill.

3. *A. psilostachya* DC. Western Ragweed. July–Oct. Dry sandy s
in fields and waste ground; occasional in n. three-fifths of Ill. (Inclu
ing var. *coronopifolia* [Torr. & Gray] Farw.) *A. coronopifolia* Torr.
Gray—J.

4. *A. artemisiifolia* L. Common Ragweed. Aug.–Oct. Fields a
waste ground; in every co. (Including var. *elatior* [L.] Desc. and v
elatior f. *villosa* Fern. & Grisc.)

5. *A. tomentosa* Nutt. False Ragweed. July–Oct. Adventive from
Waste ground, rare; La Salle and McHenry cos. *Franseria disco*
Nutt.—F, G, J.

Reported hybrids: *A. artemisiifolia* X *A. trifida*, Champaign C
A. bidentata X *A. trifida*, Clay, Jersey, Sangamon, St. Clair, and V
milion cos.

6. Xanthium L. Cocklebur

1. Leaves lanceolate, tapering to base, with 3-parted axillary spine ....
.... 1. *X. spinosum*
1. Leaves ovate, base cordate or deltoid, without axillary spine ....
.... 2. *X. strumarium*

1. X. spinosum L. Spiny Cocklebur. Aug.–Oct. Adventive from trop. Am. Waste ground, rare; Alexander, Cook, and Pulaski cos.
2. X. strumarium L. Common Cocklebur. Our plants belong to the following varieties:
var. canadensis (Mill.) Torr. & Gray. Aug.–Oct. Body of fruit and prickles glandular-hispidulous. Waste ground and along rivers; common throughout Ill. (Including *X. commune* Britt., *X. italicum* Moretti, *X. oviforme* Wallr., and *X. variens* Greene.)
var. glabratum (DC.) Cronq. Aug.–Oct. Body of fruit glabrous or merely glandular. Waste ground and along rivers; common throughout Ill. (Including *X. chasei* Fern., *X. chinense* Mill., *X. globosum* Shull, *X. inflexum* Mack. & Bush, and *X. pennsylvanicum* Mill.)

7. Heliopsis Pers. Ox-eye

1. H. helianthoides (L.) Sweet. False Sunflower. July–Aug. Open woods and prairies; common throughout Ill. (Including var. *scabra* [Dunal] Fern.)

8. Eclipta L.

1. E. alba (L.) Hassk. Yerba de Tajo. July–Oct. Wet ground and waste ground; common in the s. three-fourths of the state, rare or absent elsewhere.

9. Rudbeckia L. Coneflower

. Leaves clasping .... 1. *R. amplexicaulis*
. Leaves not clasping 2
. Disc greenish-yellow .... 2. *R. laciniata*
. Disc brown or purple 3
. Lower stem leaves 3-lobed or 3-parted 4
. Lower stem leaves merely toothed or entire 5
. Chaff of receptacle glabrous .... 3. *R. triloba*
. Chaff of receptacle pubescent toward tip .... 4. *R. subtomentosa*
. Pappus absent; style-branches elongate, acute .... 5. *R. hirta*
. Pappus present, but consisting of a minute short crown, commonly toothed on angles; style-branches short, blunt .... 6. *R. fulgida*

1. R. amplexicaulis Vahl. Coneflower. June–Aug. Waste ground along railroads, rare; Cook and Jackson cos.
2. R. laciniata L. Goldenglow. July–Sept. Moist ground; occasional throughout Ill.
3. R. triloba L. Brown-eyed Susan. Aug.–Oct. Woods; common throughout Ill.
4. R. subtomentosa Pursh. Fragrant Coneflower. Aug.–Sept. Prairies and low open woods; occasional throughout Ill.

5. R. hirta L. Black-eyed Susan. June–Sept. Open woods and fields
common throughout Ill. Occasional specimens with pale yellow ray
having white to cream tips (f. *flavescens* Clute) are observed. (Includ
ing R. *bicolor* Nutt., R. *serotina* Nutt., and R. *serotina* var. *lanceolat*
[Bisch.] Fern. & Schub.)

6. R. fulgida Ait. Orange Coneflower. July–Oct. Cauline leaves ellip
tic, sessile or narrowed to winged petiole. Woods, rare; Pope Co
(Including R. *spathulata* Michx. and R. *tenax* Boynt. & Beadle.)

var. sullivantii (Boynt. & Beadle) Cronq. Sullivant's Orange Conefl
wer. July–Oct. Cauline leaves ovate, sharply toothed, petiolate. Moi
ground; occasional in e. Ill. (Including R. *deamii* Blake, R. *palustr*
Eggert, and R. *speciosa* Boynt. & Beadle.) R. *sullivantii* Boynt.
Beadle—J.

var. missouriensis (Engelm.) Cronq. Missouri Orange Coneflowe
July–Oct. Cauline leaves linear-spatulate, entire. Dry open groun
rare; Monroe and Randolph cos. R. *missouriensis* Engelm.—F, G.

10. Echinacea Moench.   Coneflower

1. Leaves ovate, rounded at base, serrate or dentate, upper frequently e
tire .... 1. *E. purpure*
1. Leaves lanceolate to linear, attenuate to base, all entire ... 2. *E. pallic*

1. E. purpurea (L.) Moench. Purple Coneflower. July–Aug. Ope
woods and prairies; occasional throughout Ill.

2. E. pallida (Nutt.) Nutt. Pale Coneflower. July–Aug. Open woo
and prairies; occasional throughout Ill.

E. simulata McGregor, a species similar to *E. pallida,* except f
being a diploid and having yellow pollen while *E. pallida* is a t
raploid and has white pollen, has been reported from Ill. Its status
a distinct species is dubious.

11. Ratibida Raf.   Prairie Coneflower

1. Disc columnar, equalling or exceeding rays, 2–4 times as long as wi
.... 1. *R. columnife*
1. Disc ellipsoid, shorter than rays, 1.0–1.5 times as long as wide ......
.... 2. *R. pinna*

1. R. columnifera (Nutt.) Woot. & Standl. Long-headed Coneflow
July–Aug. Adventive from w. U.S. Along railroads; scattered in the
two-thirds of Ill.

2. R. pinnata (Vent.) Barnh. Drooping Coneflower; Yellow Cor
flower. July–Aug. Prairies; common in the n. three-fourths of t
state, occasional elsewhere.

12. Helianthus L.   Sunflower

1. Leaves with conspicuously bristly ciliate margins; rays less than 1
long .... 1. *H. cilia*
1. Leaves without conspicuously bristly ciliate margins; rays normally ov
1 cm long ....
2. Disc red or purple ....
2. Disc yellow ....

3. Leaves linear 4
3. Leaves lanceolate to cordate 5
4. Stem glabrous, usually glaucous .... 2. *H. salicifolius*
4. Stem strigose to pilose .... 3. *H. angustifolius*
5. Receptacle flat or nearly so 6
5. Receptacle convex or low-conic 7
6. Involucral bracts ovate, abruptly contracted above; receptacular bracts not bearded at apex .... 4. *H. annuus*
6. Involucral bracts lanceolate, tapering; receptacular bracts conspicuously bearded at apex .... 5. *H. petiolaris*
7. Leaves tapering to short thick petiole, or sessile .... 6. *H. rigidus*
7. Leaves abruptly slender petiolate .... 7. *H. silphioides*
8. Stems scapose, or with 3–5 pairs of cauline leaves smaller than the basal leaves .... 8. *H. occidentalis*
8. Stems leafy to inflorescence 9
9. Stems glabrous or nearly so below inflorescence, often glaucous .. 10
9. Stems pubescent 14
10. Heads 1.5–3.0 cm wide, disc 0.4–1.0 cm wide, rays 1.0–1.5 cm long .... 9. *H. microcephalus*
10. Heads 4–9 cm wide, disc 1.0–2.5 cm wide, rays 2–4 cm long 11
11. Leaves thin, membranaceous .... 10. *H. decapetalus*
11. Leaves thick, firm 12
12. All leaves opposite except uppermost 13
12. Middle and upper leaves alternate, lower opposite or whorled .... 13. *H. grosseserratus*
13. Leaves sessile or, if petiolate, the petioles not over 0.5 cm long .... 11. *H. divaricatus*
13. Leaves evidently petiolate, the petioles 0.5–1.0 cm long 12. *H. strumosus*
14. Leaves densely grayish-pubescent on both surfaces. ..... 14. *H. mollis*
14. Leaves not densely grayish-pubescent on both surfaces 15
15. Leaves broadly lanceolate to ovate; petioles 2–8 cm long .... 15. *H. tuberosus*
15. Leaves lanceolate; petioles, when present, not over 1.5 cm long 16
16. All leaves opposite except uppermost .... 16. *H. hirsutus*
16. Upper and middle leaves alternate, lower opposite 17
17. Stem with spreading hirsute hairs .... 17. *H. giganteus*
17. Stem with short appressed or subappressed hairs .. 18. *H. maximilianii*

1. H. ciliaris DC. Ciliate Sunflower. Aug.–Sept. Adventive from w. Waste ground, rare; St. Clair Co.

2. H. salicifolius A. Dietr. Willow-leaved Sunflower. Aug.–Oct. Adventive from w. Waste ground, rare; Cook Co.

3. H. angustifolius L. Narrow-leaved Sunflower. Aug.–Oct. Moist ground, rare; Massac and Pope cos.

4. H. annuus L. Common Sunflower; Garden Sunflower. Aug–Sept. Adventive and introduced from w., often escaped from cultivation. Fields and waste ground; occasional throughout Ill.

5. H. petiolaris Nutt. Petioled Sunflower. June–Sept. Adventive from w. Sandy soil in waste ground and fields; occasional throughout Ill.

6. H. rigidus (Cass.) Desf. Prairie Sunflower. July–Sept. Prairies; occasional throughout Ill. (Including var. *subrhomboideus* [Rydb.] Fern.) *H. laetiflorus* Pers.—F, G.

7. H. silphioides Nutt. Silphium Sunflower. July–Oct. Prairies, rare; Alexander and St. Clair cos.

8. H. occidentalis Riddell. Western Sunflower. July–Sept. Sand
soil; common in n. half of Ill., extending s. to St. Clair Co.

9. H. microcephalus Torr. & Gray. Small Wood Sunflower. Aug–Oc
Dry open woods; occasional in s. Ill.

10. H. decapetalus L. Pale Sunflower; Thin-leaved Sunflower. Aug.
Oct. Dry, open woods; occasional throughout Ill.

11. H. divaricatus L. Woodland Sunflower. July–Sept. Open woods
common throughout Ill.

12. H. strumosus L. Pale-leaved Sunflower. July–Sept. Open woods
common throughout Ill.

13. H. grosseserratus Martens. Sawtooth Sunflower. July–Oc
Prairies, edges of woods, fencerows; common throughout Ill.

14. H. mollis Lam. Downy Sunflower. Aug.–Sept. Prairies; commo
nearly throughout Ill. (Including var. *cordatus* S. Wats.)

15. H. tuberosus L. Jerusalem Artichoke. Aug.–Oct. Upper and mid
dle leaves alternate, moderately and inconspicuously pubesce
below with mostly appressed hairs. Moist ground; common throug
out Ill.

var. subcanescens Gray. Aug.–Oct. All leaves opposite except u
permost, densely pubescent below with loose or spreading hair
Moist ground; occasional throughout Ill. *H. tomentosus* Michx.—J.

16. H. hirsutus Raf. Bristly Sunflower. Aug.–Sept. Open woods ar
fields; occasional throughout Ill. (Including var. *trachyphyllus* Torr.
Gray and var. *stenophyllus* Torr. & Gray.)

17. H. giganteus L. Tall Sunflower. Aug.–Oct. Moist ground, rar
Kankakee and Tazewell cos. *H. tomentosus* Michx.—F.

18. H. maximilianii Schrader. Maximillian's Sunflower. July–Aug. A
ventive from w. Waste ground; occasional in Ill. (Including *H. da*
Britt.)

Reported hybrids: *H. X doronicoides* Lam. (*H. mollis* X *H. giga*
teus). Sangamon Co.; *H. X luxurians* E. E. Wats. (*H. giganteus* X
grosseserratus), Cook, Effingham, Lake, and Will cos.

13. Melanthera Rohr.

1. M. nivea (L.) Small. June–Oct. Floodplain woods, rare; Mass
and Pulaski cos. *M. hastata* Michx.—J.

14. Verbesina L. Crownbeard. Wingstem

1. Stem usually winged, at least in the upper part; leaves sessile ......
1. Stem wingless; leaves petiolate .... 4. *V. encelioid*
2. Rays yellow, 2–3 cm long ....
2. Rays white, 4–7 mm long .... 3. *V. virgini*
3. Stem simple; heads few (2–8); involucral bracts erect 1. *V. helianthoid*
3. Stem usually branched toward top; heads many (10–100); involuc
bracts deflexed .... 2. *V. alternifo*

1. V. helianthoides Michx. Yellow Crownbeard. May–July. Op
woods and prairies; common in s. three-fifths of Ill., absent els
where.

2. V. alternifolia (L.) Britt. Yellow Ironweed. Aug.–Sept. Moist soi
open woods; occasional throughout Ill. *Actinomeris alternifolia* (
DC.—F.

3. V. virginica L. Tickweed; Frostweed. July–Aug. Dry, open woods, rare; confined to the extreme se. cos.

4. V. encelioides (Cav.) Benth. & Hook. Golden Crownbeard. June–Aug. Adventive from w. Waste ground, rare; St. Clair Co. (Including var. *exauriculata* Robins. & Greenm.)

15. Coreopsis L.   Tickseed

1. Leaves undivided or rarely with 1–2 short lateral lobes 2
1. Leaves 3- to 5-lobed or undivided 3
2. Leaves mostly basal, linear to oblanceolate .... 1. *C. lanceolata*
2. Leaves produced to middle of stem or higher, ovate to elliptic-lanceolate
.... 2. *C. pubescens*
3. Leaves sessile, deeply 3-lobed to or below middle .... 3. *C. palmata*
3. Leaves usually petiolate, divided into 3–5 segments 4
4. Ligules of ray-flowers reddish-brown at base or throughout; disc-flowers reddish-brown 5
4. Ligules of ray-flowers yellow; disc flowers yellow or reddish-brown .. 6
5. Achenes linear, wingless; leaf segments linear to linear-lanceolate .....
.... 4. *C. tinctoria*
5. Achenes obovate, cartilaginous-margined; leaf segments lanceolate to orbicular .... 5. *C. basalis*
6. Leaf segments elliptic-lanceolate; achenes 5–7 mm long; disc-flowers yellow or reddish-brown .... 6. *C. tripteris*
6. Leaf segments linear to linear-lanceolate; achenes 1–4 mm long; disc-flowers yellow .... 7. *C. grandiflora*

1. C. lanceolata L. Tickseed Coreopsis; Sand Coreopsis. June–July. Dry sandy or rocky soils; scattered throughout Ill., but not common. (Including var. *villosa* Michx.)

2. C. pubescens Ell. Star Tickseed. June–Sept. Open woods; rare in s. Ill.

3. C. palmata Nutt. Prairie Coreopsis. June–July. Prairies and open woods; common in the n. three-fourths of Ill., occasional elsewhere.

4. C. tinctoria Nutt. Golden Coreopsis. July–Sept. Adventive from w. and escaped from cultivation. Waste ground; scattered throughout Ill., but not common. (Including f. *atropurpurea* [Hook.] Fern.)

5. C. basalis (Otto & Dietr.) Blake. June–Sept. Introduced from sw. and escaped from cultivation. Waste ground, rare; Lake Co.

6. C. tripteris L. Tall Tickseed. Aug.–Sept. Open woods; common throughout Ill. (Including var. *deamii* Standl.)

7. C. grandiflora Hogg. Large-flowered Coreopsis. June–Aug. Adventive from west and escaped from cultivation. Waste ground; occasional throughout Ill.

16. Bidens L.   Beggar-ticks. Sticktight

1. Plants aquatic, with submerged leaves finely dissected .... 1. *B. beckii*
1. Plants subaquatic or terrestrial, without submerged, finely dissected leaves 2
2. Rays 1 cm or more long 3
2. Rays less than 0.5 cm long or absent 5
3. Leaves simple .... 2. *B. cernua*
3. Leaves pinnately dissected 4

4. Achenes cuneate, 2 mm or less wide, margins ciliate   ... 3. *B. coronata*
4. Achenes obovate, more than 2 mm wide, margins scarious or less commonly only ciliate .... 4. *B. aristosa*
5. Leaves simple or sometimes 3-lobed 6
5. Leaves pinnately dissected. 7
6. Outer involucral bracts rarely much exceeding disc; achenes (3-) 4-angled, tuberculate, sparsely antrorsely appressed-pubescent; corolla 5 lobed; stamens exserted .... 5. *B. connata*
6. Outer involucral bracts 2–5 times as long as disc; achenes flat, smooth glabrous or rarely sparsely pubescent; corolla 4-lobed; stamens in cluded .... 6. *B. comosa*
7. Achenes linear .... 7. *B. bipinnata*
7. Achenes flat 8
8. Achenes downwardly barbed (upwardly barbed in a variety of *B. frondosa*); outer involucral bracts with ciliate margins 9
8. Achenes upwardly barbed; outer involucral bracts with smooth margins .... 10. *B. discoidea*
9. Outer involucral bracts 5–8; achenes sparsely pubescent 8. *B. frondosa*
9. Outer involucral bracts 10–20; achenes glabrous or nearly so ....
.... 9. *B. vulgata*

1. B. beckii Torr. Water Marigold. Aug.–Sept. Ponds and slow streams, rare; Cook, Lake, and St. Clair cos. *Megalodonta beckii* (Torr.) Greene—F.

2. B. cernua L. Sticktight; Nodding Bur Marigold. July–Oct. Wet ground; occasional throughout Ill. (Including var. *integra* Wieg. and var. *elliptica* Wieg.)

3. B. coronata (L.) Britt. Tall Swamp Marigold; Beggar-ticks; Sticktight. July–Oct. Wet ground; occasional throughout Ill. (Including var. *tenuiloba* [Gray] Sherff and var. *trichosperma* [Michx.] Fern.)

4. B. aristosa L. Swamp Marigold; Tickseed Sunflower. Aug.–Oct. Outer involucral bracts 8–10, entire, merely finely ciliate. Wet ground; common in the s. two-thirds of the state, occasional elsewhere. Occasionally specimens with awns downwardly barbed (f. *fritchey* [Fern.] Wunderlin) or with awns absent or minute (f. *mutica* [Gray] Wunderlin) are observed.

var. retrorsa (Sherff) Wunderlin. Swamp Marigold; Tickseed Sunflower. Aug.–Oct. Outer involucral bracts 12–20, coarsely ciliate, appearing almost serrulate. Wet ground; occasional throughout Ill. (Our plants belong to f. *involucrata* [Nutt.] Wunderlin.) *B. polylepis* Blake—F, G, J.

5. B. connata Muhl. Swamp Beggar-ticks. Sept.–Oct. Wet ground; occasional throughout Ill. (Including var. *petiolata* [Nutt.] Farw.) *B. tripartita* L. (in part)—G.

6. B. comosa (Gray) Wieg. Beggar-ticks. Sept.–Oct. Wet ground; common throughout Ill. *B. tripartita* L. (in part)—G.

7. B. bipinnata L. Spanish Needles. Aug.–Sept. Waste ground and open woods; occasional in the s. three-fourths of the state, rare or absent elsewhere.

8. B. frondosa L. Common Beggar-ticks; Sticktight. Aug.–Oct. Moist waste ground; common throughout Ill.

9. B. vulgata Greene. Tall Beggar-ticks; Sticktight. Aug.–Oct. Moist waste ground; occasional throughout Ill. (Including f. *puberula* [Wieg.] Fern.)

10. B. discoidea (Torr. & Gray) Britt. Swamp Beggar-ticks; Sticktight. Aug.–Oct. Wet ground; occasional in the s. two-thirds of Ill.

17. *Cosmos* Cav. *Cosmos*

1. C. sulphureus Cav. Cosmos. Aug.–Oct. Adventive from sw. Waste ground, rare; Will Co.

18. *Galinsoga* R. & P. *Peruvian Daisy*

1. Pappus of disc-flowers slightly fringed, tapering to awn-tips; pappus of ray-flowers well developed .... 1. *G. ciliata*
1. Pappus of disc-flowers conspicuously fringed, not tapering to awn-tips; pappus of ray-flowers absent or minute .... 2. *G. parviflora*

1. G. ciliata (Raf.) Blake. Peruvian Daisy; Quickweed. June–Oct. Naturalized from trop. Am. Gardens and waste ground; occasional throughout Ill.
2. G. parviflora Cav. Peruvian Daisy. June–Oct. Naturalized from trop. Am. Gardens and waste ground; rare in Ill.

19. *Hymenopappus* L'Hér.

1. H. scabiosaeus L'Hér. May–June. Open sandy woods and prairies, rare; Cass, Iroquois, Kankakee, and Mason cos.

20. *Hymenoxys* Cass.

1. H. acaulis (Pursh) Parker. Lakeside Daisy; Four-nerved Star-flower. May–July. Dry gravelly banks and fields, rare; Mason and Will cos. (Our plants belong to var. *glabra* [Nutt.] Parker.) *Actinea herbacea* (Greene) B. L. Robins.—F, J.

21. *Helenium* L. *Sneezeweed*

1. Leaves linear to linear-filiform, not decurrent .... 1. *H. amarum*
1. Leaves linear-lanceolate to ovate, decurrent 2
2. Disc depressed-globose; disc-flowers yellow .... 2. *H. autumnale*
2. Disc globose; disc-flowers purplish. .... 3. *H. flexuosum*

1. H. amarum (Raf.) Rock. Bitterweed. Aug.–Oct. Fields and waste ground; occasional in s. Ill., extending n. to Pike and Champaign cos. *H. tenuifolium* Nutt.—F, G, J.
2. H. autumnale L. Autumn Sneezeweed. Aug.–Oct. Wet ground; common throughout Ill. (Including var. *canaliculatum* [Nutt.] Torr. & Gray and var. *parviflorum* [Nutt.] Fern.)
3. H. flexuosum Raf. Purple-headed Sneezeweed. June–Sept. Fields; common in s. half of Ill., rare in n. Ill. *H. nudiflorum* Nutt.—F, j.

22. *Gaillardia* Foug. *Gaillardia. Blanket-flower*

1. Ray- and disc-flowers yellow; chaff-like development on the receptacle among the disc-flowers reduced to soft short scales, or obsolete ....
.... 1. *G. aestivalis*

1. Ray-flowers yellow and purple, disc-flowers purple, reddish-brown, or purplish-brown; chaff-like bristles on the receptacle among the disc-flowers firm and well developed, equalling or longer than achenes . . . .
.... 2. *G. pulchella*

1. *G. aestivalis* (Walt.) Rock. Gaillardia. July–Oct. Prairies, very rare; a single collection from Alexander Co., made in 1874. *G. lutea* Greene—F; *G. lanceolata* Michx. var. *flavovirens* C. Mohr.—G.
2. *G. pulchella* Foug. Blanket-flower. June–July. Introduced from w and escaped from cultivation. Waste ground; rarely escaped.

23. *Dyssodia* Cav.   Fetid Marigold

1. *D. papposa* (Vent.) Hitchc. Fetid Marigold. Sept.–Oct. Waste ground and fields; occasional throughout Ill.

24. *Grindelia* Willd.   Gumweed. Tarweed

1. *G. squarrosa* (Pursh) Dunal. Gumweed; Tarweed. July–Sept. Upper and middle leaves 2–4 times as long as wide, ovate to oblong. Fields and waste ground; occasional in the n. half of the state, rare elsewhere.
var. *serrulata* (Rydb.) Steyerm. Gumweed; Tarweed. July–Sept. Upper and middle leaves 5–8 times as long as wide, linear-oblong to oblanceolate. Fields and waste ground; occasional in the n. half of the state, rare elsewhere.

25. *Xanthocephalum* Willd.   Broomweed

1. Pappus of disc-flowers composed of 10–12 minute scales   1. *X. texanum*
1. Pappus of disc-flowers composed of 5–8 linear strap-shaped projections equalling corolla . . . . 2. *X. dracunculoides*

1. *X. texanum* (DC.) Shinners. Broomweed. July–Oct. Adventive from w. Waste ground, rare; St. Clair Co.
2. *X. dracunculoides* (DC.) Shinners. Broomweed. July–Oct. Adventive from w. Waste ground, rare; Tazewell Co. *Gutierrezia dracunculoides* (DC.) Blake—F, G, J.

26. *Heterotheca* Cass.   Golden Aster

1. Achenes of ray-flowers with capillary pappus . . . . 1. *H. villosa*
1. Achenes of ray-flowers without pappus . . . . 2. *H. latifolia*

1. *H. villosa* (Pursh) Shinners. Golden Aster. June–Sept. Sandy soil, occasional throughout Ill. (Our plants belong to var. *camporum* [Greene] Wunderlin.) *Chrysopsis camporum* Greene—F, J; *Chrysopsis villosa* (Pursh) Nutt. var. *camporum* (Greene) Cronq.—G.
2. *H. latifolia* Buckl. Camphorweed; Golden Aster. July–Oct. Sandy soil, rare; Alexander, Henry, and Union cos. (Our plants belong to var. *macgregoris* Wagenkn.) *H. subaxillaris* (Lam.) Britt. & Rusby—F, G, J.

27. Solidago L. Goldenrod

1. Inflorescence corymbose 2
1. Inflorescence paniculate, racemose, or axillary 5
2. Leaves glandular-punctate, basal and lower soon deciduous ....
   .... 1. *S. graminifolia*
2. Leaves not glandular-punctate, basal and lower persistent 3
3. Leaves ovate to elliptic, scabrous .... 2. *S. rigida*
3. Leaves linear to lanceolate, glabrous 4
4. Inflorescence puberulent; leaves conduplicate .... 3. *S. riddellii*
4. Inflorescence glabrous; leaves flat .... 4. *S. ohioensis*
5. Inflorescence racemose or axillary 6
5. Inflorescence paniculate 11
6. Stem glabrous; achenes pubescent at maturity 7
6. Stem pubescent; achenes glabrous at maturity 9
7. Leaves unequal in size, basal and lower cauline larger than middle and
   upper .... 5. *S. sciaphila*
7. Leaves nearly equal in size, or basal and lower cauline smaller than
   middle and upper 8
8. Leaves lanceolate, sessile or nearly so .... 6. *S. caesia*
8. Leaves ovate, with winged petioles .... 7. *S. flexicaulis*
9. Leaves unequal in size, basal and lower cauline larger than middle and
   upper .... 8. *S. bicolor*
9. Leaves nearly uniform in size, or basal and lower cauline smaller than
   middle and upper 10
10. Leaves thin, mostly sharply serrate .... 9. *S. buckleyi*
10. Leaves thick and firm, entire or occasionally remotely serrate ....
    .... 10. *S. petiolaris*
11. Basal and lower cauline leaves cordate; pappus much shorter than
    achene .... 11. *S. sphacelata*
11. Basal and lower cauline leaves not cordate; pappus equal or longer than
    achene 12
12. Stem below inflorescence glabrous or nearly so 13
12. Stem below inflorescence pubescent 21
13. Branches of the inflorescence glabrous or nearly so 14
13. Branches of the inflorescence pubescent 15
14. Leaves distinctly 3-nerved; achenes glabrous to sparsely pubescent ....
    .... 12. *S. missouriensis*
14. Leaves 1-nerved or indistinctly 3-nerved; achenes pubescent ....
    .... 13. *S. juncea*
15. Branches of the panicle erect or ascending 16
15. Branches of the panicle spreading or recurved 17
16. Basal and lower cauline leaves tapering to long petiole with sheathing
    base .... 14. *S. uliginosa*
16. Basal and lower cauline leaves broad and abruptly petiolate without
    sheathing base .... 15. *S. speciosa*
17. Stem angled or winged; leaves rugose, scabrous above; achenes gla-
    brous or minutely pubescent .... 16. *S. patula*
17. Stem round; leaves not rugose, smooth above; achenes pubescent . 18
18. Leaves 3-nerved .... 17. *S. gigantea*
18. Leaves 1-nerved 19
19. Leaves lanceolate, entire, glabrous below .... 18. *S. sempervirens*
19. Leaves elliptic, sharply serrate, pubescent below 20
20. Basal leaves present at flowering .... 19. *S. arguta*

20. Basal leaves usually not present at flowering .... 20. *S. ulmifolia*
21. Leaves 3-nerved 22
21. Leaves 1-nerved 24
22. Leaves ovate to elliptic 23
22. Leaves lanceolate or oblanceolate .... 23. *S. canadensis*
23. Middle and upper leaves sessile or nearly so, not rough to touch .....
.... 21. *S. drummondi*
23. All except the uppermost leaves evidently petiolate, rough to touch ...
.... 22. *S. radula*
24. Stem grayish puberulent; leaves oblanceolate .... 24. *S. nemoralis*
24. Stem hirsute, occasionally subglabrous below; leaves lanceolate to ellip
tic .... 25. *S. rugosa*

1. S. **graminifolia** (L.) Salisb. Grass-leaved Goldenrod. Aug.–Oct.
Leaves 3- to 5-nerved, glabrous or nearly so to hirtellous, not resinous
or conspicuously glandular-punctate, thin; heads glomerate, rarely
pedunculate; involucre 3–5 mm high. Moist ground; common
throughout Ill. (Including var. *nuttallii* [Greene] Fern. [= *S. hirtella*
(Greene) Bush] and var. *media* [Greene] Harris [= *S. media* (Greene)
Bush.]

var. **gymnospermoides** (Greene) Croat. Grass-leaved Goldenrod.
Aug.–Oct. Leaves 1-nerved, rarely faintly 3-nerved, glabrous, usually
resinous and conspicuously glandular-punctate, firm; heads pedunculate, rarely glomerate; involucre 4.5–6.0 mm high. Open sandy
ground; occasional in w. and n. Ill. *S. gymnospermoides* (Greene)
Fern.—F, G.

var. **remota** (Greene) Harris. Grass-leaved Goldenrod. Aug.–Sept.
Leaves 1-nerved, rarely faintly 3-nerved, glabrous or nearly so, more
or less resinous, not conspicuously glandular-punctate; involucre
3.0–4.5 mm high. Open sandy soil near Lake Michigan, rare; Cook,
Kankakee, and Lake cos. *S. remota* (Greene) Friesner—F, G.

2. S. **rigida** L. Rigid Goldenrod. Aug.–Sept. Prairies; common
throughout Ill., except for the extreme s. tip.

3. S. **riddellii** Frank. Riddell Goldenrod. Aug.–Sept. Moist ground;
occasional in the n. half of the state, rare in the s. half.

4. S. **ohioensis** Riddell. Ohio Goldenrod. Aug.–Oct. Moist ground;
occasional in ne. Ill., extending s. to Peoria and Woodford cos.

5. S. **sciaphila** Steele. Cliff Goldenrod. Aug.–Oct. Calcareous cliffs,
rare; Carroll, JoDaviess, LaSalle, and Ogle cos.

6. S. **caesia** L. Woodland Goldenrod; Blue-stem Goldenrod. Aug.–
Oct. Woods; occasional throughout Ill. (Including f. *axillaris* [Pursh]
House.)

7. S. **flexicaulis** L. Broadleaf Goldenrod. Aug.–Oct. Woods; occasional nearly throughout Ill. *S. latifolia* L.—F, G.

8. S. **bicolor** L. White Goldenrod. Aug.–Oct. Flowers white to cream.
Dry open woods; scattered throughout Ill., but not common.

var. **concolor** Torr. & Gray. Hispid Goldenrod. Aug.–Oct. Flowers
yellow-orange. Dry open woods; Alexander and Jackson cos. *S. hispida* Muhl.—F, G, J.

9. S. **buckleyi** Torr. & Gray. Buckley's Goldenrod. Sept.–Oct.
Woods; rare in s. Ill.

10. S. **petiolaris** Ait. Goldenrod. Sept.–Oct. Open rocky woods; occasional in sw. Ill. (Including var. *wardii* [Britt.] Fern.)

11. S. **sphacelata** Raf. Goldenrod. Aug.–Oct. River bluffs, rare; Pope
Co.

12. S. missouriensis Nutt. Missouri Goldenrod. Aug.–Sept. Prairies; common throughout Ill. (Including var. *fasciculata* Holz and var. *glaberrima* [Martens] Rosend. & Cronq.) S. *glaberrima* Martens—J.

13. S. juncea Ait. Early Goldenrod. July–Aug. Open woods, fields, and waste ground; common throughout Ill. (Including f. *scabrella* [Torr. & Gray] Fern.)

14. S. uliginosa Nutt. Swamp Goldenrod. Aug.–Sept. Swamps and bogs; occasional in n. one-half of Ill. (Including var. *linoides* [Torr. & Gray] Fern.)

15. S. speciosa Nutt. Showy Goldenrod. Aug.–Oct. Open woods and prairies; common throughout Ill. (Including var. *rigidiuscula* Torr. & Gray and var. *jejunifolia* [Steele] Cronq.)

16. S. patula Muhl. Spreading Goldenrod. Aug.–Oct. Wet ground; occasional throughout Ill.

17. S. gigantea Ait. Late Goldenrod. Aug.–Oct. Moist ground; common throughout Ill. (Including var. *serotina* [Kuntze] Cronq.)

18. S. sempervirens L. Seaside Goldenrod. Aug.–Oct. Adventive from e. coast. Waste ground, rare; Cook Co.

19. S. arguta Ait. Goldenrod. July–Oct. Dry open woods, rare; Union Co. S. *strigosa* Small (in part)—F, G, J.

20. S. ulmifolia Muhl. Elm-leaved Goldenrod. Aug.–Oct. Woods; common throughout Ill.

21. S. drummondii Torr. & Gray. Drummond's Goldenrod. Sept.–Oct. River bluffs; occasional in sw. Ill.

22. S. radula Nutt. Rough Goldenrod. Aug.–Oct. Dry open woods; occasional in w. and s. Ill. (Including var. *laeta* [Greene] Fern. and var. *stenolepis* Fern.)

23. S. canadensis L. Tall Goldenrod. Aug.–Oct. Open woods and fields; common throughout Ill. (Including var. *scabra* [Muhl.] Torr. & Gray, S. *altissima* L., and var. *hargeri* Fern.)

24. S. nemoralis Ait. Field Goldenrod. Aug.–Oct. Open woods and fields; common throughout Ill. (Including var. *longipetiolata* [Mack. & Bush] Palmer & Steyerm, and var. *decemflora* [DC.] Fern.)

25. S. rugosa Mill. Rough-leaved Goldenrod. Aug.–Oct. Moist ground, rare; Jackson, Johnson, Lawrence, Pope, and Randolph cos.

28. *Haplopappus Cass.*

1. H. ciliatus (Nutt.) DC. Aug.–Sept. Adventive from west. Waste ground; occasional in sw. Ill.

29. *Bellis L.   English Daisy*

1. B. perennis L. English Daisy. May–June. Introduced from Europe and escaped from cultivation. Waste ground, rare; Champaign and McLean cos.

30. *Boltonia L'Hér.   Boltonia*

. Leaves lanceolate to oblanceolate, 5–20 mm wide; disc 6–10 mm wide
.... 1. *B. asteroides*
. Leaves linear, 1–5 mm wide; disc 3–6 mm wide .... 2. *B. diffusa*

1. B. asteroides (L.) L'Hér. False Aster. July–Oct. Leaves not decurrent. Moist ground; common throughout Ill. (Including var. *recognita*

[Fern. & Grisc.] Cronq., *B. recognita* [Fern. & Grisc.] G. N. Jones, var. *latisquama* [Gray] Cronq., *B. latisquama* Gray, and *B. latisquama* var. *microcephala* Fern. & Grisc.)

var. decurrens (Torr. & Gray) Engelm. False Aster. July–Oct. Leaves decurrent. Moist ground; occasional along Illinois River. *B. latisquama* var. *decurrens* (Torr. & Gray) Fern. & Grisc.—F; *B. decurrens* (Torr. & Gray) Wood—G, J.

2. B. diffusa Ell. False Aster. July–Sept. Moist to dry open woods and fields; occasional in s. half of Ill. (Including var. *interior* Fern. & Grisc., *B. interior* [Fern. & Grisc.] G. N. Jones.)

31. Aster L. Aster

1. Basal or lower leaves cordate or subcordate 2
1. Basal or lower leaves not cordate or subcordate 9
2. Inflorescence corymbose; outer involucral bracts 1–2 mm wide 3
2. Inflorescence paniculate; outer involucral bracts less than 1 mm wide 5
3. Inflorescence glandular .... 1. *A. macrophyllus*
3. Inflorescence not glandular 4
4. Leaves distinctly pubescent below .... 2. *A. furcatus*
4. Leaves glabrous or nearly so .... 3. *A. schreberi*
5. Involucral bracts with reflexed tips .... 4 *A. anomalus*
5. Involucral bracts erect, appressed 6
6. Leaves entire or subentire 7
6. Leaves serrate 8
7. Involucral bracts puberulent; rays 10–12 mm long .... 5. *A. shortii*
7. Involucral bracts glabrous or occasionally with ciliolate margins; rays 6–8 mm long .... 6. *A. azureus*
8. Involucral bracts with conspicuous diamond-shaped tips ....
.... 7. *A. cordifolius*
8. Involucral bracts with a median green line .... 8. *A. sagittifolius*
9. Stem leaves clasping and frequently auriculate at base 10
9. Stem leaves not clasping or auriculate at base (occasionally with sheathing base) 11
10. Stem evenly pubescent, at least above 13
10. Stem glabrous or upper part pubescent in decurrent lines 14
11. Involucral bracts more or less glandular 12
11. Involucral bracts glabrous .... 12. *A. puniceus*
12. Heads numerous, subcorymbose ....
12. Heads few, scattered, solitary on slender bracteate peduncles ....
.... 11. *A. patens*
13. Leaves strongly clasping and auriculate .... 9. *A. novae-angliae*
13. Leaves slightly or not at all clasping, not auriculate 10. *A. oblongifolius*
14. Stem glabrous, often glaucous .... 13. *A. laevis*
14. Stem pubescent in decurrent lines above (if glabrous, not glaucous) 15
15. Stem leaves sharply serrate, abruptly contracted into a winged, clasping auriculate petiole .... 14. *A. prenanthoides*
15. Stem leaves entire or nearly so, not contracted into a winged, clasping auriculate petiole 16
16. Involucral bracts, at least inner ones, long-acuminate to attenuate ...
.... 12. *A. puniceus*
16. Involucral bracts obtuse to acute, rarely attenuate ... 15. *A. novi-belgii*
17. Leaves silvery-silky on both sides .... 16. *A. sericeus*
17. Leaves glabrous or pubescent, but not silvery-silky ....

1. *A. macrophyllus* L. Big-leaved Aster. Aug.–Sept. Dry open woods, rare; Boone, Cook, and Lake cos. (Including var. *pinguifolius* Burgess, var. *velutinus* Burgess, var. *sejunctus* Burgess, and var. *ianthinus* [Burgess] Fern.)

2. *A. furcatus* Burgess. Forked Aster. Aug.–Oct. Dry woods; occasional in n. half of Ill. (Including f. *erythractis* Benke.)

3. *A. schreberi* Nees. Schreber's Aster. Aug.–Oct. Woods, rare; Marshall, Peoria, and Tazewell cos. *A. chasei* G. N. Jones—J.

4. A. anomalus Engelm. Blue Aster. Sept.–Oct. Dry woods; occasional in sw. Ill., n. to Peoria and Woodford cos.

5. A. shortii Lindl. Short's Aster. Aug.–Oct. Dry woods; common throughout Ill. Rarely rose–flowered specimens (f. *gronemanni* Benke) rather than blue-flowered ones are observed.

6. A. azureus Lindl. Sky-blue Aster. Sept.–Oct. Prairies and dry open woods; occasional throughout Ill. (Including f. *laevicaulis* Fern and var. *poaceus* [Burgess] Fern.)

7. A. cordifolius L. Blue Wood Aster; Heart-leaved Aster. Aug.–Oct Dry woods; occasional throughout Ill. (Including var. *moratus* Shinners.)

8. A. sagittifolius Wedem. ex Willd. Arrow-leaved Aster. Aug.–Oct Stem glabrous or sparsely pubescent along decurrent lines on upper part. Dry woods; common throughout Ill.

var. drummondii (Lindl.) Shinners. Drummond's Aster. Aug.–Oct Stem evenly and usually densely short-pubescent throughout. Dry open woods; common throughout Ill. (Including A. *texanus* Burgess A. *sagittifolius* f. *hirtellus* [Lindl.] Shinners, and A. *drummondii* var *rhodactis* Benke.) A. *drummondii* Lindl.—G, J.

9. A. novae-angliae L. New England Aster. Aug.–Oct. Moist ground common throughout Ill. Rarely rose-flowered (f. *rosarius* House) c white-flowered (f. *geneseensis* House) specimens rather than the normal violet-purple ones are found.

10. A. oblongifolius Nutt. Aromatic Aster. Aug.–Oct. Dry open woods occasional throughout Ill., except for the e. cent. cos. (Including var *angustatus* Shinners and f. *roseoligulatus* [Benke] Shinners.)

11. A. patens Ait. Spreading Aster. Aug.–Oct. Open woods; occasional in s. third of Ill. (Including var. *phlogifolius* Muhl.)

12. A. puniceus L. Swamp Aster. Aug.–Oct. Stem evenly hisp above. Moist ground; occasional in the n. two-thirds of the state, ra elsewhere.

var. lucidulus Gray. Shiny Swamp Aster. Aug.–Oct. Stems puber lent in lines above. Moist ground; occasional in n. half of Ill. A. *lucid lus* (Gray) Wieg.—F, G. J.

13. A. laevis L. Smooth Aster. Aug.–Oct. Moist sandy soil in wood moist prairies; common throughout Ill. Rarely white-flowered spec mens (f. *beckwithiae* House) are observed.

14. A. prenanthoides Muhl. Aster. Sept.–Oct. Moist ground; rare in two-fifths of Ill. absent elsewhere.

15. A. novi-belgii L. New Belgium Aster. Aug.–Sept. Moist groun rare; Cook, Lake, and McHenry cos. A. *longifolius* Lam.—J.

16. A. sericeus Vent. Silky Aster. Sept.–Oct. Prairies; sand barrer occasional in the n. half of the state, rare or absent elsewhere.

17. A. umbellatus Mill. Flat-top Aster. Aug.–Oct. Moist ground; c casional in n. half of Ill., absent elsewhere.

18. A. ptarmicoides (Nees) Torr. & Gray. Aster. Aug.–Sept. R flowers white. Dry sandy soil; occasional in n. half of Ill., absent el where.

var. lutescens (Lindl.) Gray. Aster. Aug.–Sept. Ray-flowers yell Dry prairies, rare; Cook Co.

19. A. brachyactis Blake. Rayless Aster. July–Sept. Native of the U.S. Waste ground, rare; Cook Co.

20. A. parviceps (Burgess) Mack. & Bush. Aug.–Oct. Prairies a open woods; occasional in n. three-fifths of Ill., absent elsewhere.

21. A. pilosus Willd. Hairy Aster. Aug.–Oct. Stems and often leaves pilose. Waste ground; common throughout Ill. (Including var. *platyphyllus* [Torr. & Gray] Blake.)

var. pringlei (Gray) Blake. Pringle's Aster. Aug.–Oct. Stems and leaves glabrous or nearly so. Dry sandy soil, rare; Lake Co. (Including var. *demotus* Blake.) *A. pringlei* (Gray) Britt.—J.

22. A. ericoides L. Heath Aster. July–Oct. Stem with appressed or ascending short hairs. Prairies and dry open ground; occasional throughout Ill. Rarely blue-flowered (f. *caeruleus* [Benke] Blake) or rose-flowered (f. *gramsii* Benke) specimens are observed.

var. prostratus (Ktze.) Blake. Heath Aster. Sept.–Oct. Stems with spreading or slightly reflexed hairs. Prairies and dry open ground; common throughout Ill. (Including f. *prostratus* [Ktze.] Fern. and f. *exiguus* Fern.) *A. exiguus* (Fern.) Rydb.—J.

23. A. vimineus Lam. Sept.–Oct. Moist open ground; occasional throughout Ill. (Including var. *subdumosus* Wieg.)

24. A. dumosus L. Rice Button Aster; Bushy Aster. Aug.–Sept. Moist sandy soil, rare; Champaign, Cook, Kankakee, McDonough, Monroe and Will cos. (Including var. *cordifolius* [Michx.] Torr. & Gray and var. *strictior* Torr. & Gray.)

25. A. praealtus Poir. Willow Aster. Sept.–Oct. Moist ground; common throughout Ill. (Including var. *angustior* Wieg. and var. *subasper* [Lindl.] Wieg.)

26. A. linariifolius L. Flax-leaved Aster. Sept.–Oct. Dry open soil; occasional in n. half of Ill., extending s. to St. Clair Co.

27. A. junciformis Rydb. Rush Aster. Aug.–Sept. Bogs and calcareous fens; occasional in ne. Ill., extending s. to Peoria Co.

28. A. turbinellus Lindl. Aster. Sept.–Oct. Prairies, dry open woods; occasional in s. three-fifths of Ill.

29. A. tataricus L. f. Tartarian Aster. Sept.–Oct. Introduced from e. Asia and escaped from cultivation. Waste ground, rare; Jersey, St. Clair, and Sangamon cos.

30. A. ontarionis Wieg. Ontario Aster. Sept.–Oct. Moist ground; common throughout Ill.

31. A. lateriflorus (L.) Britt. Side-flowered Aster. Sept.–Oct. Woods; common throughout Ill.

32. A. simplex Willd. Paniculed Aster. Aug.–Oct. Moist ground; common throughout Ill. (Including var. *interior* [Wieg.] Cronq. and var. *ramosissimus* [Torr. & Gray] Cronq.)

Reported hybrids: *A.* X *amythystinus* Nutt. (*A. ericoides* X *A. novae-angliae*). Champaign and Peoria cos.; *A. ericoides* var. *prostratus* X *A. pilosus,* Champaign, Cook, and Du Page cos.; *A. laevis* X *A. simplex,* McHenry Co.

32. Erigeron L. Fleabane

Rays conspicuous, more than 5 mm long 2
Rays, inconspicuous, less than 3 mm long 5
Leaves clasping 3
Leaves not clasping 4
Heads few, 2.5–3.5 cm in diameter; rays 50–100, about 1 mm wide . . . . .
. . . . 1. *E. pulchellus*
Heads several, 1.5–2.0 cm in diameter; rays 150–200, about 0.5 mm wide
. . . . 2. *E. philadelphicus*

4. Cauline leaves many; basal leaves ovate, coarsely dentate; middle par
of stem with long, spreading hairs .... 3. *E. annuu*
4. Cauline leaves few; basal leaves spatulate, entire or nearly so; middle
part of stem usually with short, appressed hairs .... 4. *E. strigosu*
5. Stem diffusely branched from near the base; rays purplish ....
.... 5. *E. divaricatu*
5. Stem simple or nearly so up to inflorescence; rays white ....
.... 6. *E. canadensi*

1. E. pulchellus Michx. Robin's Plantain. April–June. Open woods
occasional throughout Ill.

2. E. philadelphicus L. Marsh Fleabane. May–June. Open woods
fields, and waste ground; common throughout Ill. (Including f. *sca
turicola* [Fern.] Cronq. and f. *angustatus* Vict. & Rosseau.)

3. E. annuus (L.) Pers. Daisy Fleabane; White Top. June–Oct. Ope
woods, fields , and waste ground; common throughout Ill.

4. E. strigosus Muhl. Daisy Fleabane; White Top. May–July. Ope
woods, fields, waste ground; common throughout Ill. (Including va
septentrionalis [Fern. & Wieg.] Fern. and var. *beyrichii* [Fisch. & Mey
Gray.)

5. E. divaricatus Michx. Dwarf Fleabane. June–Sept. Waste ground
occasional throughout Ill. *Conyza ramosissima* Cronq.—G.

6. E. canadensis L. Horseweed; Muleweed. May–Oct. Wast
ground; in every co. *Conyza canadensis* (L.) Cronq.—G.

33. Anthemis L.   Dogfennel. Chamomile

1. Ray-flowers yellow .... 1. *A. tinctor*
1. Ray-flowers white
2. Achenes glandular-tuberculate .... 2. *A. cotu*
2. Achenes not glandular-tuberculate ....
3. Achenes obtusely 3-angled .... 3. *A. nobi*
3. Achenes 10-ribbed .... 4. *A. arvens*

1. A. tinctoria L. Yellow Chamomile. June–Sept. Introduced fro
Europe and escaped from cultivation. Waste ground, rare; Cook, C
Page, and Winnebago cos.

2. A. cotula L. Dogfennel; Mayweed. May–Sept. Naturalized fro
Europe. Waste ground; common throughout Ill.

3. A. nobilis L. Garden Chamomile. June–Sept. Introduced fro
Europe and escaped from cultivation, rare; Cook and DuPage cos.

4. A. arvensis L. Field Chamomile; Corn Chamomile. May–Aug. N
uralized from Europe; occasional throughout Ill. (Including var. *agre
tis* [Wallr.] DC.)

34. Achillea L.   Yarrow. Milfoil

1. A. millefolium L. Common Yarrow; Milfoil. May–Aug. Leaves a
stems arachnoid to glabrescent; corymb flat-topped. Naturalized fr
Europe. Fields and waste ground; common throughout Ill. Rar
rose-purple flowered specimens (f. *rosea* Rand & Redf.) are observe

subsp. lanulosa (Nutt.) Piper. Western Yarrow. June–Aug. Lea
and stems densely woolly; corymb round-topped. Adventive from
Fields and waste ground; occasional throughout Ill. *A. lanulo
Nutt.—F. J.

35. *Matricaria* L. Wild Chamomile

1. Heads with evident white rays 2
1. Heads rayless .... 3. *M. matricarioides*
2. Achenes with 2 marginal and 1 ventral, strongly callous-thickened, almost winglike ribs .... 1. *M. maritima*
2. Achenes with 2 marginal and 3 ventral, raised, but not winglike ribs ....
.... 2. *M. chamomilla*

     1. M. maritima L. Scentless Chamomile. June–Sept. Naturalized from Europe. Waste ground, rare; DuPage Co. (Our plants belong to var. *agrestis* [Knaf] Wilmott.)
     2. M. chamomilla L. Chamomile. May–Oct. Naturalized from Europe. Roadsides and waste ground; occasional in s. half of Ill. extending n. to Woodford Co.
     3. M. matricarioides (Less.) Porter. Pineapple-weed. May–Aug. Naturalized from Pacific coast. Roadsides and waste ground; common throughout Ill.

36. *Chrysanthemum* L. Chrysanthemum

1. Heads solitary or few, 3–5 cm broad .... 1. *C. leucanthemum*
1. Heads several or numerous, 0.5–1.5 cm broad 2
2. Leaves pinnatifid or bipinnatifid; rays 3–8 mm long .. 2. *C. parthenium*
2. Leaves crenate or with few reduced basal pinnae; rays less than 1 mm long, or absent .... 3. *C. balsamita*

     1. C. leucanthemum L. Ox-eye Daisy. May–Aug. Naturalized from Europe and Asia. Fields and waste ground; common throughout Ill. (Including var. *pinnatifidum* Lecog. & Lamotte.)
     2. C. parthenium (L.) Bernh. Feverfew. June–Sept. Escaped from cultivation or naturalized from Europe. Roadsides and waste ground; occasional in n. half of Ill.
     3. C. balsamita L. Costmary. Sept–Oct. Escaped from cultivation or naturalized from Europe. Waste ground, not common. (Including var. *tanacetoides* Boiss.)

37. *Tanacetum* L. Tansy. Golden-buttons

     1. T. vulgare L. Tansy; Golden-buttons. July–Sept. Naturalized from Europe. Fields and waste ground; occasional throughout Ill.

38. *Artemisia* L. Wormwood

. Leaves white-tomentose, at least below 2
. Leaves glabrous or nearly so 7
. Leaves lanceolate to linear 3
. Leaves pinnatifid 4
. Leaves regularly serrate .... 1. *A. serrata*
. Leaves entire or irregularly few-toothed .... 2. *A. ludoviciana*
. Leaves green and glabrous above, densely white-tomentose below .....
.... 3. *A. vulgaris*
Leaves silvery-sericeous on both sides, or eventually subglabrous above in age, or the basal leaves grayish-hairy 5
. Basal leaves not grayish-hairy 6

5. Basal leaves grayish-hairy .... 7. *A. campestri*
6. Divisions of leaves oblong to lanceolate, 1.5–5.0 mm wide ....
.... 4. *A. absinthiur*
6. Divisions of leaves linear to filiform, 1 mm wide or less ... 5. *A. frigid*
7. Leaves or their divisions linear to filiform, entire margined ....
7. Leaves or their divisions lanceolate, sharply toothed 1
8. Involucral bracts glabrous ....
8. Involucral bracts tomentose .... 8. *A. abrotanu*
9. Leaves mostly entire, occasionally irregularly lobed or lower 3-
5-lobed .... 6. *A. dracunculu*
9. Leaves mostly pinnatifid .... 7. *A. campestr*
10. Inflorescence dense, spicate .... 9. *A. bienn*
10. Inflorescence loose, paniculate .... 10. *A. annu*

1. A. serrata Nutt. Aug.–Sept. Moist ground; occasional in n. half
Ill.
2. A. ludoviciana Nutt. Western Mugwort; White Sage. July–Sep
Leaves glabrate above, tomentose below. Waste ground; occasion
in the n. three-fifths of Ill.
var. gnaphalodes (Nutt.) T. & G. White Sage. July–Sept. Leav
white-tomentose on both surfaces. Waste ground; occasion
throughout Ill. (Including var. *latifolia* [Besser] Torr.) *A. gnaphalod*
Nutt.—J
3. A. vulgaris L. Common Mugwort. July–Oct. Introduced fro
Europe and escaped from cultivation. Waste ground, rare; Cha
paign, Cook, and Henry cos. (Including var. *latiloba* Ledeb.)
4. A. absinthium L. Common Wormwood; Absinth. July–Sept. I
troduced from Europe and escaped from cultivation. Waste groun
rare; Cook, DeKalb, Kane, Lake, McHenry, Peoria, and Wabash cos
5. A. frigida Willd. Wormwood. July–Sept. Adventive from w. Was
ground, rare; Cook Co.
6. A. dracunculus L. False Tarragon. July–Sept. Prairies; occasior
in n. half of Ill. *A. dracunculoides* Pursh—J; *A. glauca* Pall.—F.
7. A. campestris L. Beach Wormwood. July–Sept. Sandy soil; c
casional in n. half of Ill. (Our plants belong to ssp. *caudata* [Mich
Hall & Clem.) (Including *A. caudata* var. *calvens* Lunell.) *A. caudа*
Michx.—F. G. J.
8. A. abrotanum L. Southern Wormwood. Aug.–Sept. Introduc
from Europe and rarely persisting from cultivation. Waste groun
rare; Cook Co.
9. A. biennis Willd. Biennial Wormwood. Aug.–Oct. Adventive fr
nw. U.S. Waste ground; occasional throughout Ill.
10. A. annua L. Annual Wormwood. Aug.–Oct. Introduced fr
Europe and naturalized. Waste ground; occasional throughout Ill.

39. Pluchea Cass.   Marsh Fleabane. Stinkweed

1. P. camphorata (L.) DC. Camphorweed; Stinkweed. July–C
Swamps and sloughs; occasional in s. Ill., extending n. to DeWitt

40. Antennaria Gaertn.   Pussytoes. Ladies'-tobacco

1. Basal leaves prominently 1-nerved or obscurely 3-nerved, usually
than 1.5 cm wide .... 1. *A. negle*

1. Basal leaves prominently 3- to 5-nerved, usually 1.5 cm or more wide ..
.... 2. *A. plantaginifolia*

1. A. neglecta Greene. Pussytoes; Ladies'-tobacco. April–May. Basal leaves cuneate-spatulate, gradually tapering to the sessile base. Fields and open woods; common throughout Ill. (Including f. *simplex* [Peck] Fern., *A. campestris* Rydb., *A. petaloides* var. *scariosa* Fern., and *A. petaloides* var. *subcorymbosa* Fern.)

var. attenuata (Fern.) Cronq. Pussytoes; Ladies'-tobacco. May–June. Basal leaves obovate, abruptly contracted below the middle into a petiole-like base. Fields and open woods, rare; DeKalb, Henry, and Kendall cos. (Including *A. brainerdii* Fern., *A. rupicola* Fern., *A. neodioica* Greene, *A. neodioica* var. *interjecta* Fern., *A. neodioica* var. *chlorophylla* Fern., and *A. neodioica* var. *grandis* Fern.) *A. neodioica* Greene—J; *A. neodioica* var. *attenuata* Fern.—F.

2. A. plantaginifolia (L.) Richards. Pussytoes; Ladies'-tobacco. April–May. Upper surface of leaves arachnoid at first, tardily glabrate in age; involucre of pistillate plants 5–7 mm high; upper part of stem without purple glands. Open woods; common throughout Ill. (Including var. *petiolata* [Fern.] Heller.)

var. arnoglossa (Greene) Cronq. Pussytoes; Ladies'-tobacco. April–June. Upper surface of leaves glabrous or nearly so at first; involucre of pistillate plants 7–10 mm high; upper part of stem usually with purple glands. Open woods; occasional throughout Ill. *A. parlinii* Fern.—F, J; *A. parlinii* var. *arnoglossa* (Greene) Fern.—F.

var. ambigens (Greene) Cronq. Pussytoes; Ladies'-tobacco. May–June. Upper surface of leaves arachnoid at first, tardily glabrate in age; involucre of pistillate plants 7–10 mm high; upper part of stem without purple glands. Fields and open woods; common throughout Ill. (Including *A. fallax* Greene, *A. fallax* var. *calophylla* [Greene] Fern., *A. farwellii* Greene, and *A. munda* Fern.)

41. *Gnaphalium* L. Cudweed. Everlasting

1. Inflorescence a narrow spike-like panicle; pappus bristles united in a ring at base and deciduous as a whole .... 1. *G. purpureum*
1. Inflorescence a cymose or paniculate cluster; pappus bristles not united in a ring at base and deciduous as a whole 2
2. Involucral bracts white or whitish; achenes smooth 3
2. Involucral bracts brown; achenes scabrous .... 4. *G. uliginosum*
3. Leaves decurrent .... 2. *G. macounii*
3. Leaves not decurrent .... 3. *G. obtusifolium*

1. G. purpureum L. Early Cudweed. May–July. Fields and open woods; occasional throughout Ill., except ne. counties, more common in s. third of state.

2. G. macounii Greene. Western Cudweed; Clammy Cudweed. July–Sept. Sandy soil in open woods and fields, rare; Clark Co.

3. G. obtusifolium L. Sweet Everlasting; Catsfoot. Aug.–Oct. Meadows and open woods; common throughout Ill.

4. G. uliginosum L. Low Cudweed. June–Sept. Naturalized from Europe. Roadsides and pastures, rare; Cook and Lake cos.

42. *Inula* L.   Elecampane

1. I. helenium L. Elecampane. July–Aug. Introduced from Europe and escaped from cultivation. Fields, waste ground, and open woods; not common in Ill.

43. *Erechtites* Raf.   Fireweed

1. E. hieracifolia (L.) Raf. Fireweed. Aug.–Oct. Moist woods, bogs, waste ground, and recently burned areas; common throughout Ill. (Including var. *praealta* [Raf.] Fern.)

44. `Cacalia` L.   Indian-plantain

1. Heads 20- to 40-flowered; involucral bracts 12–15 .... 1. *C. suaveolens*
1. Heads 5-flowered; involucral bracts 5 ....
2. Lower leaves reniform, lobed or coarsely angulate-dentate ....
2. Lower leaves lance-ovate, entire, shallowly dentate, or crenate ....
.... 4. *C. tuberosa*
3. Leaves glaucous below; stem round or nearly so .... 2. *C. atriplicifolia*
3. Leaves not glaucous below; stem conspicuously grooved. ....
.... 3. *C. muhlenbergii*

1. C. suaveolens L. Sweet Indian-plantain. July–Aug. Wet ground, not common in n. Ill., rare in s. Ill.
2. C. atriplicifolia L. Pale Indian-plantain. July–Oct. Moist or dry open woods and prairies; occasional throughout Ill.
3. C. muhlenbergii (Sch.-Bip.) Fern. Great Indian-plantain. July–Sept. Woods; not common, but scattered throughout Ill.
4. C. tuberosa Nutt. Prairie Indian-plantain. June–July. Wet prairies, marshes, and bogs; not common, but scattered throughout the state.

45. *Senecio* L.   Butterweed. Ragwort. Groundsel. Squaw-weed

1. Leaves chiefly basal, basal leaves crenate, dentate, or entire, median stem leaves often pinnatifid ....
1. Stems nearly equally leafy throughout; leaves all pinnatifid or coarsely sinuate-dentate ....
2. Basal leaves with winged petioles .... 1. *S. obovatus*
2. Basal leaves without winged petioles ....
3. Basal leaves cordate to subcordate .... 2. *S. aureus*
3. Basal leaves oblanceolate to oval, not cordate ....
4. Leaves and stems, especially nodes, floccose-tomentose at maturity; peduncles tomentose .... 3. *S. plattensis*
4. Leaves and stems glabrous or nearly so at maturity; peduncles glabrous or nearly so .... 4. *S. pauperculus*
5. Involucral bracts black-tipped; rays absent .... 5. *S. vulgaris*
5. Involucral bracts not black-tipped; rays present, 3–10 mm long .....
6. Stem glabrous or obscurely floccose-tomentose at nodes; rays 6–10 mm long .... 6. *S. glabellus*
6. Stem glandular hairy; rays about 3 mm long .... 7. *S. viscosus*

1. S. obovatus Muhl. Round-leaved Groundsel. April–June. Open rich woods and rocky outcroppings, rare; Champaign, Clark, and V

milion cos. (Including var. *rotundus* Britt. and f. *elongatus* [Pursh] Fern.)

2. S. aureus L. Golden Ragwort; Squaw-weed. April–May. Wet ground; occasional throughout Ill. (Including var. *intercursus* Fern., var. *gracilis* [Pursh] Hook., and var. *aquilonius* Fern.)

3. S. plattensis Nutt. Prairie Groundsel; Prairie Ragwort. May–June. Dry woods and prairies; common in n. and w. Ill., occasional or rare elsewhere.

4. S. pauperculus Michx. Northern Ragwort. May–June. Open woods, and meadows; common in n. half of Ill., extending s. to Macoupin Co. (Including var. *balsamitae* [Muhl.] Fern.)

5. S. vulgaris L. Common Groundsel. June–July. Naturalized from Europe. Waste ground; occasional in n. half of Ill.

6. S. glabellus Poir. Butterweed. April–June. Moist shady ground along streams; common in s. half of Ill., extending n. to Peoria Co.

7. S. viscosus L. Sticky Groundsel. July–Sept. Naturalized from Europe. Waste ground, rare; Cook and Jackson cos.

46. *Eupatorium* L.   Thoroughwort

Leaves whorled 2
Leaves opposite, or the upper ones alternate 4
Heads 9- to 22-flowered; inflorescence flat-topped .... 1. *E. maculatum*
Heads 5- to 7-flowered; inflorescence convex 3
Stems solid, purple only at nodes .... 2. *E. purpureum*
Stems hollow, purple throughout .... 3. *E. fistulosum*
Leaves distinctly petiolate 5
Leaves sessile, subsessile, or connate-perfoliate 8
Involucral bracts acuminate, filiform-tipped .... 4. *E. coelestinum*
Involucral bracts rounded to acute, not filiform-tipped 6
Flowers pink to pale purple .... 5. *E. incarnatum*
Flowers white 7
Involucral bracts broadly rounded, densely pubescent . 6. *E. serotinum*
Involucral bracts acute, glabrous or sparsely pubescent . 7. *E. rugosum*
Leaves tapering to base .... 8. *E. altissimum*
Leaves cuneate, subcordate, truncate, or connate-perfoliate 9
Leaves connate-perfoliate (truncate in one form) .... 9. *E. perfoliatum*
Leaves cuneate to subcordate .... 10 *E. sessilifolium*

1. E. maculatum L. Spotted Joe-Pye-weed. July–Sept. Moist ground; common in n. Ill., rare in s. Ill.

2. E. purpureum L. Purple Joe-Pye-weed; Green-stemmed Joe-Pye-weed; Sweet Joe-Pye-weed. July–Aug. Open woods; common throughout Ill.

3. E. fistulosum Barratt. Hollow Joe-Pye-weed; Trumpet-weed. July–Sept. Low, wet ground; Alexander, Coles, Jackson, and Pope cos.

4. E. coelestinum L. Mistflower; Blue Boneset; Wild Ageratum. July–Oct. Moist ground; common in s. two-thirds of Ill., absent elsewhere, except for Cook and DuPage cos. Rarely white-flowered specimens (f. *album* E. J. Alex.) or reddish-purple-flowered specimens (f. *illinoense* Benke) rather than the normal blue- to violet-flowered specimens are observed.

5. E. incarnatum Walt. Aug.–Oct. Wet woods and swamps, rare; Alexander, Pulaski, and Union cos.

6. E. serotinum Michx. Late Boneset. Aug–Oct. Moist open woo(
and clearings; common throughout Ill.

7. E. rugosum Houtt. White Snakeroot. July–Sept. Woods; commo
throughout Ill. (Including var. *tomentellum* [Robinson] Blake.)

8. E. altissimum L. Tall Boneset; Tall Thoroughwort. Aug.–O(
Open woods and prairies; common throughout Ill.

9. E. perfoliatum L. Common Boneset; Thoroughwort. Aug.–O(
Wet ground; common throughout Ill. Rarely plants with flowers pu
ple and involucral bracts purplish-tinged (f. *purpureum* Britto(
leaves three at a node rather than two (f. *trifolium* Fassett), and le
bases truncate (f. *truncatum* [Muhl.] Fassett), are observed.

10. E. sessilifolium L. Upland Boneset. Aug.–Oct. Woods; occasio(
in the s. half of Ill., uncommon elsewhere. (Our plants belong to v
brittonianum Porter.)

Reported hybrids: *E.* X *polyneuron* (F. J. Herm.) Wunderlin (
perfoliatum X *E. serotinum*). *E. perfoliatum* L. var. *cuneat(*
(Engelm. ex Torr. & Gray) Engelm.—F, G (in part); *E. serotin(*
Michx. var. *polyneuron* F. J. Herm.—G (in part).

47. *Mikania* Willd.   Climbing Hempweed

1. M. scandens (L.) Willd. Climbing Hempweed. Aug.–Oct. L
woods, swamps, and banks of streams; not common and confined(
s. Ill.

48. *Brickellia* Ell.

1. B. eupatorioides (L.) Shinners. False Boneset. Aug.–Oct. Prai(
and open woods; occasional to common throughout Ill. (Includ(
var. *corymbulosa* [Torr. & Gray] Shinners and var. *ozarkana* [Sh
ners] Shinners.) *Kuhnia eupatorioides* L.—F, G, J.

49. *Liatris* Schreb.   Blazing-star

1. Pappus evidently plumose .... (

1. Pappus merely barbellate .... (

2. Heads 10- to 60-flowered; corolla lobes pubescent within .... (

2. Heads 4- to 6-flowered; corolla lobes glabrous within ... 3. *L. punc(*

3. Involucral bracts rounded to mucronate, appressed ... 1. *L. cylindra*

3. Involucral bracts acuminate, squarrose .... 2. *L. squarr*

4. Heads cylindrical, 3- to 18-flowered ....

4. Heads campanulate to hemispherical, 18- to 100-flowered (rarely 12–

....

5. Stem pubescent in inflorescence .... 4. *L. pycnostac(*

5. Stem glabrous in inflorescence .... 5. *L. spi(*

6. Involucral bracts herbaceous, green, ciliate margined, puberulent ..

.... 6. *L. sc(*

6. Involucral bracts scarious margined, usually colored, glabrous or ne

so ....

7. Heads uniform in size; corolla pubescent toward base within ......

.... 7. *L. as(*

7. Terminal head distinctly larger than others; corolla glabrous within

.... 8. *L. ligulis*

1. L. cylindracea Michx. Blazing-star. Aug.–Sept. Prairies; occasional throughout Ill. except for the southernmost cos. where it is rare or absent.

2. L. squarrosa (L.) Michx. Blazing-star. June–Sept. Dry open woods and prairies; occasional in s. Ill. Rarely white-flowered specimens (f. *alba* Evers & Thieret) are observed. (Including var. *hirsuta* [Rydb.] Gaiser.)

3. L. punctata Hook. Blazing-star. Aug.–Oct. Adventive from w. Along railroads, rare; DuPage Co. (Including var. *nebraskana* Gaiser.)

4. L. pycnostachya Michx. Prairie Blazing-star. July–Sept. Prairies; common throughout Ill.

5. L. spicata (L.) Willd. Marsh Blazing-star. July–Sept. Prairies and wet meadows; not common, but scattered throughout Ill. Rarely white-flowered specimens (f. *albiflora* Britt.) are observed.

6. L. scabra (Greene) K. Schum. Blazing-star. Sept.–Oct. Prairies and open woods; occasional in the s. one-third of Ill.; also Coles Co.

7. L. aspera Michx. Rough Blazing-star. Sept.–Oct. Prairies; common throughout Ill. Rarely white-flowered specimens (f. *benkei* [Macb.] Fern.) are observed. (Including var. *intermedia* [Lunell] Gaiser and *L.* X *sphaeroidea* Michx.)

8. L. ligulistylis (A. Nels.) K. Schum. Blazing-star. Aug.–Sept. Mesic to moist prairies, rare; Cook and Montgomery cos. (Including *L.* X *nieuwlandii* [Lunell] Gaiser.)

Reported hybrids: *L.* X *ridgewayi* Standl. (*L. pycnostachya* X *L. squarrosa*), Richland Co.; *L.* X *steelei* Gaiser (*L. aspera* X *L. spicata*), Lake Co.; *L.* X *gladewitzii* (Farwell) Shinners (*L. aspera* X *L. cylindracea*), Lake Co.; *L. cylindracea* X *L. spicata, Lake Co.*

50. *Vernonia* Schreb.   Ironweed

1. Tips of involucral bracts long-filiform; heads 12–20 mm wide .... 1. *V. arkansana*
1. Tips of involucral bracts obtuse to abruptly acuminate; heads 4–12 mm wide 2
2. Leaves tomentose to tomentellous beneath 3
2. Leaves glabrous or scabrous-hirtellous beneath (sometimes tomentellous on veins) 4
3. Tips of involucral bracts obtuse to mucronate, erect or slightly spreading .... 2. *V. missurica*
3. Tips of involucral bracts abruptly acuminate, squarrose . 3. *V. baldwinii*
4. Leaves glabrous, conspicuously punctate beneath; outer pappus of short capillary bristles .... 4. *V. fasciculata*
4. Leaves scabrous-hirtellous, scarcely punctate beneath; outer pappus of scale-like bristles .... 5. *V. gigantea*

1. V. arkansana DC. Aug.–Sept. Low open woods and prairies, very rare; Champaign Co. (original site since destroyed). *V. crinita* Raf.— F, G.

2. V. missurica Raf. Missouri Ironweed. July–Sept. Low open woods and prairies; common throughout Ill. Rarely rose-flowered (f. *carnea* Standl.) and white-flowered (f. *swinkii* Steyerm.) specimens are observed.

3. V. baldwinii Torr. Baldwin's Ironweed. July–Sept. Prairies and open ground; occasional in s. one-half of Ill. (Including var. *interior* (Small) Schub.).

4. V. fasciculata Michx. Common Ironweed. July–Aug. Low ground and prairies; common in n. half of Ill. occasional in s. half.

5. V. gigantea (Walt.) Trel. Tall Ironweed. July–Sept. Low woods and open ground; common throughout most of Ill. (Including var. taeniotricha Blake). V. altissima Nutt.—F, G, J.

51. Elephantopus L.   Elephant's-foot

1. E. carolinianus Willd. Elephant's-foot. Aug.–Sept. Dry woods; occasional in s. third of Ill.

52. Echinops L.   Globe Thistle

1. E. sphaerocephalus L. Globe Thistle. July–Sept. Naturalized from Europe. Waste ground, rare; Coles, Cook, DuPage, Kankakee, and Will cos.

53. Arctium L.   Burdock

1. Heads more or less corymbose, long-peduncled; petioles strongly angled ....
1. Heads more or less racemose, sessile or short-peduncled; petioles slightly angled .... 3. A. minus
2. Petioles solid; involucre glabrous or nearly so .... 1. A. lappa
2. Petioles hollow; involucre tomentose .... 2. A. tomentosum

1. A. lappa L. Great Burdock. July–Oct. Naturalized from Eurasia. Waste ground; occasional in n. Ill.; also Coles Co.

2. A. tomentosum Mill. Cotton Burdock. June–Sept. Naturalized from Europe. Fields and waste ground; Cook and Morgan cos.

3. A. minus (Hill) Bernh. Common Burdock. July–Sept. Naturalized from Eurasia. Waste ground; occasional to common throughout Ill. Occasional white-flowered (f. pallidum Farw.) or laciniate-leaved (f. laciniatum Clute) specimens are observed. (Including A. nemorosum Lej. & Court.)

54. Carduus L.   Plumeless Thistle. Musk Thistle

1. Heads solitary, nodding, 3–5 cm wide; stem unwinged immediately below head .... 1. C. nutans
1. Heads clustered, not nodding, 1.5–2.5 cm wide; stem winged immediately below head .... 2. C. acanthoides

1. C. nutans L. Musk Thistle; Nodding Thistle. June–Oct. Naturalized from Europe. Fields and waste ground; occasional throughout Ill. (Our plants belong to var. leiophyllus [Petronic] Arenes.)

2. C. acanthoides L. Plumeless Thistle. July–Sept. Naturalized from Europe. Fields and waste ground; rare in n. Ill., absent elsewhere.

55. Cirsium Mill.   Thistle

1. Outer and middle involucral bracts conspicuously spine-tipped ......
1. Outer and middle involucral bracts without or with at most a short spine less than 1 mm long ....

1. C. vulgare (Savi) Tenore. Bull Thistle. July–Aug. Naturalized from Europe. Fields and waste ground; common throughout Ill.

2. C. pitcheri (Torr.) Torr. & Gray. Dune Thistle. June–July. Sand dunes near Lake Michigan, rare; Cook and Lake cos.

3. C. discolor (Mulh.) Spreng. Field Thistle. Aug.–Sept. Fields, open woods, and waste ground; common throughout Ill. Rarely white-flowered specimens (f. *albiflorum* [Britt.] House) are observed.

4. C. altissimum (L.) Spreng. Tall Thistle. Aug.–Sept. Woods; occasional throughout Ill.

5. C. carolinianum (Walt.) Fern. & Schub. Carolina Thistle. June–July. Dry open woods, very rare; Pope Co.

6. C. pumilum (Nutt.) Spreng. Small Prairie Thistle; Hill's Thistle. June–July. Dry prairies; occasional to rare in n. three-fifths of Ill. *C. hillii* (Canby) Fern.—F, G, J.

7. C. arvense (L.) Scop. Canada Thistle. June–Oct. Naturalized from Europe. Leaves merely toothed or shallowly lobed, weakly prickly. Fields and waste ground; abundant throughout Ill. (Including var. *mite* Wimm. & Grab., var. *intergrifolium* Wimm. & Grab., and var. *vestitum* Wimm. & Grab.) *C. setosum* (Willd.) Bieb.—J.

var. horridum Wimm. & Grab. Canada Thistle. June–Oct. Naturalized from Europe. Leaves deeply pinnatifid, strongly prickly. Fields and waste ground; occasional throughout Ill., more common than var. *arvense.*

8. C. muticum Michx. Swamp Thistle. Aug.–Sept. Wet ground; calcareous fens; occasional in n. Ill., extending s. to Wabash Co.

56. *Onopordum* L. Scotch Thistle

1. O. acanthium L. Scotch Thistle. July–Aug. Naturalized from Europe. Waste ground, rare; Champaign and Cook cos.

57. *Centaurea* L. Star Thistle. Knapweed

3. Stems not winged; flowers purple .... 2. *C. calcitrapa*
4. Leaves pinnatifid with linear segments .... 4. *C. maculosa*
4. Leaves entire, toothed, or larger ones few lobed ....
5. Involucral bracts entire margined ....
5. Involucral bracts with pectinate or erose margin ....
6. Outer involucral bracts with scarious tip, inner with plumose hairy ti
.... 5. *C. repens*
6. All involucral bracts similar, coriaceous, and without scarious margin .
.... 6. *C. moschata*
7. Involucre 2–4 cm high; pappus 6–12 mm long .... 7. *C. americana*
7. Involucre 1.0–1.5 cm high; pappus 3 mm long or less ....
8. Outer involucral bracts with pectinate or erose margins to middle o
below ....
8. Outer involucral bracts with pectinate margins only near tip 11. *C. dubia*
9. Involucral bracts lanceolate, without enlarged tip .... 8. *C. cyanus*
9. Involucral bracts with enlarged tip 1
10. Outer involucral bracts with pectinate margin, inner with erose margi
.... 9. *C. nigra*
10. All involucral bracts with erose margins .... 10. *C. jacea*

    1. C. solstitalis L. Yellow Star Thistle; Barnaby's Thistle. July–Sep
Naturalized from Mediterranean region. Waste ground; Jackson Co
    2. C. calcitrapa L. Star Thistle. June–Oct. Naturalized from Medite
ranean region. Waste ground, rare; Grundy Co.
    3. C. diffusa Lam. July–Sept. Naturalized from se. Europe. Wast
ground, rare; Marshall and Will cos.
    4. C. maculosa Lam. Spotted Knapweed. July–Sept. Naturalize
from Europe. Waste ground; occasional in the n. half of the state, a
sent elsewhere.
    5. C. repens L. Russian Knapweed. June–Aug. Naturalized fro
Caucasus region. Waste ground, rare; Winnebago Co.
    6. C. moschata L. Sweet Sultan. July–Sept. Introduced from sv
Asia and occasionally escaped from cultivation. Waste ground, rar
Champaign Co.
    7. C. americana Nutt. American Basket Flower. July–Sept. Adve
tive from sw. U.S. and occasionally escaped from cultivation. Was
ground, rare; Lawrence and Wabash cos.
    8. C. cyanus L. Bachelor's Button; Corn Flower. July–Sept. I
troduced from Europe and escaped from cultivation. Waste groun
occasional throughout Ill.
    9. C. nigra L. Black Knapweed. July–Sept. Naturalized from Europ
Waste ground, rare; Boone and Cook cos.
    10. C. jacea L. Brown Knapweed. July–Sept. Naturalized fro
Europe. Waste ground, rare; Champaign, DuPage, and Winneba
cos.
    11. C. dubia Suter. Tyrol Knapweed. July–Sept. Naturalized fro
Europe. Waste ground, rare; DuPage, Gallatin, Lake, and Stephens
cos. (Our plants belong to subsp. *vochinensis* [Bernh.] Heyek.)
vochinensis Bernh.—F, J.

58. Cnicus L. Blessed Thistle

    1. C. benedictus L. Blessed Thistle. May–Sept. Introduced fro
Europe and escaped from cultivation. Waste ground, rare; Cha
paign Co.

59. *Cichorium L. Chicory*

1. C. intybus L. Chicory. June–Oct. Naturalized from Europe. Waste ground; common throughout Ill. (Including f. *roseum* Neum. and f. *album* Neum.)

60. *Lapsana L. Nipplewort*

1. L. communis L. Nipplewort. June–Sept. Naturalized from Europe. Waste ground; not common.

61. *Microseris D. Don Prairie Dandelion*

1. M. cuspidata (Pursh) Sch.-Bip. Prairie Dandelion. April–June. Dry prairies, rare; n. half of Ill. *Agoseris cuspidata* (Pursh) D. Dietr.—F, J.

62. *Krigia Schreb. Dwarf Dandelion*

1. Pappus with capillary bristles 2
1. Pappus minute or absent .... 4. *K. oppositifolia*
2. Pappus with 20–40 bristles 3
2. Pappus of 5–7 ovate scales and 5 (rarely 10) alternating bristles ....
.... 3. *K. virginica*
3. Plants scapose; heads solitary .... 1. *K. dandelion*
3. Plants with a few cauline leaves; heads several .... 2. *K. biflora*

1. K. dandelion (L.) Nutt. Potato Dandelion; Dwarf Dandelion. April–June. Open woods; occasional in s. third of Ill.
2. K. biflora (Walt.) Blake. False Dandelion. May–Sept. Open woods and prairies; common throughout Ill. (Including f. *glandulifera* Fern.)
3. K. virginica (L.) Willd. Dwarf Dandelion. Apr.–Aug. Sandy soil; occasional in n. and w. cent. Ill., rare elsewhere.
4. K. oppositifolia Raf. Dwarf Dandelion. March–April Sandy soil; occasional in s. third of Ill. *Serinia oppositifolia* (Raf.) Kuntze—F, J.

63. *Hypochaeris L. Cat's-ear*

. Leaves hispid; all achenes beaked .... 1. *H. radicata*
. Leaves glabrous or pubescent only on midrib; outer achenes beakless, inner beaked .... 2. *H. glabra*

1. H. radicata L. Cat's-ear. May–Aug. Naturalized from Europe. Waste ground, very rare; Champaign, Cook, and McHenry cos.
2. H. glabra L. Smooth Cat's-ear. May–Aug. Naturalized from Europe. Waste ground, very rare; Jackson Co.

64. *Leontodon L. Hawkbit*

. Pappus of all flowers with a single row of plumose bristles ....
.... 1. *L. autumnalis*
. Pappus of two types, inner double, with plumose and setiform bristles, outer reduced to a short irregular crown .... 2. *L. leysseri*

1. L. autumnalis L. Fall Dandelion. June–Oct. Naturalized fro
Europe. Waste ground, rare; Champaign Co., not collected sinc
1891.
2. L. leysseri (Wallr.) G. Beck. Hawkbit. June–Sept. Naturalized fro
Europe. Waste ground, rare; Cook Co.

65. *Picris* L. *Bitterweed*

1. Outer involucral bracts ovate, 3.5–8.0 mm wide; achenes with distin
   slender beak .... 1. *P. echioide*
1. Outer involucral bracts linear-lanceolate, less than 3 mm wide; achen
   with little or no beak .... 2. *P. hieracioide*

1. P. echioides L. Bristly Ox-tongue. July–Sept. Naturalized fro
Europe. Waste ground, rare; Hancock Co.
2. P. hieracioides L. Cat's-ear. July–Sept. Naturalized from Europ
Waste ground, rare; Menard Co.

66. *Tragopogon* L. *Goat's-beard*

1. Flowers purple .... 1. *T. porrifoli*
1. Flowers yellow ....
2. Involucral bracts longer than flowers; peduncles enlarged below he
   .... 2. *T. dubi*
2. Involucral bracts equalling or shorter than flowers; peduncles not
   scarcely enlarged below head .... 3. *T. pratens*

1. T. porrifolius L. Salsify; Vegetable-oyster. June–Aug. Naturaliz
from Europe. Fields and waste ground; occasional throughout Ill.
2. T. dubius Scop. Sand Goat's-beard. May–June. Naturalized fro
Europe. Fields and waste ground; common throughout Ill. *T. maj*
Jacq.—F.
3. T. pratensis L. Common Goat's-beard. May–Sept. Naturaliz
from Europe. Fields and waste ground; common throughout Ill.
Reported hybrid: (*T. pratensis* X *T. porrifolius*), Kane Co.

67. *Taraxacum* Zinn *Dandelion*

1. Achenes greenish-brown .... 1. *T. officina*
1. Achenes red to reddish-brown .... 2. *T. laevigatu*

1. T. officinale Weber. Common Dandelion. March–Nov. Naturaliz
from Europe. Fields, lawns, and waste ground; very comm
throughout Ill.
2. T. laevigatum (Willd.) DC. Red-seeded Dandelion. April–July. N
uralized from Europe. Waste ground; common throughout Ill.
erythrospermum Andrz.—F, J.

68. *Sonchus* L. *Sow Thistle*

1. Heads 3.5 cm in diameter; flowers bright yellow .... 1. *S. arven*
1. Heads 1.0–2.5 cm in diameter; flowers pale yellow ....
2. Basal auricles of leaves acute; achenes longitudinally ribbed and pap
   late .... 2. *S. olerace*

2. Basal auricles of leaves rounded; achenes longitudinally ribbed, otherwise smooth .... 3. *S. asper*

    1. S. arvensis L. Field Sow Thistle. July–Sept. Involucres and peduncles with coarse gland-tipped hairs. Naturalized from Europe. Fields and waste ground; occasional in n. half of Ill.

    var. glabrescens Guenth., Gram., & Winn. Smooth Sow Thistle. July–Sept. Involucres and peduncles glabrous or nearly so. Naturalized from Europe. Fields and waste ground; abundant in n. half of Ill., extending s. to Clay Co. *S. uliginosus* Bieb.—F, G, J.

    2. S. oleraceus L. Common Sow Thistle. July–Oct. Naturalized from Europe. Fields and waste ground; common throughout Ill. (Including f. *integrifolium* [Wallr.] Beck and f. *lacerus* [Willd.] Beck.)

    3. S. asper (L.) Hill. Spiny Sow Thistle. July–Oct. Naturalized from Europe. Fields and waste ground; occasional throughout Ill. (Including f. *inermis* [Bisch.] Beck and f. *glandulosus* Beck.)

69. *Lactuca* L.   Lettuce

. Leaves broadly ovate .... 1. *L. sativa*
. Leaves elliptic to linear-lanceolate 2
. Achenes with 1 (–3) prominent median nerves on each face 3
. Achenes with several prominent nerves on each face 4
. Involucre 10–15 mm high in fruit; achenes 4.5–6.0 mm long (beak included) .... 2. *L. canadensis*
. Involucre 15–22 mm high in fruit; achenes 7–10 mm long (beak included) .... 3. *L. ludoviciana*
. Flowers yellow; achenes with a filiform beak nearly as long as or longer than the body 5
. Flowers blue to white; achenes with a short stout beak less than half to nearly as long as body or beak absent 6
. Leaves prickly margined; achenes spinulose above. ..... 4. *L. serriola*
. Leaves not prickly margined; achenes not spinulose above 5. *L. saligna*
. Pappus brown .... 6. *L. biennis*
. Pappus white 7
. Leaves toothed, commonly pinnately lobed; involucre 10–13 mm high; heads 0.6–1.0 cm in diameter .... 7. *L. floridana*
. Leaves entire margined, lower commonly pinnately lobed; involucre 16–20 mm high; heads 1–2 cm in diameter .... 8. *L. tatarica*

    1. L. sativa L. Garden Lettuce. July–Sept. Introduced from Eurasia and escaped from cultivation, not naturalized. Waste ground; occasional throughout Ill.

    2. L. canadensis L. Wild Lettuce. June–Aug. Waste ground; common throughout Ill. (Including f. *angustata* Wieg.; var. *obovata* Wieg.; var. *obovata* f. *steelei* [Britt.] Fern.; var. *obovata* f. *stenopoda* Wieg.; var. *longifolia* [Michx.] Farw.; var. *longifolia* f. *angustipes* Wieg.; var. *latifolia* Ktze.; var. *latifolia* f. *villicaulis* Fern.; and var. *latifolia* f. *exauriculata* Wieg.)

    3. L. ludoviciana (Nutt.) DC. Prairie Lettuce. July–Sept. Dry prairies; rare in n. half of Ill., extending s. to Monroe Co. Blue-flowered specimens (f. *campestris* [Greene] Fern.) are generally found more frequently than the typical yellow-flowered specimens.

    4. L. serriola L. Prickly Lettuce. July–Sept. Naturalized from Europe.

Waste ground; common throughout Ill. (Including var. *integra*
[Gren. & Godr.] Farw.) *L. scariola* L.—F, J.

5. L. saligna L. Willow Lettuce. July–Aug. Naturalized from Europ
Waste ground; common in s. two-thirds of Ill., absent elsewhere. (I
cluding f. *ruppiana* [Wallr.] Beck.)

6. L. biennis (Moench) Fern. Tall Blue Lettuce. Aug.–Sept. Moi
open woods; occasional throughout Ill. (Including f. *integrifolia* [To
& Gray] Fern.)

7. L. floridana (L.) Gaertn. Woodland Lettuce. July–Sept. Wood
common throughout Ill. (Including f. *leucantha* Fern. and var. *villos*
[Jacq.] Cronq.)

8. L. tatarica (L.) C. A. Mey. Blue Lettuce. July–Aug. Adventive fro
w. U.S. Waste ground; Cook, DuPage, Kane, and Lake cos. (O
plants belong to ssp. *pulchella* [Pursh] Steb.). *L. pulchella* [Purs
DC.—F, G, J.

Reported hybrids: *L. canadensis* X *L. ludoviciana*, Cook Co.; *L.
morssii* Robins. (*L. canadensis* X *L. biennis*), Ill., fide Fernald.

70. *Pyrrhopappus DC. False Dandelion*

1. P. carolinianus (Walt.) DC. False Dandelion. May–June. D
woods and prairies; occasional in s. half of Ill.

71. *Crepis L. Hawksbeard*

1. C. capillaris (L.) Wallr. Hawksbeard. June–July. Naturalized fro
Europe. Waste ground, rare; Cook, DuPage, and Lake cos.

72. *Prenanthes L. Rattlesnake Root*

1. Involucral bracts pubescent ....
1. Involucral bracts glabrous ....
2. Inflorescence narrow and elongate ....
2. Inflorescence corymbose-paniculate .... 3. *P. crepidin*
3. Flowers purplish; stems and leaves glabrous and usually glaucous ..
.... 1. *P. racemo*
3. Flowers cream-colored; stems and leaves pubescent ......2. *P. aspe*
4. Involucral bracts 7–10, mostly 8; flowers 8–15, mostly 13 per he
pappus reddish-borwn .... 4. *P. a*
4. Involucral bracts 4–6, mostly 5; flowers 5–6 per head; pappus str
colored .... 5. *P. altissi*

1. P. racemosa Michx. Glaucous White Lettuce. Aug.–Sept. Invo
cral bracts 7–10, mostly 8; flowers 9–16 per head, mostly 13. M
ground; occasional throughout Ill.

subsp. multiflora Cronq. White Lettuce. Aug.–Sept. Involucral bra
10–14, mostly 13; flowers 17–26 per head, mostly 21. Moist grou
occasional throughout Ill.

2. P. aspera Michx. Rough White Lettuce. Aug.–Sept. Prairies;
casional throughout Ill.

3. P. crepidinea Michx. Great White Lettuce. Aug.–Oct. Moist o
woods; occasional throughout Ill.

4. P. alba L. White Lettuce; Rattlesnake Root. Aug.–Sept. Woo
occasional in n. half of Ill., absent elsewhere.

5. *P. altissima* L. Tall White Lettuce. Aug.–Sept. Woods; occasional in e. and s. Ill.; also Cook and Lake cos. (Our plants belong to var. *cinnamomea* Fern.)

73. *Hieracium* L. Hawkweed

1. Plants with well-developed cluster of basal leaves at flowering time . 2
1. Plants without well-developed cluster of basal leaves at flowering time 4
2. Basal leaves rounded to cordate at base; pappus white. 1. *H. murorum*
2. Basal leaves tapering to base; pappus sordid 3
3. Flowers yellow .... 2. *H. pratense*
3. Flowers red-orange .... 3. *H. aurantiacum*
4. Stem leafy nearly to summit; inflorescence leafy bracted 5
4. Stem leafless or nearly so 6
5. Lower leaves largest, others progressively reduced upwards, becoming sessile .... 4. *H. scabrum*
5. Middle and lower leaves similar in size and shape .... 5. *H. canadense*
6. Leaves and lower part of stem with hairs 1–2 cm long 6. *H. longipilum*
6. Leaves and lower part of stem with hairs less than 1 cm long ....
.... 7. *H. gronovii*

1. H. murorum L. Golden Lungwort. June–Aug. Naturalized from Europe. Waste ground; rare; Sangamon Co.

2. H. pratense Tausch. King Devil. June–Aug. Naturalized from Europe. Waste ground, rare; Cook, Lake, and McHenry cos.

3. H. aurantiacum L. Orange Hawkweed; Devil's Paint Brush. June–July. Naturalized from Europe. Waste ground, rare; Lake and Ogle cos.

4. H. scabrum Michx. Hairy Hawkweed. Aug.–Sept. Open ground and dry woods; occasional throughout Ill. (Including var. *intonsum* Fern. & St. John.)

5. H. canadense Michx. Canada Hawkweed. Aug.–Sept. Dry woods, sand barrens; occasional in n. one-fourth of Ill., absent elsewhere. (Including var. *fasciculatum* [Prush] Fern.)

6. H. longipilum Torr. Hairy Hawkweed. July–Sept. Field and open woods; occasional throughout Ill.

7. H. gronovii L. Hairy Hawkweed. July–Sept. Dry open woods; occasional in the s. three-fourths of Ill., rare elsewhere.

Sequence and Summary
of the Families

Family	Genera	Species	Lesser Taxa	Hybrids
1. Equisetaceae	1	8		4
2. Lycopodiaceae	1	6	1	1
3. Selaginellaceae	1	2		
4. Isoetaceae	1	2		
5. Ophioglossaceae	2	8	2	
6. Osmundaceae	1	3		
7. Hymenophyllaceae	1	1		
8. Polypodiaceae	18	40	4	10
9. Marsileaceae	1	1		
0. Salviniaceae	1	2		
1. Taxaceae	1	1		
2. Pinaceae	2	10		
3. Taxodiaceae	1	1		
4. Cupressaceae	2	4	1	
5. Typhaceae	1	2		
6. Sparganiaceae	1	5		
7. Ruppiaceae	1	1		
8. Zannichelliaceae	1	1		
9. Najadaceae	1	5		
0. Potamogetonaceae	1	20		2
1. Juncaginaceae	2	3		
2. Alismaceae	3	12		
3. Butomaceae	1	1		
4. Hydrocharitaceae	3	5		
5. Poaceae	87	287	42	5
6. Cyperaceae	13	225	18	2
7. Araceae	5	6	1	
8. Lemnaceae	4	14		
9. Xyridaceae	1	2		

Family	Genera	Species	Lesser Taxa	Hybrids
30. Commelinaceae	2	8		
31. Pontederiaceae	3	4		
32. Juncaceae	2	24	4	
33. Liliaceae	31	63	3	
34. Smilacaceae	1	9	2	
35. Dioscoreaceae	1	2		
36. Iridaceae	4	14	1	1
37. Burmanniaceae	1	1		
38. Orchidaceae	6	43	2	2
39. Saururaceae	1	1		
40. Salicaceae	2	31	5	4
41. Myricaceae	1	1		
42. Juglandaceae	2	12	5	
43. Betulaceae	5	11		2
44. Fagaceae	3	22	1	15
45. Ulmaceae	3	10	5	
46. Moraceae	5	7	1	
47. Urticaceae	5	9	1	
48. Santalaceae	1	1		
49. Loranthaceae	1	1		
50. Aristolochiaceae	2	3	2	
51. Polygonaceae	3	43	5	
52. Chenopodiaceae	9	29	2	
53. Amaranthaceae	4	17		
54. Nyctaginaceae	1	4		
55. Phytolaccaceae	1	1		
56. Aizoaceae	1	1		
57. Portulacaceae	3	6		
58. Caryophyllaceae	17	48		
59. Ceratophyllaceae	1	2		
60. Nymphaeaceae	2	3	1	
61. Nelumbonaceae	1	1		
62. Cabombaceae	2	2		
63. Ranunculaceae	17	57	9	
64. Berberidaceae	4	6		
65. Menispermaceae	3	3		
66. Magnoliaceae	2	2		
67. Annonaceae	1	1		
68. Lauraceae	2	2	2	
69. Papaveraceae	11	21		
70. Capparidaceae	2	4	1	

Family	Genera	Species	Lesser Taxa	Hybrids
71. Cruciferae	36	76	11	
72. Resedaceae	1	1		
73. Sarraceniaceae	1	1		
74. Droseraceae	1	2		
75. Crassulaceae	1	6		
76. Saxifragaceae	11	23		
77. Hamamelidaceae	2	2		
78. Platanaceae	1	1		
79. Rosaceae	23	132	9	1
80. Leguminosae	46	128	9	2
81. Linaceae	1	6		
82. Oxalidaceae	1	5		
83. Geraniaceae	2	8		
84. Zygophyllaceae	2	2		
85. Rutaceae	2	2	1	
86. Simaroubaceae	1	1		
87. Polygalaceae	1	7	2	
88. Euphorbiaceae	9	34	4	
89. Callitrichaceae	1	3		
90. Limnanthaceae	1	1		
91. Anacardiaceae	2	6	2	
92. Aquifoliaceae	1	3		
93. Celastraceae	2	8		
94. Staphyleaceae	1	1		
95. Aceraceae	1	7	3	
96. Hippocastanaceae	1	4	1	
97. Sapindaceae	2	2		
98. Balsaminaceae	1	2		
99. Rhamnaceae	3	8	2	
100. Vitaceae	3	12	3	
101. Tiliaceae	1	2	1	
102. Sterculiaceae	1	1		
103. Malvaceae	11	24	1	
104. Hypericaceae	3	23	3	
105. Elatinaceae	2	2		
106. Cistaceae	3	9	2	
107. Violaceae	2	25	5	1
108. Passifloraceae	1	2		
109. Loasaceae	1	3		
110. Cactaceae	1	2		
111. Thymelaeaceae	1	1		

Family	Genera	Species	Lesser Taxa	Hybrid
112. Elaeagnaceae	2	3		
113. Lythraceae	6	8		
114. Nyssaceae	1	2	1	
115. Melastomaceae	1	2		
116. Onagraceae	6	32	2	
117. Haloragidaceae	2	6		
118. Hippuridaceae	1	1		
119. Araliaceae	3	6		
120. Umbelliferae	34	46	5	
121. Cornaceae	1	10	1	
122. Ericaceae	12	22	1	
123. Primulaceae	9	21	1	1
124. Sapotaceae	1	2		
125. Ebenaceae	1	1		
126. Styracaceae	2	3		
127. Oleaceae	4	8	3	
128. Loganiaceae	2	2		
129. Gentianaceae	6	14		
130. Menyanthaceae	2	2		
131. Apocynaceae	4	7	4	1
132. Asclepiadaceae	3	23	1	
133. Convolvulaceae	6	21	3	
134. Polemoniaceae	5	10	4	
135. Hydrophyllaceae	4	9		
136. Boraginaceae	14	32	1	
137. Verbenaceae	2	9	1	7
138. Phrymaceae	1	1		
139. Labiatae	31	77	13	4
140. Solanaceae	9	32	3	1
141. Scrophulariaceae	26	74	6	
142. Bignoniaceae	3	4		
143. Martyniaceae	1	1		
144. Orobanchaceae	3	6		
145. Lentibulariaceae	1	5		
146. Acanthaceae	3	7	2	
147. Plantaginaceae	1	12	1	
148. Rubiaceae	7	28	4	
149. Caprifoliaceae	7	31		1
150. Adoxaceae	1	1		
151. Valerianaceae	2	8	1	
152. Dipsacaceae	1	2		

Family	Genera	Species	Lesser Taxa	Hybrids
153. Cucurbitaceae	7	8		
154. Campanulaceae	3	14	2	1
155. Compositae	73	290	25	15
Totals	830	2699	265	83

Additional Taxa

The following taxa new to the Illinois flora have been discovered too late to be incorporated in the text.

BORAGINACEAE—BORAGE FAMILY

Asperugo procumbens L. Madwort. June–July. Native of Europe; waste ground; Kane Co.

CALYCANTHACEAE—CALYCANTHUS FAMILY

Calycanthus floridus L. Strawberry Bush. May. Native of se. U.S.; escaped from cultivation into a woods; Jackson Co.

CARYOPHYLLACEAE—PINK FAMILY

Cerastium pumilum Curtis. Dwarf Chickweed. April. Native of Europe; adventive in waste ground; Jackson Co.

CELASTRACEAE—BITTERSWEET FAMILY

Euonymus alatus (Thunb.) Sieb. Burning-bush. May–June. Native f Asia; commonly cultivated; adventive in woodland; Coles Co.

COMMELINACEAE—SPIDERWORT FAMILY

Tradescantia subaspera Ker var. montana (Shuttlew.) Anders. & oodson. June–July. Moist woods; Jackson Co.

COMPOSITAE—SUNFLOWER FAMILY

Aster undulatus L. Aug.–Oct. Dry woodlands; Saline Co.
Grindelia lanceolata Nutt. Tarweed. July–Sept. Rocky prairies; undy Co.

CYPERACEAE—SEDGE FAMILY

Eleocharis parvula (Roem. & Schultes) Link. Dwarf Spikerush
July–Sept. Sandy soil along margin of lake; Coles Co.

DIOSCOREACEAE—WILD YAM FAMILY

Dioscorea batatas Dcne. Chinese Yam. June–Sept. Native of Asia
escaped into waste ground; Jackson Co.

ERICACEAE—HEATH FAMILY

Chimaphila maculata (L.) Pursh. Spotted Wintergreen. June–July
Dry woods; Pope Co.

GENTIANACEAE—GENTIAN FAMILY

Gentiana septemfida Pall. Gentian. July–Sept. Native of Asia; es
caped from cultivation into waste ground; Cook Co.

LILIACEAE—LILY FAMILY

Lycoris radiata Herb. Surprise Lily. July–Sept. Native of Japan; es
caped from cultivation to roadsides; Jackson Co.

LINACEAE—FLAX FAMILY

Linum perenne L. ssp. lewisii (Pursh) Hult. Flax. June–Sept. Na
tive of w. U.S.; escaped from cultivation into waste ground; Grund
Co.

MARSILEACEAE—WATER CLOVER FAMILY

Pilularia americana A. Br. June–Oct. Shallow water and shorelin
of lake; Pope Co.

OLEACEAE—OLIVE FAMILY

Ligustrum obtusifolium Sieb. & Zucc. Blunt-leaved Privet. June
July. Native of Asia; escaped from cultivation into waste ground
Jackson Co.

ORCHIDACEAE—ORCHID FAMILY

Isotria medeoloides (Pursh) Raf. Small Whorled Pogonia. Ma
Dry, rocky woods; Randolph Co.

Spiranthes romanzoffiana Cham. Hooded Ladies'-tresses. Aug.–
Sept. Meadows and thickets; Coles, Cook, and Peoria cos.

POACEAE—GRASS FAMILY

Pennisetum alopecuroides (L.) Spreng. Aug–Oct. Native of Asia;
adventive along a stream; Crawford Co.

RANUNCULACEAE—BUTTERCUP FAMILY

Eranthis hyemalis Salisb. Winter Aconite. April–May. Native of
Asia; escaped from cultivation; St. Clair Co.
Ranunculus arvensis L. Spiny-fruited Buttercup. May–June. Na-
tive of Europe; escaped along a railroad; Jackson Co.

RHAMNACEAE—BUCKTHORN FAMILY

Rhamnus davurica Pall. Buckthorn. May–June. Native of Asia; escaped from cultivation; DuPage Co.

ROSACEAE—ROSE FAMILY

Filipendula ulmaria (L.) Maxim. Queen-of-the-Meadow. June–Aug.
Native of Europe; escaped from cultivation into waste ground; Kane
Co.
Potentilla canadensis L. Cinquefoil. May–June. Dry soil; DeKalb
Co.
Rosa wichuriana Crepin. Memorial Rose. June–July. Native of
Asia; escaped from cultivation into waste ground; Jackson Co.

SCROPHULARIACEAE—FIGWORT FAMILY

Verbascum pulverulenta Schrad. July–Oct. Native of Europe; ad-
ventive in waste ground; Lake Co.
Veronica agrestis L. Field Speedwell. June–Sept. Native of
Europe; adventive in waste ground; Kane Co.

UMBELLIFERAE—CARROT FAMILY

Spermolepis echinata (Nutt.) Heller. May–June. Native of s.w.
U.S.; adventive along a railroad; Jackson Co.

Glossary

Index

Glossary

Acaulescent. Seemingly without aerial stems.

Achene. A type of one-seeded, dry, indehiscent fruit with the seed coat not attached to the mature ovary wall.

Actinomorphic. Having radial symmetry; regular, in reference to a flower.

Acuminate. Gradually tapering to a point.

Acute. Sharply tapering to a point.

Adaxial. Toward the axis; when referring to a leaf, the upper surface.

Ament. A spike of unisexual, apetalous flowers; a catkin.

Anther. The terminal part of a stamen which bears pollen.

Anthesis. Flowering time.

Antrorse. Projecting forward.

Apical. At the apex.

Apiculate. Abruptly short-pointed at the tip.

Appressed. Lying flat against the surface.

Areole. A small area between leaf veins.

Aril. An appendage of the seed, usually enclosing the seed.

Arillate. Bearing an aril.

Aristate. Bearing an awn.

Articulated. Jointed.

Attenuate. Gradually becoming narrowed.

Auricle. An ear-like lobe.

Auriculate. Bearing an ear-like process.

Awn. A bristle usually terminating a structure.

Axillary. Borne from an axil.

Beard. A tuft of stiff hairs.

Berry. A type of fruit where the seeds are surrounded only by fleshy material.

Bidentate. Having two teeth.

Bifid. Two-cleft.

459

Bifoliolate. Bearing two leaflets.

Bifurcate. Forked.

Biglandular. Bearing two glands.

Bilabiate. Two-lipped.

Bipinnate. Divided once into distinct segments, with each segment in turn divided into distinct segments.

Bipinnatifid. Divided part way to the center, with each lobe again divided partway to the center.

Bisexual. Referring to a flower which contains both stamens and pistils.

Biternate. Divided into three segments two times.

Bract. An accessory structure at the base of many flowers, usually appearing leaflike.

Bracteate. Bearing one or more bracts.

Bracteole. A secondary bract.

Bractlet. A small bract.

Bristle. A stiff hair or hair-like growth; a seta.

Bulblet. A small bulb.

Callosity. Any hardened thickening.

Callus. A hard swollen area at the outside base of a lemma or palea.

Calyx. The outermost segments of the perianth of a flower, composed of sepals.

Campanulate. Bell-shaped.

Canescent. Grayish-hairy.

Capillary. Thread-like.

Capitate. Forming a head.

Capsule. A dry, dehiscent fruit composed of more than one carpel.

Carpel. A simple pistil, or one member of a compound pistil.

Cartilaginous. Firm but flexible.

Caruncle. A fleshy outgrowth near the point of attachment of a seed.

Catkin. A spike of unisexual, apetalous flowers; an ament.

Caudate. With a tail-like appendage.

Caudex. The woody base of a perennial plant.

Cauline. Belonging to a stem.

Cespitose. Growing in tufts.

Chaffy. Covered with scales.

Chartaceous. Papery.

Cilia. Marginal hairs.

Ciliate. Bearing cilia.

Ciliolate. Bearing small cilia.

Circumscissile. Usually referring to a fruit which dehisces by a horizontal, circular line.

Claw. A narrow, basal stalk, particularly of a petal.

Coherent. The growing together of like parts.

Coma. A tuft of hairs at the end of a seed.

Compressed. Flattened.

Concave. Curved on the inner surface; opposed to convex.

Connate. Union of like parts.

Connective. That portion of a stamen between the two anther halves.

Convex. Curved on the outer surface; opposed to concave.

Convolute. Rolled lengthwise.

Coralline. Having the texture of coral.

Cordate. Heart-shaped.

Coriaceous. Leathery.

Corm. An underground, vertical stem with scaly leaves, differing from a bulb by lacking fleshy leaves.

Corolla. The segments of a flower just within the calyx, composed of petals.

Corona. A crown of petal-like structures.

Corrugated. Folded or wrinkled.

Corymb. A type of inflorescence where the pedicellate flowers are arranged along an elongated axis but with the flowers all attaining about the same height.

Corymbiform. Shaped like a corymb.

Cotyledon. A seed leaf.

Crateriform. Cone-shaped but sunken in the center at the top.

Crenate. With round teeth.

Crenulate. With small, round teeth.

Crest. A ridge.

Crisped. Curled.

Cucullate. Hood-shaped.

Culm. The stem which terminates in an inflorescence.

Cuneate. Wedge-shaped; tapering to the base.

Cupular. Shaped like a small cup.

Cuspidate. Terminating in a very short point.

Cyathium. A cup-like structure enclosing flowers.

Cyme. A type of broad and flattened inflorescence in which the central flowers bloom first.

Cymose. Bearing a cyme.

Cystolith. A stone-like concretion in the epidermis of some plants.

Deciduous. Falling away.

Decumbent. Lying flat, but with the tip ascending.

Decurrent. Adnate to the petiole or stem and then extending beyond the point of attachment.

Deflexed. Turned downward.

Dehiscent. Splitting at maturity.

Deltoid. Triangular.

Dentate. With sharp teeth, the tips of which project outward.

Denticulate. With small, sharp teeth, the tips of which projec outward.

Diffuse. Loosely spreading.

Digitate. Radiating from a common point, like the fingers from hand.

Dilated. Swollen; expanded.

Dimorphic. Having two forms.

Dioecious. With staminate flowers on one plant, pistillate flower on another.

Disarticulate. To come apart; to become disjointed.

Disk. An enlarged outgrowth of the receptacle.

Divergent. Spreading apart.

Drupe. A type of fruit in which the seed is surrounded by a hard dry covering which, in turn, is surrounded by fleshy material.

Ebracteate. Without bracts.

Echinate. Spiny.

Eciliate. Without cilia.

Eglandular. Without glands.

Ellipsoid. Referring to a solid object which is broadest at th middle, gradually tapering to both ends.

Elliptic. Broadest at middle, gradually tapering to both ends.

Emarginate. Having a shallow notch at the extremity.

Epidermis. The outermost layer of cells.

Epunctate. Without dots.

Erose. With an irregularly notched margin.

Exudate. Secreted material.

Falcate. Sickle-shaped.

Fascicle. Cluster.

Ferruginous. Rust-colored.

Fetid. Ill-smelling.

Fibrous. Referring to roots borne in tufts.

Filament. That part of the stamen supporting the anther.

Filiform. Thread-like.

Fimbriate. Fringed.

Flaccid. Weak; flabby.

Flexible. Able to be bent readily.

Flexuous. Zigzag.

Floret. A small flower.

Follicle. A type of dry, dehiscent fruit which splits along one side at maturity.

Friable. Breaking easily into small particles.

Frond. The leaf of a fern; the vegetative structure in the Lemaceae.

Funnelform. Shaped like a funnel.

Galea. A hooded portion of a perianth.

Gemmae. Bulblets or buds capable of developing into new plants.

Geniculate. Bent.

Gibbous. Swollen on one side.

Glabrate. Becoming smooth.

Glabrous. Without pubescence or hairs.

Gland. An enlarged, spherical body functioning as a secretory organ.

Glandular. Bearing glands.

Glaucous. With a whitish covering which can be rubbed off.

Globose. Round; globular.

Glochidium. A process bearing barbs.

Glomerule. A small compact cluster.

Glume. A sterile scale subtending a spikelet.

Glutinous. Covered with a sticky secretion.

Grain. The fruit of most grasses.

Hastate. Spear-shaped; said of a leaf which is triangular with spreading basal lobes.

Hirsute. With stiff hairs.

Hirsutulous. With minute stiff hairs.

Hirtellous. Finely hirsute.

Hispid. With rigid hairs.

Hispidulous. With minute rigid hairs.

Hoary. Grayish-white, usually referring to pubescence.

Hood. That part of an organ, usually of a flower, which is strongly concave and arching.

Horn. An accessory structure found in certain flowers.

Hyaline. Transparent.

Hypanthium. A development of the receptacle beneath the calyx.

Hypogynium. A structure subtending the ovary in *Scleria*.

Imbricate. Overlapping.

Indehiscent. Not splitting open at maturity.

Indurate. Hard.

Indusium. An outgrowth of the blade which covers or partial covers the sori.

Inferior. Referring to the position of the ovary when it is su rounded by the adnate portion of the floral tube or is embedded the receptacle.

Inflexed. Turned inward.

Inflorescence. A cluster of flowers.

Internode. The area between two adjacent nodes.

Involucel. A cluster of bracteoles which subtends a seconda flower cluster.

Involucre. A circle of bracts which subtends a flower cluster.

Involute. Rolled inward.

Keel. A ridgelike process.

Laciniate. Divided into narrow, pointed divisions.

Lanceolate. Lance-shaped; broadest near base, gradually tap ing to the narrower apex.

Lanceoloid. Referring to a solid object which is broadest ne base, gradually tapering to the narrower apex.

Latex. Milky juice.

Leaflet. An individual unit of a compound leaf.

Legume. A dry fruit usually dehiscing along two sides at matur

Lemma. A scale subtending a floret in grasses.

Lenticel. Corky openings on bark of twigs and branches.

Lenticular. Lens-shaped.

Lepidote. Scaly.

Ligule. An elongated tongue-like process found on the leaf *Isoetes* and *Selaginella;* a structure on the inner surface of the l of grasses and sedges at the junction of the blade and the shea

Linear. Elongated and uniform in width throughout.

Lip. The lowermost, often greatly modified, petal of a flower.

Lobulate. With small lobes.

Locular. Referring to the locule, or cavity of the ovary or anther.

Locule. The cavity of an ovary or an anther.

Loculicidal. Said of a capsule which splits down the dorsal sut of each cell.

Lodicule. A small rudimentary structure at the base of so grass flowers.

Loment. A fruit divided into 1-seeded segments by transve partitions.

Lunate. Crescent-shaped.

Lustrous. Shiny.

Lyrate. Pinnatifid, with the terminal lobe much larger than the lower ones.

Megaspore. A spore produced in a megasporangium; a spore which produces a female gametophyte.

Microspore. A spore produced in a microsporangium; a spore which produces a male gametophyte.

Moniliform. Constricted at regular intervals to resemble a string of beads.

Monoecious. Bearing both sexes in separate flowers on the same plant.

Monomorphic. Having but one form.

Mucilaginous. Slimy.

Mucro. A short, abrupt tip.

Mucronate. Possessing a short, abrupt tip.

Mucronulate. Possessing a very short, abrupt tip.

Muricate. Minutely spiny.

Nectariferous. Producing nectar.

Nigrescent. Blackish.

Node. That place on the stem from which leaves and branchlets arise.

Nutlet. A small nut.

Oblanceolate. Reverse lance-shaped; broadest at apex, gradually tapering to narrow base.

Oblong. Broadest at the middle, and tapering to both ends, but broader than elliptic.

Oblongoid. Referring to a solid object which, in side view, is nearly the same width throughout.

Obovoid. Referring to a solid object which is broadly rounded at the apex, becoming narrowed below.

Obpyramidal. Referring to an upside-down pyramid.

Obtuse. Rounded at the apex.

Ocrea. A sheathing stipule, often tubular.

Ocreola. A secondary, usually tubular, sheath.

Opaque. Incapable of being seen through.

Orbicular. Round.

Oval. Broadly elliptic.

Ovary. The lower swollen part of the pistil which produces the ovules.

Ovate. Broadly rounded at base, becoming narrowed above; broader than lanceolate.

Ovoid. Referring to a solid object which is broadly rounded at the base, becoming narrowed above.

Palea. The scale opposite the lemma which encloses the flower.
Palmate. Divided radiately, like the fingers of a hand.
Pandurate. Fiddle-shaped.
Panduriform. Fiddle-shaped.
Panicle. A type of inflorescence composed of several racemes.
Papilla. A small wart.
Papillate. Bearing small warts, or papillae.
Papillose. Bearing pimple-like processes.
Pappus. The modified calyx in the Compositae.
Papule. A pimple-like projection.
Pectinate. Pinnatifid into close, narrow segments; comb-like.
Pedicel. The stalk of a flower of an inflorescence.
Pedicellate. Bearing a pedicel.
Peduncle. The stalk of an inflorescence.
Peduncled. Provided with a peduncle.
Peltate. Attached away from the margin, in reference to a leaf.
Perennial. Living more than two years.
Perfect. Bearing both stamens and pistils in the same flower.
Perfoliate. Referring to a leaf which appears to have the stem pass through it.
Perianth. Those parts of a flower including both the calyx and corolla.
Pericarp. The ripened ovary wall.
Perigynium. A sac-like covering enclosing the achene in *Carex*.
Petal. One segment of the corolla.
Petaloid. Resembling a petal in texture and appearance.
Petiolate. Bearing a petiole, or leafstalk.
Petiole. The stalk of a leaf.
Petiolulate. Bearing a petiolule, or leaflet-stalk.
Petiolule. The stalk of a leaflet.
Pilose. Bearing soft hairs.
Pinna. A primary division of a compound blade.
Pinnate. Divided once into distinct segments.
Pinnatifid. Said of a simple leaf or leaf-part which is cleft or lobed only part way to its axis.
Pinnule. The secondary segments of a compound blade.
Pistil. The ovule-producing organ of a flower normally composed of an ovary, a style, and a stigma.
Pistillate. Bearing pistils but not stamens.
Plano-convex. Flat on the inner face, rounded on the outer face.
Plicate. Folded.

Plumose. Bearing fine hairs, like the plume of a feather.
Procumbent. Lying on the ground.
Prophyll. A bracteole.
Prophyllate. Bearing a bracteole.
Pruinose. Having a waxy covering.
Puberulent. With minute hairs.
Pubescent. Bearing some kind of hairs.
Punctate. Dotted.
Pyramidal. Shaped like a pyramid.
Pyriform. Pear-shaped.

Quadrate. Four-sided.

Raceme. A type of inflorescence where pedicellate flowers are arranged along an elongated axis.
Rachilla. The flower-bearing axis in a grass spikelet.
Rachis. A primary axis.
Receptacle. That part of the flower to which the perianth, stamens, and pistils are usually attached.
Reflexed. Turned downward.
Reniform. Kidney-shaped.
Repand. Wavy along the margin.
Repent. Creeping.
Resinous. Producing a sticky secretion, or resin.
Resupinate. Upside-down.
Reticulate. Resembling a network.
Reticulum. A network.
Retrorse. Pointing downward.
Retuse. Shallowly notched at a rounded apex.
Revolute. Rolled under from the margin.
Rhizome. An underground horizontal stem, bearing nodes, buds, and roots.
Rhombic. Becoming quadrangular.
Rosette. A cluster of leaves in a circular arrangement at the base of a plant.
Rotate. Flat and circular.
Rufescent. Reddish-brown.
Rufous. Red-brown.
Rugose. Wrinkled.
Rugulose. With small wrinkles.

Saccate. Sac-shaped.
Sagittate. Shaped like an arrowhead.

Salverform. Referring to a tubular corolla which abruptly ex pands into a flat limb.

Samara. An indehiscent winged fruit.

Scaberulous. Slightly rough to the touch.

Scabrous. Rough to the touch.

Scape. A leafless stalk bearing a flower or inflorescence.

Scarious. Thin and membranous.

Scurfy. Bearing scaly particles.

Secund. Borne on one side.

Sepaloid. Resembling a sepal in texture.

Septate. With dividing walls.

Septicidal. Said of a capsule which splits between the locules.

Sericeous. Silky; bearing soft, appressed hairs.

Serrate. With teeth which project forward.

Serrulate. With very small teeth which project forward.

Sessile. Without a stalk.

Seta. Bristle.

Setaceous. Bearing bristles, or setae.

Setiform. Bristle-shaped.

Setose. Bearing setae.

Silicle. A short silique.

Silique. An elongated capsule with a central partition separatin the valves.

Sinuate. Wavy along the margins.

Sinus. The cleft between two lobes or teeth.

Sorus. An aggregation of sporangia.

Spadix. A fleshy axis in which flowers are embedded.

Spathe. A large sheathing bract subtending or usually enclosi an inflorescence.

Spatulate. Oblong, but with the basal end elongated.

Spicate. Bearing a spike.

Spike. A type of inflorescence where sessile flowers are arrang along an elongated axis.

Spikelet. A small spike.

Spinescent. Becoming spiny.

Spinose. Bearing spines.

Spinule. A small spine.

Spinulose. Bearing small pines.

Sporangium. That structure which produces spores.

Spore. That structure formed within a sporangium which gi rise to the gametophyte generation.

Sporocarp. A compound structure containing the sporangia *Marsilea* and *Azolla*.

Sporophyll. A structure subtending the sporangia.

Spur. A sac-like extension of the flower.

Stamen. The pollen-producing organ of a flower composed of a filament and an anther.

Staminate. Bearing stamens but not pistils.

Staminodium. A sterile stamen.

Standard. The upper, usually enlarged, petal of a pea-shaped flower.

Stellate. Star-shaped.

Stipe. A stalk.

Stipel. A small stipe.

Stipitate. Possessing a stipe.

Stipular. Pertaining to a stipule.

Stipule. A leaf-like or scaly structure found at the point of attachment of a leaf to the stem.

Stolon. A slender, horizontal stem on the surface of the ground.

Stoloniferous. Bearing stolons.

Stomate. An opening in the epidermis of a leaf.

Stramineous. Straw-colored.

Striate. Marked with grooves.

Strigillose. With short, appressed, straight hairs.

Strigose. With appressed, straight hairs.

Strigulose. With short, appressed, straight hairs.

Style. That elongated part of the pistil between the ovary and the stigma.

Subcuneate. Nearly wedge-shaped.

Suborbicular. Nearly spherical.

Subulate. With a very short, narrow point.

Succulent. Fleshy.

Suffused. Spread throughout; flushed.

Superior. Referring to the position of the ovary when the free floral parts arise below the ovary.

Supra-axillary. Borne above the axil.

Tendril. A spiralling, coiling structure which enables a climbing plant to attach itself to a supporting body.

Terete. Round, in cross-section.

Ternate. Divided three times.

Testa. The seed coat.

Tetrad. A cluster of four.

Tomentose. Pubescent with matted wool.

Tomentulose. Finely pubescent with matted wool.

Tometum. Woolly hair.

Torulose. With small contractions.

Translucent. Partly transparent.

Trifoliolate. Divided into three leaflets.

Trigonous. Triangular in cross-section.

Truncate. Abruptly cut across.

Tuber. An underground fleshy stem formed as a storage organ at the end of a rhizome.

Tubercle. A small, wart-like process.

Tuberculate. Warty.

Turgid. Swollen to the point of bursting.

Umbel. A type of inflorescence in which the flower stalks arise from the same level.

Undulate. Wavy.

Unisexual. Bearing either stamens or pistils in one flower, but not both.

Urceolate. Urn-shaped.

Utricle. A small, one-seeded, indehiscent fruit with a thin covering.

Verrucose. Warty.

Verticil. A whorl.

Verticillate. Whorled.

Villous. With long, soft, slender, unmatted hairs.

Virgate. Wand-like.

Viscid. Sticky.

Whorl. An arrangement of three or more structures at a point on the stem.

Zygomorphic. Bilaterally symmetrical.

Index to Families and Genera

Index to Common Names

•2284-67
1977
5-40
C